Geometry Illuminated

An Illustrated Introduction to Euclidean and Hyperbolic Plane Geometry

© 2015 by
The Mathematical Association of America (Incorporated)
Library of Congress Control Number: 2015936098
Print ISBN: 978-1-93951-211-6
Electronic ISBN: 978-1-61444-618-7
Printed in the United States of America
Current Printing (last digit):
10 9 8 7 6 5 4 3 2 1

Geometry Illuminated

An Illustrated Introduction to Euclidean and Hyperbolic Plane Geometry

Matthew Harvey
The University of Virginia's College at Wise

Published and distributed by
The Mathematical Association of America

Council on Publications and Communications
Jennifer J. Quinn, *Chair*

Committee on Books
Fernando Gouvêa, *Chair*

MAA Textbooks Editorial Board
Stanley E. Seltzer, *Editor*

Matthias Beck
Richard E. Bedient
Otto Bretscher
Heather Ann Dye
Charles R. Hampton
Suzanne Lynne Larson
John Lorch
Susan F. Pustejovsky

MAA TEXTBOOKS

Bridge to Abstract Mathematics, Ralph W. Oberste-Vorth, Aristides Mouzakitis, and Bonita A. Lawrence

Calculus Deconstructed: A Second Course in First-Year Calculus, Zbigniew H. Nitecki

Calculus for the Life Sciences: A Modeling Approach, James L. Cornette and Ralph A. Ackerman

Combinatorics: A Guided Tour, David R. Mazur

Combinatorics: A Problem Oriented Approach, Daniel A. Marcus

Complex Numbers and Geometry, Liang-shin Hahn

A Course in Mathematical Modeling, Douglas Mooney and Randall Swift

Cryptological Mathematics, Robert Edward Lewand

Differential Geometry and its Applications, John Oprea

Distilling Ideas: An Introduction to Mathematical Thinking, Brian P. Katz and Michael Starbird

Elementary Cryptanalysis, Abraham Sinkov

Elementary Mathematical Models, Dan Kalman

An Episodic History of Mathematics: Mathematical Culture Through Problem Solving, Steven G. Krantz

Essentials of Mathematics, Margie Hale

Field Theory and its Classical Problems, Charles Hadlock

Fourier Series, Rajendra Bhatia

Game Theory and Strategy, Philip D. Straffin

Geometry Illuminated: An Illustrated Introduction to Euclidean and Hyperbolic Plane Geometry, Matthew Harvey

Geometry Revisited, H. S. M. Coxeter and S. L. Greitzer

Graph Theory: A Problem Oriented Approach, Daniel Marcus

An Invitation to Real Analysis, Luis F. Moreno

Knot Theory, Charles Livingston

Learning Modern Algebra: From Early Attempts to Prove Fermat's Last Theorem, Al Cuoco and Joseph J. Rotman

The Lebesgue Integral for Undergraduates, William Johnston

Lie Groups: A Problem-Oriented Introduction via Matrix Groups, Harriet Pollatsek

Mathematical Connections: A Companion for Teachers and Others, Al Cuoco

Mathematical Interest Theory, Second Edition, Leslie Jane Federer Vaaler and James W. Daniel

Mathematical Modeling in the Environment, Charles Hadlock

Mathematics for Business Decisions Part 1: Probability and Simulation (electronic textbook), Richard B. Thompson and Christopher G. Lamoureux

Mathematics for Business Decisions Part 2: Calculus and Optimization (electronic textbook), Richard B. Thompson and Christopher G. Lamoureux

Mathematics for Secondary School Teachers, Elizabeth G. Bremigan, Ralph J. Bremigan, and John D. Lorch

The Mathematics of Choice, Ivan Niven

The Mathematics of Games and Gambling, Edward Packel

Math Through the Ages, William Berlinghoff and Fernando Gouvea

Noncommutative Rings, I. N. Herstein

Non-Euclidean Geometry, H. S. M. Coxeter
Number Theory Through Inquiry, David C. Marshall, Edward Odell, and Michael Starbird
Ordinary Differential Equations: from Calculus to Dynamical Systems, V. W. Noonburg
A Primer of Real Functions, Ralph P. Boas
A Radical Approach to Lebesgue's Theory of Integration, David M. Bressoud
A Radical Approach to Real Analysis, 2nd edition, David M. Bressoud
Real Infinite Series, Daniel D. Bonar and Michael Khoury, Jr.
Thinking Geometrically: A Survey of Geometries, Thomas Q. Sibley
Topology Now!, Robert Messer and Philip Straffin
Understanding our Quantitative World, Janet Andersen and Todd Swanson

MAA Service Center
P.O. Box 91112
Washington, DC 20090-1112
1-800-331-1MAA   FAX: 1-240-396-5647

Contents

Preface   **xv**

0 Axioms and Models   **1**
    0.1 Fano's geometry 2
    0.2 Further reading 4
    0.3 Exercises 5

I Neutral Geometry   **7**

1 The Axioms of Incidence and Order   **9**
    1.1 Incidence 11
    1.2 Order 12
    1.3 Putting points in order 14
    1.4 Exercises 16
    1.5 Further reading 17

2 Angles and Triangles   **19**
    2.1 Exercises 24
    2.2 References 25

3 Congruence Verse I: SAS and ASA   **27**
    3.1 Triangle congruence 30
    3.2 Exercises 34

4 Congruence Verse II: AAS   **37**
    4.1 Supplementary angles 37
    4.2 The alternate interior angle theorem 40
    4.3 The exterior angle theorem 42
    4.4 AAS 43
    4.5 Exercises 44

5	**Congruence Verse III: SSS**	**45**
	5.1 Exercises	49

6	**Distance, Length, and the Axioms of Continuity**	**51**
	6.1 Synthetic comparison	51
	6.2 Distance	53
	6.3 Exercises	62

7	**Angle Measure**	**63**
	7.1 Synthetic angle comparison	63
	7.2 Right angles	67
	7.3 Angle measure	70
	7.4 Exercises	71

8	**Triangles in Neutral Geometry**	**73**
	8.1 Exercises	78
	8.2 References	79

9	**Polygons**	**81**
	9.1 Definitions	81
	9.2 Counting polygons	83
	9.3 Interiors and exteriors	84
	9.4 Interior angles: two dilemmas	88
	9.5 Polygons of note	92
	9.6 Exercises	93

10	**Quadrilateral Congruence Theorems**	**95**
	10.1 Terminology	95
	10.2 Quadrilateral congruence	96
	10.3 Exercises	103

II	**Euclidean Geometry**	**105**

11	**The Axiom on Parallels**	**107**
	11.1 Exercises	113

12	**Parallel Projection**	**115**
	12.1 Parallel projection	116
	12.2 Parallel projection, order, and congruence	117
	12.3 Parallel projection and distance	120
	12.4 Exercises	124

13	**Similarity**	**125**
	13.1 Triangle similarity theorems	127

| 13.2 The Pythagorean theorem 130
| 13.3 Exercises 133

14 Circles   135
    14.1 Definitions 135
    14.2 Intersections 137
    14.3 The inscribed angle theorem 141
    14.4 Applications of the inscribed angle theorem 144
    14.5 Exercises 147
    14.6 References 148

15 Circumference   149
    15.1 A theorem on perimeters 149
    15.2 Circumference 150
    15.3 Lengths of arcs and radians 156
    15.4 Exercises 158
    15.5 References 158

16 Euclidean Constructions   159
    16.1 Exercises 177
    16.2 References 178

17 Concurrence I   179
    17.1 The circumcenter 179
    17.2 The orthocenter 181
    17.3 The centroid 184
    17.4 The incenter 186
    17.5 Exercises 188

18 Concurrence II   191
    18.1 The Euler line 191
    18.2 The nine point circle 193
    18.3 The center of the nine point circle 196
    18.4 Exercises 198

19 Concurrence III   199
    19.1 Excenters and excircles 199
    19.2 Ceva's theorem 200
    19.3 Menelaus's theorem 205
    19.4 The Nagel point 207
    19.5 Exercises 209

20 Trilinear Coordinates   211
    20.1 Trilinear coordinates 211
    20.2 Trilinears of the classical centers 215
    20.3 Exercises 222

III Euclidean Transformations 223

21 Analytic Geometry 225
21.1 Analytic geometry 225
21.2 The unit circle approach to trigonometry 230
21.3 Exercises 235

22 Isometries 239
22.1 Definitions 239
22.2 Fixed points 243
22.3 The analytic viewpoint 245
22.4 Exercises 247

23 Reflections 249
23.1 The analytic viewpoint 254
23.2 Exercises 256

24 Translations and Rotations 257
24.1 Translation 258
24.2 Rotations 261
24.3 The analytic viewpoint 263
24.4 Exercises 264

25 Orientation 267
25.1 Exercises 271

26 Glide Reflections 273
26.1 Glide reflections 273
26.2 Compositions of three reflections 275
26.3 Exercises 280

27 Change of Coordinates 281
27.1 Vector arithmetic 281
27.2 Change of coordinates 285
27.3 Exercises 290

28 Dilation 291
28.1 Similarity mappings 291
28.2 Dilations 293
28.3 Preserving incidence, order, and congruence 297
28.4 Exercises 300

29 Applications of Transformations 303
29.1 Varignon's theorem 303
29.2 Napoleon's theorem 305

29.3 The nine point circle	308
29.4 References	310
29.5 Exercises	311

30 Area I   313
30.1 The area function	313
30.2 The laws of sines and cosines	319
30.3 Heron's formula	324
30.4 References	327
30.5 Exercises	327

31 Area II   329
31.1 Areas of polygons	329
31.2 The area of a circle	336
31.3 Exercises	338

32 Barycentric Coordinates   341
32.1 The vector approach	342
32.2 The connection to area and trilinears	346
32.3 Barycentric coordinates of triangle centers	350
32.4 References	352
32.5 Exercises	352

33 Inversion   353
33.1 Stereographic projection	353
33.2 Inversion	358
33.3 Exercises	367

34 Inversion II   369
34.1 Complex numbers, complex arithmetic	369
34.2 The geometry of complex arithmetic	373
34.3 Properties of the norm and conjugate	377
34.4 Exercises	378

35 Applications of Inversion   381
35.1 Orthogonal circles	381
35.2 The arbelos	387
35.3 Steiner's porism	389
35.4 Exercises	391

IV Hyperbolic Geometry   395

36 The Search for a Rectangle   397
36.1 If there were a rectangle ...	397

        36.2 The search for a rectangle 403
        36.3 References 409
        36.4 Exercises 409

37 Non-Euclidean Parallels   411
        37.1 Exercises 419

38 The Pseudosphere   421
        38.1 Surfaces 422
        38.2 Maps between surfaces. The Gauss map 425
        38.3 Gaussian curvature 428
        38.4 The tractrix and pseudosphere 430
        38.5 Exercises 432

39 Geodesics on the Pseudosphere   433
        39.1 Geodesics, the theory 433
        39.2 Geodesics, the calculations 434
        39.3 References 442
        39.4 Exercises 442

40 The Upper Half Plane   443
        40.1 Distance 443
        40.2 Angle measure 446
        40.3 Extending the domain 450
        40.4 Exercises 452

41 The Poincaré disk   455
        41.1 To the Poincaré disk model 455
        41.2 Interpreting "undefineds" in the Poincaré disk 457
        41.3 Verifying the axioms 458
        41.4 Exercises 473

42 Hyperbolic Reflections   475
        42.1 Exercises 482

43 Orientation-Preserving Hyperbolic Isometries   483
        43.1 An important example 485
        43.2 Classification by fixed points 486
        43.3 References 490
        43.4 Exercises 490

44 The Six Hyperbolic Trigonometric Functions   493
        44.1 Exercises 498

45 Hyperbolic Trigonometry — 501
- 45.1 Pythagorean theorem … 502
- 45.2 Sine and cosine in a hyperbolic triangle … 505
- 45.3 Circumference of a hyperbolic circle … 509
- 45.4 On a small scale … 511
- 45.5 Exercises … 512

46 Hyperbolic Area — 515
- 46.1 Area on the pseudosphere … 516
- 46.2 Areas of polygons in the Poincaré disk … 522
- 46.3 Area of a circle … 525
- 46.4 Exercises … 527

47 Tiling — 529
- 47.1 Exercises … 535

Bibliography — 537
Index — 539

Preface

Several years ago, I was asked to teach a junior-level Euclidean and non-Euclidean geometry course. I of course said yes, but I was woefully underprepared—I had never taken a comparable course as a student, and the modern geometry that I had studied seemed vastly different from the classical approach I had agreed to teach. I could easily appreciate how important it was to develop the subject systematically, and to present it to my students as such, but I was often frustrated in my attempts to do so. This book is my attempt at a systematic development of Euclidean and hyperbolic geometry. It is divided into four major parts: neutral geometry, Euclidean geometry, Euclidean transformations, and hyperbolic geometry. While we never delve into great depth in any of these subjects, there is quite a bit of material even at the introductory level.

This book takes a patient approach to the subject, and I hope that most of it should be accessible to a wide spectrum of students. It does assume that the reader has a solid comprehension of the methods of mathematical proof. Beyond that, there are places where the material calls upon other areas of mathematics, including, perhaps, some that are unfamiliar. I have not shied away from these topics. As examples:

- The development of the distance function involves associating the points of a line with the real numbers on the real number line, and this involves some discussion of the idea of Dedekind cuts.
- The analytic equations for isometries require some basic linear algebra.
- Later, the equations for inversions, and much of the work in hyperbolic geometry, assumes a familiarity with complex numbers.
- The development of the pseudosphere model for hyperbolic geometry requires some multivariable calculus.
- The calculations of geodesics on the pseudosphere requires multivariable calculus and some differential equations.

For the most part, the topics are relatively self-contained (the exception is complex numbers, which are unavoidable if you wish to discuss hyperbolic geometry).

I have tried to present the material so that motivated (and prepared) readers can read through and understand this subject on their own, outside of a traditional class structure. This is why I have divided the book into many short chapters, why it is not formal in tone, and why it has so many illustrations. Many readers, however, will encounter this material within the framework of a traditional one-semester course. In this setting, it will not be possible to cover all of the material. Realizing this, there are several possible paths through this book.

- It would be possible to build a one-semester course that deals only with Euclidean geometry. This path would focus on the first three parts of the book, covering chapters 1–29. In this approach, the ultimate goal would be chapters 17–19 (the concurrence chapters) and chapter 29 (applications of transformations). Chapters 0, 16, 20, and 30–32 could be considered optional material. It might also be wise to condense chapters 6, 7, and 21, and to rely upon the student's intuition there.
- It would also be possible to build a one-semester course that quickly leads up to, and then focuses upon, hyperbolic geometry. Here the necessary material would be chapters 1–10, (neutral geometry), 22–23 (isometries), 33–34 (inversion), and 36–47 (hyperbolic geometry). The material in chapters 22 and 23 is written with Euclidean geometry in mind, but largely applies to non-Euclidean geometry as well. Depending upon how familiar students are with the basics of Euclidean geometry, it may also be necessary to fill in some material before chapter 22. Chapters 38–40 require relatively more advanced differential geometry, but provide important motivation for the Poincaré disk model that follows.
- Usually, when I teach this course, I try to do a little bit of both Euclidean and non-Euclidean geometry. I try to cover the material in chapters 1–14, 16–18, 21–24, 27, 33–34, 36–37, and 41–45. With this approach, I will skim through chapters 6 and 7, dealing with distance and angle measure. I will present the results of chapters 17 and 18 (concurrences) from the point of view of Euclidean constructions rather than formally proving them. Finally, I will motivate the Poincaré disk model by describing the pseudosphere, without getting into any details or calculations.

In all these approaches, some important material is underemphasized, but could serve as openings for student projects and further study: chapter 0, which touches on finite projective planes; chapters 20 and 32, dealing with triangular coordinate systems; chapter 35, which shows some nice applications of inversion; and the chapters dealing with area, chapters 30, 31, and 46.

I would like to thank my family, friends, and colleagues who supported me as I worked on this book. Without their support, it never would have been finished.

0
Axioms and Models

The most elementary objects of plane geometry are points and lines. Both are, of course, abstractions. You will not find them in the real world. There are point-like objects out there (atoms might be a good example), and there are line-like objects too (laser beams come to mind). But we understand the points and lines of geometry to be an idealization of these imperfect representations. That is, we feel that these physical manifestations fall short of "true" points and lines. Points and lines, in other words, are not things we can point to in the real world. In a casual setting, that may not be important. After all, the whole of human experience requires us to deal with abstraction in a variety of contexts every day. But it is a bit more problematic to try to develop a precise mathematical system from them. Consider the opening statements in Euclid's *Elements*.

> Definition 1. A point is that which has no part.
> Definition 2. A line is breadthless length.

There is a certain poetry in those definitions, but it would be difficult to build solid proofs on top of them. Euclid doesn't define a part, nor does he define breadth or length. Were he to define them, they would have to be described using other terms, which would in turn need their own definitions, and so on. It isn't that Euclid's definitions are bad—it is that this is a hopeless situation: you can't define everything.

Modern geometry takes a fundamentally different approach to the issue of elementary terms. Like its classical counterpart, modern geometry is built upon a foundation of a few basic terms, such as point and line. Unlike the classical approach, in modern geometry no effort is made to define them. In fact, they are called the *undefineds* of the system. While the undefineds themselves have no meaning, their behavior, properties, and interactions are described in a set of statements called the *axioms* of the system. No effort is made to argue for the truth of the axioms (how could you do so?—they are statements about terms that themselves have no meaning). As long as the axioms do not contradict one another, they will describe some kind of geometry. It may be quite different from the Euclidean geometry to which we are accustomed, but it will be a geometry nonetheless.

This axiomatic approach immediately calls into question a common expectation of geometry (particularly plane geometry), that it should be the kind of math where there are pictures and illustrations. Before, I could scratch a pencil across the page and say that this is basically a line here, it is just a little imperfect because I reached the edge of the page, and the pencil wobbled, so just imagine it continued on forever and was perfectly straight. In the axiomatic approach, though, a line has no definition, and so that picture of a line may not just be imperfect, it may be completely wrong.

Nevertheless, illustrations can provide valuable insight. In that case, what is the relationship between illustrations and axioms? First, we have to accept that the illustrations are imperfect. Lines printed on paper have thickness and are finite in length. We also have to accept that illustrations can mislead: for instance, if the only triangles you draw are acute ones, then you might think that the altitude from a vertex of a triangle would always intersect the opposite side, when this is not true for obtuse triangles. We also must remind ourselves that illustrations may represent only one interpretation of the axioms. Points and lines as we traditionally depict them are one way to interpret the undefined terms of point and line, and it is an interpretation that happens to be consistent with all of the standard axioms of Euclidean geometry. But there may be a completely different interpretation of the undefineds that also satisfies the Euclidean axioms. Any such interpretation is called a *model* for the geometry. A geometry may have many models, and from a theoretical point of view, no one model is more right than any other. It is important, then, to prove facts about the geometry itself, and not peculiarities of one particular model.

0.1 Fano's geometry

To see how axiomatic geometry works without having our Euclidean intuition getting in the way, let's consider a decidedly non-Euclidean geometry called Fano's geometry (named after the Italian algebraic geometer Gino Fano). In Fano's geometry there are three undefined terms, *point*, *line*, and *on*. Five axioms govern these undefined terms.

Ax 1: There exists at least one line.
Ax 2: There are exactly three points on each line.
Ax 3: Not all points are on the same line.
Ax 4: There is exactly one line on any two distinct points.
Ax 5: There is at least one point on any two distinct lines.

Fano's geometry is a simple example of what is called a finite projective geometry. It is projective because, by the fifth axiom, all lines intersect one another (lines cannot be parallel). It is finite because, as we will see, it contains only finitely many points and lines. To get a sense of how an axiomatic proof works, let's count the points and lines in Fano's geometry. The proof of theorem 0.1 is in the style of the "two-column proof" of high school geometry, clearly identifying which axiom justifies each statement. At the same time, a chart will track the established incidences of points on lines.

Theorem 0.1 *Fano's geometry has exactly seven points and seven lines.*

0.1 Fano's geometry

Proof

PT \ LN	1	2	3	4
1	•	•		
2	•		•	
3	•			•
4		•	•	•

Ax 1: There is a line ℓ_1.
Ax 2: On ℓ_1, there are three points. Label them p_1, p_2, and p_3.
Ax 3: There is a fourth point p_4 that is not on ℓ_1.
Ax 4: There are lines ℓ_2 on p_1 and p_4, ℓ_3 on p_2 and p_4, and ℓ_4 on p_3 and p_4. They are distinct.

PT \ LN	1	2	3	4
1	∘	∘		
2	∘		∘	
3	∘			∘
4		∘	∘	∘
5		•		
6			•	
7				•

Ax 2: Each of ℓ_2, ℓ_3, and ℓ_4 has a third point on it.
Ax 4: The points are distinct and different from any of the previously declared points. Label them p_5 on ℓ_2, p_6 on ℓ_3, and p_7 on ℓ_4.

PT \ LN	1	2	3	4	5
1	∘	∘			•
2	∘		∘		
3	∘			∘	
4		∘	∘	∘	
5		∘			
6			∘		•
7				∘	•

Ax 4: There must be a line ℓ_5 on p_1 and p_6.
Ax 2: The line ℓ_5 must have one more point on it.
Ax 4: That point cannot be either p_3 or p_4.
Ax 5: For ℓ_5 and ℓ_4 to intersect, the third point of ℓ_5 must be p_7.

PT \ LN	1	2	3	4	5	6
1	∘	∘			∘	
2	∘		∘			•
3	∘			∘		
4		∘	∘	∘		
5		∘				•
6			∘		∘	
7				∘	∘	•

Ax 4: There must be a line ℓ_6 on p_2 and p_5.
Ax 2: The line ℓ_6 must have a third point on it.
Ax 4: That point cannot be p_3 or p_4.
Ax 5: For ℓ_6 and ℓ_4 to intersect, the third point of ℓ_6 must be p_7.

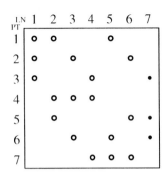

Ax 4: There must be a line ℓ_7 on p_3 and p_5.
Ax 2: The line ℓ_7 must have one more point on it.
Ax 4: That point cannot be p_2 or p_4.
Ax 5: For ℓ_7 and ℓ_3 to intersect, the third point of ℓ_7 must be p_6.

We now have seven points and seven lines as required. Could there be more? Let's suppose there were an eighth point p_8.

Ax 4: Then there would be a line ℓ_8 on p_1 and p_8.
Ax 3: Line ℓ_8 would have to have another point on it.
Ax 4: This other point would have to be distinct from each of p_2 through p_7.
Ax 5: Then ℓ_8 would not share a point with ℓ_3 (and other lines as well). Thus there cannot be an eighth point.
Ax 4: There is now a line on every pair of points. Therefore there can be no more lines.

$\square$

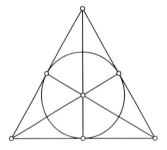

Figure 1. A model for Fano's geometry. The nodes of the graph represent the points. The six segments and the circle represent the lines.

0.2 Further reading

Euclid's *Elements* is still an interesting read. There are several editions available, both in text form and online, including, for instance, [Euc02] or [Euc56]. For more about projective geometry in general, try Coxeter's book [Cox69]. For finite projective planes, try the set of online notes by Jurgen Bierbrauer [Bie04]. At the time of this writing they are available at the web address: www.math.mtu.edu/~jbierbra/HOMEZEUGS/finitegeom04.ps.

0.3 Exercises

The following exercises deal with a finite incidence geometry called *Young's geometry*. As in Fano's geometry, there are three undefined terms: *point*, *line*, and *on*. In Young's geometry, they are governed by the following axioms.

(1) There exists at least one line.
(2) There are exactly three points on each line.
(3) Not all points are on the same line.
(4) There is exactly one line on any two distinct points.
(5) For a line L and a point P not on L, there is exactly one line on P that does not contain any points on L.

We say that two lines *intersect* if there is a point on both of them, and that they are *parallel* if there there are no points on both of them. Then the last axiom says that there is exactly one line through P that is parallel to L.

0.1. Prove that every point is on exactly four lines.

0.2. Prove that if lines L and M intersect, and L is parallel to a third line N, then lines M and N must intersect.

0.3. Prove that every line has exactly two lines parallel to it.

0.4. Prove that there are twelve lines and nine points in Young's geometry.

0.5. Construct a model for Young's geometry.

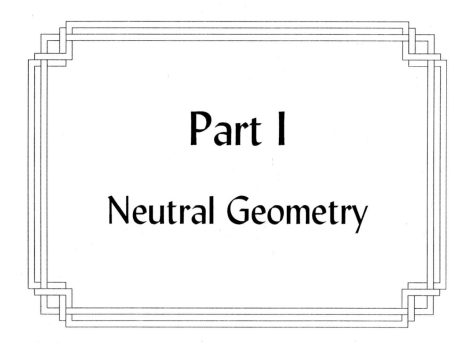

Part I
Neutral Geometry

The goal of this book is to provide a pleasant but thorough introduction to Euclidean and non-Euclidean geometry. There are many geometries that are not Euclidean, by virtue of having features that differ from Euclidean geometry. For instance, Fano's geometry (discussed in the last chapter) is a geometry that is clearly not Euclidean. For the rest of this book, however, when we speak of non-Euclidean geometry, we refer to hyperbolic geometry. Hyperbolic geometry differs from Euclidean geometry in one specific way—it satisfies all of Hilbert's axioms for Euclidean geometry except the parallel axiom.

It turns out that the parallel axiom is absolutely central to the nature of the geometry. The Euclidean geometry that can be developed from the parallel axiom is radically different from the hyperbolic geometry that can be developed from its negation. Even so, Euclidean and non-Euclidean geometry are not opposites. As different as they are in many ways, they share many basic characteristics. Neutral geometry (also known as absolute geometry in older texts) is the study of those commonalities. The first part of this book is dedicated to developing neutral geometry.

1
The Axioms of Incidence and Order

Any study of Euclidean geometry must include at least some mention of Euclid's *Elements*, a masterpiece of classical Greek mathematics, in which Euclid attempts a systematic development of geometry. The *Elements* opens with a short list of definitions. As discussed in the introductory chapter, the first few are a little problematic. If we can push past those, we get to Euclid's five postulates, the core accepted premises of his development of the subject.

Euclid's postulates.

P1: To draw a straight line from any point to any point.
P2: To produce a finite straight line continuously in a straight line.
P3: To describe a circle with any center and distance.
P4: That all right angles are equal to one another.
P5: That, if a straight line falling on two straight lines makes the interior angles on the same side less than two right angles, the two straight lines, if produced indefinitely, meet on the side on which the angles are less than two right angles.

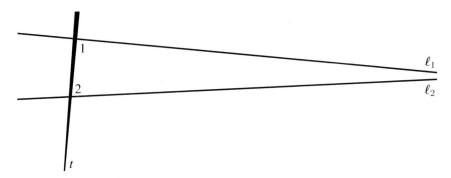

Figure 1. Euclid's fifth postulate (the parallel postulate): Because $(\angle 1) + (\angle 2) < 180°$, ℓ_1 and ℓ_2 will intersect on the *right* side of t in this illustration.

The first three postulates describe constructions. Today we would probably reinterpret them as statements about the existence of certain objects (namely lines, segments, and circles). The fourth provides a way to compare angles. As for the fifth... well, it needs to be said that Euclid's fifth postulate does not look like the other four. It is considerably longer and more complicated than the others. For that reason, generations of geometers after Euclid hoped that the fifth might actually be provable—that it could be taken as a theorem rather than a postulate. To do that, they had to work in a geometry which neither assumes nor negates it. This geometry is now called neutral geometry.

So what exactly do you give up when you decide not to use Euclid's fifth? Essentially, the postulate tells us something about the nature of parallel lines. It does so in a rather indirect way, though. It is now more common to use Playfair's axiom in place of Euclid's fifth because it addresses the issue of parallels much more directly.

Playfair's axiom. For any line ℓ and for any point P that is not on ℓ, there is exactly one line through P that is parallel to ℓ.

Playfair's axiom both implies and is implied by Euclid's fifth, so the two statements can be used interchangeably. Even without Playfair's axiom, it is relatively easy to show that in neutral geometry there must be at least one parallel through P, so what Playfair's axiom is really saying is that in Euclidean geometry there cannot be more than one parallel. The existence of a unique parallel is crucial to many of the proofs of Euclidean geometry. Without it, neutral geometry is quite limited. Still, neutral geometry is the common ground between Euclidean and non-Euclidean geometries, and it is where we begin our study.

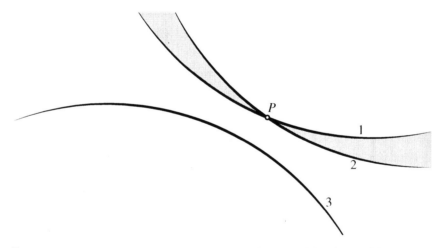

Figure 2. It looks like both lines 1 and 2 pass thorugh P and are parallel to line 3. If that is the case, this is not a possible arrangement of *Euclidean* lines.

Looking at Euclid's postulates, and his subsequent development of the subject from them, it appears that Euclid may have been attempting an axiomatic development of the subject. There is some debate about the extent to which Euclid really was trying to do that. His handling of the side-angle-side theorem, for example, is not founded upon the postulates, and not merely in a way that might be attributed to oversight. These gaps became evident as future generations of geometers tried to work within his system to prove the fifth postulate. Euclidean geometry was finally properly and completely axiomatized at the end of the nineteenth century by the German mathematician David Hilbert. His 1899 book, *The Foundations of Geometry*, gives an axiomatic description of what we think of as Euclidean geometry. Subsequently, there have been several other axiomatizations, including notably ones by Birkhoff and Tarski. The nice thing about Hilbert's approach is that proofs developed in his system "feel" like Euclid's proofs. Some of the other axiomatizations, while more streamlined, do not retain the same feel.

1.1 Incidence

In the first part of this book, we are going to develop neutral geometry following the approach of Hilbert. In Hilbert's system there are five undefined terms: *point, line, on, between*, and *congruent*. Fifteen of his axioms are needed to develop neutral plane geometry. Generally the axioms are grouped into categories to make it easier to keep track of them: the axioms of incidence, the axioms of order, the axioms of congruence, and the axioms of continuity. We will investigate them in that order over the next several chapters.

Hilbert's first set of axioms, the axioms of incidence, describe the interaction between points and lines provided by the term *on*. *On* is a binary relationship between points and lines so that, for instance, we can say that a point P is (or is not) on a line ℓ. In situations where we would want to express a line's relationship to a point, rather than saying that a line ℓ is on a point P (which is technically correct), it is much more common to say that ℓ passes through P.

The axioms of incidence.

In 1: There is a unique line on (passing through) any two distinct points.
In 2: There are at least two points on any line.
In 3: There are at least three points that do not all lie on the same line.

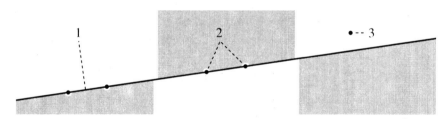

Figure 3. The axioms of incidence: a line through two points, two points on a line, but not all points are on the same line.

By themselves, the axioms of incidence do not afford a great wealth of theorems. Some notation and a few definitions are all we get. Because of the first axiom, there is only one line through any two distinct points. Therefore, for two distinct points A and B, we use the notation $\overline{AB}$ to denote the line through A and B. The first axiom guarantees that any two points share a line, and the third axiom guarantees that not all points share the same line. Between those extremes, it is generally useful to know whether points are or are not all on the same line.

Definition 1.1 Three or more points are *collinear* if they are all on the same line and are *non-collinear* if they are not.

According to the first axiom, two distinct lines can share at most one point. However, they may not share any points at all.

Definition 1.2 Two lines *intersect* if there is a point P that is on both of them. In this case, P is the *intersection* or *point of intersection* of them. Two lines that do not share a point are *parallel*.

On occasion we will use the notation $\ell_1 \parallel \ell_2$ to indicate that ℓ_1 and ℓ_2 are parallel.

1.2 Order

The axioms of order describe the undefined term *between*. Between is a relation of a point to a pair of points. We say that a point B is, or is not, between two points A and C and use the notation $A * B * C$ to indicate that B is between A and C. Closely related to this "between-ness" is the idea that a line separates the plane into two parts. This behavior, which is explained in the last of the order axioms, depends upon the following definition.

Definition 1.3 Let ℓ be a line and let A and B be two points that are not on ℓ. A and B are on the *same side* of ℓ if ℓ and $\overline{AB}$ do not intersect, or if they do but the point of intersection is not between A and B.

There are four axioms of order.

The axioms of order.

Or 1: If $A * B * C$, then A, B, and C are distinct collinear points, and $C * B * A$.
Or 2: For any two distinct points B and D, there are points A, C, and E, such that $A * B * D$, $B * C * D$, and $B * D * E$.
Or 3: Of any three distinct points on a line, exactly one lies between the other two.
Or 4: (The plane separation axiom) Given a line ℓ and points A, B, and C that are not on ℓ. If A and B are on the same side of ℓ and A and C are on the same side of ℓ, then B and C are on the same side of ℓ. If A and B are not on the same side of ℓ and A and C are not on the same side of ℓ, then B and C are on the same side of ℓ.

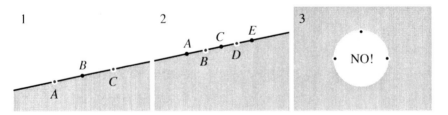

Figure 4. The first three axioms of order: three distinct points can be put in order, there are always points between and beyond two distinct points, and lines never form loops.

The last of these, the plane separation axiom is a bit more difficult to digest than the previous axioms, but it is particularly critical—it is the axiom that limits plane geometry to two dimensions. Let's take a closer look. Let ℓ be a line and let P be a point that is not on ℓ. Consider two sets of points:

S_1 : P itself and all points on the same side of ℓ as P.
S_2 : all points that are not on ℓ nor on the same side of ℓ as P.

Together, they contain all points that are not on ℓ.

1.2 Order

Exercise 1.1 How do we know that S_1 and S_2 are nonempty?

The first part of the plane separation axiom tells us is that any two points of S_1 are on the same side of ℓ; the second part tells us that any two points of S_2 are on the same side of ℓ. Hence there are two and only two sides to a line. Because of this, we can refer to points that are not on the same side of a line as being on *opposite sides*.

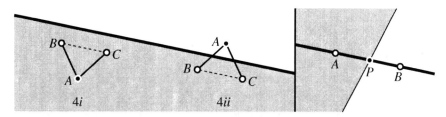

Figure 5. (left) and (center) The two parts of the fourth axiom of order (the plane separation axiom) guarantee that a line separates the plane. (right) As a consequence, any point on a line separates that line.

Just as a line separates the remaining points of the plane, a point on a line separates the remaining points on the line. If P, A, and B are collinear and if P is between A and B, then A and B are on *opposite sides* of P. Otherwise, A and B are on the *same side* of P (this separation of a line by a point could be called "line separation"). It is a direct descendent of plane separation via the following simple correspondence. For three distinct points A, B, and P on a line ℓ,

A, B are on the same side of P $\iff$ A, B are on the same side of any line through P other than ℓ,

and

A, B are on opposite sides of P $\iff$ A, B are on opposite sides of any line through P other than ℓ.

Exercise 1.2 Verify this correspondence.

Because of this, there is a counterpart to the plane separation axiom for lines. Suppose that A, B, C, and P are all on a line. If A and B are on the same side of P and A and C are on the same side of P, then B and C are on the same side of P. If A and B are on opposite sides of P and A and C are on opposite sides of P, then B and C are on the same side of P. As a result, a point divides a line into exactly two sides.

With between, we can now give a few fundamental definitions.

Definition 1.4 Given two distinct points A and B, the *line segment* between A and B is the set of points P such that $A * P * B$, together with A and B themselves. The points A and B are called the *endpoints* of the segment.

Definition 1.5 Given two distinct points A and B, the *ray* from A through B is the set containing the point A and all the points on $\overline{AB}$ that are on the same side of A as B. The point A is called the *endpoint* of the ray.

The notation for the line segment between A and B is AB. The ray with endpoint A that passes through the point B is written $\overrightarrow{AB}$.

Definition 1.6 For a ray $\overrightarrow{AB}$, the *opposite ray* $(\overrightarrow{AB})^{op}$ consists of the point A together with all the points of $\overline{AB}$ that are on the opposite side of A from B.

Exercise 1.3 Verify that an opposite ray is a ray.

1.3 Putting points in order

The order axioms describe how to put three points in order. Sometimes, though, three is not enough. It would be nice to know that more than three points on a line can be ordered in a consistent way. The axioms of order make this possible.

Theorem 1.7 *Given $n \geq 3$ collinear points, there is a labeling of them $P_1, P_2, \ldots, P_n$ so that if $1 \leq i < j < k \leq n$, then $P_i * P_j * P_k$. In that case, we write*

$$P_1 * P_2 * \cdots * P_n.$$

We will prove this by induction. Assuming that we can order n collinear points, and that we want to order $n + 1$ points, the strategy seems obvious: isolate the first point (call it Q), put the remaining points in order, and then stick Q back on the front. The problem with this is that figuring out which point is the first point essentially presupposes that you *can* put the points in order. Getting around this is a little delicate.

Proof This is a proof by induction on the number of points in the list. The initial case, where there are just three points to put in order, is an immediate consequence of the axioms of order. Now let's assume that any set of n collinear points can be put in order, and suppose that we want to put a set of $n + 1$ collinear points in order. Choose n of the $n + 1$ points. Put them in order and label them so that $p_1 * p_2 * \cdots * p_n$. Let q be the one remaining point. One of three things must happen:

$$q * p_1 * p_2 \quad \text{or} \quad p_1 * q * p_2 \quad \text{or} \quad p_1 * p_2 * q.$$

In the first case, let $Q = q$ and let $P_1 = p_1, P_2 = p_2, \ldots, P_n = p_n$. In the second and third cases, let $Q = p_1$. Then put the remaining points $p_1, \ldots, p_n$ and q in order and label them $P_1, P_2, \ldots, P_n$. Having done this, we have two pieces of an ordering

$$Q * P_1 * P_2 \quad \text{and} \quad P_1 * P_2 * \cdots * P_n.$$

The P's are all properly ordered now but the proof is not yet complete. We still need to show that Q is ordered properly with respect to the remaining P's. That is, we need to show $Q * P_i * P_j$ when $1 \leq i < j \leq n$. Let's do that (in several cases).

1.3 Putting points in order

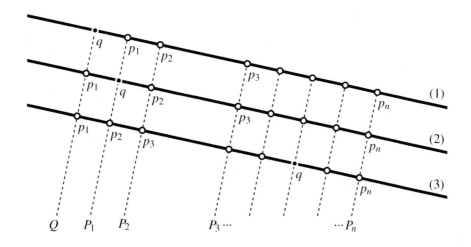

Figure 6. The three possible positions of q in relation to p_1 and p_2.

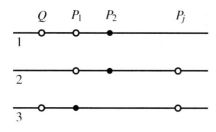

Figure 7. Case 1.

Case 1: $i = 1$. The result is given when $j = 2$, so let's suppose that $j > 2$. Then

(1) $Q * P_1 * P_2$ so Q and P_1 are on the same side of P_2.
(2) $P_1 * P_2 * P_j$ so P_1 and P_j are on opposite sides of P_2.
(3) Therefore Q and P_j are on opposite sides of P_1, so $Q * P_1 * P_j$.

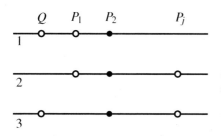

Figure 8. Case 2.

Case 2: $i = 2$.

(1) $Q * P_1 * P_2$ so Q and P_1 are on the same side of P_2.
(2) $P_1 * P_2 * P_j$ so P_1 and P_j are on opposite sides of P_2.
(3) Therefore Q and P_j are on opposite sides of P_2, so $Q * P_2 * P_j$.

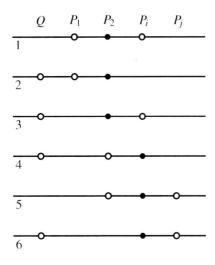

Figure 9. Case 3.

Case 3: $i > 2$.

(1) $P_1 * P_2 * P_i$ so P_1 and P_i are on opposite sides of P_2.
(2) $Q * P_1 * P_2$ so Q and P_1 are on the same side of P_2.
(3) Therefore Q and P_i are on opposite sides of P_2, so $Q * P_2 * P_i$.
(4) Consequently, Q and P_2 are on the same side of P_i.
(5) Meanwhile, $P_2 * P_i * P_j$ so P_2 and P_j are on opposite sides of P_i.
(6) Therefore, Q and P_j are on opposite sides of P_i, so $Q * P_i * P_j$.

□

There is a "uniqueness" result that goes with this theorem. It is left as an exercise.

Exercise 1.4 Prove that the ordering of points $P_1 * P_2 * \cdots * P_n$ as described in the previous theorem is one of only two possible consistent orderings of the points (the other is to list them in the reverse order).

With this consistency of ordering, it is possible to make (and prove) basic statements about containments of segments in other segments, and rays in other rays. We will prove one such result; others can be found in the exercises. Recall the following standard notation from set theory:

$X \subset Y$: X is a *subset* of Y,
$X \cup Y$: the *union* of X and Y,
$X \cap Y$: the *intersection* of X and Y.

Theorem 1.8 *If $A * B * C$, then $AB \subset AC$.*

Proof Let P be a point on the segment AB. Then, by definition, either $P = A$ or $P = B$ or $A * P * B$. If $P = A$, then clearly P is on the segment AC. If $P = B$, then we are given that $A * P * C$, so again P is on AC. Finally, if $A * P * B$, then we have $A * P * B * C$, and so $A * P * C$. Therefore, in all cases, P is on the segment AC, and so $AB \subset AC$. □

1.4 Exercises

1.5. The third axiom of incidence states that there are at least three points that are not on the same line. Extend this by proving:

- For any line, there is a point which is not on it.
- For any point, there is a line which does not pass through it.

1.6. Prove that if $A * B * C$, then $AB \cup BC = AC$ and $AB \cap BC = B$.

1.7. Prove that if $A * B * C$, then $\overrightarrow{AB} = \overrightarrow{AC}$ and $\overrightarrow{BC} \subset \overrightarrow{AC}$.

1.8. Prove that if $A * B * C$, then $(\overrightarrow{BC})^{op} = \overrightarrow{BA}$.

1.9. Prove that if $A * B * C * D$ then $AC \cup BD = AD$ and $AC \cap BD = BC$.

1.10. For two distinct points A and B, prove that $(\overrightarrow{AB}) \cup (\overrightarrow{BA}) = \overleftrightarrow{AB}$.

1.11. For two distinct points A and B, prove that $(\overrightarrow{AB}) \cap (\overrightarrow{BA}) = AB$.

1.12. Prove that there are infinitely many points on a line.

1.13. Prove that there are infinitely many lines through a point.

1.14. The familiar model for Euclidean geometry is the "Cartesian model." In that model, points are interpreted as coordinate pairs of real numbers (x, y). Lines are loosely interpreted as equations of the form

$$Ax + By = C,$$

but technically, there is a little bit more to it than that. First, A and B cannot simultaneously be zero. Second, if $A' = kA$, $B' = kB$, and $C' = kC$ for some nonzero constant k, then the equations $Ax + By = C$ and $A'x + B'y = C'$ both represent the same line (in truth then, a line is represented by an equivalence class of equations). In this model, a point (x, y) is on a line $Ax + By = C$ if its coordinates make the equation true. With this interpretation, verify the axioms of incidence.

1.15. In the Cartesian model, a point (x_2, y_2) is between two other points (x_1, y_1) and (x_3, y_3) if

- the three points are distinct and on the same line,
- x_2 is between x_1 and x_3 (either $x_1 \leq x_2 \leq x_3$ or $x_1 \geq x_2 \geq x_3$), and
- y_2 is between y_1 and y_3 (either $y_1 \leq y_2 \leq y_3$ or $y_1 \geq y_2 \geq y_3$).

With this interpretation, verify the axioms of order.

1.5 Further reading

At this stage, with few results to build from, there is very little flexibility about where to go next. Since we have adopted the axioms of Hilbert, our initial steps (in this and the next few chapters) follow fairly closely those of Hilbert in his *Foundations of Geometry* [Hil50]. In addition, there are many more contemporary books that examine the first steps in the development of the subject. Moise's *Elementary Geometry from an Advanced Standpoint* [Moi74] is one of my favorites. I have taught from both Wallace and West's *Roads to Geometry* [WW04], and Greenberg's *Euclidean and Non-Euclidean Geometries* [Gre08].

2
Angles and Triangles

Through these first few chapters, we must move with caution. It is difficult to keep intuition from making unjustified leaps. All of the theorems of this chapter are examples of this. They just seem so obvious that it would be easy to overlook them (and historically, they were at times overlooked). But these are results that we will need to use, so they need to be proved.

In the last chapter we defined ray and segment. They are the most elementary of objects, defined directly from the undefined terms. This chapter begins with the next layer: angles and triangles, which are built from rays and segments.

Definition 2.1 An *angle* consists of a (unordered) pair of non-opposite rays with the same endpoint. The mutual endpoint is called the *vertex* of the angle.

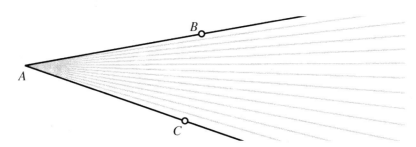

Figure 1. The angle $\angle BAC$ formed by the rays $\overrightarrow{AB}$ and $\overrightarrow{AC}$. Point A is the vertex.

The angle formed by $\overrightarrow{AB}$ and $\overrightarrow{AC}$ is written $\angle BAC$, with the vertex listed in the middle spot. There is more than one way to indicate the angle. For one, it does not matter which order the rays are taken, so $\angle CAB$ refers to the same angle as $\angle BAC$. If B' is on $\overrightarrow{AB}$ and C' is on $\overrightarrow{AC}$ (not the endpoint of course), then $\angle B'AC'$ is the same as $\angle BAC$. Frequently, it is clear from context that you only care about one angle at a particular vertex. On those occasions it is convenient to use the abbreviation $\angle A$ in place of the full $\angle BAC$. Just as a line divides the plane into two sides, so too does an angle. The two parts are called the interior and the exterior of the angle.

Definition 2.2 A point lies *in the interior* or is an *interior point* of $\angle BAC$ if it is on the same side of $\overline{AB}$ as C and same side of $\overline{AC}$ as B. A point that does not lie in the interior of the angle and does not lie on either of the rays composing the angle is *exterior* to the angle and is called an *exterior point*.

Just as an angle is formed by joining two rays at their mutual endpoint, a triangle is formed by joining three segments at mutual endpoints.

Definition 2.3 A *triangle* is a set of three non-collinear points called *vertices*, together with all the points on the segments between each pair of vertices. The segments between vertices are called the *sides* or *edges* of the triangle.

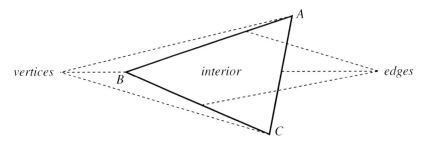

Figure 2. Parts of the triangle $\triangle ABC$.

We write $\triangle ABC$ to denote the triangle whose vertices are at points A, B, and C. The ordering of the three vertices does not matter, so there is more than one way to write a given triangle: the notations $\triangle ABC$, $\triangle ACB$, $\triangle BAC$, $\triangle BCA$, $\triangle CAB$, and $\triangle CBA$ all refer to the same triangle (although it should be noted that the ordering of the vertices matters when comparing triangles, such as the triangle congruence theorems of the next few chapters). The three angles $\angle ABC$, $\angle BCA$, and $\angle CAB$ are called the *interior angles* of $\triangle ABC$. A point that is in the interior of all three of the interior angles is said to be *inside* the triangle. Together, those points form the *interior* of the triangle. Points that are neither on the triangle nor in its interior are said to be *outside* the triangle. Together, they form the *exterior* of the triangle.

The rest of this chapter is dedicated to three fundamental theorems. The first, a result about lines crossing triangles, is called Pasch's lemma after Moritz Pasch, a nineteenth century German mathematician whose works are a precursor to Hilbert's. It is a direct consequence of the plane separation axiom. The second result, the crossbar theorem, is a bit more difficult—it deals with lines crossing through the vertex of an angle. The third says that rays from a common endpoint that all lie on one side of a line can be ordered in a consistent way.

Theorem 2.4 *(Pasch's lemma) If a line intersects a side of a triangle at a point other than a vertex, then it must intersect another side of the triangle. If a line intersects all three sides of a triangle, then it must intersect two of the sides at a vertex.*

Proof Suppose that a line ℓ intersects side AB of $\triangle ABC$ at a point P other than the endpoints. If ℓ also passes through C, then that's the other intersection; in this case ℓ passes through all three sides of the triangle, but it passes through two of them at a vertex. Now what if ℓ does not pass through C? There are only two possibilities: either C is on the same side of ℓ as A, or it is on the opposite side of ℓ from A. We can turn to the plane separation axiom. Because P is between A and B, the two points have to be on opposite sides of ℓ. Thus, if C is on the same side of ℓ as A, then it is on the opposite side of ℓ from B, and so ℓ intersects BC but not AC. On the other hand, if C is on the opposite side of ℓ from A, then it is on the same side of ℓ as B, so ℓ intersects AC but not BC. Either way, ℓ intersects two of the three sides of the triangle. □

The crossbar theorem is similar in spirit to Pasch's lemma, but its proof is rather more complicated. It is helpful to separate one small part into the following lemma.

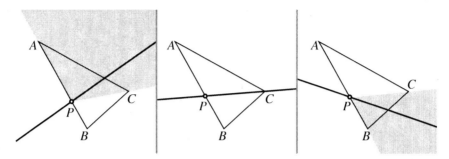

Figure 3. Pasch's lemma in action. (left) ℓ passes through AC; (center) ℓ passes through C; (right) ℓ passes through BC.

Lemma 2.5 *If A is a point on line ℓ, and B is a point that is not on ℓ, then all the points of $\overrightarrow{AB}$ (and therefore all the points of AB) except A are on the same side of ℓ as B.*

Exercise 2.1 Prove lemma 2.5.

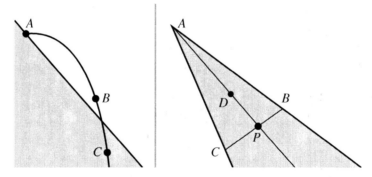

Figure 4. (l) Lemma 2.5 says that a ray cannot cross back over a line like this; (r) the crossbar theorem guarantees the existence of the intersection point P.

Theorem 2.6 *(The crossbar theorem) If D is an interior point of angle $\angle BAC$, then the ray $\overrightarrow{AD}$ intersects the segment BC.*

Although it is phrased differently, this seems close to Pasch's lemma, with $\overline{AD}$ crossing $\triangle ABC$. The problem is that it crosses the triangle at the vertex, and that is the one place where Pasch's lemma is not immediately helpful. Still, the basic idea is sound. The proof does use Pasch's lemma; we just have to bump the triangle a little bit so that $\overrightarrow{AD}$ doesn't cross through the vertex.

Proof According to the second axiom of order, there are points on the opposite side of A from C. Let A' be one of them. Now $\overline{AD}$ intersects the side $A'C$ of $\triangle A'BC$. By Pasch's lemma, $\overline{AD}$ must intersect one of the other two sides of the triangle, either $A'B$ or BC. There are questions. First, what if $\overline{AD}$ crosses $A'B$ instead of BC? Second, what if $\overline{AD}$ does cross BC, but the intersection is on $(\overrightarrow{AD})^{op}$ instead of $\overrightarrow{AD}$?

Let's start by ruling out the second possibility. If D' is any point on $(\overrightarrow{AD})^{op}$, then it is on the opposite side of A from D. Therefore D' and D are on opposite sides of $\overline{A'C}$. Since D is an interior point, it is on the same side of $\overline{A'C}$ as B, and so D' and B are on opposite sides of $\overline{A'C}$. By the previous lemma, all the points of $A'B$ and of BC are on the same side of $\overline{A'C}$ as B. Therefore they are on the opposite side of $\overline{A'C}$ from D', so no point of $(\overrightarrow{AD})^{op}$ may lie on either $A'B$ or BC.

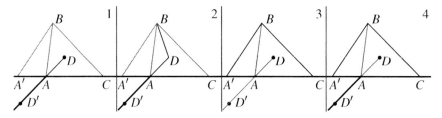

Figure 5. First steps in the proof of the crossbar theorem.

With the opposite ray ruled out entirely, we now just need to make sure that $\overrightarrow{AD}$ does not intersect $A'B$. Points A' and C are on opposite sides of $\overline{AB}$. Because D is in the interior of $\angle BAC$, D and C are on the same side of $\overline{AB}$. Therefore A' and D are on opposite sides of $\overline{AB}$. Using the preceding lemma, all the points of $A'B$ are on opposite sides of $\overline{AB}$ from all the points of $\overrightarrow{AD}$. Hence $\overrightarrow{AD}$ cannot intersect $A'B$, so it must intersect BC.

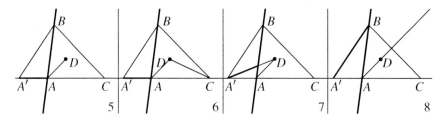

Figure 6. Last steps in the proof of the crossbar theorem.   □

The crossbar theorem provides a essential conduit between the notion of *between* for points and *interior* for angles. We can use it in the next theorem, which is the angle interior analog to the ordering of points theorem in the last chapter. First, a few lemmas.

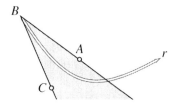

Figure 7. According to lemma 2.7, a ray cannot escape an angle in this way.

Angles and Triangles

Lemma 2.7 *Consider $\angle ABC$ and a ray r whose endpoint is B. Either all the points of r other than B lie in the interior of $\angle ABC$, or none of them do.*

Exercise 2.2 Prove lemma 2.7.

Lemma 2.8 *Suppose that two distinct rays $\overrightarrow{BC_1}$ and $\overrightarrow{BC_2}$ are both on the same side of $\overline{AB}$. Then either $\overrightarrow{BC_1}$ is in the interior of $\angle ABC_2$ or $\overrightarrow{BC_2}$ is in the interior of $\angle ABC_1$.*

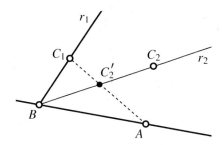

Figure 8. Assuming $\overrightarrow{BC_1}$ is not in the interior of $\angle ABC_2$, then $\overrightarrow{BC_2}$ must be in the interior of $\angle ABC_1$.

Proof Suppose that $\overrightarrow{BC_1}$ is not in the interior of $\angle ABC_2$; that is, point C_1 is not in the interior of $\angle ABC_2$. It is on the same side of $\overline{AB}$ as C_2, so it must be on the opposite side of $\overline{BC_2}$ from A. The segment AC_1 then intersects $\overline{BC_2}$ (and in fact, it intersects the ray $\overrightarrow{BC_2}$) at a point C_2'. The point C_2' is in the interior of $\angle ABC_1$, so by the previous lemma, all of $\overrightarrow{AC_2'} = \overrightarrow{AC_2}$ must be in the interior of $\angle ABC_1$. □

Exercise 2.3 Verify that C_2' in the last proof is on $\overrightarrow{BC_2}$ (rather than the opposite ray).

Theorem 2.9 *(On ordering rays) Consider $n \geq 2$ distinct rays with a common endpoint B that are all on the same side of a line $\overline{AB}$ through B. There is an ordering of them*

$$r_1, r_2, \ldots, r_n$$

so that if $i < j$ then r_i is in the interior of the angle formed by $\overrightarrow{BA}$ and r_j.

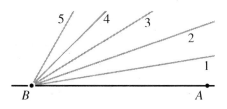

Figure 9. An ordering of five rays and five angles so that each ray is in the interior of all the subsequent angles.

Proof This is a proof by induction. The base case is handled by the previous lemma: choose points C_1 and C_2 on r_1 and r_2 so that $r_1 = \overrightarrow{BC_1}$ and $r_2 = \overrightarrow{BC_2}$. Now let's tackle the inductive step. Assume that any n rays can be put in order and consider a set of $n+1$ rays all sharing a common endpoint B and on the same side of $\overline{AB}$. Take n of them and put them in order as r_1, $r_2, \ldots, r_n$. That leaves just one more ray—call it s. We need to compare s to what is currently the "outermost" ray, r_n. One of two things can happen: either s lies in the interior of the angle formed by $\overrightarrow{BA}$ and r_n, or it doesn't, and in this case, as we saw in the proof of the base case, that means that r_n lies in the interior of the angle formed by $\overrightarrow{BA}$ and s. Our path splits now, as we consider the two cases.

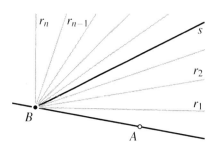

Figure 10. Case 1.

When s is in the interior of the angle formed by $\overrightarrow{BA}$ and r_n. Here r_n is the outermost ray, so let's relabel it as R_{n+1}. The remaining rays $r_1, r_2, \ldots, r_{n-1}$ and s are all in the interior of the angle formed by $\overrightarrow{BA}$ and R_{n+1}. Therefore, if C_{n+1} is any point on R_{n+1} (other than B) then according to the crossbar theorem each of $r_1, r_2, \ldots, r_{n-1}$ and s intersect the segment AC_{n+1}. We can put the intersection points in order

$$A * C_1 * C_2 * \cdots * C_n * C_{n+1}.$$

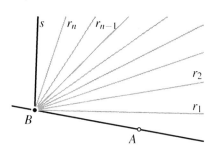

Figure 11. Case 2.

When s is not in the interior of the angle formed by $\overrightarrow{BA}$ and r_n. In this case, we will eventually see that s is the outermost ray, but all we know at the outset is that it is farther out than r_n. Let's relabel s as R_{n+1} and let C_{n+1} be a point on it. Since r_n is in the interior of the angle formed by $\overrightarrow{BA}$ and R_{n+1}, by the crossbar theorem, r_n must intersect AC_{n+1}. Let C_n be the intersection point. But we know that $r_1, r_2, \ldots, r_{n-1}$ lie in the interior of the angle formed by $\overrightarrow{BA}$ and R_n, so AC_n must intersect each of $r_1, r_2, \ldots, r_n$. We can put the intersection points in order

$$A * C_1 * C_2 * \cdots * C_n * C_{n+1}.$$

With the rays sorted and the intersections marked, the two strands of the proofs merge. Label the ray with point C_i as R_i. Then, for any $i < j$, C_i is on the same side of C_j as A, and so R_i is in the interior of the angle formed by $\overrightarrow{BA}$ and C_j. This is the ordering that we want. □

Exercise 2.4 Prove that the ordering described in the previous theorem is unique.

2.1 Exercises

2.5. Suppose that a line ℓ intersects a triangle at two points P and Q. Prove that all the points on the segment PQ other than the endpoints P and Q are in the interior of the triangle.

2.6. Prove that there are infinitely many points in the interior of an angle.

2.7. Prove that there are infinitely many points in the interior of a triangle.

2.8. Let $\overrightarrow{A_1B_1}$ and $\overrightarrow{A_1C_1}$ be two non-opposite rays with vertex A_1 and suppose that all the points on them are contained in the interior of $\angle B_2A_2C_2$. Prove that all the points in the interior of $\angle B_1A_1C_1$ are also in the interior of $\angle B_2A_2C_2$.

2.9. For any $\triangle ABC$, prove that there is another $\triangle A'B'C'$ that is entirely contained in the interior of $\triangle ABC$.

2.10. Given $\angle BAC$, label points D and E on rays $\overrightarrow{AB}$ and $\overrightarrow{AC}$ respectively so that $A*B*D$ and $A*C*E$. Prove that BC and DE do not intersect.

2.11. We have assumed plane separation as an axiom and used it to prove Pasch's lemma. Try to reverse that—in other words, assume Pasch's lemma (and all the other axioms of incidence and order) and prove the plane separation axiom.

2.12. Consider the following model for a geometry. The points of $\mathbb{Q}^2$ are coordinate pairs (x, y) where x and y are rational numbers. A line is given by an equation of the form $Ax + By = C$ as in the Cartesian model, with the added condition that in order to represent a line in $\mathbb{Q}^2$, it must pass through at least two points in $\mathbb{Q}^2$ (for instance $y = \sqrt{2}$ is not a line in this model). Incidence and order are as in the Cartesian model. Demonstrate that $\mathbb{Q}^2$ models a geometry that satisfies all the axioms of incidence and order except the plane separation axiom. Show that Pasch's lemma and the crossbar theorem do not hold in this geometry.

2.13. Consider the following model for a geometry. The points are equivalence classes of coordinates (x, y), where two coordinate pairs (x_1, y_1) and (x_2, y_2) are considered to be equivalent if $x_1 - x_2$ is an integer (you can imagine this equivalence relation as rolling up the Cartesian plane into an infinitely long cylinder with a circumference of one). Lines are still defined as in the Cartesian model, as equations of the form $Ax + By = C$. To what extent can the Cartesian interpretations of incidence and order be modified to work with this model? Which of the axioms of incidence and order will hold with those modifications?

2.2 References

I got my proof of the crossbar theorem from Moise's book on Euclidean geometry [Moi74].

3
Congruence Verse I: SAS and ASA

This is the first of three chapters dealing with the undefined term *congruence* and the axioms governing it. After Pasch's lemma and the crossbar theorem, these next few chapters begin to follow more closely Euclid's own development of the subject, and the results are more in line with a traditional beginning geometry course.

We have now defined several of the elementary objects of geometry, including segments, rays, angles, and triangles. There is an idea of equality for these objects. For instance, two segments are equal if they share all the same points—that is, if they have the same endpoints. But in general, in geometry there is a more expansive and more useful notion of equivalence. Intuitively, it is an equivalence between objects that have the same shape but that are in different positions or orientations. This is the equivalence described by the term congruence. Initially, congruence comes in two varieties—congruence of segments and congruence of angles. It describes a relation between a pair of segments or a pair of angles, so that we can say, for instance, that two segments are or are not congruent, or that two angles are or are not congruent. Later, the term is extended so that we can talk about congruence of triangles and other more general shapes. The notation used to indicate that two things (segments, angles, whatever) are congruent

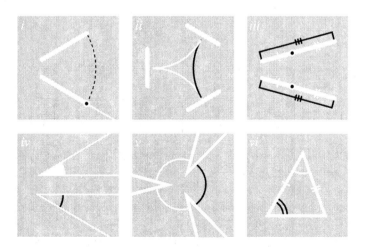

Figure 1. The six axioms of congruence. Construction (i), (iv); equivalence relation (ii), (v); segment addition (iii); and SAS (vi).

is $\simeq$. In Hilbert's system, there are six axioms of congruence. Three deal with congruence of segments, two deal with congruence of angles, and one involves both segments and angles.

The axioms of congruence.

Cg1: (The segment construction axiom) If A and B are distinct points and if A' is any point, then for each ray r with endpoint A', there is a unique point B' on r such that $AB \simeq A'B'$.

Cg2: Segment congruence is reflexive (every segment is congruent to itself), symmetric (if $AA' \simeq BB'$ then $BB' \simeq AA'$), and transitive (if $AA' \simeq BB'$ and $BB' \simeq CC'$, then $AA' \simeq CC'$).

Cg3: (The segment addition axiom) If $A*B*C$ and $A'*B'*C'$, and if $AB \simeq A'B'$ and $BC \simeq B'C'$, then $AC \simeq A'C'$.

Cg4: (The angle construction axiom) Given $\angle BAC$ and $\overrightarrow{A'B'}$, there is a unique $\overrightarrow{A'C'}$ on a given side of $\overline{A'B'}$ such that $\angle BAC \simeq \angle B'A'C'$.

Cg5: Angle congruence is reflexive (every angle is congruent to itself), symmetric (if $\angle A \simeq \angle B$, then $\angle B \simeq \angle A$), and transitive (if $\angle A \simeq \angle B$ and $\angle B \simeq \angle C$, then $\angle A \simeq \angle C$).

Cg6: (The side angle side (SAS) axiom.) Given $\triangle ABC$ and $\triangle A'B'C'$, if $AB \simeq A'B'$, $\angle B \simeq \angle B'$, and $BC \simeq B'C'$, then $\angle A \simeq \angle A'$.

Figure 2. In illustrations, hash marks indicate that segments and angles are congruent to one another.

The first and fourth axioms make it possible to construct congruent copies of segments and angles wherever we want. They are a little reminiscent of Euclid's postulates in that way. The second and fifth tell us that congruence is an equivalence relation. The third and sixth are paired as well, although not in as direct a way—both deal with three points and the segments that have them as their endpoints. In the third axiom, the points are collinear, while in the sixth they are not. There is a more direct counterpart to the third axiom though, a statement that does for angles what the segment addition axiom does for segments. It is called the angle addition theorem and we will prove it later.

We have added congruence to a system that already had incidence and order. It is important to understand how the notions interact. The axioms of congruence themselves describe some basic interactions. Another important connection is the triangle inequality, but that result is still a little ways away. In the meantime, the next theorem provides a simple connection.

Theorem 3.1 *Suppose that $A_1 * A_2 * A_3$ and that B_3 is a point on $\overrightarrow{B_1 B_2}$. If $A_1 A_2 \simeq B_1 B_2$ and $A_1 A_3 \simeq B_1 B_3$, then $B_1 * B_2 * B_3$.*

Proof Since B_3 is on $\overrightarrow{B_1B_2}$ one of three things is going to happen:

$$(1)\ B_2 = B_3,\ \text{or}\ (2)\ B_1 * B_3 * B_2,\ \text{or}\ (3)\ B_1 * B_2 * B_3.$$

The last is what we want, so it is just a matter of ruling out the other two possibilities.

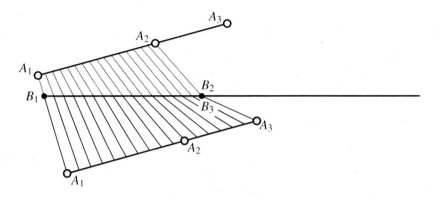

Figure 3. The case against case I.

(1) Why can't B_3 be equal to B_2? With $B_2 = B_3$, both A_1A_2 and A_1A_3 are congruent to the same segment. Therefore they are two different constructions of a segment starting from A_1 along $\overrightarrow{A_1A_2}$ and congruent to B_1B_2. The segment construction axiom says that there is only one.

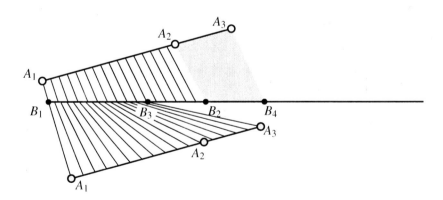

Figure 4. The case against case II.

(2) Why can't B_3 be between B_1 and B_2? By the segment construction axiom, there is a point B_4 on the opposite side of B_2 from B_1 so that $B_2B_4 \simeq A_2A_3$. Because $B_1B_2 \simeq A_1A_2$ and $B_2B_4 \simeq A_2A_3$, the segment addition axiom gives $B_1B_4 \simeq A_1A_3$. This creates the same problem as before: two different segments B_1B_3 and B_1B_4, both starting from B_1 along the same ray. They are each supposed to be congruent to A_1A_3, though, so this is a contradiction. □

3.1 Triangle congruence

Congruence of segments and angles is undefined, subject only to the axioms of congruence. But congruence of triangles is defined. It is defined in terms of the congruences of the segments and angles that make up the triangles.

Definition 3.2 Triangles $\triangle ABC$ and $\triangle A'B'C'$ are *congruent* if all their corresponding sides and angles are congruent:

- $AB \simeq A'B'$, $BC \simeq B'C'$, and $CA \simeq C'A'$,
- $\angle A \simeq \angle A'$, $\angle B \simeq \angle B'$, and $\angle C \simeq \angle C'$.

The definition suggests that there are six steps to show that two triangles are congruent. In actuality, triangles aren't that flexible, and in fact it is often sufficient to match up only half that many things. For example, the next result we will prove, the SAS triangle congruence theorem, says that you only have to match up two sides of the triangles, and the angles between them, to show that the triangles are congruent. In this chapter, we begin to investigate what are the minimum conditions necessary to guarantee that two triangles are congruent.

Before embarking, be aware that there is another useful way to think about these theorems, which is in terms of the uniqueness of its construction. The triangle congruence theorems are written as a way to compare two triangles. Another way to think of them is as a set of restrictions on the way that a triangle can be formed. To take an example, the SAS theorem says that, modulo congruence, there is only one triangle with a given pair of sides and a given intervening angle. Therefore, if you are building a triangle, and have decided on two sides and an intervening angle, then the triangle is determined.

Theorem 3.3 *(SAS triangle congruence)* In $\triangle ABC$ and $\triangle A'B'C'$, if

$$AB \simeq A'B', \quad \angle B \simeq \angle B', \quad \text{and} \quad BC \simeq B'C',$$

then $\triangle ABC \simeq \triangle A'B'C'$.

Proof Two of the three necessary side congruences are given, as is one of the angle congruences. The SAS axiom guarantees a second angle congruence, $\angle A \simeq \angle A'$. That leaves just one angle congruence and one side congruence.

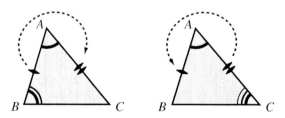

Figure 5. Two orderings of the list of congruences for the SAS axiom.

3.1 Triangle congruence

Let's do the remaining angle congruence first. The SAS axiom tells us about $\angle A$ in $\triangle ABC$. But let's not be misled by lettering. Because $\triangle ABC = \triangle CBA$ and $\triangle A'B'C' = \triangle C'B'A'$, we can reorder the given congruences:

$$CB \simeq C'B', \quad \angle B \simeq \angle B', \text{ and } BA \simeq B'A'.$$

Then the SAS axiom says that $\angle C \simeq \angle C'$ (the argument seems a little sneaky, but this is a legitimate use of the SAS axiom).

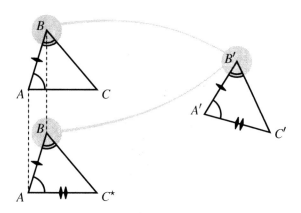

Figure 6. To show the last sides are congruent, construct a third triangle from parts of the original two. The key to the location of C is the angle at B.

That leaves just the sides AC and $A'C'$. We will construct a congruent copy of $\triangle A'B'C'$ on top of $\triangle ABC$ (Euclid's flawed proof of SAS in the *Elements* used a similar argument but without the axioms to back it up). Thanks to the segment construction axiom, there is a unique point $C^\star$ on $\overrightarrow{AC}$ so that $AC^\star \simeq A'C'$. Now if we can just show that $C^\star = C$ we will be done. Look:

$$BA \simeq B'A', \quad \angle A \simeq \angle A', \text{ and } AC^\star \simeq \angle A'C'.$$

By the SAS axiom then, $\angle ABC^\star \simeq \angle A'B'C'$. That in turn means that $\angle ABC^\star \simeq \angle ABC$. But both of those angles are constructed on the same side of $\overrightarrow{BA}$. According to the angle construction axiom, they must be the same. That is, $\overrightarrow{BC} = \overrightarrow{BC^\star}$. Both C and $C^\star$ are the intersection of this ray and $\overline{AC}$. Since a ray can intersect a line only once, C and $C^\star$ have to be the same. Therefore $AC = AC^\star \simeq A'C'$. □

Let's move immediately to another theorem describing another set of conditions that guarantee triangle congruence. Its proof shares some of the same features as the last one.

Theorem 3.4 *(ASA triangle congruence)* In $\triangle ABC$ and $\triangle A'B'C'$, if

$$\angle A \simeq \angle A', \quad AB \simeq A'B', \text{ and } \angle B \simeq \angle B',$$

then $\triangle ABC \simeq \triangle A'B'C'$.

Proof Because of the segment construction axiom, there is a point $C^\star$ on $\overrightarrow{AC}$ so that $AC^\star \simeq A'C'$. The hope is that $C^\star = C$, and that is what we need to show. To do that, observe that

$$BA \simeq B'A', \quad \angle A \simeq \angle A', \quad \text{and} \quad AC^\star \simeq A'C'.$$

By SAS, $\triangle ABC^\star \simeq \triangle A'B'C'$. In particular, look at what is happening at vertex B:

$$\angle ABC^\star \simeq \angle A'B'C' \simeq \angle ABC.$$

There is only one way to make the angle on that side of $\overrightarrow{BA}$, and so $\overrightarrow{BC^\star} = \overrightarrow{BC}$. Since C and $C^\star$ are where the ray intersects $\overline{AC}$, $C = C^\star$. Since $C = C^\star$, $AC = AC^\star \simeq A'C'$. Then

$$BA \simeq B'A', \quad \angle A \simeq \angle A', \quad \text{and} \quad AC \simeq A'C',$$

and by SAS, $\triangle ABC \simeq \triangle A'B'C'$.

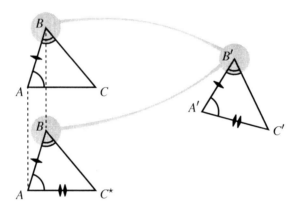

Figure 7. A familiar setup. The strategy for ASA mimics the SAS strategy. □

The naming convention for the triangle congruence theorems is elegant. The name of the theorem describes, in shorthand, exactly what segment and angle congruences are required to apply the theorem. Let's take a look at how the triangle congruence theorems can be put to work. This next theorem is the angle analog of the earlier theorem that related congruence and the order of points.

Theorem 3.5 *Suppose that $\angle ABC \simeq \angle A'B'C'$. Suppose that D is in the interior of $\angle ABC$. Suppose that D' is located on the same side of $\overline{A'B'}$ as C' so that $\angle ABD \simeq \angle A'B'D'$. Then D' is in the interior of $\angle A'B'C'$.*

Proof To generate the necessary triangle congruences, we need to reposition points along their respective rays. Let $A^\star$ be the point on $\overrightarrow{BA}$ so that $BA^\star \simeq B'A'$. Let $C^\star$ be the point on $\overrightarrow{BC}$ so that $BC^\star \simeq B'C'$. Since D is in the interior of $\angle ABC$, the crossbar theorem guarantees

3.1 Triangle congruence

that $\overrightarrow{BD}$ intersects $A^\star C^\star$. Let's call this intersection E. Then

$$A^\star B \simeq A'B', \quad \angle A^\star BC^\star \simeq \angle A'B'C', \text{ and } BC^\star \simeq B'C',$$

so by SAS, $\triangle A^\star BC^\star \simeq \triangle A'B'C'$.

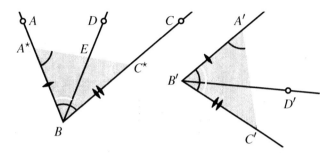

Figure 8. After repositioning points, the first use of SAS.

Let's focus on the second configuration of points—the ones with the ′ marks. According to the segment construction axiom, there is a point E' on $\overrightarrow{A'C'}$ so that $A'E' \simeq A^\star E$. Furthermore, thanks to the earlier theorem relating congruence and order, since E is between $A^\star$ and $C^\star$, E' must be between A' and C', so it is in the interior of $\angle A'B'C'$. Because

$$BA^\star \simeq B'A', \quad \angle BA^\star E \simeq \angle B'A'E', \text{ and } A^\star E \simeq A'E',$$

by SAS, $\triangle BA^\star E \simeq \triangle B'A'E'$.

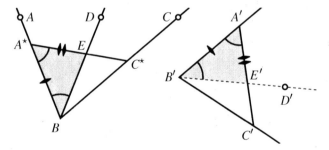

Figure 9. The second use of SAS: E' and D' are on the same ray.

In particular, this means that $\angle A^\star BE \simeq \angle A'B'E'$. But we were originally told that $\angle A^\star BE \simeq \angle A'B'D'$. Since angle congruence is transitive this means that $\angle A'B'D' = \angle A'B'E'$. According to the angle construction axiom, $\overrightarrow{B'D'}$ and $\overrightarrow{B'E'}$ must in fact be the same. Since E' is in the interior of $\angle A'B'C'$, D' must be as well. □

It should not come as a great surprise that in some triangles, two or even all three sides or angles may be congruent (although we have not yet shown, for example, that it is possible to construct a triangle with three congruent angles). Thanks to the triangle congruence theorems, we can show that these triangles are congruent to themselves in non-trivial ways, which reveal the internal symmetries of those triangles.

Definition 3.6 If all three sides of a triangle are congruent, the triangle is *equilateral*. If at least two sides of a triangle are congruent, the triangle is *isosceles*. If no pair of sides of the triangle is congruent, the triangle is *scalene*.

With this definition, every equilateral triangle is an isosceles triangle. Some texts require instead that an isosceles triangle have exactly two congruent sides. By that stricter definition, equilateral triangles would not be isosceles. Let's end this chapter with a theorem about the internal symmetries of a triangle.

Theorem 3.7 *(The isosceles triangle theorem) In an isosceles triangle, the angles opposite the congruent sides are congruent.*

Proof Suppose $\triangle ABC$ is isosceles, with $AB \simeq AC$. Then

$$AB \simeq AC, \quad \angle A \simeq \angle A, \text{ and } AC \simeq AB,$$

so by SAS, $\triangle ABC \simeq \triangle ACB$ (there's the non-trivial congruence of the triangle with itself). Comparing corresponding angles, $\angle B \simeq \angle C$. $\square$

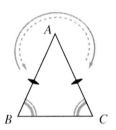

Figure 10. The internal symmetry of an isosceles triangle.

3.2 Exercises

3.1. Given a point P and a segment AB, prove that there are infinitely many points Q so that $PQ \simeq AB$.

3.2. Given a scalene triangle $\triangle ABC$ and a segment $A'B'$ that is congruent to AB, how many points C' are there so that $\triangle A'B'C' \simeq \triangle ABC$?

3.3. Verify that triangle congruence is an equivalence relation (that it is reflexive, symmetric, and transitive).

3.4. Prove the converse of the isosceles triangle theorem: that if two interior angles of a triangle are congruent, then the sides opposite them must also be congruent.

3.5. Let $\triangle ABC$ be isosceles with $AB \simeq AC$. Suppose that points M and N are located on BC so that $B * M * N * C$ and so that $BM \simeq CN$. Prove that $\triangle AMN$ is isosceles.

3.2 Exercises

3.6. Prove that the interior angles of an equilateral triangle are congruent. Conversely, prove that if the interior angles of a triangle are congruent, then the triangle is equilateral.

3.7. Prove that no two interior angles of a scalene triangle can be congruent.

3.8. Given $\triangle ABC$ and a segment $A'B'$ so that $A'B' \simeq AB$, draw rays r_1 and r_2 from A' and B' respectively, on the same side of $\overline{A'B'}$ that form angles congruent to $\angle A$ and $\angle B$. If r_A and r_B intersect, then by ASA the triangle formed is congruent to $\triangle ABC$. How do we know r_A and r_B intersect?

3.9. In the exercises in chapter 1, we looked at the Cartesian model and saw how *point*, *line*, *on*, and *between* are interpreted in that model. Let's extend the model now to include congruence. In the Cartesian model, segment congruence is defined in terms of the length of the segment, which, in turn, is defined using the distance function. If (x_a, y_a) and (x_b, y_b) are the coordinates of A and B, then the length of the segment AB, written $|AB|$, is

$$|AB| = \sqrt{(x_a - x_b)^2 + (y_a - y_b)^2}.$$

Two segments are congruent if and only if they are the same length. With this interpretation, verify the first three axioms of congruence.

3.10. Angle congruence is the most difficult of the undefined terms to interpret in the Cartesian model. Like segment congruence, angle congruence is defined via measure, in this case angle measure. You may remember from calculus that the dot product provides a way to measure the angle between two vectors: that for two vectors v and w,

$$v \cdot w = |v||w| \cos \theta,$$

where θ is the angle between v and w. That is the key here. Given an angle $\angle ABC$, its measure, written $(\angle ABC)$, is computed as follows. Let (x_a, y_a), (x_b, y_b), and (x_c, y_c) be the coordinates for points A, B, and C. Then define vectors

$$v = \langle x_a - x_b, y_a - y_b \rangle \text{ and } w = \langle x_c - x_b, y_c - y_b \rangle$$

and measure

$$(\angle ABC) = \cos^{-1}\left(\frac{v \cdot w}{|v||w|}\right).$$

Two angles are congruent if and only if they have the same angle measure. With this interpretation, verify the last three axioms of congruence.

3.11. A variation of the Cartesian model realizes points, lines, on, and between in the same way, but defines the length of the segment from $A = (x_a, y_a)$ to $B = (x_b, y_b)$ using the taxicab metric:

$$|AB| = |x_a - x_b| + |y_a - y_b|.$$

As in the Cartesian model, two segments are congruent if they are the same length. Show that the model does still satisfy the first three axioms of congruence (it fails the SAS axiom, but to properly address that would require a more careful examination of angle measurement in the metric).

4
Congruence Verse II: AAS

The ultimate objective of this chapter is to derive a third triangle congruence theorem, AAS. The technique we used in the last chapter to prove SAS and ASA does not quite work this time though, so along the way we are going to get to see a few more of the tools of neutral geometry: supplementary angles, the alternate interior angle theorem, and the exterior angle theorem.

4.1 Supplementary angles

There aren't that many letters in the alphabet, and so it is easy to burn through most of them in a proof if you aren't frugal. Even if your variables don't run the full gamut from A to Z, it can be challenging just trying to keep up with them. Some notation just can't be avoided; fortunately, some of it can. One common trick used to cut down on some notation is "relocation". Let's say you are working with a ray $\overrightarrow{AB}$. You can't change the endpoint A without changing the ray itself, but there is flexibility with the point B. If B' is any other point on the ray (other than A), then $\overrightarrow{AB}$ and $\overrightarrow{AB'}$ are the same. So rather than introduce a whole new point on the ray, we just "relocate" B to a more convenient location. The same technique can be used for angles and lines. Be warned: you must be careful not to abuse this relocation power! Yes, there is some flexibility to the placement of some points, but once you have used it up, the point has to stay put. For instance, if you relocate a point on a line to a particular intersection, and your proof depends on that intersection, then you can no longer move the point to another location.

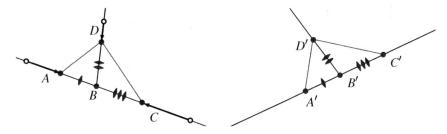

Figure 1. Relocation of points is a shortcut to cut down on notation. Illustrated here are the relocations of the points A, B, and C to make the congruences needed to prove that the supplements of congruent angles are congruent.

Three non-collinear points A, B, and C define an angle $\angle ABC$. When they are collinear, they do not define an angle, but you may want to think of them as forming a kind of degenerate

37

angle. If $A * B * C$, then A, B, and C form what is called a "straight angle". One of the most basic relationships that two angles can have is defined in terms of these straight angles.

Definition 4.1 Suppose that A, B, and C form a straight angle with $A * B * C$. Let D be a fourth point that is not on the line through A, B, and C. Then $\angle ABD$ and $\angle CBD$ are *supplementary angles*.

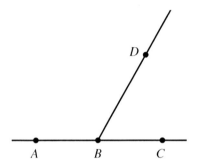

Figure 2. A pair of supplementary angles: $\angle ABD$ and $\angle CBD$.

Supplements have a healthy relationship with congruence as related in the next theorem.

Theorem 4.2 *The supplements of congruent angles are congruent: given two pairs of supplementary angles*

Pair 1: $\angle ABD$ and $\angle CBD$ and
Pair 2: $\angle A'B'D'$ and $\angle C'B'D'$,

if $\angle ABD \simeq \angle A'B'D'$, then $\angle CBD \simeq \angle C'B'D'$.

Proof First, we relocate points: reposition A, C, and D on their respective rays $\overrightarrow{BA}$, $\overrightarrow{BC}$, and $\overrightarrow{BD}$ so that
$$BA \simeq B'A', \quad BC \simeq B'C', \quad \text{and} \quad BD \simeq B'D'.$$
In all, we have to use the SAS congruence theorem three times.

- Because
$$AB \simeq A'B', \quad \angle ABD \simeq \angle A'B'D', \quad \text{and} \quad BD \simeq B'D',$$
by SAS, $\triangle ABD \simeq \triangle A'B'D'$.
- In particular, this means that $AD \simeq A'D'$ and $\angle A \simeq \angle A'$. Furthermore, by the segment addition axiom, $AC \simeq A'C'$. By SAS again, $\triangle ACD \simeq \triangle A'C'D'$.
- Hence $CD \simeq C'D'$ and $\angle C \simeq \angle C'$. This, together with the construction $BC \simeq B'C'$ and one more use of SAS, implies that $\triangle BCD \simeq \triangle B'C'D'$. Therefore $\angle CBD \simeq \angle C'B'D'$. □

Every angle has two supplements. To get a supplement of an angle, replace one of the two rays forming the angle with its opposite ray. Since there are two candidates for the

4.1 Supplementary angles

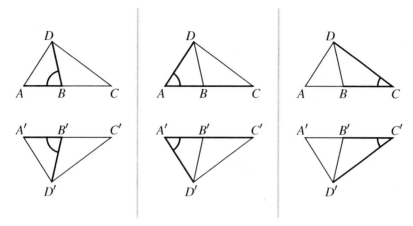

Figure 3. Three applications of SAS show that the supplements of congruent angles are congruent.

replacement, there are two supplements. There is a name for the relationship between these two supplements.

Definition 4.3 *Vertical angles* are two angles that are supplementary to the same angle.

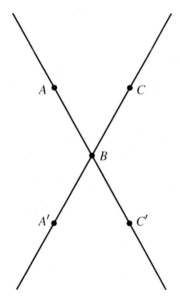

Figure 4. Two intersecting lines create two pairs of vertical angles. The first pair: $\angle ABC$ and $\angle A'BC'$. The second pair: $\angle ABA'$ and $\angle CBC'$.

Exercise 4.1 Prove that every angle is part of one and only one vertical angle pair.

The single most important property of vertical angles is the following theorem, which is an immediate consequence of the previous theorem on supplements.

Corollary 4.4 *Vertical angles are congruent.*

Proof Two vertical angles are, by definition, supplementary to the same angle. That angle is congruent to itself (because of the second axiom of congruence). Since the vertical angles are supplementary to congruent angles, they themselves must be congruent. □

4.2 The alternate interior angle theorem

The farther we go in the study of neutral geometry, the more we are going to bump into issues relating to how parallel lines behave. Many of the results we will derive are close to results of Euclidean geometry, and this can lead to several dangerous pitfalls. The alternate interior angle theorem is one of the first glimpses of that.

Definition 4.5 Given a set of lines, $\{\ell_1, \ell_2, \ldots, \ell_n\}$, a *transversal* is a line that intersects all of them.

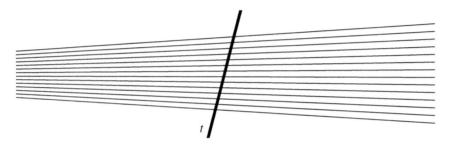

Figure 5. A transversal t of a set of lines.

Definition 4.6 Let t be a transversal to ℓ_1 and ℓ_2. *Alternate interior angles* are pairs of angles formed by ℓ_1, ℓ_2, and t that are between ℓ_1 and ℓ_2 and on opposite sides of t. *Adjacent interior angles* are pairs of angles formed by ℓ_1, ℓ_2, and t that are between ℓ_1 and ℓ_2 and on the same side of t.

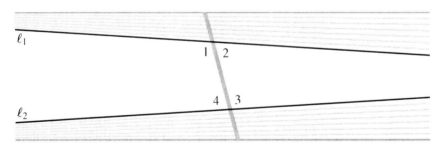

Figure 6. Angles 1 and 3 are alternate interior angles, as are angles 2 and 4. Angles 1 and 4 are adjacent interior angles, as are angles 2 and 3.

The alternate interior angle theorem tells us something about transversals and parallel lines, but read it carefully. The converse of this theorem is used often in Euclidean geometry, but in neutral geometry it is not an "if and only if" statement.

4.2 The alternate interior angle theorem

Theorem 4.7 *(The alternate interior angle theorem) Let ℓ_1 and ℓ_2 be two lines crossed by a transversal t. If the alternate interior angles formed by ℓ_1, ℓ_2, and t are congruent, then ℓ_1 and ℓ_2 are parallel.*

Proof First a small clarification of the statement of the theorem. The lines ℓ_1, ℓ_2, and t will form two pairs of alternate interior angles. However, the angles in one pair are the supplements of the angles in the other pair, so if the angles in one pair are congruent then angles in the other pair also have to be congruent. Now let's get on with the proof, a proof by contradiction. Suppose that ℓ_1 and ℓ_2 are crossed by a transversal t so that alternate interior angles are congruent, but suppose that ℓ_1 and ℓ_2 are not parallel. Label

A : the intersection of ℓ_1 and t,
B : the intersection of ℓ_2 and t, and
C : the intersection of ℓ_1 and ℓ_2.

By the segment construction axiom there are points

D : on ℓ_1 so that $D * A * C$ and so that $AD \simeq BC$, and
D' : on ℓ_2 so that $D' * B * C$ and so that $BD' \simeq AC$.

In terms of these marked points the congruent pairs of alternate interior angles are

$$\angle ABC \simeq \angle BAD \text{ and } \angle ABD' \simeq \angle BAC.$$

The first congruence, together with the fact that that we have constructed $AD \simeq BC$ and $AB \simeq BA$, is enough to use SAS: $\triangle ABC \simeq \triangle BAD$. We can focus on one pair of corresponding angles in those triangles: $\angle ABD \simeq \angle BAC$. Now $\angle BAC$ is congruent to its alternate interior pair $\angle ABD'$, so since angle congruence is transitive, this means that $\angle ABD \simeq \angle ABD'$.

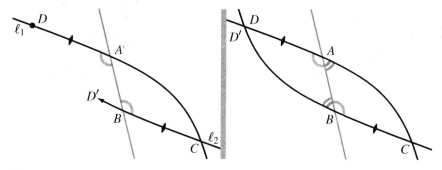

Figure 7. The key to the proof of the alternate interior angle theorem: if ℓ_1 and ℓ_2 crossed on one side of t, they would have to cross on the other side as well.

Here's the problem. There is only one way to construct this angle on that side of t, so $\overrightarrow{BD}$ and $\overrightarrow{BD'}$ must be the same. Therefore D, which we originally placed on ℓ_1, is also on ℓ_2. That would imply that ℓ_1 and ℓ_2 share two points, C and D, in violation of the first axiom of incidence. □

4.3 The exterior angle theorem

We have talked about congruent angles, but so far we have not discussed any way of saying that one angle is larger or smaller than the other. That is something that we will need to do eventually, in order to develop a system of measurement for angles. For now though, we need at least some rudimentary description, even if the more fully developed system will wait until later. Given $\angle A_1 B_1 C_1$ and $\angle A_2 B_2 C_2$, the angle construction axiom guarantees that there is a point $A^\star$ on the same side of $\overline{B_2 C_2}$ as A_2 so that $\angle A^\star B_2 C_2 \simeq \angle A_1 B_1 C_1$. If $A^\star$ is on $\overrightarrow{B_2 A_2}$, then the two angles are congruent as we have previously seen. If the angles are not congruent, we can use the position of $A^\star$ in relation to $\angle A_2 B_2 C_2$ to determine which of the two is larger.

Definition 4.8 If $A^\star$ is in the interior of $\angle A_2 B_2 C_2$, then we say that $\angle A_1 B_1 C_1$ is *smaller* than $\angle A_2 B_2 C_2$. If $A^\star$ is neither in the interior of $\angle A_2 B_2 C_2$, nor on the ray $\overrightarrow{B_2 A_2}$, then $\angle A_1 B_1 C_1$ is *larger* than $\angle A_2 B_2 C_2$.

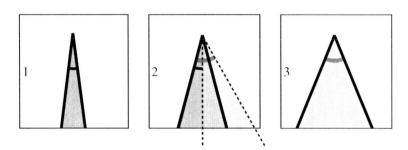

Figure 8. $\angle 1$ is smaller than $\angle 2$; $\angle 3$ is larger than $\angle 2$.

We need this notion of relative angle size to compare the interior and exterior angles of triangles.

Definition 4.9 An *exterior angle* of a triangle is an angle supplementary to one of the triangle's interior angles.

The exterior angle theorem is our first chance to compare the sizes of angles that are not congruent. It should be noted that in Euclidean geometry, there is a stronger relationship between the angles in this theorem. In fact, there is a stronger version of this theorem even in neutral geometry, but it relies on a few facts we have yet to prove. Our proof of this modest version of the exterior angle theorem relies essentially on the alternate interior angle theorem just proved.

Theorem 4.10 *(The exterior angle theorem) An exterior angle of a triangle is larger than either of its nonadjacent interior angles.*

Proof This is a proof by contradiction. Starting with $\triangle ABC$, extend the side AC past C and pick a point D so that $A * C * D$. Suppose that the interior angle at B is larger than the exterior angle $\angle BCD$. Then there is a ray r from B on the same side of BC as A so that $\overrightarrow{BC}$ and r

4.4 AAS

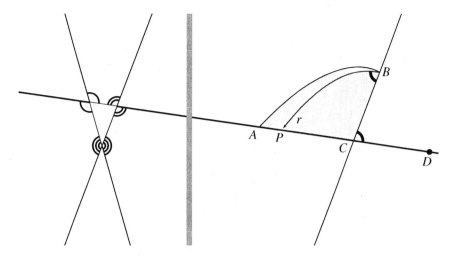

Figure 9. (l) Three vertical angle pairs of exterior angles. (r) A proof by contradiction of the exterior angle theorem.

form an angle congruent to $\angle BCD$. The ray will lie in the interior of $\angle B$, though, so by the crossbar theorem, r must intersect AC. Call the intersection point P. But then the alternate interior angles formed by the transversal $\overline{BC}$ across $\overline{BP}$ and $\overline{CD}$, namely $\angle PBC$ and $\angle BCD$, are congruent. According to the alternate interior angle theorem r and AC must be parallel, and so they can't intersect. This is a contradiction. □

4.4 AAS

That brings us to our third triangle congruence theorem, AAS. The setup of the proof of this congruence theorem is the same as the proof of ASA, but the critical step uses the exterior angle theorem.

Theorem 4.11 *(AAS triangle congruence) In $\triangle ABC$ and $\triangle A'B'C'$, if*

$$\angle A \simeq \angle A', \quad \angle B \simeq \angle B', \quad \text{and} \quad BC \simeq B'C',$$

then $\triangle ABC \simeq \triangle A'B'C'$.

Proof Locate $A^\star$ on $\overrightarrow{BA}$ so that $A^\star B \simeq A'B'$. By SAS, $\triangle A^\star BC \simeq \triangle A'B'C'$. Therefore $\angle A^\star \simeq \angle A' \simeq \angle A$. Now if $B * A * A^\star$ (as illustrated) then $\angle A$ is an exterior angle and $\angle A^\star$ is a nonadjacent interior angle of $\triangle AA^\star C$. According to the exterior angle theorem, the angles can't be congruent. If $B * A * A^\star$, then $\angle A^\star$ is an exterior angle and $\angle A$ is a nonadjacent interior angle. Again, the exterior angle theorem says the angles can't be congruent. The only other possibility, then, is that $A = A^\star$, so $AB = AB^\star \simeq A'B'$. That, together with the two congruences, $\angle B \simeq \angle B'$ and $BC \simeq B'C'$, is enough to use SAS. Then $\triangle ABC \simeq \triangle A'B'C'$. □

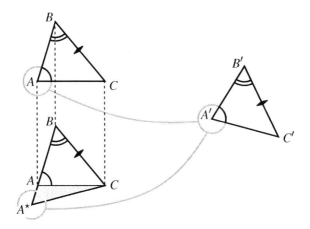

Figure 10. The proof of the AAS triangle congruence theorem.

4.5 Exercises

4.2. Prove that if $\angle A$ is smaller than $\angle B$, then the supplement of $\angle A$ is larger than the supplement of $\angle B$.

4.3. Given an angle $\angle P_0 A P_n$, let $r_1, r_2, \ldots, r_{n-1}$ be rays with endpoint A that are arranged in order in $\angle P_0 A P_n$ (as in theorem 2.9). Let P_i be the point of intersection of r_i and $P_0 P_n$, so that $P_0 * P_1 * P_2 * \ldots * P_n$. Prove that if $i > j$, then $\angle P_0 A P_i$ is larger than $\angle P_0 A P_j$.

4.4. Suppose that $A * B * C$ and that A' and C' are on opposite sides of $\overline{AC}$. Prove that if $\angle ABA' \simeq \angle CBC'$, then $A' * B * C'$.

4.5. Prove that for every segment AB there is a point M on AB so that $AM \simeq MB$. This point is called the *midpoint* of AB.

4.6. Let $\triangle ABC$ be isosceles, with $AB \simeq AC$. Let M be the midpoint of BC. Prove that $\triangle AMB \simeq \triangle AMC$.

4.7. Prove that for every angle $\angle ABC$ there is a unique ray $\overrightarrow{BD}$ in the interior of $\angle ABC$ so that $\angle ABD \simeq \angle DBC$. This ray is called the *bisector* of $\angle ABC$.

4.8. Let $\triangle ABC$ be isosceles, with $AB \simeq AC$. Let r be the angle bisector of $\angle A$. Prove that r intersects BC at its midpoint.

4.9. We used the alternate interior angle theorem to prove the exterior angle theorem. Now reverse that: assume the exterior angle theorem and derive the alternate interior angle theorem from it.

5
Congruence Verse III: SSS

The first and fourth axioms of congruence are naturally paired—one applies to segments, the other to angles. Likewise, the second and fifth axioms are paired. But the third axiom, the segment addition axiom, has no matching angle counterpart. Instead, angle addition can be proved from the other axioms, and that is one of the goals of this chapter. Let's begin with a warm-up exercise.

Theorem 5.1 *(The segment subtraction theorem) Suppose that $A*B*C$, $A'*B'*C'$, $AB \simeq A'B'$, and $AC \simeq A'C'$. Then $BC \simeq B'C'$.*

Exercise 5.1 Prove the segment subtraction theorem.

Now let us turn our attention to angles.

Theorem 5.2 *(The Angle subtraction theorem) Let D and D' be interior points of $\angle ABC$ and $\angle A'B'C'$ respectively. If*

$$\angle ABC \simeq \angle A'B'C' \text{ and } \angle ABD \simeq \angle A'B'D',$$

then $\angle DBC \simeq \angle D'B'C'$.

Proof The first step is relocation. Relocate A and C on $\overrightarrow{BA}$ and $\overrightarrow{BC}$ respectively so that $AB \simeq A'B'$ and $BC \simeq B'C'$. Since D is in the interior of $\angle ABC$, by the crossbar theorem, $\overrightarrow{BD}$ intersects AC. Relocate D to that intersection. Likewise, relocate D' to the intersection of

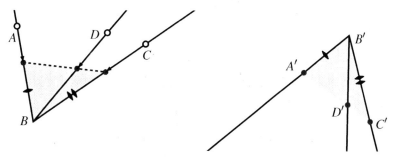

Figure 1. Relocation of points to create the congruent triangles necessary for the proof of the angle subtraction theorem.

45

$\overrightarrow{B'D'}$ and $A'C'$. We cannot yet say that BD and $B'D'$ are congruent, although that is something that we will establish in the course of the proof.

From the congruences
$$AB \simeq A'B', \quad \angle ABC \simeq \angle A'B'C', \quad \text{and} \quad BC \simeq B'C',$$
by SAS, $\triangle ABC \simeq \triangle A'B'C'$. We can draw three conclusions:

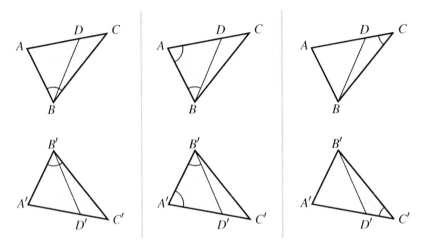

Figure 2. Using SAS, ASA, and SAS again to prove the angle subtraction theorem.

(1) $\angle A \simeq \angle A'$. That, together with $AB \simeq A'B'$ and $\angle ABD \simeq \angle A'B'D'$ implies that $\triangle ABD \simeq \triangle A'B'D'$ (by ASA), so $AD \simeq A'D'$.

(2) $AC \simeq A'C'$. From the previous step, we know that $AD \simeq A'D'$, so by segment subtraction $CD \simeq C'D'$.

(3) $\angle C \simeq \angle C'$. Put that together with $BC \simeq B'C'$ (constructed) and $CD \simeq C'D'$ (from the previous step). By SAS $\triangle BCD \simeq \triangle B'C'D'$, and in particular $\angle DBC \simeq \angle D'B'C'$. □

With angle subtraction in the toolbox, angle addition is now easy to prove.

Corollary 5.3 *(The angle addition theorem) Suppose that D is in the interior of $\angle ABC$ and that D' is in the interior of $\angle A'B'C'$. If*
$$\angle ABD \simeq \angle A'B'D' \quad \text{and} \quad \angle DBC \simeq \angle D'B'C',$$
then $\angle ABC \simeq \angle A'B'C'$.

Proof Because of the angle construction axiom, there is a ray $\overrightarrow{BC^\star}$ on the same side of $\overline{AB}$ as C so that $\angle ABC^\star \simeq \angle A'B'C'$. We can now use the angle subtraction theorem:
$$\angle ABC^\star \simeq \angle A'B'C' \text{ and } \angle ABD \simeq \angle A'B'D' \implies \angle DBC^\star \simeq \angle D'B'C'.$$

We already know that $\angle D'B'C' \simeq \angle DBC$, so $\angle DBC^\star \simeq \angle DBC$. The angle construction axiom tells us that there is but one way to construct the angle on the opposite side of $\overline{DB}$ from A, so $\overrightarrow{BC^\star}$ and $\overrightarrow{BC}$ have to be the same. Therefore, $\angle ABC = \angle ABC^\star \simeq \angle A'B'C'$. □

We end this chapter with the last of the triangle congruence theorems, the SSS triangle congruence theorem. The proofs of the previous triangle congruence theorems were all set up in

Congruence Verse III: SSS

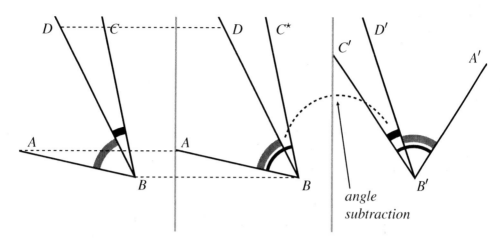

Figure 3. The angle addition theorem as a consequence of the angle subtraction theorem.

a similar manner, but it was a setup that required at least one angle congruence to be given. With SSS, no angle congruence is given, so we use a different approach that relies on the isosceles triangle theorem.

Theorem 5.4 *(SSS triangle congruence)* In $\triangle ABC$ and $\triangle A'B'C'$ if
$$AB \simeq A'B', \quad BC \simeq B'C', \quad \text{and} \quad CA \simeq C'A',$$
then $\triangle ABC \simeq \triangle A'B'C'$.

Proof The first step is to get the two triangles into a more convenient configuration. To do that, create a congruent copy of $\triangle A'B'C'$ on the opposite side of $\overline{AC}$ from B. The construction is simple enough: there is a unique point $B^\star$ on the opposite side of $\overline{AC}$ from B such that
$$\angle CAB^\star \simeq \angle C'A'B' \quad \text{and} \quad AB^\star \simeq A'B'.$$

In addition, we already know that $AC \simeq A'C'$, so by SAS, $\triangle AB^\star C$ is congruent to $\triangle A'B'C'$. Now the real question is whether $\triangle AB^\star C$ is congruent to $\triangle ABC$.

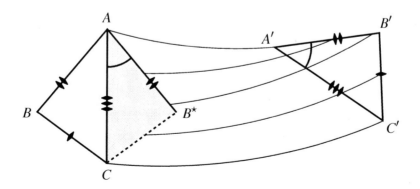

Figure 4. The first step in the proof of SSS. Create a congruent copy of the second triangle abutting the first triangle.

Since B and $B^\star$ are on opposite sides of $\overline{AC}$, $BB^\star$ intersects $\overline{AC}$. Let's call the point of intersection P. We don't know anything about where P is on $\overline{AC}$, and that opens up some options:

- P could be between A and C, or
- P could be either of the endpoints A or C, or
- P could be on $\overline{AC}$ but not the segment AC.

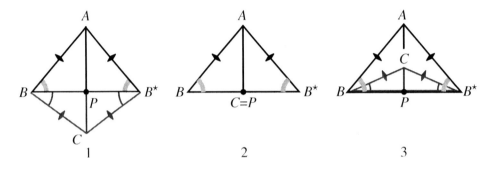

Figure 5. Three possible location of P, and the resulting isosceles triangles.

Let's consider that first possibility (the other two are left as exercises). Assuming that $A \ast P \ast C$, both $\triangle ABB^\star$ and $\triangle CBB^\star$ are isosceles:

$$AB \simeq A'B' \simeq AB^\star;$$
$$CB \simeq C'B' \simeq CB^\star.$$

According to the isosceles triangle theorem, the angles opposite the congruent sides are congruent:

$$\angle ABP \simeq \angle AB^\star P;$$
$$\angle CBP \simeq \angle CB^\star P.$$

Since we have assumed that P is between A and C, the angle addition theorem guarantees that the larger angles $\angle ABC$ and $\angle AB^\star C$ are congruent. We already know $\angle AB^\star C \simeq \angle A'B'C'$, so $\angle ABC \simeq \angle A'B'C'$. That is the needed angle congruence. By SAS, $\triangle ABC \simeq \triangle A'B'C'$. □

Exercise 5.2 Complete the proof of the SSS triangle congruence theorem by working out the details of the remaining two cases in that proof.

We have established four triangle congruence theorems: SAS, ASA, AAS, and SSS. Each required three components, some mix of sides and angles. It would be natural to wonder whether there are any other combinations of three sides and angles that guarantee congruence. There are only two other fundamentally different letter combinations: AAA and SSA. Neither is a valid congruence theorem in neutral geometry. In fact, both fail to guarantee congruence in

Euclidean geometry. In non-Euclidean geometry, the situation is a little different, but I am going to postpone that issue for the time being.

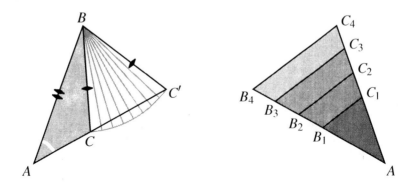

Figure 6. SSA and AAA in the Euclidean model. (l) Two non-congruent triangles sharing SSA. (r) Four non-congruent triangles sharing AAA.

5.1 Exercises

5.3. One of the conditions in the statement of the angle subtraction theorem is that D and D' must be in the interiors of their respective angles. In fact, this condition can be weakened: prove that you do not need to assume that D' is in the interior of the angle, just that it is on the same side of $A'B'$ as C'.

5.4. Suppose that A, B, C, and D are four distinct points, no three of which are collinear. Prove that if $\triangle ABC \simeq \triangle DCB$, then $\triangle BAD \simeq \triangle CDA$.

5.5. Let C and D be on opposite sides of $\overline{AB}$ so that $\triangle ABC \simeq \triangle BAD$. Let M be the intersection of CD and $\overline{AB}$. Prove that $AM \simeq BM$ and that $CM \simeq DM$.

5.6. Let C and D be points on the opposite sides of $\overline{AB}$ that are positioned so that $AC \simeq BD$ and $BC \simeq AD$. Prove that $\overline{AC}$ is parallel to $\overline{BD}$ and that $\overline{AD}$ is parallel to $\overline{BC}$.

5.7. Let $\triangle ABC$ be isosceles with $AB \simeq AC$. Let b be the midpoint of AC, let c be the midpoint of AB, and let P be the intersection of Bb and Cc.
   (1) Prove that $\triangle BPc \simeq \triangle CPb$.
   (2) Prove that $\triangle APc \simeq \triangle APb$.
   (3) Let a be the midpoint of BC. Prove that $\triangle aBP \simeq \triangle aCP$.

5.8. Suppose that $A * B * C$ and $A' * B * C'$ (but that the five points are not all on the same line), that $AB \simeq BC$, and that $A'B \simeq BC'$. Let P be a point on AA' and let Q be the intersection of $\overline{PB}$ and CC'. Prove that $PB \simeq QB$.

5.9. Suppose that $\triangle ABC$ is isosceles, with $AB \simeq AC$, and let D and E be points on the segments AB and AC, respectively, so that $AD \simeq AE$. Let P be the intersection of BE and CD. Prove that $\triangle BDP \simeq \triangle CEP$.

6

Distance, Length, and the Axioms of Continuity

Hilbert's axioms lend themselves to a synthetic, rather than analytic, development of the subject. That is, they describe a geometry that is not centrally built upon measurement. Nowadays, it is more common to take a metrical approach to geometry, to establish the geometry based upon a measurement. In the metrical approach, we begin by defining a distance function, a function d that assigns to each pair of points a real number and satisfies the following requirements

(i) $d(P, Q) \geq 0$, with $d(P, Q) = 0$ if and only if $P = Q$,

(ii) $d(P, Q) = d(Q, P)$, and

(iii) $d(P, R) \leq d(P, Q) + d(Q, R)$.

Once the distance function has been chosen, the *length* of a segment PQ is defined to be the distance between its endpoints. That is, its length is $d(P, Q)$. The length of PQ is written $|PQ|$. Then congruence is defined by saying that two segments are congruent if they have the same length. Incidence and order are also defined in terms of d: points P, Q, and R are collinear and Q is between P and R when the inequality in (iii) is an equality. In other words, the synthetic notions of incidence, order, and congruence, are defined analytically. This is essentially the perspective that we will take when we develop hyperbolic geometry. But to this point, we have been developing a synthetic geometry, and so in this chapter we will see how we can take those synthetic notions and use them to build a distance function.

6.1 Synthetic comparison

The first part of the process is to describe a way to compare non-congruent segments, and to learn a few properties of such comparisons. I prove one, and leave the rest as exercises.

Definition 6.1 Given segments AB and CD, label E on $\overrightarrow{CD}$ so that $CE \simeq AB$. If $C * E * D$, then AB is *shorter than* CD, written $AB \prec CD$. If $C * D * E$, then AB is *longer than* CD, written $AB \succ CD$.

If we replace CD in this definition with DC, things will change slightly: calculations will be done on the ray $\overrightarrow{DC}$ rather than $\overrightarrow{CD}$. That could be a problem, since CD and DC are the same

51

segment. The next theorem is a reassurance that $\prec$ and $\succ$ are defined the same way, whether using CD or DC.

Theorem 6.2 *($\prec$ and $\succ$ are well-defined) Given segments AB and CD, label*

E: *the unique point on $\overrightarrow{CD}$ so that $AB \simeq CE$ and*
F: *the unique point on $\overrightarrow{DC}$ so that $AB \simeq DF$.*

Then $C * E * D$ if and only if $D * F * C$.

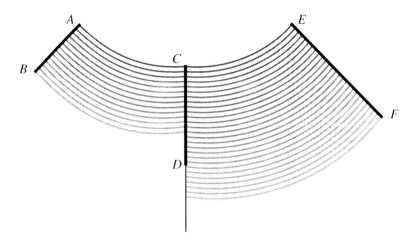

Figure 1. $AB \prec CD$ and $EF \succ CD$.

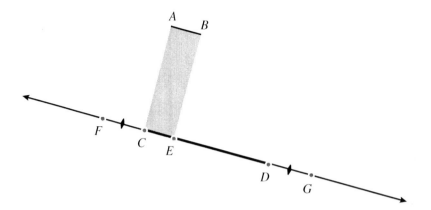

Figure 2. Segment comparison is well-defined: if $AB \prec CD$, then $AB \prec DC$ (a proof by contradiction).

Proof Suppose that $C * E * D$, but that F is not between D and C, so $D * C * F$. According to the segment construction axiom, there is a point G so that $C * D * G$ and $DG \simeq FC$. Now

$CD \simeq DC$ and $DG \simeq FC$, so by the segment addition axiom, $CG \simeq DF$. Therefore,
$$CE \simeq AB \simeq DF \simeq CG.$$
But both CE and CG are constructed along the same ray. According to the segment construction axiom (the uniqueness part this time), there is only one way to construct the segment. That would imply that $E = G$, but that isn't possible since they are on opposite sides of D. Therefore F must be between D and C. The proof of the converse statement is essentially the same. □

Here are some essential properties of $\prec$. There are corresponding properties for $\succ$.

Theorem 6.3 *(Transitivity of $\prec$) If $AB \prec CD$ and $CD \prec EF$, then $AB \prec EF$. If $AB \prec CD$ and $CD \simeq EF$, then $AB \prec EF$. If $AB \simeq CD$ and $CD \prec EF$, then $AB \prec EF$.*

Exercise 6.1 Prove theorem 6.3.

Theorem 6.4 *(Symmetry between $\prec$ and $\succ$) For any two segments AB and CD, $AB \prec CD$ if and only if $CD \succ AB$.*

Exercise 6.2 Prove theorem 6.4.

Theorem 6.5 *(Order and $\prec$) If $A * B * C * D$, then $BC \prec AD$.*

Exercise 6.3 Prove theorem 6.5.

Theorem 6.6 *(Additivity of $\prec$) Suppose that $A * B * C$ and $A' * B' * C'$. If $AB \prec A'B'$ and $BC \prec B'C'$, then $AC \prec A'C'$.*

Exercise 6.4 Prove theorem 6.6.

6.2 Distance

Now we are in position to move from synthetic comparisons of segments to something numerical, by building a distance function d. That function will have to satisfy conditions (i)-(iii) mentioned at the start of the chapter, and so it is fair to have certain expectations for d.

(1) The distance between any two distinct points should be a positive real number and the distance from a point to itself should be zero. That way, d will satisfy condition (i) above.
(2) Congruent segments should have the same length. That takes care of condition (ii) above, since $AB \simeq BA$, but it does a whole lot more too. Choose a ray r and label its endpoint O. According to the segment construction axiom, for any segment AB, there is a unique point P on r so that $AB \simeq OP$. If congruent segments are to have the same length, then that means $|AB| = d(O, P)$. Therefore, if we can work out the distance from O to the other points on r, then all other distances will follow.

54 Distance, Length, and the Axioms of Continuity

(3) If $A * B * C$, then

$$|AB| + |BC| = |AC|.$$

This is a part of property (iii) of a distance function. Since we are going to develop the distance function on r, we don't have to worry about non-collinear points just yet (that will come a little later). Since d never assigns negative values, this implies

$$AB \prec CD \implies |AB| < |CD|,$$
$$AB \succ CD \implies |AB| > |CD|.$$

It is up to us to build a distance function that meets all three requirements. The rest of this chapter is devoted to doing that. With those conditions in mind, let's start building the distance function d. The picture to keep in mind is that of a ruler (not a metric one, but rather an English one, marked in inches). It has a $1''$ mark on it. That distance is halved, and halved, and halved again to get the $1/2''$, $1/4''$, and $1/8''$ marks. Then, depending upon the precision of the ruler, there may be $1/16''$ or $1/32''$ markings as well. All the other marks on the ruler are multiples of these. That ruler is the blueprint for how we will build the skeleton of d. First, because of condition (1), $d(O, O) = 0$. Now take a step along r to another point. Any point is fine. Like the inch mark on the ruler, it sets the unit of measurement. Call the point P_0 and define $d(O, P_0) = 1$. As with the ruler, we want to repeatedly halve OP_0. That requires a little theory.

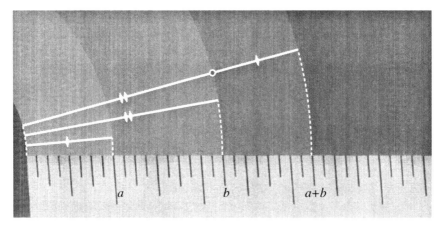

Figure 3. The additivity condition for d.

Definition 6.7 A point M on a segment AB is the *midpoint* of AB if $AM \simeq MB$.

Theorem 6.8 *Every segment has a unique midpoint.*

Proof We will prove the existence portion of the theorem, and leave the uniqueness as an exercise. Given AB, choose a point P that is not on $\overline{AB}$. According to the angle and segment construction axioms, there is a point Q on the opposite side of $\overline{AB}$ from P so that $\angle ABP \simeq \angle BAQ$ (that's the angle construction part) and so that $BP \simeq AQ$ (that's the segment construction part). Since P and Q are on opposite sides of $\overline{AB}$, the segment PQ intersects it.

6.2 Distance

Call the point of intersection M. I claim that M is the midpoint of AB. Why? Compare $\triangle MBP$ and $\triangle MAQ$. We have constructed them so that $\angle MAQ \simeq \angle MBP$ and $BP \simeq AQ$. Furthermore, the vertical angles $\angle AMQ$ and $\angle BMP$ are congruent. By AAS, $\triangle MBP \simeq \triangle MAQ$, and in particular, $AM \simeq MB$.

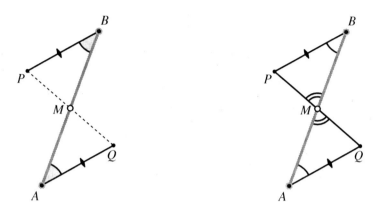

Figure 4. Constructing a midpoint.   Figure 5. Using AAS to verify that M bisects AB.

$\square$

Exercise 6.5 Prove that the midpoint of a segment AB is unique.

Let's go back to OP_0. We now know that it has a unique midpoint. Call it P_1. For the distance function d to satisfy condition (3),

$$|OP_1| + |P_1 P_0| = |OP_0|.$$

But OP_1 and $P_1 P_0$ are congruent, so for d to satisfy condition (2), they have to have the same length. Therefore $2|OP_1| = 1$ and so $|OP_1| = 1/2$. Repeat. Take OP_1, and find its midpoint. Call it P_2. Then

$$|OP_2| + |P_2 P_1| = |OP_1|.$$

Again, OP_2 and $P_2 P_1$ are congruent, so the must have the same length. Therefore $2|OP_2| = 1/2$, and so $|OP_2| = 1/4$. By repeating the process, we can identify the points P_n that are distances of $1/2^n$ from O.

With the points P_n as building blocks, we can start combining segments of lengths $1/2^n$ to get to other points. In fact, we can find a point whose distance from O is $m/2^n$ for any positive integers m and n. It is just a matter of chaining together enough congruent copies of OP_n as follows. Begin with the point P_n. By the first axiom of congruence, there is a point P_n^2 on the opposite side of P_n from O so that $P_n P_n^2 \simeq OP_n$, and there is a point P_n^3 on the opposite side of P_n^2 from P_n so that $P_n^2 P_n^3 \simeq OP_n$, and a point P_n^4 on the opposite side of P_n^3 from P_n^2 so that $P_n^3 P_n^4 \simeq OP_n$, and so on. This process can be continued until m segments are chained together stretching from O to a point that we will label P_n^m. In order for the distance function to satisfy the additivity condition (3),

$$\left|OP_n^m\right| = \left|OP_n\right| + \left|P_n P_n^2\right| + \left|P_n^2 P_n^3\right| + \cdots + \left|P_n^{m-1} P_n^m\right|.$$

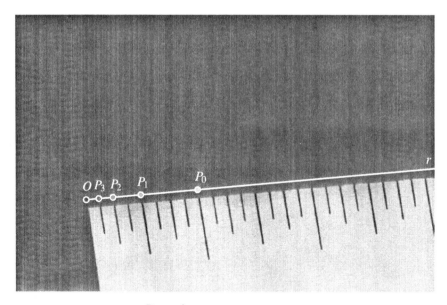

Figure 6. Subdividing a segment.

All the segments are congruent, though, so they have to be the same length (for condition (2)), so

$$|OP_n^m| = m \cdot |OP_n| = m \cdot 1/2^n = m/2^n.$$

Rational numbers whose denominator can be written as a power of two are called dyadic rationals. In that spirit, I will call these points the dyadic points of r.

Figure 7. Chaining together segments.

6.2 Distance

There are plenty of real numbers that aren't dyadic rationals though, and there are plenty of points on r that aren't dyadic points. How can we measure the distance from O to them? We now need the last two axioms of neutral geometry, the axioms of continuity. They a little more technical than the previous ones. The first essentially says that you can get to any point on a line if you take enough steps. The second, which is inspired by Dedekind's construction of the real numbers, essentially says that there are no gaps in a line.

The axioms of continuity.

Ct1: *Archimedes' axiom* If AB and CD are two segments, there is some positive integer n such that n congruent copies of CD constructed end-to-end from A along $\overrightarrow{AB}$ will pass beyond B.

Ct2: *Dedekind's axiom* Let $\mathbb{S}_<$ and $\mathbb{S}_\geq$ be two nonempty subsets of a line ℓ satisfying: (i) $\mathbb{S}_< \cup \mathbb{S}_\geq = \ell$, (ii) no point of $\mathbb{S}_<$ is between two points of $\mathbb{S}_\geq$, and (iii) no point of $\mathbb{S}_\geq$ is between two points of $\mathbb{S}_<$. Then there is a unique point O on ℓ such that for any two other points P_1 and P_2 with $P_1 \in \mathbb{S}_<$ and $P_2 \in \mathbb{S}_\geq$ then $P_1 * O * P_2$.

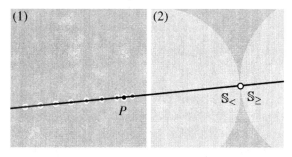

Figure 8. (l) Archimedes' axiom: given enough steps, the point P will be passed. (r) Dedekind's axiom: there is a point between any two separated partitions of a line.

These two axioms allow us to fill the gaps between the dyadic points on r. Let's consider a point P on r that is not a dyadic point. However, it is surrounded by enough dyadic points that we can use them to estimate the distance from O to P to any level of precision. For instance, suppose we consider just the dyadic points whose denominators can be written as 2^0:

$$S_0 = \{O, P_0^1, P_0^2, P_0^3, \ldots\}.$$

By Archimedes' axiom, eventually, the points in this list will pass beyond P. If we focus our attention on the one right before P, say $P_0^{m_0}$, and the one right after, $P_0^{m_0+1}$, then

$$O * P_0^{m_0} * P * P_0^{m_0+1}.$$

We can compare the relative sizes of the segments

$$OP_0^{m_0} \prec OP \prec OP_0^{m_0+1}$$

and so, if our distance is going to satisfy condition (3),

$$\left|OP_0^{m_0}\right| < |OP| < \left|OP_0^{m_0+1}\right|$$

or
$$m_0 < |OP| < m_0 + 1.$$

To get a slightly more precise approximation, replace S_0, with S_1, the set of dyadic points whose denominators can be written as 2^1:

$$S_1 = \{O, P_1, P_1^2 = P_0, P_1^3, P_1^4 = P_0^2, \ldots\}.$$

Again, Archimedes' axiom guarantees that eventually the points in S_1 will pass beyond P. Let $P_1^{m_1}$ be the last one before that happens. Then

$$O * P_1^{m_1} * P * P_1^{m_1+1}$$

so

$$\left|OP_1^{m_1}\right| < |OP| < \left|OP_1^{m_1+1}\right|$$

or

$$m_1/2 < |OP| < (m_1 + 1)/2,$$

and this gives $|OP|$ to within an accuracy of $1/2$. Continuing in this way, using S_2, dyadics whose denominators can be written as 2^2, will give an approximation of $|OP|$ to within $1/4$; using S_3, dyadics whose denominators can be written as 2^3, will give an approximation of $|OP|$ to within $1/8$. Generally speaking, the dyadic rationals in S_n will provide an upper and lower bound for $|OP|$ that differ by $1/2^n$. As n goes to infinity, $1/2^n$ goes to zero, forcing the upper and lower bounds to come together at a single number. This number is going to have to be $|OP|$. Now, we don't really need both the increasing and decreasing sequences of approximations to define $|OP|$. After all, they both end up at the same number. Here is the description of $|OP|$ using just the increasing sequence: for each positive integer n, let $P_n^{m_n}$ be the last point in the list S_n that is between O and P. For the distance function to satisfy condition (3), we must set

$$|OP| = \lim_{n \to \infty} \left|OP_n^{m_n}\right| = \lim_{n \to \infty} m_n/2^n.$$

Example 6.9 (A sequence of dyadics approaching $1/3$) In general, finding a dyadic sequence approaching a particular number can be tricky. Finding a sequence approaching $1/3$ is easy, though, as long as you remember the geometric series formula

$$\sum_{n=0}^{\infty} x^n = \frac{1}{1-x} \quad \text{if } |x| < 1.$$

With a little trial and error, I found that letting $x = 1/4$,

$$1 + \frac{1}{4} + \frac{1}{16} + \frac{1}{64} + \frac{1}{256} + \cdots = \frac{4}{3}.$$

6.2 Distance

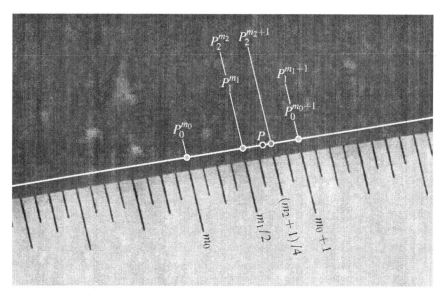

Figure 9. Capturing a non-dyadic point between two sequences of dyadic points.

Subtracting 1 from both sides gives an infinite sum of dyadics to 1/3, and we can extract the sequence from that:

$$\frac{1}{4} = 0.25,$$

$$\frac{1}{4} + \frac{1}{16} = \frac{5}{16} = 0.3125,$$

$$\frac{1}{4} + \frac{1}{16} + \frac{1}{64} = \frac{21}{64} = 0.32825.$$

Every point of r now has a distance associated with it, but is there a point at every possible distance? Do we know, for instance, that there is a point at exactly a distance of 1/3 from O? The answer is yes—it is just a matter of reversing the distance calculation process we just described and using Dedekind's axiom. Let's take as our prospective distance a positive real number x (we have already seen how to construct the dyadic point corresponding to any positive dyadic rational, so let's assume x is not a dyadic rational). For each integer $n \geq 0$, let $m_n/2^n$ be the largest dyadic rational less than x whose denominator can be written as 2^n and let $P_n^{m_n}$ be the corresponding dyadic point on r. Now we are going to define two sets of points:

$\mathbb{S}_<$: all the points of r that lie between O and any of the $P_n^{m_n}$, together with all the points of r^{op}.

$\mathbb{S}_\geq$: all the remaining points of r.

So $\mathbb{S}_<$ contains a sequence of dyadic points whose distance from O is increasing to x

$$\{P_0^{m_0}, P_1^{m_1}, P_2^{m_2}, P_3^{m_3}, \dots\},$$

and $\mathbb{S}_\geq$ contains a sequence of dyadic points whose distance from O is decreasing to x

$$\{P_0^{m_0+1}, P_1^{m_1+1}, P_2^{m_2+1}, P_3^{m_3+1}, \dots\}.$$

Together $\mathbb{S}_<$ and $\mathbb{S}_\geq$ contain all the points of the line through r, but they do not intermingle: no point of $\mathbb{S}_<$ is between two of $\mathbb{S}_\geq$ and no point of $\mathbb{S}_\geq$ is between two of $\mathbb{S}_<$. According to Dedekind's axiom, then, there is a unique point P between $\mathbb{S}_<$ and $\mathbb{S}_\geq$. Let's look at how far P is from O. For all n,

$$OP_n^{m_n} \prec OP \prec OP_n^{m_n+1}$$
$$|OP_n^{m_n}| < |OP| < |OP_n^{m_n+1}|$$
$$m_n/2^n < |OP| < (m_n+1)/2^n.$$

As n approaches infinity, the interval between the two consecutive dyadics shrinks. Ultimately, the only number left is x. So $|OP| = x$.

For two points P and Q, there is a unique point R on the ray r so that $OR \simeq PQ$. Define $d(P, Q) = |OR|$. With this setup, our distance function will satisfy conditions (1) and (2). That leaves condition (3)—a lot of effort went into trying to build d so that condition would be satisfied. Now let's make sure that it succeeded. Let's close out this chapter with two theorems that do that.

Theorem 6.10 *If P and Q are points on r, with $|OP| = x$ and $|OQ| = y$, and if P is between O and Q, then $|PQ| = y - x$.*

Proof If both P and Q are dyadic points, then this is fairly easy. First express their dyadic distances with a common denominator:

$$|OP| = m/2^n \quad |OQ| = m'/2^n.$$

Then OP is built from m segments of length $1/2^n$ and OQ is built from m' segments of length $1/2^n$. To get $|PQ|$, we take the m segments from the m' segments, so $|PQ|$ is made up of $m' - m$ segments of length $1/2^n$. That is

$$|PQ| = (m' - m) \cdot \frac{1}{2^n} = y - x.$$

If one or both of P and Q are not dyadic, then there is a bit more work to do. In this case, P and Q are approximated by a sequence of dyadics $P_n^{m_n}$ and $P_n^{m'_n}$ where

$$\lim_{n \to \infty} \frac{m_n}{2^n} = x \quad \text{and} \quad \lim_{n \to \infty} \frac{m'_n}{2^n} = y.$$

We can trap $|PQ|$ between segments that have dyadic lengths:

$$P_n^{m_n+1} P_n^{m'_n} \prec PQ \prec P_n^{m_n} P_n^{m'_n+1}$$
$$\left|P_n^{m_n+1} P_n^{m'_n}\right| < |PQ| < \left|P_n^{m_n} P_n^{m'_n+1}\right|$$
$$\frac{m'_n - m_n - 1}{2^n} < |PQ| < \frac{m'_n + 1 - m_n}{2^n}.$$

6.2 Distance

As n approaches infinity, $|PQ|$ is stuck between two values that are approaching $y - x$. Therefore $|PQ|$ must be $y - x$. □

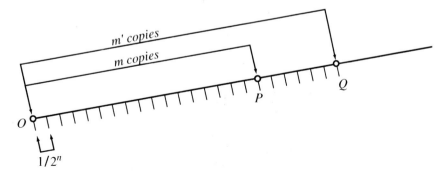

Figure 10. Measuring the distance between two dyadic points.

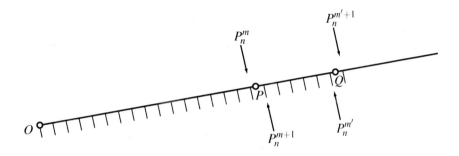

Figure 11. Measuring the distance between two non-dyadic points.

The last result of this chapter is a reinterpretation of the segment addition axiom in terms of distance. We will see that segment lengths satisfy the more general triangle inequality condition (for non-collinear points) in a few more chapters.

Theorem 6.11 *(Segment addition, the measured version)* If $P * Q * R$, then
$$|PQ| + |QR| = |PR|.$$

Proof The first step is to transfer the problem over to the ray r, where we have already established distances. So locate Q' and R' on r so that $O * Q' * R'$, $PQ \simeq OQ'$, and $QR \simeq Q'R'$. According to the segment addition axiom, $PR \simeq OR'$. Now we can use the last theorem,
$$|QR| = |Q'R'| = |OR'| - |OQ'| = |PR| - |PQ|.$$
Solving for $|PR|$ gives the result. □

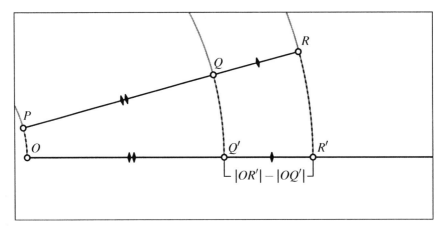

Figure 12. Proof of the measured version of segment addition.

We have established a correspondence between the points of a ray r and the positive real numbers. In many ways, though, the more natural correspondence is between points on a line and the numbers on the real number line. We can extend our correspondence from the ray r to the line containing it as follows: to any point P on r^{op}, associate the (negative) real number $-|OP|$.

Exercise 6.6 Let P_1 and P_2 be points on a line that has been put in correspondence with the real number line, so that x is the number corresponding to P_1 and y is the number corresponding to P_2. Show that $|P_1 P_2| = |x - y|$.

6.3 Exercises

6.7. Write $1/7$, $1/6$, and $1/5$ as infinite sums of dyadic rationals.

6.8. In chapter 40, we will examine a metric on the upper half plane, the portion of the plane above the real axis. In that metric, the distance between two points $P = (x, y_1)$ and $Q = (x, y_2)$ on a vertical line is defined to be

$$d_H(P, Q) = |\ln(y_1/y_2)|$$

(the distance between points that are not on a vertical line is more complicated). Show that if this metric were to be extended to the entire plane, the resulting geometry would not satisfy the Archimedes' axiom.

7
Angle Measure

The purpose of this chapter is to develop a measurement system for angles. As in the last chapter, we start by building up the properties of a synthetic comparison (larger versus smaller) of non-congruent angles. Then we define and examine right angles. Finally, we build the degree measurement system for angles. Since there are many similarities between this and what we did for segment length, much of it is just sketched, with the details left as exercises.

7.1 Synthetic angle comparison

Recall that we already described a synthetic way to compare angles (in chapter 4) that worked as follows.

Definition 7.1 Given $\angle ABC$ and $\angle A'B'C'$, label $C^\star$ on the same side of AB as C so that $\angle ABC^\star \simeq \angle A'B'C'$. If $C^\star$ is in the interior of $\angle ABC$, then $\angle A'B'C'$ is *smaller than* $\angle ABC$, written $\angle A'B'C' \prec \angle ABC$. If $C^\star$ is in the exterior of $\angle ABC$, then $\angle A'B'C'$ is *larger than* $\angle ABC$, written $\angle A'B'C' \succ \angle ABC$.

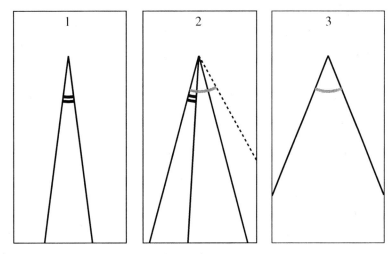

Figure 1. $\angle 1 \prec \angle 2$ and $\angle 3 \succ \angle 2$.

In chapter 2, we proved that there is an ordering of rays from a common endpoint:

Theorem 2.9. *Given $n \geq 2$ distinct rays with a common endpoint B that are all on the same side of $\overline{AB}$ through B, there is an ordering of them $r_1, r_2, \ldots, r_n$ so that if $i < j$ then r_i is in the interior of the angle formed by $\overrightarrow{BA}$ and r_j.*

In chapter 3, we proved that angle interiors work well with congruence, in the sense of the following theorem.

Theorem 3.5. *Given $\angle ABC \simeq \angle A'B'C'$ and that the point D is in the interior of $\angle ABC$. Suppose that D' is located on the same side of $\overline{A'B'}$ as C' so that $\angle ABD \simeq \angle A'B'D'$. Then D' is in the interior of $\angle A'B'C'$.*

As with the segment comparison definitions, there is a potential issue with the definitions of $\prec$ and $\succ$. What if we had decided to construct C^* off of $\overrightarrow{BC}$ instead of $\overrightarrow{BA}$? Since $\angle ABC = \angle CBA$, and since we are interested in comparing the angles themselves, this notion of larger or smaller should not depend on which ray we are building from. The next theorem tells us not to worry.

Theorem 7.2 *($\prec$ and $\succ$ are well-defined) Given $\angle ABC$ and $\angle A'B'C'$, label*

C^* : *a point on the same side of AB as C for which $\angle ABC^* \simeq \angle A'B'C'$*
A^* : *a point on the same side of BC as A for which $\angle CBA^* \simeq \angle A'B'C'$.*

Then C^* is in the interior of $\angle ABC$ if and only if A^* is.

Proof Note that $\angle ABC \simeq \angle CBA$, $\angle A^*BC \simeq \angle C^*BA$, and C^* is on the same side of AB as C. So if A^* is in the interior of $\angle ABC$, then C^* must be too. Conversely, A^* is on the same side of BC as A, so if C^* is in the interior, then A^* must be too. □

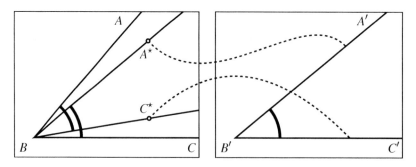

Figure 2. When comparing angles, it doesn't matter which ray is used as the base.

Let's look at some of the properties of synthetic angle comparison, focusing on the $\prec$ version of these properties. The $\succ$ version is similar. There is nothing particularly elegant about these proofs. They mainly rely upon the two propositions listed above.

7.1 Synthetic angle comparison

Theorem 7.3 *(Transitivity of $\prec$)*

$\prec\prec$: If $\angle A_1B_1C_1 \prec \angle A_2B_2C_2$ and $\angle A_2B_2C_2 \prec \angle A_3B_3C_3$, then $\angle A_1B_1C_1 \prec \angle A_3B_3C_3$.

$\simeq\prec$: If $\angle A_1B_1C_1 \simeq \angle A_2B_2C_2$ and $\angle A_2B_2C_2 \prec \angle A_3B_3C_3$, then $\angle A_1B_1C_1 \prec \angle A_3B_3C_3$.

$\prec\simeq$: If $\angle A_1B_1C_1 \prec \angle A_2B_2C_2$ and $\angle A_2B_2C_2 \simeq \angle A_3B_3C_3$, then $\angle A_1B_1C_1 \prec \angle A_3B_3C_3$.

Proof The second and third statements are easier, and so are left as exercises. For the first, most of the proof is just getting points shifted into a useful position.

1 : Copy the first angle to the second: since $\angle A_1B_1C_1 \prec \angle A_2B_2C_2$, there is a point A_1' in the interior of $\angle A_2B_2C_2$ so that $\angle A_1B_1C_1 \simeq \angle A_1'B_2C_2$.

2 : Copy the second angle to the third: since $\angle A_2B_2C_2 \prec \angle A_3B_3C_3$, there is a point A_2' in the interior of $\angle A_3B_3C_3$ so that $\angle A_2B_2C_2 \simeq \angle A_2'B_3C_3$.

3 : Copy the first angle to the third (although we don't know quite as much about this one): pick a point A_1'' on the same side of B_3C_3 as A_1 so that $\angle A_1''B_3C_3 \simeq \angle A_1B_1C_1$.

Since A_1' is in the interior of $\angle A_2B_2C_2$, A_1'' has to be in the interior of $\angle A_2'B_3C_3$. Since $\overrightarrow{B_3A_1''}$ is in the interior of $\angle A_2'B_3C_3$ and $\overrightarrow{B_3A_2'}$ is in the interior of $\angle A_3B_3C_3$, $\overrightarrow{B_3A_1''}$ has to be in the interior of $\angle A_3B_3C_3$. Therefore $\angle A_1B_1C_1 \prec \angle A_3B_3C_3$.

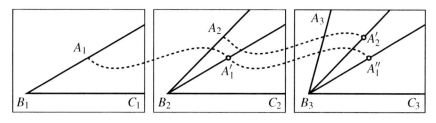

Figure 3. The transitivity of $\prec$. □

Exercise 7.1 Prove the other two cases in the previous theorem ($\simeq\prec$ and $\prec\simeq$).

Theorem 7.4 *(Symmetry between $\prec$ and $\succ$)* For two angles $\angle A_1B_1C_1$ and $\angle A_2B_2C_2$, $\angle A_1B_1C_1 \prec \angle A_2B_2C_2$ if and only if $\angle A_2B_2C_2 \succ \angle A_1B_1C_1$.

Proof Suppose that $\angle A_1B_1C_1 \prec \angle A_2B_2C_2$. Then there is a point A_1' in the interior of $\angle A_2B_2C_2$ so that $\angle A_1B_1C_1 \simeq \angle A_1'B_2C_2$. Moving back to the first angle, there is a point $A_2^\star$ on the opposite side of A_1B_1 from C_1 so that $\angle A_1B_1A_2^\star \simeq \angle A_1'B_2A_2$. By angle addition, $\angle A_2^\star B_1C_1 \simeq \angle A_2B_2C_2$, and since $A_2^\star$ is not in the interior of $\angle A_1B_1C_1$, $\angle A_2B_2C_2 \succ \angle A_1B_1C_1$. The other direction in this proof is similar. □

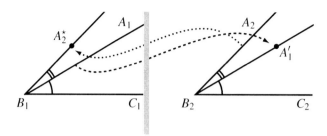

Figure 4. The relationship between $\prec$ and $\succ$.

Theorem 7.5 *If A_2 and C_2 are in the interior of $\angle A_1 B C_1$, then $\angle A_2 B C_2 \prec \angle A_1 B C_1$.*

Proof Locate $A_2^\star$ on the same side of $\overline{BC_1}$ as A_1 so that $\angle A_2^\star B C_1 \simeq \angle A_2 B C_2$. Then the question is: is $A_2^\star$ in the interior of $\angle A_1 B C_1$? Suppose that it isn't. Then

$$\angle A_2 B C_2 \prec \angle A_2^\star B C_2 \prec \angle A_2^\star B C_1.$$

Since we have established that $\prec$ is transitive, $\angle A_2 B C_2 \prec \angle A_2^\star B C_1$. But this cannot be, since the angles are supposed to be congruent. Hence $A_2^\star$ has to be in the interior of $\angle A_1 B C_1$, and so $\angle A_2 B C_2 \prec \angle A_1 B C_1$. $\square$

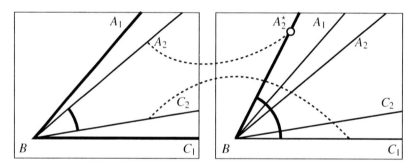

Figure 5. A proof (by contradiction) that rays inside an angle form a smaller angle.

Theorem 7.6 *(Additivity of $\prec$) Suppose that D_1 lies in the interior of $\angle A_1 B_1 C_1$ and that D_2 lies in the interior of $\angle A_2 B_2 C_2$. If $\angle A_1 B_1 D_1 \prec \angle A_2 B_2 D_2$ and $\angle D_1 B_1 C_1 \prec \angle D_2 B_2 C_2$, then $\angle A_1 B_1 C_1 \prec \angle A_2 B_2 C_2$.*

Proof Find D_1' in the interior of $\angle A_2 B_2 D_2$ so that $\angle A_2 B_2 D_1' \simeq \angle A_1 B_1 D_1$. Find C_1' on the opposite side of $\overline{B_2 D_1'}$ from A_2 so that $\angle D_1' B_2 C_1' \simeq \angle D_1 B_1 C_1$. By angle addition, $\angle A_2 B_2 C_1' \simeq \angle A_1 B_1 C_1$, so the question is whether or not C_1' is in the interior of $\angle A_2 B_2 C_2$. If it is not, then $\angle D_2 B_2 C_2$ is contained in $\angle D_1' B_2 C_1'$, and so by the previous result $\angle D_2 B_2 C_2 \prec \angle D_1' B_2 C_1'$. Therefore $\angle D_2 B_2 C_2 \prec \angle D_1 B_1 C_1$. But that is a contradiction, and so C_1' will have to lie in the interior of $\angle A_2 B_2 C_2$. That is, $\angle A_1 B_1 C_1 \prec \angle A_2 B_2 C_2$. $\square$

7.2 Right angles

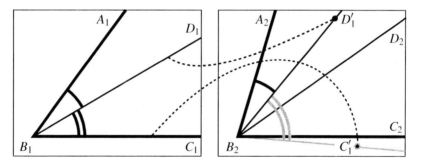

Figure 6. A proof (by contradiction) of the additivity of $\prec$.

7.2 Right angles

When we constructed distance, we used an arbitrary segment to determine unit length. Angle measure is handled differently—a specific angle is used as the baseline from which the rest is developed (although, at least in the degree measurement system, that angle is then assigned a pretty random measure). That angle is the right angle.

Definition 7.7 A *right angle* is an angle that is congruent to its supplement.

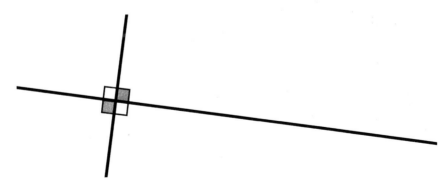

Figure 7. Two lines intersecting at right angles. In diagrams, square angle markers are often used to indicate that an angle is right.

In fact, we have already seen right angles, in the proof of the SSS theorem. But it ought to be stated again, that

Theorem 7.8 *Right angles exist.*

Proof We will prove that right angles exist by constructing one. Start with a segment AB. Choose a point P that is not on $\overline{AB}$. If $\angle PAB$ is congruent to its supplement, then it is a right angle, and that's it. If $\angle PAB$ is not congruent to its supplement, then there is more work to do. Thanks to the segment and angle construction axioms, there is a point P' on the opposite side of $\overline{AB}$ from P so that $\angle P'AB \simeq \angle PAB$ (angle construction) and $AP' \simeq AP$ (segment construction). Since P and P' are on opposite sides of $\overline{AB}$, PP' has to intersect $\overline{AB}$. Call the point of intersection Q. With that construction,

$$PA \simeq P'A, \quad \angle PAQ \simeq \angle P'AQ, \quad \text{and} \quad AQ \simeq AQ;$$

so by the SAS triangle congruence theorem, $\triangle PAQ \simeq \triangle P'AQ$. In those two triangles, the relevant congruence is between the two angles that share the vertex Q: $\angle AQP \simeq \angle AQP'$. The angles are supplements. They are congruent. By definition, they are right angles. □

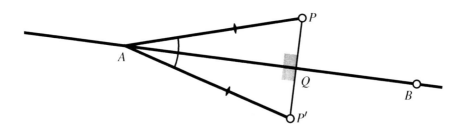

Figure 8. A proof that right angles exist, by construction.

Having established the existence of right angles, the next question is how common they are. The next result is something like a uniqueness statement—that there is really only one right angle up to congruence.

Theorem 7.9 *Suppose that $\angle ABC$ is a right angle. Then $\angle A'B'C'$ is a right angle if and only if it is congruent to $\angle ABC$.*

Proof ⟹ To start, let's mark a few more points so that we can refer to the supplements of the angles. Mark the points

D : on $\overline{BC}$ so that $D * B * C$ and
D' : on $\overline{B'C'}$ so that $D' * B' * C'$.

Therefore $\angle ABC$ and $\angle ABD$ are a supplementary pair, as are $\angle A'B'C'$ and $\angle A'B'D'$. Now suppose that $\angle ABC$ and $\angle A'B'C'$ are right angles. Thanks to the angle construction axiom, it is possible to build a congruent copy of $\angle A'B'C'$ on top of $\angle ABC$: there is a ray $\overrightarrow{BA^*}$ on the same side of BC as A so that $\angle A^*BC \simeq \angle A'B'C'$. Earlier we proved that the supplements of congruent angles are congruent, so $\angle A^*BD \simeq \angle A'B'D'$. How, though, does $\angle A^*BC$ compare to $\angle ABC$? If $\overrightarrow{BA^*}$ and $\overrightarrow{BA}$ are the same ray, then the angles are equal, meaning that $\angle ABC$ and $\angle A'B'C'$ are congruent, which is what we want. But what happens if the rays are not equal? Then one of two things can happen: either $\overrightarrow{BA^*}$ is in the interior of $\angle ABC$, or it is in the interior of $\angle ABD$. Both cases will lead to essentially the same problem, so let's focus on the first one. In that case, A^* is in the interior of $\angle ABC$, so $\angle A^*BC \prec \angle ABC$; and A^* is in the exterior of

7.2 Right angles

$\angle ABD$, so $\angle A^\star BD \succ \angle ABD$. That leads to a string of congruences and inequalities:

$$\angle A'B'C' \simeq \angle A^\star BC \prec \angle ABC \simeq \angle ABD \prec \angle A^\star BD \simeq \angle A'B'D'.$$

Because of the transitivity of $\prec$, $\angle A'B'C' \prec \angle A'B'D'$. This can't be since the supplements are supposed to be congruent. The second scenario plays out in the same way, with $\succ$ in place of $\prec$. Therefore $\overrightarrow{BA^\star}$ and $\overrightarrow{BA}$ have to be the same ray, and so $\angle A'B'C' \simeq \angle ABC$.

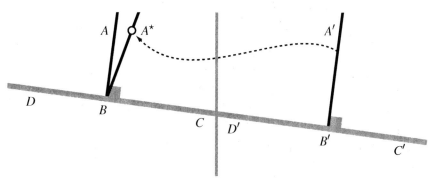

Figure 9. Any two right angles are congruent: if one right angle were larger or smaller than another, then it could not be congruent to its supplement.

$\impliedby$ The other direction is easier: suppose that $\angle A'B'C' \simeq \angle ABC$ and that $\angle ABC$ is a right angle. Let's recycle the points D and D' from the first part of the proof. Then $\angle A'B'D'$ and $\angle ABD$ are supplementary to congruent angles, so they too must be congruent. Therefore

$$\angle A'B'C' \simeq \angle ABC \simeq \angle ABD \simeq \angle A'B'D',$$

and so we can see that $\angle A'B'C'$ is congruent to its supplement: it must be a right angle.

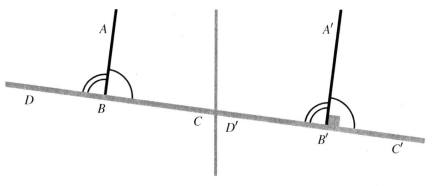

Figure 10. If an angle is congruent to a right angle, it is a right angle too. □

Definition 7.10 Two lines that intersect at right angles are said to be *perpendicular*.

We use the standard notation $\ell_1 \perp \ell_2$ to indicate that ℓ_1 and ℓ_2 are perpendicular. With $\prec$ and $\succ$ and with right angles as a point of comparison, we have a way to classify non-right angles.

Definition 7.11 An angle is *acute* if it is smaller than a right angle. An angle is *obtuse* if it is larger than a right angle.

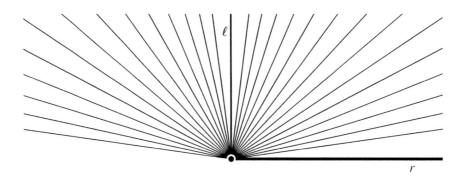

Figure 11. To the left of ℓ, rays that form an obtuse angle with r. To the right of ℓ, rays that form an acute angle with r.

Exercise 7.2 A *perpendicular bisector* to a segment AB is a line that passes through the midpoint of AB and is perpendicular to AB. Show that every segment has a unique perpendicular bisector.

7.3 Angle measure

The rest of this lesson is an outline of the construction of the degree measurement system for angles. The details are left as exercises. Just as there is a tendency to write a segment AB when you really mean its length $|AB|$, there is also a temptation to write an angle $\angle ABC$ when you really mean its measure. As with segments and lengths, this rarely leads to confusion in practice; but on principle, it seems to be a good idea to have notation that distinguishes an angle from its measure. One common approach is to write $m(\angle ABC)$ for the measure of $\angle ABC$, which I find to be burdensome. My notation is an abbreviation: I write $(\angle ABC)$ for the measure of $\angle ABC$.

Now let's talk about what we will want in a system of angle measurement. We want, analogous to measures for distance:

(1) The measure of an angle should be a positive real number.
(2) Congruent angles should have the same measure (because of the angle construction axiom, which allows us to focus our investigation on angles that are built off of just one fixed ray).
(3) If D is in the interior of $\angle ABC$, then

$$(\angle ABC) = (\angle ABD) + (\angle DBC).$$

Therefore, since the measure of an angle has to be positive,

$$\angle ABC \prec \angle A'B'C' \implies (\angle ABC) < (\angle A'B'C')$$
$$\angle ABC \succ \angle A'B'C' \implies (\angle ABC) > (\angle A'B'C').$$

Mirroring the distance approach from the last chapter, we first need to prove that it is possible to divide an angle in half.

Theorem 7.12 *For an angle* $\angle ABC$, *there is a unique ray* $\overrightarrow{BD}$ *in the interior of* $\angle ABC$ *so that* $\angle ABD \simeq \angle DBC$. *This ray is called the* angle bisector *of* $\angle ABC$.

Exercise 7.3 Prove theorem 7.12.

In the degree measurement system, a right angle is assigned a measure of 90°. As with segments, the next step is to repeatedly subdivide the right angle, and to chain together those pieces to form the dyadic angles. A fundamental difference between angles and segments is that segments can be extended arbitrarily, but angles cannot be put together to exceed a straight angle. Therefore segments can be arbitrarily long, but all angles must measure less than 180° (since a straight angle is made up of two right angles). It is true that the unit circle in trigonometry shows how you can loop back around to define angles with any real measure, positive or negative, sometimes a useful extension, but one that creates some problems (the measure of an angle is not uniquely defined, for instance).

Exercise 7.4 Describe the construction of angles that have measures of the form $\theta = 90° \cdot m/2^n$, where m and n are integers and $0 < \theta < 180°$.

Finally, we need to fill in the gaps between the dyadic angles: both to see that every angle has a real number measure, and to see that every real number between 0° and 180° is the measure of an angle. As before, we need to use a sequence of approximating dyadics to do this. This time the key word "interior" will replace the key word "between." We previously used Dedekind's axiom as a way to fill the gaps. That axiom is still useful, but it cannot be applied as directly in this situation.

Exercise 7.5 Describe the process of finding the measure of a non-dyadic angle.

Exercise 7.6 Show that for every real number θ between 0° and 180°, there is an angle whose measure is θ.

Finally, with angles measured in this way, we may now prove a measured version of the angle addition theorem.

Theorem 7.13 *(Angle addition, the measured version) If D is in the interior of* $\angle ABC$, *then* $(\angle ABC) = (\angle ABD) + (\angle DBC)$.

Exercise 7.7 Prove theorem 7.13.

7.4 Exercises

7.8. Verify that the supplement of an acute angle is an obtuse angle and that the supplement of an obtuse angle is an acute angle.

7.9. Prove that an acute angle cannot be congruent to an obtuse angle.

7.10. For a line ℓ and point P, prove that there is a unique line through P that is perpendicular to ℓ. There are two cases: P may or may not be on ℓ.

7.11. Suppose that A_2 is the midpoint of the segment $A_1 A_3$. Let B_1, B_2, and B_3 be on the perpendiculars to $A_1 A_3$ through A_1, A_2, and A_3 respectively. Suppose furthermore that B_1 and B_3 are on the same side of $\overline{A_1 A_3}$. Prove that $B_1 B_2 \simeq B_2 B_3$ (there are two cases to consider, depending upon whether B_2 is on the same side, or the opposite side of $\overline{A_1 A_3}$ from B_1 and B_3).

7.12. Consider two distinct isosceles triangles with a common side: $\triangle ABC$ and $\triangle A'BC$ with $AB \simeq AC$ and $A'B \simeq A'C$. Prove that $\overline{AA'}$ is perpendicular to $\overline{BC}$.

7.13. Two angles are *complementary* if together they form a right angle. That is, if D is in the interior of a right angle $\angle ABC$, then $\angle ABD$ and $\angle DBC$ are complementary angles. Prove that every acute angle has a complement. Prove that if $\angle ABC$ and $\angle A'B'C'$ are congruent acute angles, then their complements are also congruent.

7.14. Verify that if ℓ_1 is perpendicular to ℓ_2 and ℓ_2 is perpendicular to ℓ_3, then either $\ell_1 = \ell_3$, or ℓ_1 and ℓ_3 are parallel.

7.15. Prove that if D is in the interior of $\triangle ABC$, then $\triangle ABD$ cannot be congruent to $\triangle ABC$.

7.16. We have defined equilateral triangles as those whose three sides are all congruent, and proved some properties of them (for instance, that they are equiangular). We have not yet proved, though, that equilateral triangles exist. Describe now a construction in neutral geometry that will create an equilateral triangle (warning: in neutral geometry, you cannot assume that the interior angles of a triangle measure $60°$).

7.17. Given an equilateral triangle $\triangle ABC$, label points along its edges A_1, A_2, B_1, B_2, C_1, and C_2 so that

$$A * A_1 * A_2 * B$$
$$B * B_1 * B_2 * C$$
$$C * C_1 * C_2 * A$$

and

$$AA_1 \simeq A_1 A_2 \simeq A_2 B$$
$$BB_1 \simeq B_1 B_2 \simeq B_2 C$$
$$CC_1 \simeq C_1 C_2 \simeq C_2 A$$

(in essence trisect the edges of the triangle). Show that the lines $\overline{AB_1}$, $\overline{BC_1}$, and $\overline{CA_1}$ intersect to form an equilateral triangle.

8
Triangles in Neutral Geometry

In this chapter we will see what our two newly-created measurement systems, segment length and angle measure, can tell us about triangles. We will derive three of the most fundamental results of neutral geometry: the Saccheri-Legendre theorem, the scalene triangle theorem, and the triangle inequality. Let's begin with the Saccheri-Legendre theorem, a theorem about the measures of the interior angles of a triangle.

Definition 8.1 The *angle sum* $s(\triangle ABC)$ of $\triangle ABC$ is the sum of its three interior angles. That is

$$s(\triangle ABC) = (\angle A) + (\angle B) + (\angle C).$$

It is a well known fact of Euclidean geometry that the angle sum of a triangle is $180°$. That is not necessarily the case in neutral geometry, though, so we will have to be content with a less restrictive result.

Theorem 8.2 *(The Saccheri-Legendre theorem) The angle sum of a triangle is at most* $180°$.

The proof of this important theorem is in three parts—two preparatory lemmas followed by the proof of the theorem.

Figure 1. (l) a Euclidean triangle, (c) a non-Euclidean triangle, (r) not a valid triangle in neutral geometry.

Lemma 8.3 *The sum of the measures of two angles in a triangle is less than* $180°$.

73

Proof Consider ∠A and ∠B of △ABC. Label one more point: choose D so that D ∗ A ∗ C. Then, by the exterior angle theorem, (∠ABC) < (∠BAD), and so

$$(\angle BAC) + (\angle ABC) < (\angle BAC) + (\angle BAD) = 180°. \qquad \square$$

This means that a triangle cannot support more than one right or obtuse angle—if a triangle has a right angle, or an obtuse angle, then the other two angles have to be acute. That leads to some more terminology.

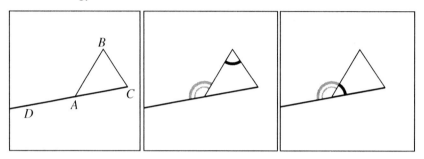

Figure 2. The proof of lemma 8.3. (l) Labels, (c) an exterior angle is larger than a non-adjacent interior angle, and (r) supplementary to the adjacent interior angle.

Definition 8.4 A triangle is *acute* if all three of its angles are acute. A triangle is *right* if it has a right angle. A triangle is *obtuse* if it has an obtuse angle.

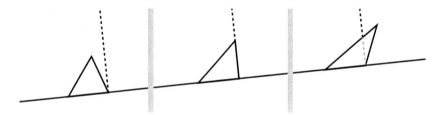

Figure 3. (l) An acute triangle, (c) a right triangle, (r) an obtuse triangle.

The real key to this proof of the Saccheri-Legendre theorem, the clever mechanism that makes it work, is the next lemma.

Lemma 8.5 *For any triangle △ABC, there is another triangle △A'B'C' so that*

(1) $s(\triangle A'B'C') = s(\triangle ABC)$, *and*
(2) $(\angle A') \leq (\angle A)/2$.

Proof We will describe how to build a triangle that meets both requirements. First we need to label a few more points:

D: the midpoint of BC, and
E: on $\overrightarrow{AD}$, so that A ∗ D ∗ E and AD ≃ DE.

I claim that △ACE satisfies (1) and (2). Showing that it does involves comparing angle measures, and for that it is helpful to abbreviate some of the angles:

∠1 for ∠BAD,   ∠2 for ∠DAC,   ∠3 for ∠DCE,   ∠4 for ∠ACD.

Because

$$BD \simeq CD, \quad \angle BDA \simeq \angle CDE, \quad \text{and} \quad DA \simeq DE,$$

by SAS, $\triangle BDA \simeq \triangle CDE$. Matching the two remaining pairs of angles in those triangles $\angle 1 \simeq \angle E$ and $\angle B \simeq \angle 3$.

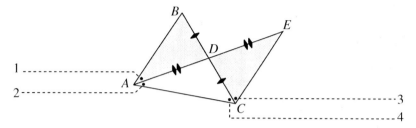

Figure 4. Labels in the proof of lemma 8.5.

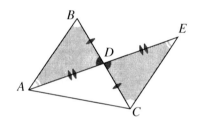

Figure 5. Use SAS to establish congruences.

Compare the angle sums of $\triangle ABC$ and $\triangle ACE$:

$$s(\triangle ABC) = (\angle A) + (\angle B) + (\angle 4) = (\angle 1) + (\angle 2) + (\angle B) + (\angle 4)$$
$$s(\triangle ACE) = (\angle 2) + (\angle ACE) + (\angle E) = (\angle 2) + (\angle 3) + (\angle 4) + (\angle E).$$

Since $(\angle 1) = (\angle E)$ and $(\angle B) = (\angle 3)$, the angle sums are the same. Now note that

$$(\angle BAC) = (\angle 1) + (\angle 2) = (\angle E) + (\angle 2).$$

Therefore it is not possible for both $\angle E$ and $\angle 2$ to measure more than $(\angle BAC)/2$. If $\angle E$ is the smaller angle, use the labels $A' = E$, $B' = C$, and $C' = A$. If $\angle 2$ is the smaller angle, use the labels $A' = A$, $B' = C$, and $C' = E$. Either way, $s(\triangle A'B'C') = s(\triangle ABC)$, and $(\angle A') \leq (\angle A)/2$. □

Now we can combine the lemmas into a proof of the Saccheri-Legendre theorem.

Proof Suppose that there is a triangle $\triangle ABC$ whose angle sum is more than $180°$. To keep track of the excess, write $s(\triangle ABC) = (180+x)°$. According to lemma 8.5, there are triangles

$\triangle A_1 B_1 C_1$: with the same angle sum but $(\angle A_1) \leq \frac{1}{2}(\angle A)$,
$\triangle A_2 B_2 C_2$: with the same angle sum but $(\angle A_2) \leq \frac{1}{2}(\angle A_1) \leq \frac{1}{4}(\angle A)$,
$\triangle A_3 B_3 C_3$: with the same angle sum but $(\angle A_3) \leq \frac{1}{2}(\angle A_2) \leq \frac{1}{8}(\angle A)$, and so on.

After going through this procedure n times, we will end up with a triangle $\triangle A_n B_n C_n$ whose angle sum is still $(180 + x)°$ but with one very tiny angle— $(\angle A_n) \leq \frac{1}{2^n}(\angle A)$. No matter how big $\angle A$ is or how small x is, there is a large enough value of n so that $\frac{1}{2^n}(\angle A) < x$. In that case, the remaining two angle measures of the triangle, $(\angle B_n)$ and $(\angle C_n)$, have to add up to more than $180°$. According to lemma 8.3, this cannot happen. Therefore there cannot be a triangle with an angle sum greater than $180°$. □

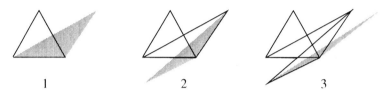

Figure 6. Starting from an equilateral triangle, the first three iterations.

The scalene triangle theorem relates the measures of the angles of a triangle to the measures of its sides. Essentially, it guarantees that the largest angle is opposite the longest side and that the smallest angle is opposite the shortest side. More precisely,

Theorem 8.6 *(The scalene triangle theorem)* In $\triangle ABC$ suppose that $|BC| > |AC|$. Then $(\angle BAC) > (\angle ABC)$.

Proof First we construct an isosceles triangle in $\triangle ABC$. That requires one additional point: since $|BC| > |AC|$, there is a point D between B and C so that $CA \simeq CD$. Then, since D lies in the interior or $\angle BAC$, $(\angle BAC) > (\angle DAC)$; by the isosceles triangle theorem, $\angle DAC \simeq \angle ADC$; and by the exterior angle theorem, $(\angle ADC) > (\angle ABC)$ (note that $\angle ADC$ is an exterior angle to $\triangle ABD$ and $\angle ABC$ is a non-adjacent interior angle). Putting it all together,

$$(\angle BAC) > (\angle DAC) = (\angle ADC) > (\angle ABC).$$ □

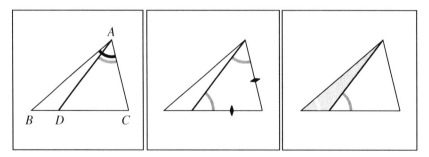

Figure 7. The proof of the scalene triangle theorem: (l) D is in the interior of $\angle BAC$, (c) the isosceles triangle theorem, (r) the exterior angle theorem.

In spite of the theorem's name, it applies to any two unequal-length sides of a triangle. In other words, it applies to the non-congruent sides of an isosceles triangle as well.

Triangles in Neutral Geometry

The last result of this chapter is the triangle inequality. The triangle inequality relates the lengths of the three sides of a triangle, providing upper and lower bounds for one side in terms of the other two. The upper bound is what people usually think of when they think of the triangle inequality, and it is the one that we will prove here. The other is left as an exercise.

Theorem 8.7 *(The triangle inequality) In $\triangle ABC$, the length of one side AC is bounded above and below by the lengths of the other two sides AB and BC:*

$$||AB| - |BC|| < |AC| < |AB| + |BC|.$$

Proof We will prove the second of the inequalities, that $|AC| < |AB| + |AC|$. This is obviously true if AC is not the longest side of the triangle, so let's focus our attention on the only interesting case, when AC is the longest side. As in the proof of the scalene triangle theorem, we build an isosceles triangle inside $\triangle ABC$. To do that, label D between A and C so that $AD \simeq AB$. According to the isosceles triangle theorem, $\angle ADB \simeq \angle ABD$. From the Saccheri-Legendre theorem, we know that the angles can't both be right or obtuse, so they have to be acute. Therefore, $\angle BDC$, which is supplementary to $\angle ADB$, is obtuse. Using the Saccheri-Legendre theorem again, $\triangle BDC$ will only support one obtuse angle, so $\angle BDC$ has to be the largest angle in that triangle. According to the scalene triangle theorem, BC has to be the longest side of $\triangle BDC$. Hence $|DC| < |BC|$. Now let's put it together:

$$|AC| = |AD| + |DC| < |AB| + |BC|. \qquad \square$$

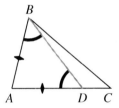

Figure 8. In $\triangle ABD$, $\angle B$ and $\angle D$ are congruent, so they must be acute.

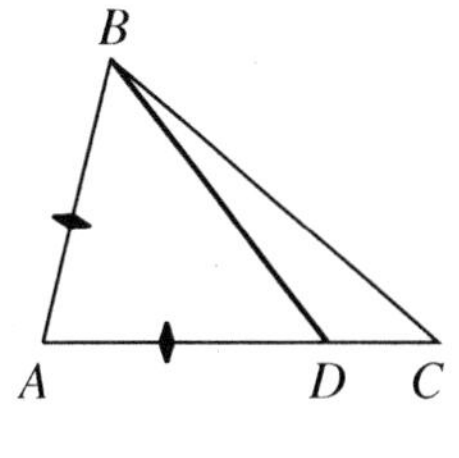

Figure 9. In $\triangle BDC$, $\angle D$ is obtuse. Opposite it, BC is the longest side.

For proper triangles, the triangle inequality promises strict inequalities with $<$ instead of $\leq$. When the three points A, B, and C collapse into a straight line, they no longer form a proper triangle, and that is when the inequalities become equalities:

if $C * A * B$, then $|AC| = |BC| - |AB|$;
if $A * C * B$, then $|AC| = |AB| - |BC|$;
if $A * B * C$, then $|AC| = |AB| + |BC|$.

So for any configuration of points, A, B, and C, we can say that

$$||AB| - |BC|| \leq |AC| \leq |AB| + |BC|.$$

The inequalities become equalities only when the three points are collinear (and this can be a useful way to determine if points are collinear):

Exercise 8.1 Prove that if $|AC| = |AB| + |BC|$, then $A * B * C$.

8.1 Exercises

8.2. Prove the converse of the scalene triangle theorem: in $\triangle ABC$, if $(\angle BAC) > (\angle ABC)$ then $|BC| > |AC|$.

8.3. Prove the other half of the triangle inequality (that $||AB| - |BC|| < |AC|$).

8.4. Given $\triangle ABC$, consider the interior and exterior angles at a vertex, say vertex A. Prove that their bisectors are perpendicular.

8.5. Prove that for a point P and line ℓ, there are points on ℓ that are arbitrarily far away from P.

8.6. Prove a strengthened form of the exterior angle theorem: for any triangle, the measure of an exterior angle is greater than or equal to the sum of the measures of the two nonadjacent interior angles.

8.7. Prove that if a triangle is acute, then the line that passes through a vertex and is perpendicular to the opposite side will intersect that side (the segment, that is, not just the line containing the segment).

8.8. Given $\triangle ABC$, let C' be a point on $\overrightarrow{AC}$ so that $A * C * C'$. Prove that the angle sum of $\triangle ABC'$ cannot be greater than the angle sum of $\triangle ABC$.

8.9. Given $\triangle ABC$, let B' and C' be points on $\overrightarrow{AB}$ and $\overrightarrow{AC}$, respectively, so that $A * B * B'$ and $A * C * C'$. Prove that the angle sum of $\triangle AB'C'$ cannot be greater than the angle sum of $\triangle ABC$.

8.10. Given $\triangle ABC$, and points A' in the interior of $\triangle ABC$, and B' and C' so that $B * B' * C' * C$, prove that the angle of sum of $\triangle A'B'C'$ cannot be less than the angle sum of $\triangle ABC$.

Recall that SSA is not a valid triangle congruence theorem. If you know just a little bit more about the triangles in question, though, SSA can be enough to prove triangles congruent. The next exercises look at some of those situations.

8.11. In a right triangle, the side opposite the right angle is called the *hypotenuse*. By the scalene triangle theorem, it is the longest side of the triangle. The other two sides are called the *legs* of the triangle. Consider two right triangles $\triangle ABC$ and $\triangle A'B'C'$ with right angles at C and C', respectively. Suppose in addition that

$$AB \simeq A'B' \quad \& \quad AC \simeq A'C'$$

(the hypotenuses are congruent, as are one set of legs). Prove that $\triangle ABC \simeq \triangle A'B'C'$. This is called the HL congruence theorem for right triangles.

8.12. Suppose that $\triangle ABC$ and $\triangle A'B'C'$ are acute triangles and that
$$AB \simeq A'B', \quad BC \simeq B'C', \text{ and } \angle C \simeq \angle C'.$$
Prove that $\triangle ABC \simeq \triangle A'B'C'$.

8.2 References

The proof that I give for the Saccheri-Legendre Theorem is the one I learned from Wallace and West's book [WW04].

9
Polygons

We have spent a lot of time talking about triangles, and while we are certainly not done with them, in this chapter we will broaden the focus, and look at polygons with more than three sides.

9.1 Definitions

The first step is to get a definition for the term *polygon*. This may not be as straightforward as you think. Remember that three non-collinear points P_1, P_2, and P_3 define a triangle. The triangle itself consists of all the points on the segments $P_1 P_2$, $P_2 P_3$, and $P_3 P_1$. At the least, a definition of polygons (as we think of them) involves a list of points and segments connecting each point to the next in the list, and then the last point back to the first:

The Vertices: $P_1, P_2, P_3, \ldots, P_n$
The Sides: $P_1 P_2, P_2 P_3, P_3 P_4, \ldots, P_{n-1} P_n, P_n P_1$.

Now the non-trivial problem is, what conditions should we place on the points? With triangles, we insisted that the points be non-collinear. What is the appropriate way to extend that beyond $n = 3$? This is not an easy question to answer.

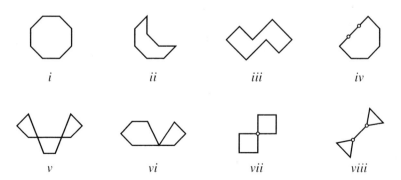

Figure 1. "Polygons" on configurations of eight not necessarily distinct points. Which of these should be considered octagons (polygons with eight sides and eight vertices)?

81

Before settling on a definition, let's discuss notation. No matter the definition, the vertices will cycle around: $P_1, P_2, \ldots, P_n$, and then back to the start P_1. Because polygons loop back around like this, we often end up crossing from P_n back to P_1. For example, look at the listing of the sides of the polygon—all but one of them can be written in the form $P_i P_{i+1}$, but the last side, $P_n P_1$, doesn't fit that pattern. We can sidestep this issue: rather than using integer subscripts for the vertices, use integers modulo n, where n is the number of vertices. That way, for instance, in a polygon with eight vertices, P_9 and P_1 would stand for the same point since $9 \equiv 1 \mod 8$, and the sides of the polygon would be $P_i P_{i+1}$ for $1 \leq i \leq 8$.

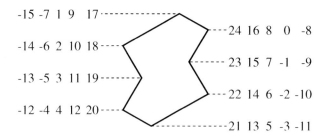

Figure 2. Cyclic labeling of vertices of a polygon.

Now let's return to the definition. There is a range of opinion. Which definition is most appropriate probably depends on the situation. For instance, in *Are your polyhedra the same as my polyhedra?* [Grü03], Grünbaum argues for a more inclusive definition. The most inclusive definition would be:

Definition 9.1 (Inclusive version) An ordered list of points $\{P_i | 1 \leq i \leq n\}$ defines a *polygon*, written $P_1 P_2 \cdots P_n$, with vertices P_i, $1 \leq i \leq n$, and sides $P_i P_{i+1}$, $1 \leq i \leq n$.

This is an easy definition, but it allows for all manner of degeneracies. To be more particular, we might want to insist that the vertices be distinct, and that two sides intersect at most once. That eliminates shapes (vii) and ($viii$) in figure 1. We might also want to prohibit a polygon from intersecting itself, as shapes (v) and (vi) do. Thinking back to the non-collinearity requirement for the vertices of a triangle, it might be tempting to exclude shape (iv) as well, or at least to consider it as a polygon with fewer than eight sides. For our purposes, though, the essential behavior of a polygon that we will almost always want is this: a polygon should enclose a single connected region. In figure 1, shapes (i)-(iv) do this, while shapes (v)-($viii$) do not. To distinguish these from the more general polygons as defined above, they are called *simple polygons*.

What is it that prevents shapes (v)-($viii$) from enclosing a single connected region? In a word, self-intersections. Sides of simple polygons must intersect one another only at the endpoints where they are joined to their adjacent sides. That is encapsulated in the following definition.

Definition 9.2 (Simple polygon) An ordered list of distinct points $\{P_i | 1 \leq i \leq n\}$ that satisfies the condition

if $i \neq j$ and $P_i P_{i+1}$ intersects $P_j P_{j+1}$ then either $i = j + 1$ and the intersection is at $P_i = P_{j+1}$, or $j = i + 1$ and the intersection is at $P_{i+1} = P_j$

defines a *simple polygon*, written $P_1 P_2 \cdots P_n$, with vertices P_i, $1 \leq i \leq n$, and sides $P_i P_{i+1}$, $1 \leq i \leq n$.

For our purposes, almost all polygons will be simple, and so it is lazy but convenient to leave off the adjective "simple". This will be our definition for the word *polygon* for the remainder of this text. On the rare occasions that we encounter polygons that are not simple, we will just point them out as non-simple polygons. In any event, no matter how you choose to define a polygon, the definition of one important invariant does not change:

Definition 9.3 The *perimeter* of a polygon is the sum of the lengths of its sides. The perimeter of the polygon $P_1 P_2 \cdots P_n$ is

$$\sum_{i=1}^{n} |P_i P_{i+1}|.$$

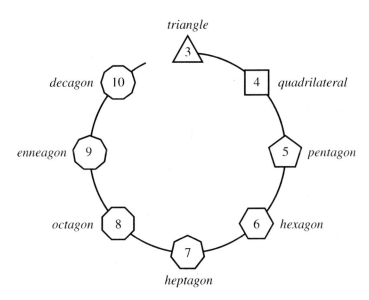

Figure 3. Polygons names based upon their number of sides (and vertices). Note the greek prefixes hepta- and ennea- for the seven-sided and nine-sided polygons. The Latin prefixes septa- and nona- are also common for these. In general, an n-sided polygon is called an n-gon.

9.2 Counting polygons

Two polygons are the same if they have all the same vertices and the same edges. That means that the order that you list the vertices generally matters—different orders can lead to different sets of sides. Not all rearrangements of the list lead to new polygons though. For instance, the listings $P_1 P_2 P_3 P_4$, $P_3 P_4 P_1 P_2$, and $P_4 P_3 P_2 P_1$ all define the same polygon: one with sides $P_1 P_2$, $P_2 P_3$, $P_3 P_4$, and $P_4 P_1$. More generally, any two listings that differ either by a cycling of the vertices or by a reversal of the order of a cycling will describe the same polygon.

So how many possible polygons are there on n points? That depends on what definition of polygon you are using. The inclusive definition leads to the easiest calculation, for in that

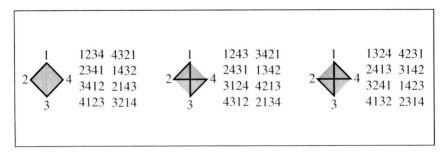

Figure 4. The 24 permutations of 1,2,3,4 and the corresponding (not necessarily simple) polygons on four points.

case, any configuration on n points results in a polygon. There are $n!$ orderings of n distinct elements. However for each ordering there are n cyclings of the entries and n reversals of them, leading to a total of $2n$ listings that correspond to the same polygon. Therefore, there are $n!/(2n) = (n-1)!/2$ possible polygons that can be built on n vertices. When $n = 3$, there is only one possibility, and that is why this was not an issue when we were dealing with triangles. The situation is more complicated if you insist that the polygons be simple. That condition throws the problem from the relatively comfortable world of combinatorics into a much murkier geometric realm. Not much is known about this interesting problem.

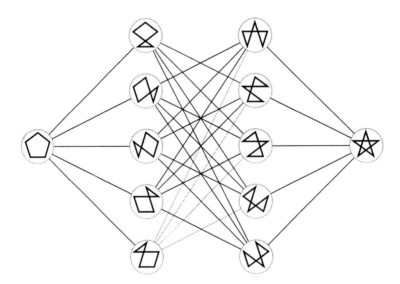

Figure 5. The twelve polygons on a configuration of five points (only one of which is simple). In this illustration, segments connect two polygons that differ by a swap of two adjacent vertices.

9.3 Interiors and exteriors

One characteristic of the triangle is that it separates the rest of the plane into two sets, an interior and an exterior. Our definition of a simple polygon was inspired by this—the intent was that a

9.3 Interiors and exteriors

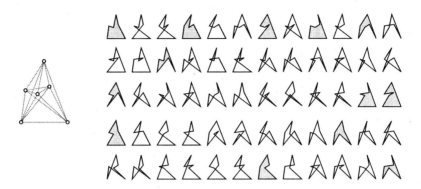

Figure 6. Thirteen of the sixty polygons on this (Euclidean) configuration of six points are simple.

simple polygon should also enclose a single connected region (and that would be its interior). We now need to prove that this does in fact happen. This is actually a special case of the celebrated Jordan curve theorem, which states that every simple closed curve in the plane separates the plane into an interior and an exterior. The Jordan curve theorem is one of those notorious results that seems easy to prove, but is actually brutally difficult. In the special case of simple polygons, our case, there are simpler proofs. One such proof is from *What is Mathematics?* by Courant and Robbins [CR41]. I will paraphrase it here.

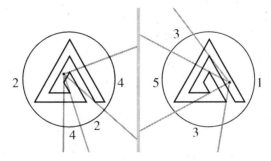

Figure 7. Rays from a point. The number of intersections with a polygon (in black) depends on which ray is chosen, but its parity (even or odd) does not.

Theorem 9.4 *(Polygonal plane separation) Every simple polygon separates the remaining points of the plane into two connected regions.*

Proof Let $\mathcal{P}$ be a simple polygon, and let P be a point that is not on $\mathcal{P}$. Now look at a ray R_P whose endpoint is P. As long as R_P does not run exactly along an edge, it will intersect the edges of $\mathcal{P}$ a finite number of times (perhaps none). You want to think of each such intersection as a crossing of R_P into or out of $\mathcal{P}$.

Since there are only finitely many intersections, they are all within a finite distance of P. That means that eventually R_P will pass beyond all the points of $\mathcal{P}$. This is the essence of the argument: eventually the ray is outside of the polygon, so by counting back the intersections crossing into and out of the polygon, we can figure out whether the beginning of the ray is inside or outside of $\mathcal{P}$. The one situation where we have to be careful is when R_P intersects a vertex of $\mathcal{P}$. Here is the way to count those intersections:

once: if R_P separates the two neighboring edges, and
twice: if R_P does not separate them.

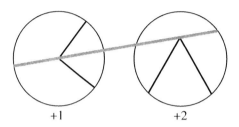

Figure 8. Procedure for counting intersections at a vertex.

When we count intersections this way, the number of intersections depends not just on the point P, but also on the direction of R_P. The key, though, is that whether the number of intersections is odd or even, the "parity" of P, does not depend on the direction. To see why, you have to look at what happens as you move the ray R_P around, and in particular what causes the number of intersections to change. Changes can happen only when R_P crosses one of the vertices of $\mathcal{P}$. Each vertex crossing corresponds to either an increase by two in the number of crossings, a decrease by two in the number of crossings, or no change in the number of crossings. In each case, the parity is not changed. Therefore $\mathcal{P}$ separates the remaining points of the plane into two sets, those with even parity and those with odd parity.

Figure 9. As the ray shifts across a vertex, the intersection count changes by $+2$, -2, or 0, all even numbers.

Now we need to show that each of those sets is connected. Let P and Q be points in one of the sets, call it S. We need to show that there is a path connecting P to Q that is entirely contained in S. Start by considering $\overrightarrow{PQ}$. Let n_P be the number of intersections of $\overrightarrow{PQ}$ with $\mathcal{P}$ (this determines the parity of P), and let n_Q be the number of intersections of $\overrightarrow{PQ}$ with $\mathcal{P}$

9.3 Interiors and exteriors

that occur beyond Q (this determines the parity of Q). Then the segment PQ and $\mathcal{P}$ intersect $n_P - n_Q$ times. Since P and Q have the same parity, this difference is an even number. If $n_P - n_Q = 0$, then PQ itself is the connecting path from P to Q, and we are done; in general, the straight line path PQ will not be entirely contained in S. If that is the case, then scanning along the segment PQ, from P to Q, there is an even number, say m, of intersections with $\mathcal{P}$ (to simplify the argument, let's assume that none of the intersections happen at vertices). Label them A_i so that

$$P * A_1 * A_2 * \cdots * A_m * Q.$$

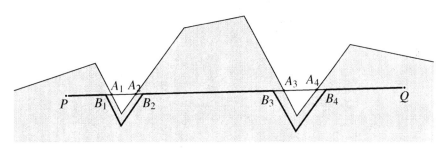

Figure 10. If P and Q are in the interior of $\mathcal{P}$, then we can construct a path in $\mathcal{P}$ that connects P to Q.

The points between P and A_1 are all in S, the points between A_1 and A_2 are all outside of S, the points between A_2 and A_3 are all inside S, and so on. The alternating in-and-out pattern continues until we reach the other end, where all the points between A_m and Q are in S. Next, we need to replace all the points A_i with points that are just slightly inside S. Choose B_1 just a bit before A_1, choose B_2 just a bit after A_2, choose B_3 just a bit before A_3, and so on. Finally, we can construct a path that lies entirely in S, connecting P to B_1, B_1 to B_2, and so on until finally it joins B_m to Q. This path is made up of the pieces

- σ_1 : the segment from P to B_1
- σ_2 : a path that runs just right beside the edge of $\mathcal{P}$ from B_1 to B_2
- σ_3 : the segment from B_2 to B_3
- σ_4 : a path that runs just right beside the edge of $\mathcal{P}$ from B_3 to B_4

and so on. At the end, σ_{m-1} is a path that runs just right beside the edge of $\mathcal{P}$ from B_{m-1} to B_m and σ_m is the segment from B_m to Q. □

In the previous proof, I have been vague about how close the B_is need to be to the A_is and how close the paths σ_i (where i is even) need to be to the edge of $\mathcal{P}$. The key is to move the σ_i pieces slightly to one side of $\mathcal{P}$ to put them in S, but to be sure that nudging them does not cause them to intersect any other edges of $\mathcal{P}$. Since $\mathcal{P}$ is simple, it does not self-intersect. Therefore, there is a small but positive minimum distance ϵ between any two nonadjacent edges. As long as the σ_i path stays within ϵ of the edges it is shadowing, it cannot intersect another side of $\mathcal{P}$. In Euclidean geometry, the shadowing path can be constructed with segments that are parallel to the edges; in non-Euclidean geometry it gets trickier, because in non-Euclidean geometry parallel lines diverge from another. In any event, it can be done. The exact details of this are left to the reader.

Definition 9.5 For a simple polygon $\mathcal{P}$, the set of points with odd parity (as described in the last proof) is called the *interior* of $\mathcal{P}$. The set of points with even parity is called the *exterior* of $\mathcal{P}$.

Exercise 9.1 Prove that a polygon's interior is always a bounded region and that its exterior is always an unbounded region.

9.4 Interior angles: two dilemmas

Now let's look more carefully at the angles of a simple polygon. Consider the three marked angles in the polygons in figure 11. The first, $\angle 1$, is the interior angle of a triangle. You can see that the entire interior of the triangle is contained in the interior of the angle, and that seems proper, that close connection between the interiors of the interior angles and the interior of the triangle. Now look at $\angle 2$, and you can see that for a general simple polygon, things do not work quite as well: the entire polygon does not lie in the interior of this angle. But at least the part of the polygon interior which is closest to that vertex is in the interior of that angle. Finally look at $\angle 3$: the interior of $\angle 3$ encompasses exactly none of the interior of the polygon—it is actually pointing away from the polygon.

Figure 11. Angle interiors and polygon interiors.

First consider the issue involving $\angle 3$. We have said that two non-opposite rays define a single angle, and later established a measure for that angle, some number between 0 and $180°$. But really, two rays like this divide the plane into two regions, and correspondingly, they should form two angles. One is the proper angle that we have already dealt with. The other angle is what is called a *reflex angle*. Together, the measures of the proper angle and the reflex angle formed by any two rays should add up to $360°$. There does not seem to be a standard terminology to describe this relationship between angles; I have seen the term "conjugate" as well as the term "explementary". So the problem with $\angle 3$ is that the interior angle isn't the proper angle, but instead, that it is its explement.

As long as the polygon is not particularly convoluted, this is all fairly clear, but suppose that we were looking at an angle $\angle P_i$ in a much more elaborate polygon. Should the interior angle at P_i be the proper angle $\angle P_{i-1} P_i P_{i+1}$ or its conjugate? To answer that question, you need to look at the segment $P_{i-1} P_{i+1}$. It may cross into and out of the interior of the polygon, but if the interior angle is the proper angle, then the first and last points of $P_{i-1} P_{i+1}$ (the ones closest to P_{i-1} and P_{i+1}) will be in the interior of the polygon. If the interior angle is the reflex angle, then the first and last points of $P_{i-1} P_{i+1}$ will not be in the interior of the polygon.

9.4 Interior angles: two dilemmas

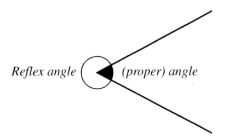

Figure 12. A pair of "explementary angles", one proper and the other reflex.

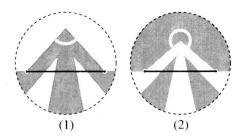

Figure 13. The dark region shows the polygon interior around a vertex. In (1), the connecting segment begins in the interior, so the interior angle is the proper angle. In (2), the connecting segment begins in the exterior, so the interior angle is the reflex angle.

With the interior angles of a polygon now properly accounted for, we can define what it means for two polygons to be congruent.

Definition 9.6 Two polygons $\mathcal{P} = P_1 P_2 \cdots P_n$ and $\mathcal{Q} = Q_1 Q_2 \cdots Q_n$ are *congruent*, written $\mathcal{P} \simeq \mathcal{Q}$, if corresponding sides and interior angles are congruent:

$$P_i P_{i+1} \simeq Q_i Q_{i+1} \text{ and } \angle P_i \simeq \angle Q_i, \text{ for all } i.$$

Now let's take a look at the issue involving $\angle 2$, where not all of the interior of the polygon lies in the interior of the angle. The problem here is more intrinsic and cannot be avoided by fiddling with definitions. There is, though, a class of simple polygon for which the polygon interior always lies in the interior of each interior angle. These are the convex polygons.

Definition 9.7 A polygon $\mathcal{P}$ is *convex* if, for two points P and Q in the interior of $\mathcal{P}$, the entire line segment PQ is in the interior of $\mathcal{P}$.

Convexity is a big word in geometry and it comes up in a wide variety of contexts. Our treatment here will be very elementary, and just touch on the most basic properties of a convex polygon.

Theorem 9.8 *If $\mathcal{P} = P_1 P_2 \cdots P_n$ is a convex polygon, then all the points of the interior of $\mathcal{P}$ lie on the same side of each of the lines $\overline{P_i P_{i+1}}$.*

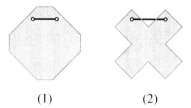

(1)   (2)

Figure 14. (1) A convex polygon, and (2) a non-convex one. In the second, a segment joins two points in the interior, but passes outside of the polygon.

The fundamental mechanism that makes this proof work is the way that we defined the interior and exterior of a polygon by drawing a ray and counting how many times it intersects the sides of $\mathcal{P}$. In particular, suppose that P and Q lie on opposite sides of a segment $P_i P_{i+1}$, so that PQ intersects $P_i P_{i+1}$, and suppose further that PQ intersects no other sides of the polygon. Then $\overrightarrow{PQ}$ intersects $\mathcal{P}$ one more time than $(\overrightarrow{QP})^{op}$ does. Therefore P and Q have different parities, and so P and Q cannot both be interior points.

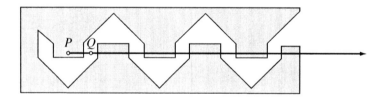

Figure 15. A single side of the polygon comes between P and Q, they cannot both be in the interior.

Pursuing a proof by contradiction, we might suppose that P and Q are in the interior of a convex polygon, but that they are on the opposite sides of $\overline{P_i P_{i+1}}$. After the previous discussion, it is tempting to draw a picture that looks like figure 16. We could then choose points R_1 and R_2 between P and Q close enough to the segment $P_i P_{i+1}$ so that no other edges intersect $R_1 R_2$. Their parities would differ by one, and so one would have to be outside of $\mathcal{P}$. Then PQ would go outside $\mathcal{P}$, which would mean that $\mathcal{P}$ is not convex, and that would be the contradiction.

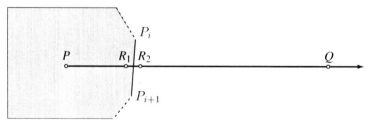

Figure 16. When $\overrightarrow{PQ}$ intersects segment $P_i P_{i+1}$.

9.4 Interior angles: two dilemmas

Unfortunately, that would miss an important (and indeed likely) case, where PQ intersects $\overline{P_iP_{i+1}}$ but not P_iP_{i+1}. To deal with that, we are going to have to maneuver the intersection so that it occurs on the segment, which requires a more delicate argument.

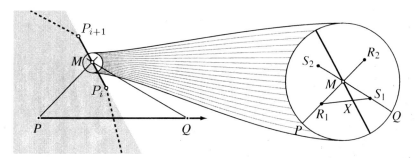

Figure 17. When $\overrightarrow{PQ}$ intersects the line $\overline{P_iP_{i+1}}$ (but not between P_i and P_{i+1}).

Proof Let M be the midpoint of P_iP_{i+1}, and let d be the distance from the midpoint to either P_i or P_{i+1}. The idea is to relay the interior/exterior information from P and Q back to points that are in close proximity to M. We need to choose points R_1, R_2, S_1, and S_2 so that

- $P * R_1 * M * R_2$ and $Q * S_1 * M * S_2$,
- both $|R_1M|$ and $|S_1M|$ are less than $d/2$, and
- R_1, R_2, S_1, and S_2 are all close enough to M that no other side of $\mathcal{P}$ intersects R_1R_2, S_1S_2, or R_1S_1 (this is possible because $\mathcal{P}$ is simple).

Because of the first statement, R_1 and R_2 are separated by the side P_iP_{i+1}, and because of the third statement, P_iP_{i+1} is the only side of $\mathcal{P}$ that separates them. Therefore one of the two is in the interior of $\mathcal{P}$ and the other is not. If R_2 were the one in the interior, then PR_2 would connect two interior points of $\mathcal{P}$, but would contain R_1, a point not in $\mathcal{P}$. Since $\mathcal{P}$ is convex, that cannot be. Therefore R_1 must be the point inside $\mathcal{P}$. Working the same argument, but using Q instead, we see that S_1 must also be an interior point of $\mathcal{P}$.

Now consider S_1R_1. Its endpoints S_1 and R_1 are on opposite sides of $\overline{P_iP_{i+1}}$, so S_1R_1 intersects that line at a point X. In fact, though, the second statement guarantees that X must lie on the segment P_iP_{i+1}. To see why, we need to use the triangle inequality in three triangles:

$\triangle MS_1X : |MX| \leq |MS_1| + |XS_1|,$
$\triangle MR_1X : |MX| \leq |MR_1| + |XR_1|,$ and
$\triangle MS_1R_1 : |S_1R_1| \leq |MS_1| + |MR_1|.$

Add the first two inequalities, use segment addition, substitute in the third inequality, and use the second statement above to get

$$2|MX| \leq |MS_1| + |MR_1| + |XS_1| + |XR_1|$$
$$\leq |MS_1| + |MR_1| + |S_1R_1|$$
$$\leq 2|MS_1| + 2|MR_1|$$
$$\leq 2(d/2) + 2(d/2)$$
$$\leq 2d.$$

Therefore $|MX| < d$, and so the intersection point X is closer to the midpoint than either P_i or P_{i+1}. That is, it occurs between P_i and P_{i+1}. Therefore R_1 and S_1 are separated by a single side of $\mathcal{P}$, which means that they cannot both be interior points. This is a contradiction, so P and Q cannot lie on opposite sides of $\overline{P_i P_{i+1}}$. □

Exercise 9.2 Prove: If $\mathcal{P}$ is a convex polygon, then none of its interior angles is a reflex angle.

Exercise 9.3 Prove that if $\mathcal{P}$ is a convex polygon, then the interior of $\mathcal{P}$ lies in the interior of each interior angle $\angle P_i$.

Exercise 9.4 Prove that if P is an interior point of a convex polygon $\mathcal{P}$, then any ray whose endpoint is at P will intersect $\mathcal{P}$ exactly once.

Exercise 9.5 Let $\overline{\mathcal{P}}$ denote the set of all points that are either on $\mathcal{P}$ or in the interior of $\mathcal{P}$. Prove that if $\mathcal{P}$ is convex, and P and Q are two points on the polygon $\mathcal{P}$, then all the points of the segment PQ are in $\overline{\mathcal{P}}$.

9.5 Polygons of note

We end this chapter by defining a few particularly well-behaved types of polygons.

Definition 9.9 An *equilateral* polygon is one in which all sides are congruent. A *cyclic* polygon is one in which all vertices are equidistant from a fixed point (hence, all vertices lie on a circle, to be discussed later). A *regular* polygon is one in which all sides are congruent and all interior angles are congruent.

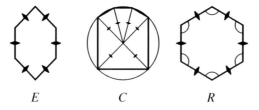

Figure 18. Types of polygons. E: equilateral, C: cyclic, R: regular.

The third of these types is actually a combination of the previous two as the next theorem shows.

Theorem 9.10 *A polygon $\mathcal{P}$ that is both equilateral and cyclic is regular.*

Exercise 9.6 Prove the previous theorem.

There are also non-simple polygons that satisfy the regularity condition. These are called *star polygons*. A few of these are shown in Figure 19.

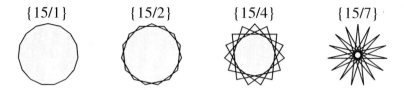

Figure 19. There is a regular star n-gon for each integer p between 1 and $n/2$ that is relatively prime to n. For instance, there are four star polygons when $n = 15$. The notation $\{n/p\}$ is called the Schläfli symbol.

9.6 Exercises

9.7. Verify that a triangle is a convex polygon.

9.8. A diagonal of a polygon is a segment connecting nonadjacent vertices. How many diagonals does an n-gon have?

9.9. Prove that a regular convex polygon is cyclic (to find that equidistant point, you may have to consider the odd and even cases separately).

9.10. Prove that if a polygon is convex, then its diagonals lie entirely in the interior of the polygon (except for the endpoints).

9.11. Prove that if a polygon is not convex, then at least one of its diagonals does not lie entirely in the interior of the polygon.

9.12. Verify that the perimeter of a polygon is more than twice the length of its longest side.

9.13. Prove that the sum of the interior angles of a convex n-gon is at most $180°(n-2)$.

9.14. Let $\mathcal{P} = P_1 P_2 \ldots P_n$ be a convex polygon. Prove that if $3 \leq j \leq n$, then the polygon $\mathcal{Q} = P_1 P_2 \ldots P_j$ is convex.

9.15. Let $\mathcal{P}$ be a convex polygon and let A and B be two points on $\mathcal{P}$. Prove that the segment AB divides $\mathcal{P}$ into two polygons, both of which are convex.

9.16. Let $\mathcal{P} = P_1 P_2 \ldots P_n$ and $\mathcal{Q} = Q_1 Q_2 \ldots Q_m$ be convex polygons that intersect at two points A and B. Show that A, B, the vertices of $\mathcal{P}$ that are in the interior of $\mathcal{Q}$, and the vertices of $\mathcal{Q}$ that are in the interior of $\mathcal{P}$, are the vertices of a convex polygon.

9.17. Prove that if a polygon $P_1 P_2 \ldots P_n$ is convex, then there is no other simple polygon that has $P_1, P_2, \ldots, P_n$ as its vertices.

10
Quadrilateral Congruence Theorems

This is the last chapter in neutral geometry. After this, we will allow ourselves one more axiom dealing with parallel lines, and that is the axiom that turns neutral geometry into Euclidean geometry. Before turning down the Euclidean path, let's spend a little time looking at quadrilaterals. The primary goal of this section will be to develop quadrilateral congruence theorems similar to the triangle congruence theorems in earlier chapters.

10.1 Terminology

Before we start working on congruence theorems, though, let's quickly run through the definitions of a few particular types of quadrilaterals.

Table 1. Classification of types of quadrilaterals.

Name	A quadrilateral with...
Trapezoid	a pair of parallel sides
Parallelogram	two pairs of parallel sides
Rhombus	four congruent sides
Rectangle	four right angles
Square	four congruent sides and four right angles

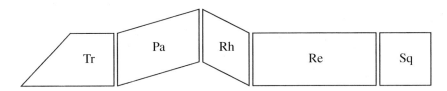

Figure 1. Types of Quadrilaterals.

When you define an object by requiring it to have certain properties, the risk is that you may define something that cannot be, something like an equation with no solution. The quadrilaterals defined above are all such common shapes in everyday life that we usually don't question their existence. In neutral geometry, though, there is no construction that guarantees you can make a quadrilateral with four right angles. That is, neutral geometry does not guarantee the existence

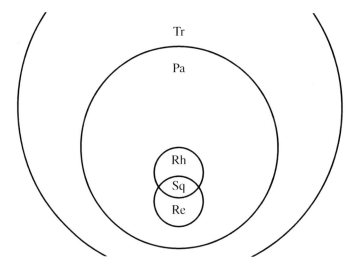

Figure 2. Venn diagram of special quadrilaterals. Rhombuses and rectangles are parallelograms. A square is both a rhombus and a rectangle.

of rectangles or squares. At the same time, it does nothing to prohibit the existence of squares or rectangles either. You can make a quadrilateral with three right angles easily, but once you have done that, you have no control over the fourth angle, and the axioms of neutral geometry are not sufficient to prove whether or not the fourth angle is a right angle. This is one of the fundamental differences that separates Euclidean geometry from non-Euclidean geometry. In Euclidean geometry, the fourth angle is a right angle, so there are rectangles. In non-Euclidean geometry, the fourth angle cannot be a right angle, so there are no rectangles. When we eventually turn our attention to non-Euclidean geometry, we will begin that study with a more thorough investigation of quadrilaterals that try to be like rectangles, but fail.

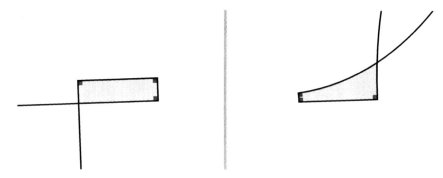

Figure 3. Quadrilaterals with three right angles. On the left, in Euclidean geometry, the fourth angle is a right angle. On the right, in non-Euclidean geometry, it is acute.

10.2 Quadrilateral congruence

By definition two triangles are congruent if all their corresponding sides and angles are congruent. Sometimes congruence can be guaranteed with certain subsets of that information.

10.2 Quadrilateral congruence

Likewise, two quadrilaterals are congruent if all their corresponding sides and angles (eight total) are congruent, but again, certain subsets of that information are often sufficient. The triangle congruence theorems generally require three given congruences. For quadrilaterals, it seems that the magic number is five, but as with triangles, some combinations of five congruences work, while others do not. The quadrilateral congruence theorems describe which ones work.

The first step is some basic combinatorics. Each of the theorems has a five letter name consisting of some mix of Ss and As. When forming it, there are two choices, S and A, for each of the five letters, and so there are a total of $2^5 = 32$ possible names. Two, SSSSS and AAAAA, don't make any sense in the context of quadrilateral congruences, since a quadrilateral doesn't have five sides or five angles. That leaves thirty different words. It is important to notice that not all of them represent fundamentally different information about the quadrilaterals. For instance, SSASA and ASASS both represent the same information, just listed in reverse order. Similarly, SSASS and SSSSA both represent the same information, four sides and one angle. Once equivalences are taken into consideration, we are left with ten potential quadrilateral congruence theorems.

Table 2. Possible quadrilateral congruences.

Name	Variations		Valid theorem?
SASAS			yes
ASASA			yes
AASAS	SASAA		yes
SSSSA	SSSAS	SSASS	no(*)
	SASSS	ASSSS	
ASAAS	SAASA		no
ASASS	SSASA		no
ASSAS	SASSA		no
AAAAS	AAASA	AASAA	no
	ASAAA	SAAAA	
SSSAA	AASSS	ASSSA	no
	SAASS	SSAAS	
AAASS	SAAAS	SSAAA	yes
	ASSAA	AASSA	

(*) SSSSA is a valid congruence theorem for convex quadrilaterals.

10.2.1 SASAS, ASASA, and AASAS

Each of these is a valid congruence theorem for simple quadrilaterals, both convex and non-convex. The basic strategy for their proofs is to use a diagonal of the quadrilateral to separate it into two triangles, and then to use the triangle congruence theorems. It complicates things to allow both convex and non-convex quadrilaterals in this discussion, because only one diagonal passes through the interior of a non-convex quadrilateral. Let's start by examining the nature of the interior angles and diagonals of a quadrilateral.

Exercise 10.1 Prove that a quadrilateral can have at most one reflex interior angle.

Consider □$ABCD$ (here the square symbol denotes a simple quadrilateral). Now let's look at the position of the point D relative to $\triangle ABC$. Each of $\overline{AB}$, $\overline{BC}$, and $\overline{AC}$ separate the plane into two pieces. It is not possible, though, for any point of the plane to simultaneously be

- on the opposite side of AB from C,
- on the opposite side of AC from B, and
- on the opposite side of BC from A.

Thus the lines of $\triangle ABC$ divide the plane into seven ($2^3 - 1$) regions. For D in a region, we can determine whether the diagonals AC and BD are in the interior of □$ABCD$. This is always an all-or-nothing proposition:

Exercise 10.2 Prove that either the entire diagonal AC lies in the interior of □$ABCD$ (excepting the endpoints A and C) or none of it does.

Additionally, in each case, a diagonal lies in the interior of a quadrilateral if and only if it lies in the interior of both the angles formed by □$ABCD$ at its endpoints.

Exercise 10.3 Prove that if AC is in the interior of □$ABCD$, then AC will be in the interior of both $\angle DAB$ and $\angle BCD$. Prove that if AC is not in the interior of □$ABCD$, then AC will not be in the interior of either $\angle DAB$ or $\angle BCD$.

All that information is gathered in table 3, and with it we can now address the congruence theorems directly. Probably the most useful of them is SASAS.

Table 3. Diagonals of a quadrilateral □$ABCD$.

D is in region	is □$ABCD$ simple?	D is on the same side of:			Reflex angle	Interior diagonal:	
		BC as A	AC as B	AB as C		AC	BD
I	✓	✓			A	✓	
II	✓		✓		B		✓
III	✓			✓	C	✓	
IV		✓	✓		—	—	—
V	✓	✓		✓	none	✓	✓
VI			✓	✓	—	—	—
VII	✓	✓	✓	✓	D		✓

Theorem 10.1 *(SASAS quadrilateral congruence)* Given quadrilaterals □$ABCD$ and □$A'B'C'D'$ with

$$AB \simeq A'B', \quad \angle B \simeq \angle B', \quad BC \simeq B'C', \quad \angle C \simeq \angle C', \text{ and } CD \simeq C'D',$$

then □$ABCD \simeq$ □$A'B'C'D'$.

10.2 Quadrilateral congruence

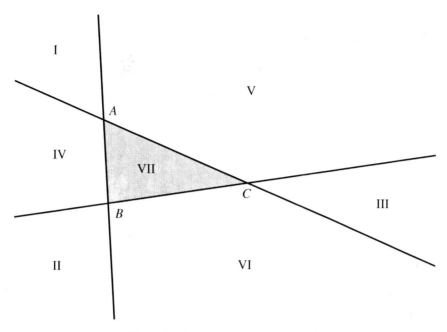

Figure 4. The seven "sides" of a triangle.

The diagonals AC and $A'C'$ are the keys to turning this into a problem of triangle congruence. Unfortunately, we do not know whether the diagonals are in the interiors of their quadrilaterals. That means we have to tread somewhat carefully.

Proof Because of SAS, $\triangle ABC \simeq \triangle A'B'C'$. Observe what is happening at vertex C. If AC is in the interior of the quadrilateral, then it is in the interior of $\angle BCD$ so $(\angle BCA) < (\angle BCD)$. Then, since $\angle B'C'A' \simeq \angle BCA$ and $\angle B'C'D' \simeq \angle BCD$, $(\angle B'C'A') < (\angle B'C'D')$. Therefore $A'C'$ must be in the interior of $\angle B'C'D'$ and in the interior of $\square A'B'C'D'$. With the same reasoning, we can argue that if AC is not in the interior of $\square ABCD$, then $A'C'$ cannot be in the interior of $\square A'B'C'D'$. So there are two cases, and the assembly of the quadrilateral from the triangles depends upon the case. See figure 5.

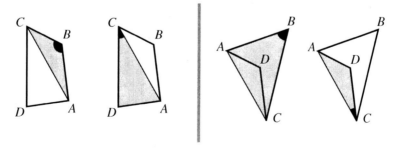

Figure 5. The proof of the SASAS by two applications of SAS (only $\square ABCD$ is shown in these illustrations). There are two cases depending on whether the quadrilaterals are convex or not.

Case 1. If AC is in the interior of the quadrilateral, then it is in the interior of $\angle BCD$ and that means $(\angle BCA) < (\angle BCD)$. Then, since $\angle B'C'A' \simeq \angle BCA$ and $\angle B'C'D' \simeq \angle BCD$, $(\angle B'C'A') < (\angle B'C'D')$. Therefore $A'C'$ must be in the interior of $\angle B'C'D'$ and in the interior of $\square A'B'C'D'$. Now by angle subtraction, $\angle ACD \simeq \angle A'C'D'$. We also know that $CD \simeq C'D'$ and $AC \simeq A'C'$ so again by SAS, $\triangle ACD \simeq \triangle A'C'D'$. That immediately yields the remaining side congruence, $AD \simeq A'D'$, and one more angle congruence, $\angle D \simeq \angle D'$. Finally, note that the diagonals AC and $A'C'$ are in the interiors of $\angle A$ and $\angle A'$ respectively. Therefore, by angle addition, $\angle A \simeq \angle A'$.

Case 2. The second case is essentially the same, with a few small modifications. First, if AC is not in the interior of the quadrilateral, then $(\angle BCA) > (\angle BCD)$. To show that $\angle ACD \simeq \angle A'C'D'$, we still use angle subtraction, but we subtract the angles in the opposite order. Second, note that $\angle D$ in $\triangle ACD$ is a proper angle, but $\angle ACD$ in the quadrilateral is reflex. Third, AC is not in the interior of $\angle A$ (and $A'C'$ is not in the interior of $\angle A'$), so to show $\angle A \simeq \angle A'$, we have to use angle subtraction rather than angle addition. □

The proofs of ASASA and AASAS are similar; however these theorems do not hold for quadrilaterals with a straight angle (in other words, quadrilaterals that look like a triangle).

Exercise 10.4 Prove the ASASA quadrilateral congruence theorem (for quadrilaterals with no straight angles).

Exercise 10.5 Prove the AASAS quadrilateral congruence theorem (for quadrilaterals with no straight angles).

Exercise 10.6 Give Euclidean examples that demonstrate that ASASA and AASAS fail to be valid congruence theorems for quadrilaterals with straight angles.

10.2.2 SSSSA

The SSSSA condition is almost enough to guarantee quadrilateral congruence. Suppose that you know the lengths of all four sides of $\square ABCD$, and you also know $\angle A$. Then $\triangle BAD$ is completely determined (SAS) and from that $\triangle BCD$ is completely determined (SSS). That still does not mean that $\square ABCD$ is completely determined, though, because there are two ways to assemble $\triangle BAD$ and $\triangle BCD$ (as illustrated in figure 6). One assembly creates a convex quadrilateral, the other a non-convex one. There will be times when you know the quadrilaterals are convex, and then SSSSA can be used to show that convex quadrilaterals are congruent.

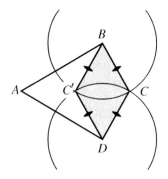

Figure 6. Two non-congruent quadrilaterals with matching SSSSA.

10.2.3 ASAAS, ASASS, ASSAS, AAAAS, and SSSAA

None of these provide sufficient information to guarantee congruence and counterexamples can be found in Euclidean geometry. For example, in figure 7, $\square ABCD$ and $\square ABC'D'$ have corresponding SSSAA but are not congruent.

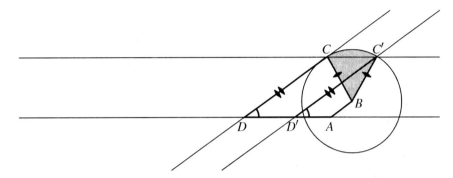

Figure 7. Two non-congruent quadrilaterals with matching SSSAA.

Exercise 10.7 Give Euclidean counterexamples that demonstrate that ASAAS, ASASS, ASSAS, and AAAAS are not valid quadrilateral congruence theorems.

10.2.4 AAASS

This is the intriguing one. The idea of splitting the quadrilateral into triangles along the diagonal doesn't work. It fails to give enough information about either triangle. Yet, (as we will see) in Euclidean geometry, the angle sum of a quadrilateral has to be 360°. Since three angle congruences are given, that means that in the Euclidean realm the fourth angles must be congruent as well. In that case, this set of congruences is equivalent to ASASA, which is a valid congruence theorem. The problem is that in neutral geometry the angle sum of a quadrilateral does not have to be 360°. Because of the Saccheri-Legendre theorem, the angle sum of a quadrilateral cannot be more than 360°, but that is all we can say. It turns out that this is a valid congruence theorem in neutral geometry, though the proof is a little difficult. To simplify matters, we will only consider the situation where the quadrilaterals are convex. The argument that we will use requires us to "drop a perpendicular".

Lemma 10.2 *For a line ℓ and point P not on ℓ, there is a unique line through P that is perpendicular to ℓ.*

The intersection of ℓ and the perpendicular line is often called the *foot* of the perpendicular. The process of finding the foot is called *dropping a perpendicular*. It was an exercise to prove this lemma at the end of chapter 7. This is a common technique, so if you did not prove it then, it is worth proving now. One particularly relevant property of the foot of the perpendicular is the following.

Lemma 10.3 *Let ℓ be a line, P a point not on ℓ, and Q the foot of the perpendicular to ℓ through P. Then P is closer to Q than it is to any other point on ℓ.*

Exercise 10.8 Prove lemma 10.3.

Lemma 10.4 *Suppose that ℓ_1 and ℓ_2 are parallel lines, and that P is a point on the opposite side of ℓ_1 from all the points of ℓ_2. Let F_1 be the foot of the perpendicular from P to ℓ_1, and let F_2 be the foot of the perpendicular from P to ℓ_2. Then $|PF_1| < |PF_2|$.*

Exercise 10.9 Prove lemma 10.4 (warning: it is not necessarily true that $P * F_1 * F_2$).

With those preliminaries out of the way, we can now tackle AAASS.

Theorem 10.5 *(AAASS quadrilateral congruence) Given quadrilaterals $\square ABCD$ and $\square A'B'C'D'$ with*

$$\angle A \simeq \angle A', \quad \angle B \simeq \angle B', \quad \angle C \simeq \angle C', \quad CD \simeq C'D', \text{ and } DA \simeq D'A',$$

then $\square ABCD \simeq \square A'B'C'D'$.

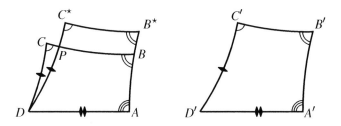

Figure 8. The setup of a proof of AAASS. Suppose that $\square ABCD$ and $\square A'B'C'D'$ are not congruent. Construct a conguent copy of $\square A'B'C'D'$ on top of $\square ABCD$.

Proof As mentioned above, we will prove this only for convex quadrilaterals. Suppose that the convex quadrilaterals $\square ABCD$ and $\square A'B'C'D'$ share the corresponding pieces as described in the statement of the theorem, but that they are not congruent. Then AB and $A'B'$ must not be congruent (if they were, then the quadrilaterals would be congruent, by AASAS). We may assume, without loss of generality, that AB is the shorter segment. That is, that $|AB| < |A'B'|$. Label $B^\star$ on $\overrightarrow{AB}$ so that $A * B * B^\star$ and so that $AB^\star \simeq A'B'$. Next label $C^\star$ on the same side of $\overline{AB}$ as C so that $B^\star C^\star \simeq B'C'$ and $\angle AB^\star C^\star \simeq \angle AB'C'$. Now, we have constructed $\square AB^\star C^\star D$ so that it will be congruent to $\square A'B'C'D'$ (by SASAS). In particular, this means that

$$\angle B^\star \simeq \angle B, \quad \angle C^\star \simeq \angle C, \text{ and } CD \simeq C^\star D.$$

According to the alternate interior angle theorem, the congruent angles $\angle B$ and $\angle B^\star$ require $\overline{BC}$ and $\overline{B^\star C^\star}$ to be parallel. Now look at the congruent angles $\angle C$ and $\angle C^\star$.

If they are right angles, we get an immediate contradiction: the sides CD and $C^\star D$ are congruent, but lemma 10.4 says that CD should be shorter than $C^\star D$. It is quite likely that $\angle C$ and $\angle C^\star$ are not right angles. In this case, drop the perpendicular from D to each of the lines

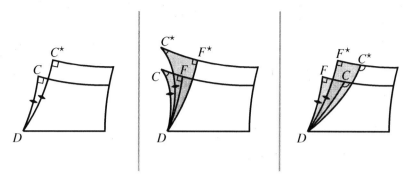

Figure 9. Three cases: (l) ∠C and ∠C* are right angles; (c) ∠C and ∠C* are acute angles; (r) ∠C and ∠C* are obtuse angles.

$\overline{BC}$ and $\overline{B^\star C^\star}$ Label the respective feet as F and $F^\star$. By lemma 10.4, DF must be shorter than $DF^\star$. But compare $\triangle CDF$ and $\triangle C^\star DF^\star$. In these triangles,

- ∠F and ∠F* are right angles,
- ∠C and ∠C* are congruent (they are either the congruent angles of the quadrilateral or their supplements), and
- the sides CD and $C^\star D$ are congruent.

By AAS, $\triangle CDF$ and $\triangle C^\star DF^\star$ are congruent. This would imply that $DF \simeq DF^\star$, which contradicts our previous finding that $|DF| < |DF^\star|$. The assumption that AB and $A'B'$ are not congruent is false, so by AASAS, $\square ABCD \simeq \square A'B'C'D'$. □

10.3 Exercises

10.10. A convex quadrilateral with two pairs of congruent adjacent sides is called a *kite*. Prove that the diagonals of a kite are perpendicular to one another.

10.11. Prove the SSSSA quadrilateral congruence theorem for convex quadrilaterals. Identify exactly where in the proof you use the convex condition.

10.12. Let $\square ABCD$ be a convex quadrilateral with right angles at B and C and congruent sides AB and CD (such a quadrilateral is called a Saccheri quadrilateral). Let E be the midpoint of BC and F be the midpoint of AD. Prove that EF is perpendicular to both AD and BC.

10.13. Let $\triangle ABC$ be an equilateral triangle. Let a be the midpoint of BC, b be the midpoint of AC and c be the midpoint of AB. On $\overrightarrow{aA}$, $\overrightarrow{bB}$, and $\overrightarrow{cC}$, label A', B', and C' so that

$$a * A * A', \quad b * B * B', \quad c * C * C',$$

and $AA' \simeq BB' \simeq CC'$. Prove that $\triangle A'B'C'$ is equilateral.

10.14. Let $\triangle ABC$ be an isosceles triangle with $AB \simeq AC$. Let a be the midpoint of BC, b be the midpoint of AC, and c be the midpoint of AB. Finally, let P be the intersection of Bb and Cc. Prove that $\square BaPc \simeq \square CaPb$.

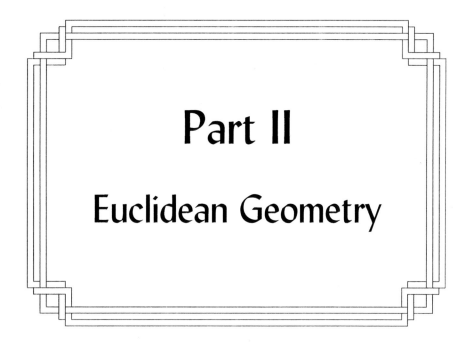

Part II

Euclidean Geometry

The goal of this book is to provide an introduction to both Euclidean and non-Euclidean (hyperbolic) geometry. The two geometries share many features, but they have fundamental and radical differences. Neutral geometry is the part of the path they have in common and that is what we have been studying so far, but we have now come to the fork in the path. That fork comes when you try to answer this question:

> Given a line ℓ and a point P that is not on ℓ, how many lines pass through P and are parallel to ℓ?

Using just the axioms of neutral geometry, we can prove that there is always at least one such parallel. We can also prove that if there is more than one parallel, then there must be infinitely many. But that is the extent of what the neutral axioms can say. The neutral axioms are not enough to determine whether there is one parallel or many. A single axiom is what separates Euclidean and non-Euclidean geometry: the final axiom of Euclidean geometry calls for exactly one parallel, the final axiom of non-Euclidean geometry calls for more than one parallel.

The next several chapters are devoted to Euclidean geometry. Euclidean geometry is very old, and so there is a lot of it. We only cover the fundamentals in this book, but there are many excellent books that do much more, such as *Geometry Revisited* [CG67] by Coxeter and Greitzer.

11
The Axiom on Parallels

Euclidean geometry is distinguished from neutral geometry by one axiom, one that describes the nature of parallel lines. There are many formulations of the parallel axiom for Euclidean geometry, but the one that gets right to the heart of the matter is *Playfair's axiom*, named after the Scottish mathematician John Playfair.

> *(Playfair's axiom) Let ℓ be a line, and let P be a point that is not on ℓ. Then there is exactly one line through P that is parallel to ℓ.*

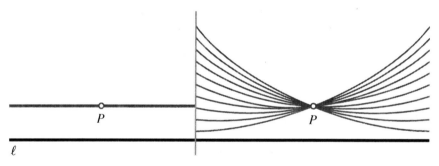

Figure 1. On the left, Euclidean geometry: a single line through P parallel to ℓ. On the right, more than one line through P parallel to ℓ.

In this chapter, we look at a small collection of theorems that are almost immediate consequences of Playfair's axiom, and as such are at the very core of Euclidean geometry. The first of these is Euclid's fifth postulate. This is the controversial postulate in the *Elements*, but also the one that guarantees the same parallel behavior that Playfair's axiom provides. Euclid's fifth is a little unwieldy, particularly when compared to Playfair's axiom, but it is the historical impetus for much of what followed. So let's use Playfair's axiom to prove Euclid's fifth postulate.

Theorem 11.1 *(Euclid's fifth postulate) If lines ℓ_1 and ℓ_2 are crossed by a transversal t, and the sum of adjacent interior angles on one side of t measure less than $180°$, then ℓ_1 and ℓ_2 intersect on that side of t.*

Proof First some labels.

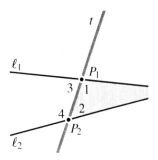

Figure 2. The labels.

Start with lines ℓ_1 and ℓ_2 crossed by transversal t. Label P_1 and P_2, the points of intersection of t with ℓ_1 and ℓ_2 respectively. On one side of t, the two adjacent interior angles should add up to less than 180°. Label the one at P_1 as ∠1 and the one at P_2 as ∠2. Label the supplement of ∠1 as ∠3 and label the supplement of ∠2 as ∠4.

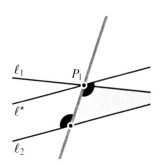

Figure 3. Constructing the unique parallel.

This theorem asserts that the lines intersect, and says where the intersection occurs. We will tackle those two statements in that order. Note that ∠1 and ∠4 are alternate interior angles, but they are not congruent: if they were, their supplements ∠2 and ∠3 would be too, and then

$$(\angle 1) + (\angle 2) = (\angle 1) + (\angle 3) = 180°.$$

There is, however, another line ℓ^* through P_1 that forms an angle congruent to ∠4 (because of the angle construction axiom), and by the alternate interior angle theorem, ℓ^* must be parallel to ℓ_2. Because of Playfair's axiom, ℓ^* is the only parallel to ℓ_2 through P_1. Therefore ℓ_1 intersects ℓ_2.

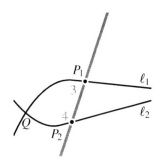

Figure 4. An impossible triangle on the wrong side of t.

The second part of the proof is to figure out on which side of t that ℓ_1 and ℓ_2 cross. Let's see what would happen if they intersected at a point Q on the wrong side of t, the side with ∠3 and ∠4. Then $\triangle P_1 P_2 Q$ would have two interior angles, ∠3 and ∠4, whose measures add up to more than 180°. This would violate the Saccheri-Legendre theorem. So ℓ_1 and ℓ_2 cannot intersect on the side of t with ∠3 and ∠4 and so they must intersect on the side with ∠1 and ∠2.

□

One of the truly useful theorems of neutral geometry is the alternate interior angle theorem. In fact, we just used it in the last proof. You may recall from high school geometry that the converse of the theorem is often even more useful. The problem is that the converse of the alternate interior angle theorem can't be proved using just the axioms of neutral geometry. It depends on the Euclidean behavior of parallel lines.

The Axiom on Parallels

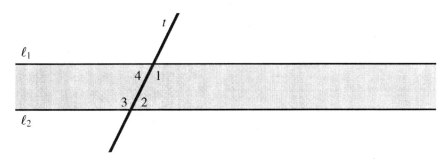

Figure 5. The converse of the alternate interior angle theorem: if $\ell_1 \parallel \ell_2$, then $\angle 1 \simeq \angle 3$ and $\angle 2 \simeq \angle 4$.

Theorem 11.2 *(Converse of the alternate interior angle theorem) If ℓ_1 and ℓ_2 are parallel, then the pairs of alternate interior angles formed by a transversal t are congruent.*

Proof Consider two parallel lines crossed by a transversal. Label adjacent interior angles: $\angle 1$ and $\angle 2$, and $\angle 3$ and $\angle 4$, so that $\angle 1$ and $\angle 4$ are supplementary and $\angle 2$ and $\angle 3$ are supplementary. That means that the pairs of alternate interior angles are $\angle 1$ and $\angle 3$, and $\angle 2$ and $\angle 4$. Now, we just have to do a little arithmetic. From the two pairs of supplementary angles:

(i): $(\angle 1) + (\angle 4) = 180°$
(ii): $(\angle 2) + (\angle 3) = 180°$.

Adding all four angles together,

$$(\angle 1) + (\angle 2) + (\angle 3) + (\angle 4) = 360°.$$

Now we turn to Euclid's fifth postulate, or rather its contrapositive. The lines ℓ_1 and ℓ_2 are parallel, which means that they do not intersect on either side of t. Therefore Euclid's fifth says that on neither side of t may the sum of adjacent interior angles be less than 180°:

- $(\angle 1) + (\angle 2) \geq 180°$
- $(\angle 3) + (\angle 4) \geq 180°$.

If either sum was greater than 180°, though, the sum of all four angles would have to be more than 360°. We saw above that is not the case, so the inequalities are actually equalities:

(iii): $(\angle 1) + (\angle 2) = 180°$
(iv): $(\angle 3) + (\angle 4) = 180°$.

Now we have two systems of equations with four unknowns, and it is basic algebra from here. Subtract *(iv)* from *(i)* to get

$$(\angle 1) - (\angle 3) = 0 \implies (\angle 1) = (\angle 3).$$

Subtract *(i)* from *(iii)* to get

$$(\angle 2) - (\angle 4) = 0 \implies (\angle 2) = (\angle 4).$$

Therefore the alternate interior angles are congruent. □

One of the key theorems we proved in the neutral geometry section was the Saccheri-Legendre theorem: that the angle sum of a triangle is at most 180°. That is all you can say with the axioms of neutral geometry, but in a world with Playfair's axiom and the converse of the alternate interior angle theorem, we can make a more bold statement about angle sums of triangles.

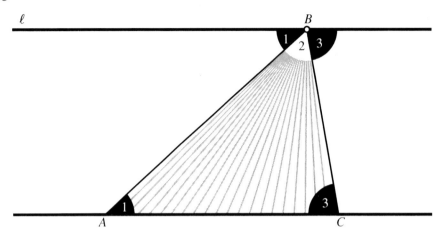

Figure 6. Use the converse of the alternate interior angle theorem to find the angle sum of a triangle.

Theorem 11.3 *The angle sum of a triangle is* 180°.

Proof Given $\triangle ABC$, by Playfair's axiom, there is a unique line ℓ through B that is parallel to $\overline{AC}$. That line and the rays $\overrightarrow{BA}$ and $\overrightarrow{BC}$ form three angles, $\angle 1$, $\angle 2$, and $\angle 3$ as shown in figure 6. By the converse of the alternate interior angle theorem, two pairs of alternate interior angles are congruent: $\angle 1 \simeq \angle A$ and $\angle 3 \simeq \angle C$. Therefore, the angle sum of $\triangle ABC$ is

$$s(\triangle ABC) = (\angle A) + (\angle B) + (\angle C) = (\angle 1) + (\angle 2) + (\angle 3) = 180°.$$ □

So, the three angles in an equilateral triangle all measure 60° and the measures of the two acute angles in a right triangle must add up to 90°.

In the last chapter on quadrilaterals we talked a little bit about the uncertain status of rectangles in neutral geometry, saying that it is easy to make a convex quadrilateral with three right angles, but that once you have done that, there is no guarantee that the fourth angle will be a right angle. Here it is now in the Euclidean context:

Theorem 11.4 *Let $\angle ABC$ be a right angle. Let r_A and r_C be rays so that r_A has endpoint A, is on the same side of $\overline{AB}$ as C, and is perpendicular to $\overline{AB}$; r_C has endpoint C, is on the same side of $\overline{BC}$ as A, and is perpendicular to $\overline{BC}$. Then r_A and r_C intersect at a point D, and the angle formed at this intersection, $\angle ADC$, is a right angle. Therefore $\square ABCD$ is a rectangle.*

Proof First we need to make sure that r_A and r_C intersect. Let ℓ_A and ℓ_C be the lines containing r_A and r_C respectively. By the alternate interior angle theorem, the right angles at A and B imply that ℓ_A and $\overline{BC}$ are parallel. So $\overline{BC}$ is the one line parallel to ℓ_A through C, and that implies that ℓ_C cannot be parallel to ℓ_A: it has to intersect ℓ_A. Let's call that point of intersection D. This

The Axiom on Parallels

theorem claims that it is the rays, not the lines, that intersect. Therefore we need to rule out the possibility that the intersection of ℓ_A and ℓ_C might happen on one (or both) of the opposite rays. Since ℓ_A is parallel to $\overline{BC}$, all the points of ℓ_A are on the same side of $\overline{BC}$ as A. None of the points of r_C^{op} is on that side of BC, so D cannot be on r_C^{op}. Likewise, all the points of ℓ_C are on the same side of $\overline{AB}$ as C. None of the points of r_A^{op} is on that side of AB, so D cannot be on r_A^{op}.

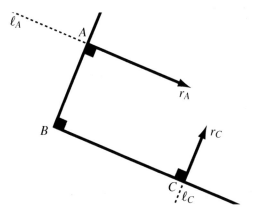

Figure 7. Constructing a rectangle.

So now $\square ABCD$ has three right angles, $\angle A$, $\angle B$, and $\angle C$. It is actually a convex quadrilateral too (verifying this is left as an exercise), so the diagonal AC divides $\square ABCD$ into two triangles $\triangle ABC$ and $\triangle ADC$ (see figure 8). Then, since the angle sum of a triangle is $180°$,

$$s(\triangle ABC) + s(\triangle ADC) = 180° + 180°$$
$$(\angle CAB) + (\angle B) + (\angle ACB) + (\angle CAD) + (\angle D) + (\angle ACD) = 360°$$
$$(\angle A) + (\angle B) + (\angle C) + (\angle D) = 360°$$
$$90° + 90° + 90° + (\angle D) = 360°$$
$$(\angle D) = 90°.$$

Therefore $\angle D$ is a right angle. □

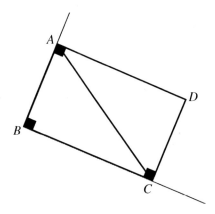

Figure 8. Verifying that the fourth angle is right.

Exercise 11.1 Fill in the gap in the proof by explaining why $\square ABCD$ must be convex.

We see then that, yes, rectangles exist in Euclidean geometry. Two essential properties of rectangles are provided in the following (easy) exercises.

Exercise 11.2 Let $\square ABCD$ be a rectangle. Prove that $\overline{AB}$ is parallel to $\overline{CD}$ and $\overline{AD}$ is parallel to $\overline{BC}$.

Exercise 11.3 Let $\square ABCD$ be a rectangle. Prove that $AB \simeq CD$ and $AD \simeq BC$ and $AC \simeq BD$.

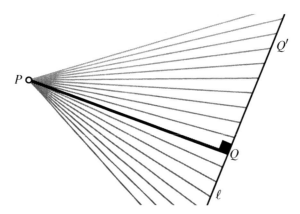

Figure 9. The distance from P to ℓ is measured along the segment from P to ℓ that is perpendicular to ℓ.

For the last result of this section, we return to parallel lines. One of the things that we will see when we study non-Euclidean geometry is that parallel lines tend to diverge from each other. That doesn't happen in Euclidean geometry. This is one of the key differences between the two geometries. Let's make this more precise. Suppose that P is a point that is not on a line ℓ. Define the distance from P to ℓ to be the minimum distance from P to a point on ℓ:

$$d(P, \ell) = \min\left\{|PQ| \,\Big|\, Q \text{ is on } \ell\right\}.$$

Exercise 11.4 Prove that the minimum actually occurs when Q is the foot of the perpendicular to ℓ through P (this can be proved using solely the axioms of neutral geometry).

For a pair of parallel lines, that distance as measured along perpendiculars does not change.

Theorem 11.5 *If ℓ and ℓ' are parallel lines, then the distance from a point on ℓ to ℓ' is constant. In other words, if P and Q are points on ℓ, then*

$$d(P, \ell') = d(Q, \ell').$$

11.1 Exercises

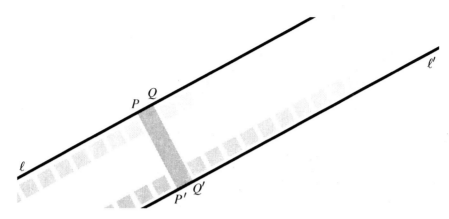

Figure 10. Perpendicular segments PP' and QQ' between parallel lines have the same length. That is, parallel lines are equidistant.

Proof Let P' and Q' be the feet of the perpendiculars on ℓ' from P and Q respectively. That way, by the previous exercise,

$$d(P, \ell') = |PP'| \text{ and } d(Q, \ell') = |QQ'|.$$

Then $\angle PP'Q'$ and $\angle QQ'P'$ are right angles. By the converse of the alternate interior angle theorem, $\angle P$ and $\angle Q$ are right angles too, so $\square PQQ'P'$ is a rectangle. According to exercise 11.2, PP' and QQ', which are opposite sides of a rectangle, are congruent. □

11.1 Exercises

11.5. Let ℓ be a line and P a point not on ℓ. Using only the results of neutral geometry, prove that there is at least one line through P that is parallel to ℓ.

11.6. Suppose that ℓ_1, ℓ_2, and ℓ_3 are three distinct lines such that ℓ_1 and ℓ_2 are parallel, and ℓ_2 and ℓ_3 are parallel. Prove that ℓ_1 and ℓ_3 are parallel.

11.7. Suppose that ℓ_1 and ℓ_2 are parallel lines, and that t is a line that intersects (but is distinct from) ℓ_1. Prove that t intersects ℓ_2.

11.8. Given three distinct parallel lines, prove that there is a labeling of them as ℓ_1, ℓ_2, and ℓ_3 so that

- no point of ℓ_1 lies between a point of ℓ_2 and a point of ℓ_3
- every point of ℓ_2 lies between a point of ℓ_1 and of ℓ_3
- no point of ℓ_3 lies between a point of ℓ_1 and a point of ℓ_2.

In this case, we can say that ℓ_2 is *between* ℓ_1 and ℓ_3.

11.9. This exercise is an extension of the previous one. Given n distinct parallel lines, prove that there is an ordering of them as $\ell_1, \ell_2, \ldots, \ell_n$ so that if $i < j < k$ then ℓ_j is between ℓ_i and ℓ_k.

11.10. Find the angle sum of a convex n-gon as a function of n.

11.11. What is the measure of an interior angle of a regular n-gon (as a function of n)?

11.12. Consider a convex quadrilateral $\square ABCD$. Prove that the two diagonals of $\square ABCD$ bisect each other if and only if $\square ABCD$ is a parallelogram.

11.13. Show that a parallelogram $\square ABCD$ is a rectangle if and only if $AC \simeq BD$.

11.14. Suppose that the diagonals of a convex quadrilateral $\square ABCD$ intersect one another at a point P and that
$$AP \simeq BP \simeq CP \simeq DP.$$
Prove that $\square ABCD$ is a rectangle.

11.15. Suppose that the diagonals of a convex quadrilateral bisect one another at right angles. Prove that the quadrilateral must be a rhombus.

11.16. Let $\triangle ABC$ be isosceles with $AB \simeq AC$, and let a be the midpoint of BC. Let b and c be points on the line through A that is parallel to BC so that $b * A * c$ and so that $bA \simeq cA \simeq Ba$. Prove that $\triangle ABC \simeq \triangle abc$.

11.17. Verify that the Cartesian model (as developed through the exercises in chapters 1–3) satisfies Playfair's axiom.

12
Parallel Projection

In the course of this chapter, we will need to use a few properties of parallelograms. It is rather common to confuse the properties of a parallelogram with its definition, so first, recall the definition of a parallelogram.

Definition 12.1 A *parallelogram* is a simple quadrilateral whose opposite sides are parallel.

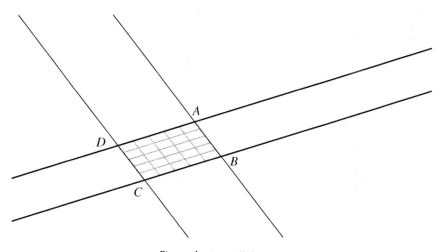

Figure 1. A parallelogram.

Now on to the properties of parallelograms that we will need for this chapter.

Exercise 12.1 Prove that in a parallelogram, the two pairs of opposite sides are congruent and the two pairs of opposite angles are congruent.

Exercise 12.2 Prove that if a convex quadrilateral has one pair of opposite sides that are both parallel and congruent, then it is a parallelogram.

Exercise 12.3 Let $\square ABB'A'$ be a simple quadrilateral. Verify that if AA' and BB' are parallel, but AB and $A'B'$ are not, then AA' and BB' cannot be congruent.

12.1 Parallel projection

The purpose of this chapter is to introduce a mechanism called *parallel projection*, a particular kind of mapping from the points on one line to the points on another. Parallel projection is the piece of machinery that you have to have in place to understand similarity, which is in turn essential for much of what we will be doing in the next chapters. The primary goal at this point is to understand how distances between points may be distorted by the parallel projection mapping. Once that is figured out, we will be able to turn our attention to the geometry of similarity.

Definition 12.2 A *parallel projection* from one line ℓ to another ℓ' is a map $\Phi : \ell \to \ell'$ that assigns to each point P on ℓ a point $\Phi(P)$ on ℓ' so that all the lines connecting a point and its image are parallel to one another.

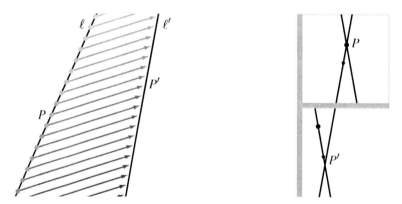

Figure 2. The path from a point P on ℓ to a point P' on ℓ' defines a parallel projection as long as neither P nor P' is the intersection of ℓ and ℓ' (as shown at right).

It is easy to construct parallel projections. Any point P on ℓ and its image $\Phi(P)$ on ℓ' completely determines the projection: for any other point Q on ℓ there is a unique line that passes through Q and is parallel to $\overline{P\,\Phi(P)}$. Wherever the line intersects ℓ' will have to be $\Phi(Q)$. There are only two cases where this construction will not work: if P is the intersection of ℓ and ℓ', then the lines of projection run parallel to ℓ' and so fail to provide a point of intersection; and if $\Phi(P)$ is the intersection of ℓ and ℓ', then the lines of projection actually coincide rather than being parallel.

Theorem 12.3 *A parallel projection is both one-to-one and onto.*

12.2 Parallel projection, order, and congruence

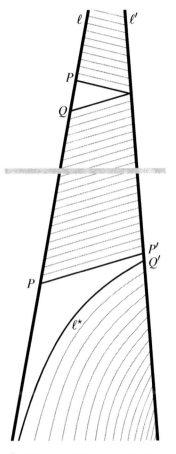

Figure 3. Using proofs by contradiction, we can show that parallel projection is both one-to-one (top) and onto (bottom).

Proof Let $\Phi : \ell \to \ell'$ be a parallel projection. First let's see why Φ is one-to-one. Suppose that it is not. That is, suppose that P and Q are two distinct points on ℓ but that $\Phi(P) = \Phi(Q)$. Then the two projecting lines $\overline{P\,\Phi(P)}$ and $\overline{Q\,\Phi(Q)}$, which ought to be parallel, actually share a point. This can't happen.

Now let's see why Φ is onto. Suppose that Q' is a point on ℓ' that is not in the image of Φ. To get a sense of how Φ is casting points from ℓ to ℓ', consider a point P on ℓ and its image $P' = \Phi(P)$ on ℓ'. By Playfair's axiom, there is a unique line $\ell^\star$ that passes through Q' and is parallel to $\overline{PP'}$. If $\ell^\star$ intersects ℓ at a point Q, then $\Phi(Q) = Q'$, contradicting the initial supposition that Q' is not in the image of Φ. If $\ell^\star$ does not intersect ℓ, then both ℓ and $\overline{PP'}$ are lines that pass through P and are parallel to $\ell^\star$, in violation of Playfair's axiom. In either case, we arrive at a contradiction. Therefore, an arbitrary point Q' of ℓ must be in the image of Φ, and so Φ is onto. □

Now that we have established that parallel projection is a bijection, it helps to use a naming convention to makes things more readable. I will use a prime mark ′ to indicate the parallel projection of a point, so $\Phi(P) = P'$, $\Phi(Q) = Q'$, and so on.

12.2 Parallel projection, order, and congruence

So far we have seen that parallel projection establishes a correspondence between the points of one line and the points of another. What about the order of the points? Can points get shuffled up in the process of a parallel projection? No.

Theorem 12.4 *Let $\Phi : \ell \to \ell'$ be a parallel projection. If A, B, and C are points on ℓ and $A * B * C$, then $A' * B' * C'$.*

Proof Because B is between A and C, A and C must be on opposite sides of $\overline{BB'}$. But, $\overline{AA'}$ does not intersect $\overline{BB'}$, so A' has to be on the same side of $\overline{BB'}$ as A; and $\overline{CC'}$ does not

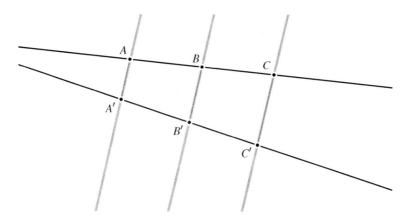

Figure 4. Parallel projection keeps points in order.

intersect $\overline{BB'}$, so C' has to be on the same side of $\overline{BB'}$ as C. Therefore A' and C' have to be on opposite sides of $\overline{BB'}$, and so the intersection of $\overline{BB'}$ and $A'C'$, which is B', must be between A' and C'. □

Now what about congruence?

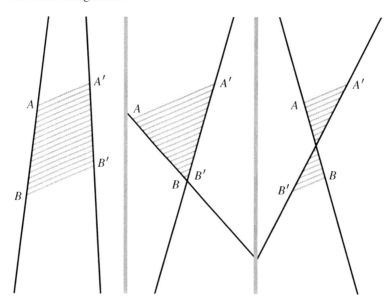

Figure 5. There are three positions for A and B relative to the image line: both on the same side, one on the image line, or one on each side. Likewise, there are three positions for a and b. There are nine cases in all.

Theorem 12.5 *Let $\Phi : \ell \to \ell'$ be a parallel projection. If a, b, A, and B are points on ℓ and if $ab \simeq AB$, then $a'b' \simeq A'B'$.*

Proof There are several cases, depending on the positions of ab and AB relative to ℓ'. They could lie on the same side of ℓ', they could lie on opposite sides of ℓ', one or both could straddle

ℓ', or one or both could have an endpoint on ℓ'. Each case must be handled differently. We will look at the case where both segments are on the same side of ℓ'. The others are left as exercises. There are still two cases to consider, depending on whether or not ℓ and ℓ' are parallel.

Case 1: when ℓ and ℓ' are parallel (see figure 6). This is the simpler case, but it helps illuminate the more general case. Because aa' is parallel to bb' and ab is parallel to $a'b'$, $\square aa'b'b$ is a parallelogram; AA' is parallel to BB' and AB is parallel to $A'B'$ so $\square AA'B'B$ is also a parallelogram. Because the opposite sides of a parallelogram are congruent (exercise 12.1), $a'b' \simeq ab$ and $AB \simeq A'B'$. Since $ab \simeq AB$, $a'b' \simeq A'B'$.

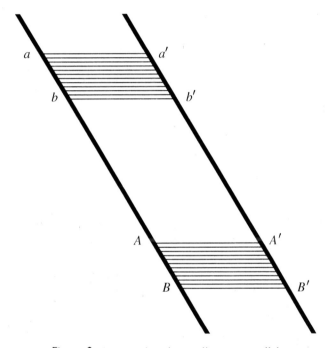

Figure 6. Case 1, when the two lines are parallel.

Case 2: when ℓ and ℓ' are not parallel (see figure 7). This is the more likely case, in which $\square aa'b'b$ and $\square AA'B'B$ will not be parallelograms. We can takes a similar approach to case 1, though, by building some parallelograms into the problem. Because ℓ and ℓ' are not parallel, the segments aa' and bb' cannot have the same length (exercise 12.3), and the segments AA' and BB' cannot have the same length. Let's assume that aa' is shorter than bb' and that AA' is shorter than BB'. If this is not the case, then it is just a matter of switching some labels to make it so. Then

- there is a point c between b and b' so that $bc \simeq aa'$, and
- there is a point C between B and B' so that $BC \simeq AA'$.

This creates four shapes of interest: the quadrilaterals $\square a'abc$ and $\square A'ABC$ that are parallelograms (exercise 12.2), and the triangles $\triangle a'b'c$ and $\triangle A'B'C$. The key here is to prove that $\triangle a'b'c \simeq \triangle A'B'C$.

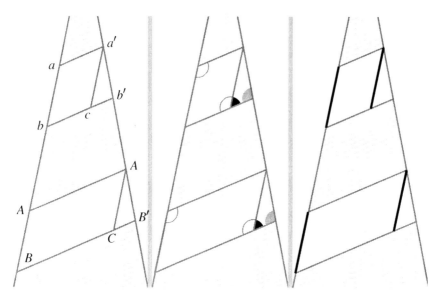

Figure 7. The non-parallel case. (l) First construct parallelograms. (c) The two angles and (r) the sides needed for AAS to show that $\triangle a'b'c' \simeq \triangle A'B'C'$.

- **A:** $\angle b' \simeq \angle B'$. The lines $\overline{cb'}$ and $\overline{CB'}$ are parallel (they are two of the projecting lines) and they are crossed by the transversal ℓ'. By the converse of the alternate interior angle theorem, $\angle a'b'c$ and $\angle A'B'C$ are congruent.
- **A:** $\angle c \simeq \angle C$. The opposite angles of the two parallelograms are congruent. Therefore $\angle a'cb \simeq \angle a'ab$ and $\angle A'AB \simeq \angle A'CB$. But $\overline{aa'}$ and $\overline{AA'}$ are parallel lines cut by the transversal ℓ, so $\angle a'ab \simeq \angle A'AB$, so $\angle a'cb \simeq \angle A'CB$, and so their supplements $\angle a'cb'$ and $\angle A'CB'$ are also congruent.
- **S:** $a'c \simeq A'C$. The opposite sides of the two parallelograms are congruent too. Therefore $a'c \simeq ab$ and $AB \simeq A'C$, and since $ab \simeq AB$, we get $a'c \simeq A'C$.

By AAS, then, $\triangle a'b'c \simeq \triangle A'B'C$. The corresponding sides $a'b'$ and $A'B'$ have to be congruent. □

Exercise 12.4 Fill in some gaps:

(1) Prove the theorem in the case when both a and A lie on ℓ' (at the intersection of ℓ and ℓ').
(2) Prove the theorem in the case when a and A lie on opposite sides of ℓ' from b and B.

12.3 Parallel projection and distance

That brings us to the question at the very heart of parallel projection. If Φ is a parallel projection and A and B are two points on ℓ, how do the lengths $|AB|$ and $|A'B'|$ compare? In case 1 of the last proof, AB and $A'B'$ ended up being congruent, but that was because ℓ and ℓ' were parallel. In general, AB and $A'B'$ do not have to be congruent. But it turns out that in the process of parallel projecting from one line to another, all distances are scaled by a constant multiple.

12.3 Parallel projection and distance

Theorem 12.6 *If $\Phi : \ell \to \ell'$ is a parallel projection, then there is a constant k such that*

$$|A'B'| = k|AB|$$

for all points A and B on ℓ.

It is helpful to get a little perspective on this before the proof. First note that the previous theorem gives us a way to narrow the scope of the problem. Fix a point O on ℓ and let r be one of the two rays along ℓ with O as its endpoint. The segment construction axiom says that every segment AB on ℓ is congruent to a segment OP where P is some point on r. We have just seen that parallel projection maps congruent segments to congruent segments. So if Φ scales all segments of the form OP by a factor of k, then it must scale all the segments of ℓ by the same factor.

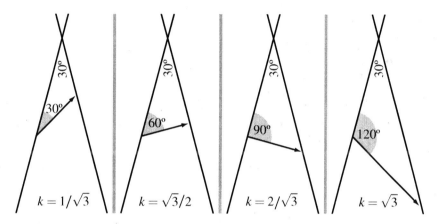

Figure 8. Some parallel projections and their scaling constants.

Second, we can extend that by considering what happens when we parallel project end-to-end congruent copies of a segment. For this, let me introduce another convenient notation convention: for the rest of this argument, let P_d denote the point on r that is a distance d from O. Now, pick a positive real value x, and let

$$k = |O'P'_x|/|OP_x|,$$

so that Φ scales the segment OP_x by a factor of k. Eventually we will have to show that Φ scales all segments by that factor, but for now let's restrict our attention to the segments OP_{nx}, where n is a positive integer. Between O and P_{nx} are $P_x, P_{2x}, \ldots P_{(n-1)x}$ in order:

$$O * P_x * P_{2x} * \cdots * P_{(n-1)x} * P_{nx}.$$

We have seen that parallel projection preserves the order of points, so

$$O' * P'_x * P'_{2x} * \cdots * P'_{(n-1)x} * P'_{nx}.$$

Each segment $P_{ix}P_{(i+1)x}$ is congruent to OP_x and consequently each parallel projection $P'_{ix}P'_{(i+1)x}$ is congruent to $O'P'_x$. Add their lengths:

$$|O'P'_{nx}| = |O'P'_x| + |P'_x P'_{2x}| + |P'_{2x} P'_{3x}| + \cdots + |P'_{(n-1)x} P'_{nx}|$$
$$= kx + kx + kx + \cdots + kx \quad (n \text{ times})$$
$$= k \cdot nx$$

and so Φ scales OP_{nx} by a factor of k.

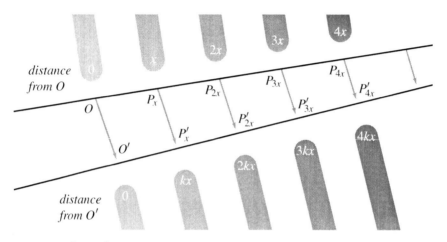

Figure 9. Scaling of segments of length kx, where k is an integer.

Sadly, no matter what x is, the points P_{nx} account for an inconsequential portion of the set of all points of r. However, if OP_x and OP_y were to have two different scaling factors we could use this end-to-end copying to magnify the difference between them. Third, then, consider an example that demonstrates how this works, and which shows how this ultimately prevents there being two different scaling factors. In this example, let's suppose that $|O'P'_1| = 2$, so all integer length segments on ℓ are scaled by a factor of 2. Let's take a look at what this means for $P_{3.45}$. Let k be the scaling factor for $OP_{3.45}$. Figure 10 shows how the first few end-to-end copies of $OP_{3.45}$ begin to limit the possible values of k.

Fourth, note that the floor function, $f(x) = \lfloor x \rfloor$, assigns to each real number x the largest integer that is less than or equal to it. The ceiling function, $f(x) = \lceil x \rceil$, assigns to each real number x the smallest integer that is greater than or equal to it.

It is finally time to prove that parallel projection scales distance.

Proof Let $k = |O'P'_1|$ so that k is the scaling factor for the segment of length 1 (and consequently for all integer length segments). Now take a point P_x on ℓ and let k' be the scaling factor for the segment OP_x. We want to show that $k' = k$ and we can do that by following the same strategy as in the example above, by capturing k' in an increasingly narrow band around k by looking at the parallel projection of P_{nx} as n increases. Start by working back and forth between

12.3 Parallel projection and distance

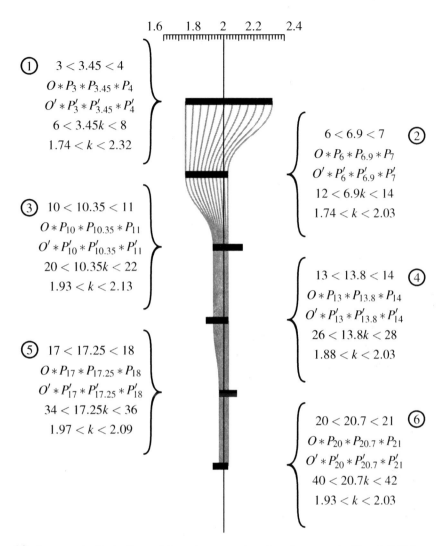

Figure 10. An example. Projecting end-to-end congruent copies of a segment of length 3.45 by a parallel projection that scales all integer length segments by a factor of two.

inequalities of real numbers and orders of points:

$$\lfloor nx \rfloor \leq nx \leq \lceil nx \rceil$$
$$O * P_{\lfloor nx \rfloor} * P_{nx} * P_{\lceil nx \rceil}$$
$$O' * P'_{\lfloor nx \rfloor} * P'_{nx} * P'_{\lceil nx \rceil}$$
$$k\lfloor nx \rfloor \leq k'nx \leq k\lceil nx \rceil.$$

(The notation here is a bit casual—if nx is an integer, then $\lfloor nx \rfloor = nx$, so $P_{\lfloor nx \rfloor}$ is not actually between O and P_{nx}. This does not cause any problems for the proof.) Because $(nx - 1) \leq \lfloor nx \rfloor$

and $\lceil nx \rceil \leq (nx + 1)$ we can move to a less precise but more manageable set of inequalities:

$$k(nx - 1) \leq k'nx \leq k(nx + 1).$$

Divide by nx to get

$$k\left(\frac{nx - 1}{nx}\right) \leq k' \leq k\left(\frac{nx + 1}{nx}\right).$$

As n increases, the ratios $(nx - 1)/(nx)$ and $(nx + 1)/(nx)$ both approach 1. In the limit as n goes to infinity, they are 1. Since the inequalities have to be true for all n, the only possible value for k', then, is k. $\square$

12.4 Exercises

12.5. Suppose that Φ is a parallel projection from ℓ to ℓ'. If ℓ and ℓ' intersect at P, prove that $\Phi(P) = P$.

12.6. Prove that if ℓ and ℓ' are parallel, then the scaling factor of a parallel projection between them must be one, but that if ℓ and ℓ' are not parallel, then there is a parallel projection with every possible scaling factor k where $0 < k < \infty$.

12.7. Let $\Phi : \ell \to \ell'$ be a parallel projection. Show that the inverse map $\Phi^{-1} : \ell' \to \ell$ is also a parallel projection.

12.8. Let $\Phi_1 : \ell \to \ell'$ and $\Phi_2 : \ell' \to \ell''$ be parallel projections. Is the composition $\Phi_2 \circ \Phi_1 : \ell \to \ell''$ necessarily a parallel projection?

12.9. In chapter 6, we constructed a distance function, and one of the keys to the construction was locating the points on a ray that were a distance of $m/2^n$ from its endpoint. In Euclidean geometry, there is a construction that locates all the points on a ray that are any rational distance m/n from its endpoint. Take two non-opposite rays r and r' with a common endpoint O. Along r, lay out m congruent copies of a segment of length 1, ending at the point P_m. Along r', lay out n congruent copies of a segment of length 1, ending at the point Q_n. Mark the point Q_1 on r' that is a distance 1 from O. Verify that the line that passes through Q_1 and is parallel to $P_m Q_n$ intersects r a distance of m/n from O.

13
Similarity

In the chapters on neutral geometry, we spent a lot of effort to gain an understanding of polygon congruence, particularly congruence of triangles and quadrilaterals. Remember that polygon congruence is an equivalence relation (it is reflexive, symmetric, and transitive). It turns out that congruence is not the only important equivalence relation between polygons, though, and the purpose of this chapter is to investigate another: similarity. Similarity is a less demanding relation than congruence. Intuitively, we tend to think of congruent polygons as exactly the same, just positioned differently. Similar polygons are not just positioned differently, but are scaled versions of one another: the same shape, but possibly different sizes.

Definition 13.1 Two n-sided polygons $P_1 P_2 \ldots P_n$ and $Q_1 Q_2 \ldots Q_n$ are *similar* to one another if they meet two sets of conditions:

(1) corresponding interior angles are congruent:
$$\angle P_i \simeq \angle Q_i, \quad 1 \leq i \leq n.$$

(2) corresponding side lengths differ by the same constant multiple k:
$$|P_i P_{i+1}| = k \cdot |Q_i Q_{i+1}|, \quad 1 \leq i \leq n.$$

Figure 1. An arrangement of similar triangles.

To indicate that two polygons $P_1 P_2 \ldots P_n$ and $Q_1 Q_2 \ldots Q_n$ are similar, write $P_1 P_2 \ldots P_n \sim Q_1 Q_2 \ldots Q_n$. There are a few things worth noting here. First, if polygons are congruent, they will be similar as well with the scaling constant 1. Second, it is easy to see that similarity is reflexive, symmetric and transitive and so is an equivalence relation. Third, when you jump from one polygon to another similar polygon, the corresponding segment lengths are scaled by the same amount. That behavior echoes the work we did in the last chapter, and for good reason: parallel projection underlies everything that we are going to do in this chapter.

Much of the time, when working with either parallel projection or similarity, the scaling constant is not important. The only thing that matters is that there is one. Fortunately, the existence of a scaling constant can be indicated without ever mentioning what it is. The key to doing this is ratios. Consider a parallel projection from line ℓ to line ℓ'. Let A, B, a, and b be points on ℓ and let A', B', a', and b' be their respective images on ℓ'. The main result of the last chapter was that there is a scaling constant k so that

$$|A'B'| = k \cdot |AB| \text{ and } |a'b'| = k \cdot |ab|.$$

The ratios are only a step away from this pair of equations.

Ratio 1: Solve for k in both equations and set them equal to each other:

$$\frac{|A'B'|}{|AB|} = \frac{|a'b'|}{|ab|}.$$

Ratio 2: Starting from the first ratio, multiply by $|AB|$ and divide by $|a'b'|$:

$$\frac{|A'B'|}{|a'b'|} = \frac{|AB|}{|ab|}.$$

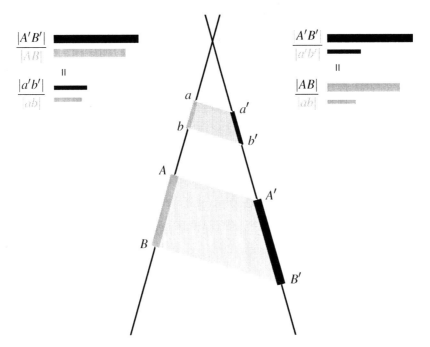

Figure 2. Two invariant ratios of a parallel projection.

13.1 Triangle similarity theorems

Let's begin with a few theorems that deal with similarity of triangles. I like to think of them as degenerations of the triangle congruence theorems, where the strict condition of side congruence, $A'B' \simeq AB$, is replaced with the more flexible condition of constant scaling, $|A'B'| = k|AB|$. The first is the SAS similarity theorem.

Theorem 13.2 *(SAS similarity)* In $\triangle ABC$ and $\triangle A'B'C'$, *if* $\angle A \simeq \angle A'$ *and if there is a constant k so that*

$$|A'B'| = k \cdot |AB| \text{ and } |A'C'| = k \cdot |AC|,$$

then $\triangle ABC \sim \triangle A'B'C'$.

Proof First note that as with parallel projection, the second condition in the SAS similarity theorem can be recast in terms of ratios:

$$\frac{|A'B'|}{|AB|} = \frac{|A'C'|}{|AC|} \text{ or } \frac{|A'B'|}{|A'C'|} = \frac{|AB|}{|AC|}.$$

With that said, what we need to do in the proof is to establish two more angle congruences, that $\angle B \simeq \angle B'$ and $\angle C \simeq \angle C'$, and one more ratio of sides, that $|B'C'| = k|BC|$. Two parallel projections form the backbone of this proof.

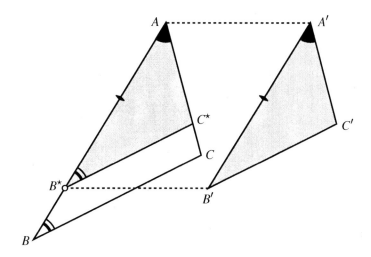

Figure 3. Proving SAS similarity. The first parallel projection.

The first parallel projection. The primary purpose of the first projection is to build a transitional triangle that is congruent to $\triangle A'B'C'$ but positioned on top of $\triangle ABC$. Begin by locating the point $B^\star$ on $\overrightarrow{AB}$ so that $AB^\star \simeq A'B'$. We cannot know the exact location of $B^\star$ relative to B on this ray. It depends on whether $A'B'$ is shorter or longer than AB. For this argument, assume that $A'B'$ is shorter than AB, which will place $B^\star$ between A and B (the other case is not substantially different). Consider the parallel projection

$$\Phi_1 : (\overline{AB}) \longrightarrow (\overline{AC})$$

that maps B to C. Because A is the intersection of these two lines, $\Phi_1(A) = A$. Label $C^\star = \Phi_1(B^\star)$. Let's see how the newly formed $\triangle AB^\star C^\star$ compares with $\triangle A'B'C'$. By parallel projection,

$$\frac{|AC^\star|}{|AC|} = \frac{|AB^\star|}{|AB|},$$

but we chose $B^\star$ so that $|AB^\star| = |A'B'|$. Therefore

$$\frac{|AC^\star|}{|AC|} = \frac{|A'B'|}{|AB|} = \frac{|A'C'|}{|AC|},$$

and so $|AC^\star| = |A'C'|$. Put that together with what we already know, that $AB^\star \simeq A'B'$ and $\angle A \simeq \angle A'$, and by SAS, $\triangle A'B'C'$ and $\triangle AB^\star C^\star$ are congruent. In particular, that means $\angle B' \simeq \angle B^\star$ and $\angle C' \simeq \angle C^\star$. Now let's turn back to see how $\triangle AB^\star C^\star$ relates to $\triangle ABC$. To locate $C^\star$, we used a projection that was parallel to $\overline{BC}$. Hence $\overline{B^\star C^\star}$ and $\overline{BC}$ are parallel, and so, by the converse of the alternate interior angle theorem, $\angle B^\star \simeq \angle B$ and $\angle C^\star \simeq \angle C$. Since angle congruence is transitive, we now have the required angle congruences, $\angle B \simeq \angle B'$ and $\angle C \simeq \angle C'$.

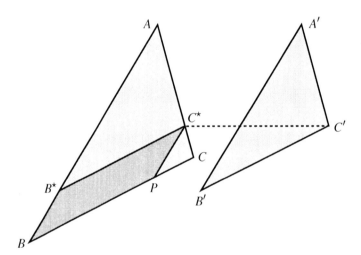

Figure 4. Proving SAS similarity. The second parallel projection.

The second parallel projection. Now consider the parallel projection

$$\Phi_2 : (\overline{AC}) \longrightarrow (\overline{BC})$$

that maps A to B. Since C is the intersection of $\overline{AC}$ and $\overline{BC}$, $\Phi_2(C) = C$. The other point of interest this time is $C^\star$. Let $P = \Phi_2(C^\star)$. In doing so, we have effectively carved out a parallelogram $\square BB^\star C^\star P$. Recall (from the last chapter) that the opposite sides of a parallelogram are congruent, so $B^\star C^\star \simeq BP$, and therefore $|B'C'| = |BP|$. The parallel projection Φ_2 establishes

13.1 Triangle similarity theorems

equal ratios $|BP|/|BC| = |AC^\star|/|AC|$, and so

$$\frac{|B'C'|}{|BC|} = \frac{|BP|}{|BC|} = \frac{|AC^\star|}{|AC|} = \frac{|A'C'|}{|AC|} = k.$$

Thus, $|B'C'| = k|BC|$, as needed. □

Back in the neutral geometry chapters, after SAS we next encountered ASA and AAS. Unlike SAS, both theorems reference only one pair of sides in the triangles. Let's look at what happens when we try to modify those congruence conditions into similarity conditions.

ASA congruence	ASA similarity (?)
$\angle A \simeq \angle A'$	$\angle A \simeq \angle A'$
$AB \simeq A'B'$	$|A'B'| = k \cdot |AB|$
$\angle B \simeq \angle B'$	$\angle B \simeq \angle B'$

AAS congruence	AAS similarity (?)
$\angle A \simeq \angle A'$	$\angle A \simeq \angle A'$
$\angle B \simeq \angle B'$	$\angle B \simeq \angle B'$
$BC \simeq B'C'$	$|B'C'| = k \cdot |BC|$

In each of these conversions, the condition on the one side is automatically satisfied: there will always be a real value of k that makes the equation true. That suggests that it may take only two angle congruences to guarantee similarity.

Theorem 13.3 *(AA similarity) In $\triangle ABC$ and $\triangle A'B'C'$, if $\angle A \simeq \angle A'$ and $\angle B \simeq \angle B'$, then $\triangle ABC \sim \triangle A'B'C'$.*

We have plenty of information about the angles, so what we need is some information about ratios of sides. If we can show that

$$\frac{|A'B'|}{|AB|} = \frac{|A'C'|}{|AC|},$$

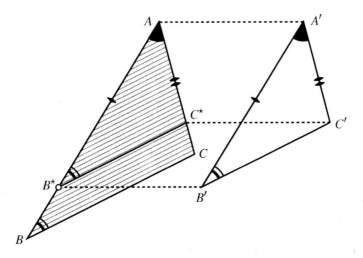

Figure 5. Proof of AA similarity, by parallel projection.

then that, along with the congruence $\angle A \simeq \angle A'$, will be enough to use the SAS similarity theorem.

Proof Start by constructing a transition triangle: one that is positioned on top of $\triangle ABC$ but that is congruent to $\triangle A'B'C'$. To do that, locate $B^\star$ on $\overrightarrow{AB}$ so that $AB^\star \simeq A'B'$, and $C^\star$ on $\overrightarrow{AC}$ so that $AC^\star \simeq A'C'$. By SAS, $\triangle AB^\star C^\star$ and $\triangle A'B'C'$ are congruent. Consider the congruent angles

$$\angle B^\star \simeq \angle B' \simeq \angle B.$$

According to the alternate interior angle theorem, $\overline{B^\star C^\star}$ and $\overline{BC}$ must be parallel. Therefore the parallel projection from $\overline{AB}$ to $\overline{AC}$ that maps B to C and A to itself will also map $B^\star$ to $C^\star$ and so

$$\frac{|A'B'|}{|AB|} = \frac{|AB^\star|}{|AB|} = \frac{|AC^\star|}{|AC|} = \frac{|A'C'|}{|AC|}.$$

The first and last terms give the ratio we need. That, together with the known congruence $\angle A \simeq \angle A'$, is enough for SAS similarity, so $\triangle ABC \sim \triangle A'B'C'$. □

While AAA was not enough to guarantee congruence, thanks to the result above, we now know that it is (more than) enough to guarantee similarity. The last of the triangle similarity theorems is SSS similarity (SSA, which just misses as a congruence theorem, is done in again by the same counterexample).

Theorem 13.4 *(SSS similarity)* In $\triangle ABC$ and $\triangle A'B'C'$, if there is a constant k so that

$$|A'B'| = k \cdot |AB|, \quad |B'C'| = k \cdot |BC|, \quad \text{and} \quad |C'A'| = k \cdot |CA|,$$

then $\triangle ABC \sim \triangle A'B'C'$.

Exercise 13.1 Prove the SSS triangle similarity theorem.

13.2 The Pythagorean theorem

Before we close this chapter, let's meet one of the real celebrities of the subject, the Pythagorean theorem. There are many, many proofs of this theorem. This one involves dividing the triangle into two smaller triangles, each similar to the initial triangle, and then working with ratios.

Theorem 13.5 *(The Pythagorean theorem)* Let $\triangle ABC$ be a right triangle whose right angle is at the vertex C. Let

$$a = |BC|, \quad b = |AC|, \quad \text{and} \quad c = |AB|.$$

Then $c^2 = a^2 + b^2$.

13.2 The Pythagorean theorem

Proof Let D be the foot of the perpendicular to AB through C. The segment CD divides $\triangle ABC$ into two smaller triangles: $\triangle ACD$ and $\triangle BCD$. Label the lengths of their sides as $c_1 = |AD|$, $c_2 = |BD|$, and $d = |CD|$, and note that $c = c_1 + c_2$. Now $\triangle ADC$ shares $\angle A$ with $\triangle ACB$, and they both have a right angle, so by the AA similarity theorem, $\triangle ADC \sim \triangle ACB$. Similarly, $\triangle BDC$ shares $\angle B$ with $\triangle ACB$, and they both have a right angle as well, so again by AA similarity, $\triangle BDC \sim \triangle BCA$. From these similarities, there are many ratios, but the two that we need are

$$\frac{a}{c} = \frac{c_2}{a} \implies a^2 = c \cdot c_2 \quad \text{and} \quad \frac{b}{c} = \frac{c_1}{b} \implies b^2 = c \cdot c_1.$$

All that is left is to add the equations and simplify to get the Pythagorean theorem:

$$a^2 + b^2 = c \cdot c_2 + c \cdot c_1 = c(c_2 + c_1) = c^2. \qquad \square$$

Here is a first application of the Pythagorean theorem. The triangle inequality says that if a, b, and c are the lengths of the sides of a triangle, then

$$|a - b| < c < a + b.$$

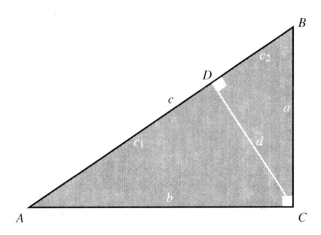

Figure 6. A proof of the Pythagorean theorem via similarity.

What about the converse? If a, b, and c are positive reals satisfying the triangle inequality conditions, can we put together a triangle with sides of those lengths?

Theorem 13.6 *Let a, b, and c be positive real numbers. Suppose that c is the largest of them and that $c < a + b$. Then there is a triangle with sides of length a, b, and c.*

Proof Start with a segment AB whose length is c. We need to place a third point C so that it is a distance a from B and b from A (according to SSS, there is only one such triangle up to congruence, so this may not be too easy). The strategy is to build the triangle out of two right triangles. Mark D on $\overrightarrow{AB}$ and label $d = |AD|$. Mark C on one of the rays with endpoint D

that is perpendicular to AB and label $e = |CD|$. Then $\triangle ACD$ and $\triangle BCD$ are right triangles. Furthermore, by sliding D and C along their respective rays, we can make d and e any positive numbers.

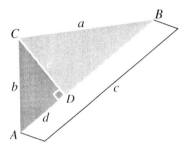

Figure 7. Building a triangle out of two right triangles.

We need to see if it is possible to position the two so that $|AC| = b$ and $|BC| = a$. To get $|AC| = b$, we will need
$$d^2 + e^2 = b^2.$$
To get $|BC| = a$, we will need
$$(c - d)^2 + e^2 = a^2.$$
With a little algebra, we can find d and e. Solve for e^2 in the equations and set them equal:
$$b^2 - d^2 = e^2 = a^2 - (c - d)^2$$
$$b^2 - d^2 = a^2 - c^2 + 2cd - d^2$$
$$b^2 = a^2 - c^2 + 2cd$$
$$(b^2 - a^2 + c^2)/2c = d.$$
Since we required $c > a$, d will be positive. Substitute to find e:
$$e^2 = b^2 - d^2 = b^2 - \left(\frac{b^2 - a^2 + c^2}{2c}\right)^2.$$
Here is the essential part. Because we will have to take a square root to find e, the right hand side of this equation has to be positive; if not the equation has no solution and the triangle cannot be built. Let's go back to see if the triangle inequality condition on the three sides will help:
$$c < a + b$$
$$c - b < a$$
$$(c - b)^2 < a^2$$
$$c^2 - 2bc + b^2 < a^2$$
$$c^2 - a^2 + b^2 < 2bc$$
$$(c^2 - a^2 + b^2)/2c < b$$
$$((c^2 - a^2 + b^2)/2c)^2 < b^2$$
$$0 < b^2 - ((c^2 - a^2 + b^2)/2c)^2$$

which is exactly what we want. As long as $c < a + b$, then, a value for e can be found, and the triangle can be built. □

13.3 Exercises

13.2. Prove that similarity of polygons is an equivalence relation.

13.3. State and prove the SASAS similarity theorem for convex quadrilaterals.

13.4. State and prove the ASASA similarity theorem for convex quadrilaterals.

13.5. State and prove the SSSSA similarity theorem for convex quadrilaterals.

13.6. Is AAAA a valid similarity theorem for quadrilaterals?

13.7. Suppose that $P_1 P_2 \cdots P_n$ and $Q_1 Q_2 \cdots Q_n$ are similar polygons, and $P_1 P_j$ is a diagonal that lies entirely in the interior of $P_1 P_2 \cdots P_n$ (other than its endpoints). Prove that $Q_1 Q_j$ lies entirely in the interior of $Q_1 Q_2 \cdots Q_n$ (other than its endpoints).

13.8. Suppose that $P_1 P_2 \cdots P_n$ and $Q_1 Q_2 \cdots Q_n$ are similar polygons, and $P_1 P_j$ is a diagonal that lies entirely in the interior of $P_1 P_2 \cdots P_n$ as in the last exercise. Prove that $P_1 P_2 \cdots P_j \sim Q_1 Q_2 \cdots Q_j$.

13.9. For $\triangle ABC$, the lines

- through A, parallel to BC;
- through B, parallel to AC; and
- through C, parallel to AB

intersect to form a triangle. Prove that the triangle is similar to $\triangle ABC$.

13.10. Let $\square ABCD$ be a trapezoid, with $AB \parallel CD$. Let M_1 be the midpoint of AB, M_2 the midpoint of CD, N_1 the midpoint of BC, and N_2 the midpoint of AD. Then the segments $M_1 M_2$ and $N_1 N_2$ cut $\square ABCD$ into four pieces. Under what conditions (on the original $\square ABCD$) are the four pieces each similar to the original $\square ABCD$?

13.11. Let $\triangle ABC$ be an isosceles triangle with $AB \simeq AC$. Label D so that $B * C * D$ and so that $AC \simeq CD$ (so that $\triangle ACD$ is isosceles). What angle measure or measures must $\angle B$ have in order for the large triangle $\triangle DAB$ to be similar to $\triangle ABC$?

13.12. Consider a rectangle $\square ABCD$ with $|AB| < |BC|$, and suppose that it has the property that if a square $\square ABEF$ is constructed inside $\square ABCD$, then the remaining rectangle $\square ECDF$ is similar to the original $\square ABCD$. A rectangle with this property is called a golden rectangle. Find the value of $|BC|/|AB|$, a value known as the golden ratio.

13.13. Let $\triangle ABC$ be a right triangle with right angle at C, and suppose that $|AB| = a$ and $|BC| = a/2$. Let M be the midpoint of AB. The line through M that is perpendicular to AB intersects AC at a point N. What is $|MN|$ in terms of a?

13.14. Let $\triangle ABC$ be an equilateral triangle with sides of length a. Prove that the three segments that connect each vertex to the midpoint of the opposite side intersect at a point P. What is the distance from a vertex to P in terms of a?

The six trigonometric functions can be defined, for values of θ between 0 and 90°, as ratios of pairs of sides of a right triangle with an interior angle θ. If the length of the hypotenuse is h, the length of the leg adjacent to θ is a, and the length of the leg opposite θ is o, then these functions are defined as

$$\text{sine: } \sin(\theta) = o/h$$
$$\text{cosine: } \cos(\theta) = a/h$$
$$\text{tangent: } \tan(\theta) = o/a$$
$$\text{cotangent: } \cot(\theta) = a/o$$
$$\text{secant: } \sec(\theta) = h/a$$
$$\text{cosecant: } \csc(\theta) = h/o.$$

In chapter 21, we will extend these functions to accept all real values. In the meantime, we will have need of these functions, and some of the basic identities relating them as listed in the exercises below.

13.15. Verify that the six trigonometric functions are well-defined. That is, show that it does not matter which right triangle with interior angle θ you choose—these six ratios will not change.

13.16. Verify the Pythagorean identities (for values of θ between 0 and 90°).

$$\sin^2\theta + \cos^2\theta = 1$$
$$1 + \tan^2\theta = \sec^2\theta$$
$$1 + \cot^2\theta = \csc^2\theta$$

13.17. Verify the cofunction identities (for values of θ between 0 and 90°).

$$\sin(90° - \theta) = \cos\theta$$
$$\cos(90° - \theta) = \sin\theta$$
$$\tan(90° - \theta) = \cot\theta$$
$$\cot(90° - \theta) = \tan\theta$$
$$\sec(90° - \theta) = \csc\theta$$
$$\csc(90° - \theta) = \sec\theta$$

14

Circles

This is the first of two chapters that deal with circles. This chapter gives some basic definitions and some elementary theorems, the most important of which is the inscribed angle theorem. In the next, we will tackle the important issue of circumference and see how that leads to the radian angle measurement system.

14.1 Definitions

Circles are basic geometric objects, but are making a late entrance in this book. In fact, circles can be defined very early in neutral geometry, the only prerequisites being points, segments, and congruence. However, most of the interesting theorems involving circles are specific to Euclidean geometry (or, in the case of the circumference formula, are different in Euclidean and non-Euclidean geometry). In any case, now is the time for a proper definition.

Definition 14.1 For a point O and positive real number r, the *circle* with *center* O and *radius* r is the set of points that are a distance r from O.

A few observations.

(1) A circle is a set, but we say that a point in the set is *on* the circle, rather than *in* the circle. The term *in* is reserved for points in the interior of the circle.
(2) In the definition, the radius is a number. We often talk about the radius as a geometric entity though, as one of the segments from the center to a point on the circle.
(3) We tend to think of the center of a circle as a fundamental part of it, but it is not a point on the circle.
(4) It is not common to talk about circles as congruent or not congruent. If you did, you would say that two circles are congruent if and only if they have the same radius.

Before we get into anything complicated, let's make some related definitions.

Definition 14.2 A line segment with both endpoints on a circle is called a *chord* of the circle. A chord that passes through the center of the circle is called a *diameter* of the circle.

Just like the term radius, the term diameter plays two roles, a numerical one and a geometric one. The diameter in the numerical sense is just the length of the diameter in the geometric sense (and is twice the radius).

Figure 1. Twelve chords, three diameters, and four central angles.

Definition 14.3 An angle with its vertex at the center of a circle is called a *central angle* of the circle.

We will see in the next section that a line intersects a circle at most twice. Therefore, if AB is a chord of a circle, then all the points of the circle other than A and B are on one side or the other of $\overline{AB}$. Thus $\overline{AB}$ separates the points into two sets, called *arcs* of the circle. There are three types of arcs—semicircles, major arcs, and minor arcs—depending on where the chord crosses the circle.

Definition 14.4 Let AB be a diameter of a circle $\mathcal{C}$. All the points of $\mathcal{C}$ that are on one side of $\overline{AB}$, together with the endpoints A and B, form a *semicircle*.

Each diameter divides the circle into two semicircles, overlapping at the endpoints A and B.

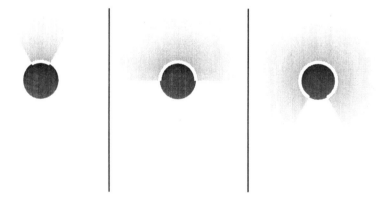

Figure 2. (l) A minor arc, (c) a semicircle, and (r) a major arc.

Definition 14.5 Let AB be a chord of a circle $\mathcal{C}$ that is not a diameter, and let O be the center of the circle. The points of $\mathcal{C}$ that are on the same side of $\overline{AB}$ as O, together with the endpoints A and B, form a *major arc*. The points of $\mathcal{C}$ that are on the opposite side of $\overline{AB}$ from O, together with the endpoints A and B, form a *minor arc*.

Like the two semicircles defined by a diameter, the major arc and minor arc defined by a chord overlap only at the endpoints A and B. We will use the notation $\overparen{AB}$ for arcs, including diameters. The notation is ambiguous, as it does not specify whether we are referring to the minor arc or to the major arc. In practice, this ambiguity rarely causes problems. Furthermore, we are usually interested in the minor arc.

There is a simple, direct, and important correspondence between arcs and central angles. Recall from chapter 9, that two rays with a common endpoint define not one, but two angles, a proper angle and a reflex angle, that are related to the minor and major arcs as described in the next exercise.

Exercise 14.1 Let AB be a chord of a circle $\mathcal{C}$ with center O that is not a diameter. Prove that the points of the minor arc $\widehat{AB}$ are A, B, and the points of $\mathcal{C}$ in the interior of the proper angle $\angle AOB$. Prove that the points of the major arc $\widehat{AB}$ are A, B, and the points of $\mathcal{C}$ in the interior of the reflex angle $\angle AOB$ (that is, the points exterior to the proper angle).

14.2 Intersections

Circles are different from the shapes we have studied up this point because they are not built out of lines or line segments. Circles share at least one characteristic with simple polygons though: they have an interior and an exterior.

Definition 14.6 For a circle $\mathcal{C}$ with center O and radius r, and for a point P that is not on $\mathcal{C}$,

- if $|OP| < r$, then P is inside $\mathcal{C}$,
- if $|OP| > r$, then P is outside $\mathcal{C}$.

The set of points inside the circle is the *interior* of the circle. The set of points outside the circle is its *exterior*.

Just like simple polygons, circles separate their interior and exterior. To get a better sense of that, we need to look at how circles intersect other geometric objects.

Theorem 14.7 *A line will intersect a circle in 0, 1, or 2 points.*

Proof Let O be the center of a circle $\mathcal{C}$ of radius r, and let ℓ be a line. It is easy to find points on ℓ that are far from $\mathcal{C}$, but are there points on ℓ that are close to $\mathcal{C}$? The easiest way to figure out how close ℓ gets to $\mathcal{C}$ is to look at the closest point on ℓ to the center O. We have seen

Figure 3. A circle and line that do not intersect.

(lemma 10.3) that the closest point to O on ℓ is the foot of the perpendicular, call it Q. There are three possibilities.

Zero intersections: If $|OQ| > r$, then the other points of ℓ are even farther from O, so none of the points on ℓ can be on $\mathcal{C}$.

One intersection: If $|OQ| = r$, then Q is an intersection, but it is the only intersection because the other points on ℓ are farther away from O.

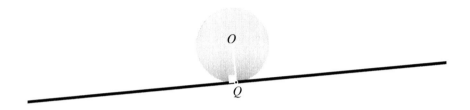

Figure 4. A circle and line that intersect at a single point.

Two intersections: If $|OQ| < r$, then the line lies both inside and outside the circle. We just need to find where the line crosses into, and then back out of, the circle. The idea is to relate a point's distance from O to its distance from Q, and we can do that with the Pythagorean theorem. If P is a point on ℓ other than Q, then $\triangle OQP$ will be a right triangle with side lengths that are related by the Pythagorean theorem:

$$|OQ|^2 + |QP|^2 = |OP|^2.$$

If P is on the circle, $|OP| = r$. Therefore $|PQ| = \sqrt{r^2 - |OQ|^2}$. Since $|OQ| < r$, this expression is a positive real number. There are then exactly two points on ℓ, one on each side of Q, that are this distance from Q.

Figure 5. A circle and line that intersect at two points. □

A line that intersects a circle once (at the foot of the perpendicular) is called a *tangent* line to the circle. A line that intersects a circle twice is called a *secant* line of the circle.

14.2 Intersections

There is a important corollary that turns this last theorem about lines into a statement about segments.

Corollary 14.8 *If a point P is inside a circle, and a point Q is outside it, then the segment PQ intersects the circle (exactly once).*

Proof Label the center of the circle O. From the last theorem, we know that $\overline{PQ}$ intersects the circle twice, and that the two intersections are separated by F, the foot of the perpendicular to PQ through O. The important intersection here is the one that is on the same side of the foot of the perpendicular as Q, call it R. According to the Pythagorean theorem (with $\triangle OFQ$ and $\triangle OFR$),

$$|FQ| = \sqrt{|OQ|^2 - |OF|^2} \text{ and } |FR| = \sqrt{|OR|^2 - |OF|^2}.$$

Since $|OQ| > |OR|$, $|FQ| > |FR|$, which places R between F and Q. We cannot say whether P and Q are on the same side of F, though. If they are on opposite sides of F, then $P * F * R * Q$, so R is between P and Q as needed. If P and Q are on the same side of F, then we need to look at the right triangles $\triangle OFP$ and $\triangle OFR$. They tell us that

$$|FP| = \sqrt{|OP|^2 - |OF|^2} \text{ and } |FR| = \sqrt{|OR|^2 - |OF|^2}.$$

Since $|OP| < |OR|$, $|FP| < |FR|$, which places P between F and R. Finally, if $F * P * R$ and $F * R * Q$, then R has to be between P and Q. □

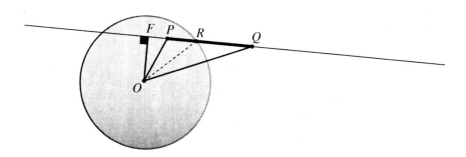

Figure 6. Intersection of a circle and a segment.

There is another important question of intersections: how can two circles intersect?

Theorem 14.9 *Two circles intersect at 0, 1, or 2 points.*

Proof Three factors come in to play here: the radii of the circles and the distance between their centers. Label

O_1, O_2 : the centers of the two circles,
r_1, r_2 : the radii of the two circles, and
c : the distance between the centers.

An intersection point X will have to be a distance r_1 from O_1 and r_2 from O_2. We know that $|O_1 O_2| = c$, so the question is: can we build a triangle with sides of lengths r_1, r_2, and c? We studied this question at the end of the last chapter. There are three possibilities.

Two intersections: when $|r_1 - r_2| < c < r_1 + r_2$.
In this case, the proposed side lengths satisfy the triangle inequality conditions. Therefore, there are exactly two triangles, $\triangle O_1 X O_2$ and $\triangle O_1 Y O_2$, one on each side of $O_1 O_2$, with sides of the required lengths. There are then exactly two intersections of the circles.

Figure 7. Two intersections.

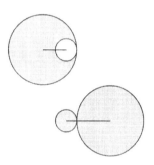

One intersection: when $c = |r_1 - r_2|$ or $c = r_1 + r_2$.
In these limiting cases, the triangle devolves into a line segment and the two intersections merge. In the first, either $O_1 * O_2 * X$ or $X * O_1 * O_2$, depending on which radius is larger. In the second, $O_1 * X * O_2$.

Figure 8. One intersection.

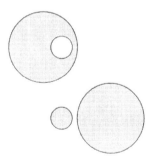

Zero intersections: when $c < |r_1 - r_2|$ or $c > r_1 + r_2$.
In this case, the proposed side lengths do not satisfy the triangle inequality conditions. No triangle can be formed with sides of these lengths, so the two circles cannot intersect. In the first case, one circles lies entirely inside the other. In the second, they are separated from one another.

Figure 9. No intersections. □

As you established in exercise 14.1, there is a correspondence between central angles and arcs. In the next chapter we will establish a metrical connection between the measure of the central angle and the length of the corresponding arc (which is the basis for radian measure). In the meantime, the correspondence is a helpful way to simplify illustrations. Using an arc

14.3 The inscribed angle theorem

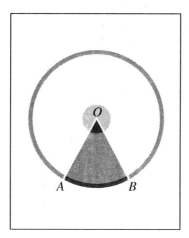

Figure 10. The major arc corresponds to the reflex angle $\angle AOB$. The minor arc corresponds to the proper angle $\angle AOB$.

to indicate a central angle keeps a picture from getting too crowded around the center of the circle.

14.3 The inscribed angle theorem

In this section we will prove the inscribed angle theorem, a result that is indispensable when working with circles. I suspect that this theorem is the most elementary result of Euclidean geometry that is generally not known to the average calculus student. Before stating the theorem, we must define an inscribed angle, the subject of the theorem.

Definition 14.10 If A, B, and C are points on a circle, then $\angle ABC$ is an *inscribed angle* on the circle.

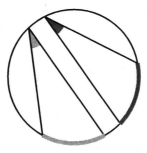

Figure 11. Two angles inscribed in a circle.

Given an inscribed angle $\angle ABC$, either proper or reflex, the points A and C are the endpoints of two arcs, either a minor and a major arc or two semicircles. Excluding the endpoints, one of the arcs will be contained in the interior of ∠ABC. We say, then, that ∠ABC is inscribed on the arc $\overset{\frown}{AC}$. The inscribed angle theorem describes the relationship between an inscribed angle and the central angle on the same arc.

Theorem 14.11 *(The inscribed angle theorem) If ∠ABC is an inscribed angle on a circle with center O, then*

$$(\angle ABC) = \frac{1}{2}(\angle AOC).$$

The proof of the inscribed angle theorem is a good demonstration of the benefits of starting with an easy case. There are three cases in the proof, depending on the location of the vertex B relative to the lines $\overline{OA}$ and $\overline{OC}$. The first case establishes the theorem for two particular locations of B, where things work out easily. But the special cases are the key that unlocks everything else.

Proof Case 1. *Where B is the intersection of $\overrightarrow{OC}^{op}$ with the circle, or where B is the intersection of $\overrightarrow{OA}^{op}$ with the circle.* We will prove the result where B is on $\overrightarrow{OC}^{op}$. To deal with the other possibility, where B is on $\overrightarrow{OA}^{op}$, just switch the letters A and C. Label ∠AOB as ∠1 and ∠AOC as ∠2. These angles are supplementary, so

(i) $(\angle 1) + (\angle 2) = 180°.$

The angle sum of △AOB is 180°, but in that triangle ∠A and ∠B are opposite congruent segments (radii), so by the isosceles triangle theorem they are congruent. Therefore

(ii) $2(\angle B) + (\angle 1) = 180°,$

and if we subtract (ii) from (i), we get $(\angle 2) - 2(\angle B) = 0$, so $(\angle AOC) = 2(\angle ABC)$.

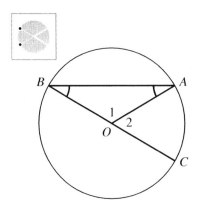

Figure 12. The inscribed angle theorem, case 1: an inscribed angle with one ray along a diameter.

14.3 The inscribed angle theorem

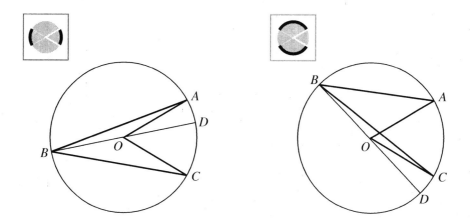

Figure 13. The inscribed angle theorem, case 2.   Figure 14. The inscribed angle theorem, case 3.

Case 2. Where B is in the interior of $\angle AOC$, or where B is in the interior of the angle formed by $\overrightarrow{OA^{op}}$ and $\overrightarrow{OC^{op}}$, or when $A * O * C$. There are three cases here—in the first the central angle is reflex, in the second it is proper, and in the third it is a straight angle—but the proof is the same for all of them. In each case, $\overline{OB}$ splits both the inscribed and the central angles. To identify the four angles, label one more point: D is the second intersection of $\overline{OB}$ with the circle (so BD is a diameter of the circle). Using angle addition in conjunction with the previous results,

$$(\angle AOC) = (\angle AOD) + (\angle DOC)$$
$$= 2(\angle ABD) + 2(\angle DBC)$$
$$= 2((\angle ABD) + (\angle DBC))$$
$$= 2(\angle ABC).$$

Case 3. Where B is in the interior of the angle formed by $\overrightarrow{OA}$ and $\overrightarrow{OC^{op}}$, or where B is in the interior of the angle formed by $\overrightarrow{OC}$ and $\overrightarrow{OA^{op}}$. As in the last case, label D so that BD is a diameter. The difference this time is that we need to use angle subtraction instead of angle addition. Since subtraction is less symmetric than addition, the two cases will differ slightly (in terms of lettering). In the first,

$$(\angle AOC) = (\angle AOD) - (\angle DOC)$$
$$= 2(\angle ABD) - 2(\angle DBC)$$
$$= 2((\angle ABD) - (\angle DBC))$$
$$= 2(\angle ABC).$$

To get the second, just switch A and C. □

There are two important and immediate corollaries to this theorem. First, because all inscribed angles on a given arc share the same central angle,

Corollary 14.12 *All inscribed angles on a given arc are congruent.*

Second, the special case where the central angle $\angle AOC$ is a straight angle, so that the inscribed angle $\angle ABC$ is a right angle, is important enough to earn its own name.

Corollary 14.13 *(Thales' theorem) If C is a point on a circle with diameter AB (and C is neither A nor B), then $\triangle ABC$ is a right triangle.*

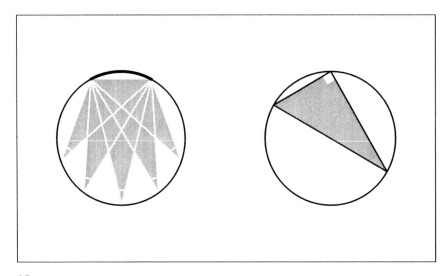

Figure 15. (l) Five congruent angles inscribed on the same arc. (r) A right angle inscribed on a semicircle.

Thales' theorem is often presented as an if-and-only-if statement, and both directions are quite useful.

Exercise 14.2 Prove that if $\triangle ABC$ is a right triangle with right angle at C, then C is a point on the circle with diameter AB.

14.4 Applications of the inscribed angle theorem

Using the inscribed angle theorem, we can establish several nice relationships between chords, secants, and tangents associated with a circle. We will look at two of these results to end this chapter.

Theorem 14.14 *(The chord-chord formula) Let $\mathscr{C}$ be a circle with center O. Suppose that AC and BD are chords of $\mathscr{C}$, and suppose that they intersect at a point P. Label the angle of intersection, $\theta = \angle APD \simeq \angle BPC$. Then*

$$(\theta) = \frac{(\angle AOD) + (\angle BOC)}{2}.$$

14.4 Applications of the inscribed angle theorem

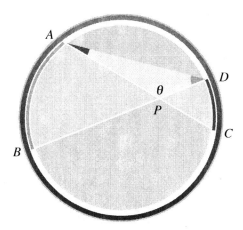

Figure 16. Two intersecting chords in a circle.

Proof The angle θ is an interior angle of $\triangle APD$, so

$$(\theta) = 180° - (\angle A) - (\angle D).$$

$\angle A$ and $\angle D$ are inscribed angles: $\angle A$ is inscribed on the arc $\widehat{CD}$ and $\angle D$ is inscribed on the arc $\widehat{AB}$. According to the inscribed angle theorem, they are half the size of the corresponding central angles, so

$$\begin{aligned}(\theta) &= 180° - \tfrac{1}{2}(\angle COD) - \tfrac{1}{2}(\angle AOB) \\ &= \tfrac{1}{2}(360° - (\angle COD) - (\angle AOB)).\end{aligned}$$

This relates θ to central angles, but alas, not to the central angles in the formula. If we add all four central angles around O, though,

$$(\angle AOB) + (\angle BOC) + (\angle COD) + (\angle DOA) = 360°$$

and so

$$(\angle BOC) + (\angle DOA) = 360° - (\angle COD) - (\angle AOB).$$

Now substitute to get the desired formula:

$$(\theta) = \tfrac{1}{2}((\angle BOC) + (\angle DOA)). \qquad \square$$

According to the chord-chord formula, as long as the intersection point P is inside the circle, θ can be computed as the average of two central angles. What would happen if P moved outside the circle? Then we would not be talking about chords, since chords stop at the circle boundary, but about the secant lines containing them.

Theorem 14.15 *(The secant-secant formula) Suppose that A, B, C, and D are points on a circle arranged so that □ABCD is a simple quadrilateral, and that the secant lines AB and CD intersect at a point P that is outside the circle. Label the angle of intersection, ∠APD, as θ. Let ∠AOD and ∠BOC be taken so that P is in their interiors (that is, the angles may be reflex). If P occurs on the same side of AD as B and C, then*

$$(\theta) = \frac{(\angle AOD) - (\angle BOC)}{2}.$$

If P occurs on the same side of BC as A and D, then

$$(\theta) = \frac{(\angle BOC) - (\angle AOD)}{2}.$$

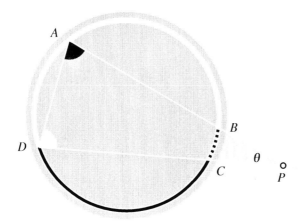

Figure 17. Two lines that intersect outside of the circle.

Proof There is obviously a great deal of symmetry between the two cases, so let me just address the first. The same principles apply here as in the last proof. Angle θ is an interior angle of △APD, so

$$(\theta) = 180° - (\angle A) - (\angle D).$$

∠A and ∠D are inscribed angles— ∠A is inscribed on arc $\widehat{BD}$ and ∠D is inscribed on arc $\widehat{AC}$. We need to use the inscribed angle theorem to relate the angles to central angles, and in this case, the central angles overlap, so we will need to break them down further, but the rest is straightforward.

$$\begin{aligned}(\theta) &= 180° - \tfrac{1}{2}(\angle BOD) - \tfrac{1}{2}(\angle AOC) \\ &= \tfrac{1}{2}[360° - (\angle BOD) - (\angle AOC)] \\ &= \tfrac{1}{2}[360° - (\angle BOC) - (\angle COD) - (\angle AOB) - (\angle BOC)] \\ &= \tfrac{1}{2}[[360° - (\angle AOB) - (\angle BOC) - (\angle COD)] - (\angle BOC)] \\ &= \tfrac{1}{2}[(\angle AOD) - (\angle BOC)].\end{aligned}$$

□

14.5 Exercises

14.3. Prove that no line is entirely contained in any circle.

14.4. Prove that a circle is convex. That is, prove that if points P and Q are inside a circle, then all the points on the segment PQ are inside the circle.

14.5. Prove that for any circle there is a triangle entirely contained in it, so all the points of the triangle are inside the circle.

14.6. Prove that for any circle there is a triangle which entirely contains it, so all the points of the circle are in the interior of the triangle.

14.7. Let P be a point on a line ℓ, and let Q be a point that is not on ℓ. Prove that there is a unique circle $\mathcal{C}$ that passes through Q and is tangent to ℓ at P.

14.8. Suppose that $\square ABCD$ is cyclic, so there is a circle through all four of its vertices. Prove that $(\angle A) + (\angle C) = 180°$ and $(\angle B) + (\angle D) = 180°$.

14.9. Given a circle $\mathcal{C}$ with center O and points A and C on $\mathcal{C}$, prove that:
- if a point B is inside $\mathcal{C}$, then $(\angle ABC) > \frac{1}{2}(\angle AOC)$, and
- if a point B is outside $\mathcal{C}$, then $(\angle ABC) < \frac{1}{2}(\angle AOC)$.

Note that you will have to think about when $\angle ABC$ and $\angle AOC$ should be taken as the reflex angle.

14.10. Let $\angle ABC$ be an inscribed angle on a circle. Prove that, excluding the endpoints, exactly one of the two arcs $\widehat{AC}$ lies in the interior of $\angle ABC$.

14.11. Let $\mathcal{C}$ be a circle and P be a point outside of it. Prove that there are exactly two lines that pass through P and are tangent to $\mathcal{C}$. Let Q and R be the points of tangency for the two lines. Prove that PQ and PR are congruent.

14.12. Let A be a point outside of a circle $\mathcal{C}$ with center O. Let B be the point on $\mathcal{C}$ so that $A * O * B$. There are two lines through A that are tangent to $\mathcal{C}$. Label their intersections with $\mathcal{C}$ as P and Q. What is the relationship between the angle measures $(\angle PAQ)$ and $(\angle PBQ)$?

14.13. Let $\mathcal{P}$ be a regular polygon. Prove that there is a unique circle $\mathcal{C}$ contained in $\mathcal{P}$ so that $\mathcal{C}$ is tangent to each of the sides of $\mathcal{P}$. This is called the circle *inscribed* in $\mathcal{P}$.

14.14. Suppose that a parallelogram is inscribed in a circle. Prove that the parallelogram must be a rectangle.

14.15. Let $\square ABCD$ be a trapezoid with exactly one set of parallel sides that is inscribed in a circle. Prove that the non-parallel sides of $\square ABCD$ must be congruent.

14.16. Suppose that circles $\mathcal{C}_1$ with center O_1 and $\mathcal{C}_2$ with center O_2 intersect at two points P and Q. Prove that $\angle O_1 P O_2 \simeq \angle O_1 Q O_2$.

14.17. Let $\triangle ABC$ be equilateral, with sides of length a. Let $\mathcal{C}_A$, $\mathcal{C}_B$, and $\mathcal{C}_C$ be three circles, each with radius $a/2$, centered at A, B, and C. The three circles are tangent to one another and determine a curved but approximately triangular region in the center of $\triangle ABC$. Show that a circle can be inscribed in the region, tangent to $\mathcal{C}_A$, $\mathcal{C}_B$, and $\mathcal{C}_C$. What is the radius of this circle? Hint: see exercise 13.14.

14.18. The "tangent-tangent" formula. Let P be a point that is outside of a circle $\mathcal{C}$. Consider the two tangent lines to $\mathcal{C}$ that pass through P and let A and B be the points of tangency between those lines and the circle. Prove that
$$(\angle APB) = \frac{(\angle 1) - (\angle 2)}{2}$$
where $\angle 1$ is the reflex central angle corresponding to the major arc $\overgroup{AB}$ and $\angle 2$ is the proper central angle corresponding to the minor arc $\overgroup{AB}$.

14.19. Let AC and BD be two chords of a circle that intersect at a point P inside the circle. Prove that
$$|AP| \cdot |CP| = |BP| \cdot |DP|.$$

14.6 References

I learned of the chord-chord, secant-secant, and tangent-tangent formulas in the Wallace and West book *Roads to Geometry* [WW04]. They use the names two-chord angle theorem, two-secant angle theorem, and two-tangent angle theorem.

15
Circumference

15.1 A theorem on perimeters

In chapter 9 on polygons, we defined the perimeter of a polygon $\mathcal{P} = P_1 \cdots P_n$ as

$$|\mathcal{P}| = \sum_{i=1}^{n} |P_i P_{i+1}|,$$

but then left it at that. In this lesson we are going to use perimeters of cyclic polygons to find the circumference of the circle. Along the way, we will need to use the following result that compares the perimeters of two convex polygons when one is contained in the other. We will prove the theorem by a process of "shaving down" the sides of the larger polygon until we reach the smaller one.

Theorem 15.1 *If $\mathcal{P}$ and $\mathcal{Q}$ are convex polygons and all the points of $\mathcal{P}$ are on or inside $\mathcal{Q}$, then $|\mathcal{P}| \leq |\mathcal{Q}|$.*

Proof Some of the edges of $\mathcal{P}$ may run along the edges of $\mathcal{Q}$, but unless $\mathcal{P} = \mathcal{Q}$, at least one edge of $\mathcal{P}$ must pass through the interior of $\mathcal{Q}$. Let s be one of those interior edges. The line containing s intersects $\mathcal{Q}$ twice—call the intersections a and b—dividing $\mathcal{Q}$ into two smaller polygons that share the side ab, one on the same side of s as $\mathcal{P}$, the other on the opposite side. Define a new polygon $\mathcal{Q}_1$ that consists of

- points of $\mathcal{Q}$ on the same side of s as $\mathcal{P}$, and
- points on the segment ab.

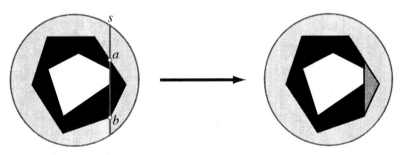

Figure 1. Shaving a corner off of a polygon.

There are two things to notice about Q_1. First, Q_1 and P have one more coincident side (the side s) than Q and P had. Second, the portions of Q and Q_1 on the side of s with P are identical, so the segments making up that part contribute the same amount to their perimeters. On the other side, though, the path that Q takes from a to b is longer than the direct route along the segment ab of Q_1 (because of the triangle inequality). Combining the two parts, we get $|Q_1| \leq |Q|$.

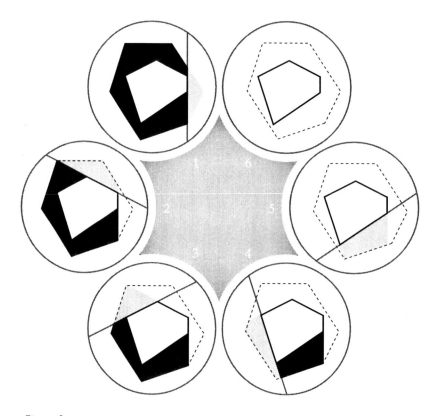

Figure 2. One at a time, shave the edges of the outer polygon down to the inner one.

Now we can repeat this process with P and Q_1, generating Q_2 with even smaller perimeter than Q_1 and another coincident side with P. And again, to get Q_3. Eventually, though, after say m steps, we run out of sides that pass through the interior, at which point $P = Q_m$. Then

$$|P| = |Q_m| \leq |Q_{m-1}| \leq \cdots \leq |Q_2| \leq |Q_1| \leq |Q|.$$ □

15.2 Circumference

The purpose of this chapter is to understand the connection between the radius of a circle, and the distance around it, its circumference. The final result, the formula $C = 2\pi r$, ranks with the Pythagorean theorem in celebrity, although I think it is an often misunderstood celebrity. Let's

15.2 Circumference

be clear about what this equation is not. It is not an equation comparing two known quantities C and $2\pi r$. Instead, it is saying that the ratio $C/(2r)$ is a constant. The constant is what we have come to call π.

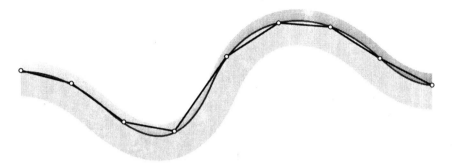

Figure 3. An approximation of an arc by segments.

To define the circumference of a circle, let's borrow an idea from calculus, the idea of approximating a curve by straight line segments, and then refining the approximation by increasing the number of segments. For a circle $\mathcal{C}$, the approximating line segments will be the edges of a simple cyclic polygon $\mathcal{P}$ inscribed in the circle. We will want the circumference of $\mathcal{C}$ to be bigger than the perimeter of $\mathcal{P}$. We should also expect that by adding in additional vertices to $\mathcal{P}$, we should be able to get the perimeter of $\mathcal{P}$ as close as we want to the circumference of $\mathcal{C}$. All this suggests that to get the circumference of $\mathcal{C}$, we need to find out how large the perimeters of inscribed polygons can be.

Definition 15.2 The circumference of a circle $\mathcal{C}$, written $|\mathcal{C}|$, is

$$|\mathcal{C}| = \sup\left\{|\mathcal{P}|\,\Big|\,\mathcal{P} \text{ is a simple cyclic polygon inscribed in } \mathcal{C}\right\}.$$

There is nothing in the definition to guarantee that the supremum exists. It is conceivable that the lengths of the approximating perimeters might grow with bound. An example is the *Koch snowflake*. It is constructed as follows. Take an equilateral triangle with sides of length 1. The perimeter of this triangle is 3. Divide each side into thirds. On each middle third, build an equilateral triangle by adding two more sides then remove the original side. You have made a shape with $3 \cdot 4$ sides, each with length $1/3$, for a perimeter of 4. Now iterate: divide the sides into thirds; build equilateral triangles on each middle third, and remove the base. That will make $3 \cdot 16$ sides of length $1/9$, for a perimeter of $16/3$. After the next step, there are $3 \cdot 64$ sides of length $1/27$ for a perimeter of $64/9$. Generally, after n iterations, there are $3 \cdot 4^n$ sides of length $1/3^n$ for a total perimeter of $4^n/3^{n-1}$, and

$$\lim_{n\to\infty} \frac{4^n}{3^{n-1}} = \lim_{n\to\infty} 3\left(\frac{4}{3}\right)^n = \infty.$$

The Koch snowflake, which is the limiting shape in this process, has infinite perimeter! The first thing we need to do, then, is to make sure that circles are better behaved than this.

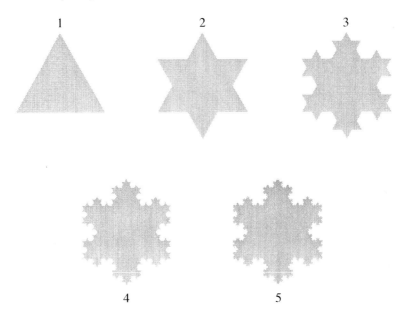

Figure 4. The first few steps in the construction of the Koch snowflake.

Theorem 15.3 *If $\mathcal{C}$ is a circle of radius r, then $|\mathcal{C}| \leq 8r$.*

Proof The first step is to build a circumscribing square around $\mathcal{C}$, the smallest square that still contains $\mathcal{C}$. Begin by choosing two perpendicular diameters d_1 and d_2. Each will intersect $\mathcal{C}$ twice, for a total of four intersections, P_1, P_2, P_3, and P_4. For each i between 1 and 4, let t_i be the tangent line to $\mathcal{C}$ at P_i. The tangents intersect to form the circumscribing square. The length of each side of the square is equal to the diameter of $\mathcal{C}$, so the perimeter of the square is $4 \cdot 2r = 8r$.

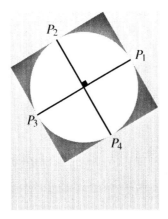

Figure 5. The circumscribing square.

15.2 Circumference

Now we turn to the theorem we proved to start this chapter. Each simple cyclic polygon inscribed in $\mathcal{C}$ is a convex polygon contained in the circumscribing square. Therefore the perimeter of any such approximating polygon is bounded above by $8r$. Remember that we have defined $|\mathcal{C}|$ to be the supremum of all the approximating perimeters, so it cannot exceed $8r$ either. □

Now that we know that a circle has a circumference, the next step is to find a way to calculate it. The key to that is the next theorem.

Theorem 15.4 *The ratio of the circumference of a circle to its radius is a constant.*

Proof Let's suppose that the ratio is not a constant, so that there are two circles $\mathcal{C}_1$ and $\mathcal{C}_2$ with centers O_1 and O_2 and radii r_1 and r_2, but with unequal ratios

$$|\mathcal{C}_1|/r_1 > |\mathcal{C}_2|/r_2.$$

As we have defined circumference, there are approximating cyclic polygons to $\mathcal{C}_1$ whose perimeters are arbitrarily close to its circumference. In particular, there is an approximating cyclic polygon $\mathcal{P} = P_1 P_2 \ldots P_n$ for $\mathcal{C}_1$ so that

$$|\mathcal{P}|/r_1 > |\mathcal{C}_2|/r_2.$$

We can construct a cyclic polygon $\mathcal{Q}$ on $\mathcal{C}_2$ that is similar to $\mathcal{P}$. Intuitively, we just need to scale $\mathcal{P}$ so that it fits in the circle $\mathcal{C}_2$. The formal construction is:

- Begin by placing a point Q_1 on circle $\mathcal{C}_2$.
- Locate Q_2 on $\mathcal{C}_2$ so that $\angle P_1 O_1 P_2$ is congruent to $\angle Q_1 O_2 Q_2$ (there are two choices for Q_2).
- Locate Q_3 on $\mathcal{C}_2$ and on the opposite side of $O_2 Q_2$ from Q_1 so that $\angle P_2 O_1 P_3$ is congruent to $\angle Q_2 O_2 Q_3$.
- Continue placing points on $\mathcal{C}_2$ in this fashion until Q_n has been placed to form the polygon $\mathcal{Q} = Q_1 Q_2 \ldots Q_n$. At the end of the process $\angle P_n O_1 P_1$ is congruent to $\angle Q_n O_2 Q_1$ by angle subtraction.

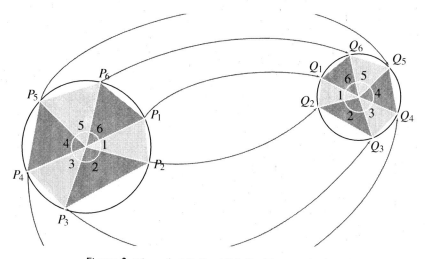

Figure 6. The ratio $|Q_i Q_{i+1}|/|P_i P_{i+1}|$ is a constant.

Then

$$\frac{|O_2 Q_i|}{|O_1 P_i|} = \frac{r_2}{r_1} = \frac{|O_2 Q_{i+1}|}{|O_1 P_{i+1}|} \text{ and } \angle Q_i O_2 Q_{i+1} \simeq \angle P_i O_1 P_{i+1},$$

so by SAS similarity, $\triangle Q_i O_2 Q_{i+1}$ is similar to $\triangle P_i O_1 P_{i+1}$. That gives us the ratio of the third sides of the triangle as

$$\frac{|Q_i Q_{i+1}|}{|P_i P_{i+1}|} = \frac{r_2}{r_1}.$$

Thus we can describe the perimeter of $\mathcal{Q}$ as

$$|\mathcal{Q}| = \sum_{i=1}^{n} |Q_i Q_{i+1}| = \sum_{i=1}^{n} \frac{r_2}{r_1} |P_i P_{i+1}| = \frac{r_2}{r_1} \sum_{i=1}^{n} |P_i P_{i+1}| = \frac{r_2}{r_1} |\mathcal{P}|.$$

That would mean that

$$\frac{|\mathcal{Q}|}{r_2} = \frac{|\mathcal{P}|}{r_1} > \frac{|\mathcal{C}_2|}{r_2}.$$

That is, $|\mathcal{Q}| > |\mathcal{C}_2|$, when in fact the circumference of $\mathcal{C}_2$ is supposed to be greater than the perimeter of any of its approximating cyclic polygons. □

Definition 15.5 The constant π is the ratio of the circumference of a circle to its diameter: $\pi = |\mathcal{C}|/(2r)$.

It is more convenient to express this definition in terms of a limit rather than a supremum. Fortunately, the supremum can be reached as a limit of perimeters of a sequence of regular polygons as follows. Arrange n angles each measuring $360°/n$ around the center of a circle $\mathcal{C}$. The rays of the angles intersect $\mathcal{C}$ n times, and the points P_i are the vertices of a regular n-gon, $\mathcal{P}_n = P_1 P_2 \ldots P_n$. The tangent lines to $\mathcal{C}$ at the neighboring points P_i and P_{i+1} intersect at a point Q_i. Taken together, the n points are the vertices of another regular n-gon $\mathcal{Q}_n = Q_1 Q_2 \ldots Q_n$. The polygon $\mathcal{P}_n$ is just one of the many cyclic polygons inscribed in $\mathcal{C}$ so $|\mathcal{P}_n| \leq |\mathcal{C}|$. The polygon $\mathcal{Q}_n$ circumscribes $\mathcal{C}$, and every cyclic polygon inscribed on $\mathcal{C}$ lies inside $\mathcal{Q}_n$, so $|\mathcal{Q}_n| \geq |\mathcal{C}|$.

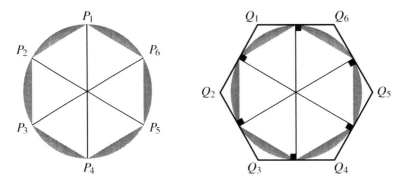

Figure 7. Regular inscribed (l) and circumscribed (r) hexagons.

15.2 Circumference

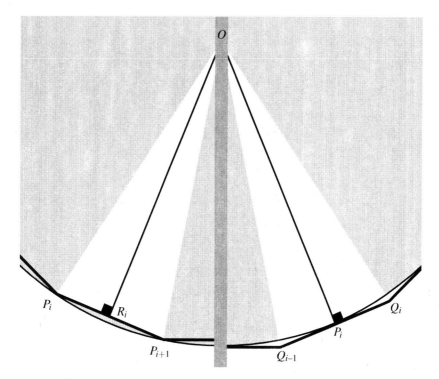

Figure 8. Approximation by inscribed and circumscribed polgons.

The lower bound prescribed by $\mathcal{P}_n$. $\overrightarrow{OQ_i}$ is a perpendicular bisector of P_iP_{i+1}, intersecting it at a point R_i and dividing $\triangle OP_iP_{i+1}$ in two. By the HL congruence theorem for right triangles, the two parts, $\triangle OR_iP_i$ and $\triangle OR_iP_{i+1}$, are congruent. Then $\mathcal{P}_n$ is built from $2n$ segments of length $|P_iR_i|$. Now for a little trigonometry:

$$\sin(360°/2n) = \frac{|P_iR_i|}{r} \implies |P_iR_i| = r\sin(360°/2n)$$

so $|\mathcal{P}_n| = 2nr\sin(360°/2n)$.

The upper bound prescribed by $\mathcal{Q}_n$. $\overrightarrow{OP_i}$ is a perpendicular bisector of $Q_{i-1}Q_i$, intersecting it at P_i and dividing $\triangle OQ_{i-1}Q_i$ in two. By SAS, the two parts, $\triangle OP_iQ_{i-1}$ and $\triangle OP_iQ_i$, are congruent. Then $\mathcal{Q}_n$ is built from $2n$ segments of length $|P_iQ_i|$. Because

$$\tan(360°/2n) = |P_iQ_i|/r \implies |P_iQ_i| = r\tan(360°/2n),$$

$|\mathcal{Q}_n| = 2nr\tan(360°/2n)$.

Let's compare $|\mathcal{P}_n|$ and $|\mathcal{Q}_n|$ as n increases (the key to the calculation is that as x approaches 0, $\cos(x)$ approaches 1):

$$\lim_{n\to\infty}\frac{|\mathcal{P}_n|}{|\mathcal{Q}_n|} = \lim_{n\to\infty}\frac{2nr\tan(360°/2n)}{2nr\sin(360°/2n)}$$
$$= \lim_{n\to\infty}\frac{\sin(360°/2n)}{\cos(360°/2n)} \cdot \frac{1}{\sin(360°/2n)}$$
$$= \lim_{n\to\infty}\frac{1}{\cos(360°/2n)}$$
$$= 1.$$

Therefore $\lim_{n\to\infty} |\mathcal{P}_n| = \lim_{n\to\infty} |\mathcal{Q}_n|$. Since $|\mathcal{C}|$ is trapped between $|\mathcal{P}_n|$ and $|\mathcal{Q}_n|$ for all n, and since they approach the same number as n goes to infinity, $|\mathcal{C}|$ also approaches this number. That gives a more comfortable equation for circumference as

$$|\mathcal{C}| = \lim_{n\to\infty} 2nr \sin(360°/2n),$$

and since $|\mathcal{C}| = 2\pi r$, we can make a definition of π as

$$\pi = \lim_{n\to\infty} n \sin(360°/2n).$$

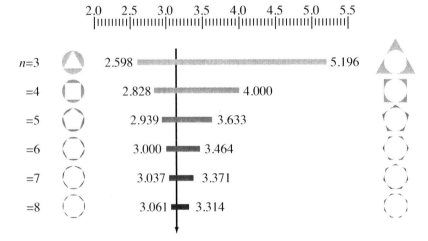

Figure 9. Using perimeters of regular inscribed and circumscribed polygons to determine upper and lower bounds for π.

15.3 Lengths of arcs and radians

It doesn't take much modification to get a formula for the length of an arc. The 360° in the formula for $|\mathcal{C}|$ is the measure of the central angle corresponding to an arc that goes completely around the circle. To get the measure of any other arc, we can replace the 360° with the measure of the corresponding central angle.

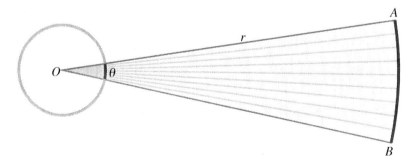

Figure 10. An arc.

15.3 Lengths of arcs and radians

Theorem 15.6 *If $\widehat{AB}$ is the arc of a circle with radius r, and if θ is the measure of the central angle $\angle AOB$, then*

$$|\widehat{AB}| = \frac{\pi}{180°}\theta \cdot r.$$

Proof To start, replace the 360° in the circumference formula with θ:

$$|\widehat{AB}| = \lim_{n\to\infty} 2nr \sin(\theta/2n) = 2r \cdot \lim_{n\to\infty} n \sin(\theta/2n).$$

This limit is clearly related to the one that defines π. We can absorb the difference between the two into the variable with the substitution $n = m \cdot \theta/360°$. As n approaches infinity, m will as well, so

$$\begin{aligned}
|\widehat{AB}| &= 2r \cdot \lim_{m\to\infty} \frac{m \cdot \theta}{360°} \sin\left(\frac{\theta}{2m \cdot \theta/360°}\right) \\
&= \frac{2r\theta}{360°} \cdot \lim_{m\to\infty} m \sin(360°/2m) \\
&= \frac{\theta}{180°} r\pi.
\end{aligned}$$
□

There is one more thing to notice before the end of this chapter. The arc length formula provides a direct connection between angle measure (of the central angle) and distance (along the arc). And yet, the $\frac{\pi}{180}$ factor in that formula suggests that distance and the degree measurement system are a little out of sync with one another. This can be fixed by modernizing our method of angle measurement. The preferred angle measurement system, and the one that we will use, is *radian* measurement.

Definition 15.7 One radian is $180°/\pi$.

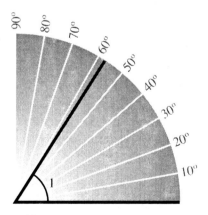

Figure 11. One radian is approximately 57.296°.

Then the measure of a straight angle is π radians. The measure of a right angle is $\pi/2$ radians. One complete turn of the circle is 2π radians. If $\theta = (\angle AOB)$ is measured in radians, then $|\widehat{AB}| = r \cdot \theta$.

15.4 Exercises

15.1. Let A and B be points on a circle $\mathcal{C}$ with radius r. Let θ be the measure of the central angle corresponding to the minor arc (or semicircle) $\widehat{AB}$. What is the relationship (in the form of an equation) between θ, r, and $|AB|$?

15.2. Let A and B be points on a circle $\mathcal{C}$ with center O and radius r. If $(\angle BAO) = \theta$, what is $|\widehat{AB}|$ (in terms of r and θ)?

15.3. Let AB be a diameter of a circle $\mathcal{C}$, and let P be a point on AB. Let $\mathcal{C}_1$ be the circle with diameter AP and let $\mathcal{C}_2$ be the circle with diameter BP. Show that the sum of the circumferences of $\mathcal{C}_1$ and $\mathcal{C}_2$ is equal to the circumference of $\mathcal{C}$ (the shape formed by the three semicircles on one side of AB is called an *arbelos*).

15.4. What is the radius of the circle that passes through the three vertices of a equilateral triangle with sides of length 1?

15.5. Let $\mathcal{C}_1$ be the circle inscribed in an equilateral triangle $\triangle ABC$ with sides of length 1. There is a line that is parallel to AB and tangent to $\mathcal{C}_1$ that divides $\triangle ABC$ into a trapezoid and a smaller equilateral triangle. Let $\mathcal{C}_2$ be the circle inscribed in the smaller equilateral triangle. Continuing in this way, we can form a stack of smaller and smaller circles. Is the sum of their circumferences finite, and if so, what is it?

15.6. In the construction of the Koch snowflake, the middle third of each segment is replaced with two-thirds of an equilateral triangle. Suppose, instead, that middle third was replaced with three of the four sides of a square. What is the perimeter of the n-th stage of this operation? Would the limiting perimeter still be infinite?

15.7. This problem deals with the possibility of angle measurement systems other than degrees or radians. Let $\mathscr{A}$ be the set of angles in the plane. Consider a function

$$\star : \mathscr{A} \to (0, \infty) : \angle A \to (\angle A)^\star$$

that satisfies

- if $\angle A \simeq \angle B$, then $(\angle A)^\star = (\angle B)^\star$
- if D is in the interior of $\angle ABC$, then

$$(\angle ABC)^\star = (\angle ABD)^\star + (\angle DBC)^\star.$$

Prove that the $\star$ measurement system is a constant multiple of the degree measurement system (or, for that matter, the radian measurement system). That is, prove that there is a $k > 0$ such that for all $\angle A \in \mathscr{A}$,

$$(\angle A)^\star = k \cdot (\angle A).$$

15.5 References

The Koch snowflake is an example of a fractal. Gerald Edgar's book *Measure, Topology, and Fractal Geometry* [Edg90], deals with these objects and their measures.

16
Euclidean Constructions

This chapter is a diversion from our projected path, but I maintain that it is a pleasant and worthwhile diversion. We get a break from the heavy proofs, and we get a more tactile approach to the subject. Many of the concurrences we will see in the next few chapters are best explored using the compass and straightedge constructions described in this chapter.

Figure 1. A modern compass and straightedge.

At the beginning of this book, we talked briefly about Euclid's postulates. The first three were

P1 : To draw a straight line from any point to any point.
P2 : To produce a finite straight line continuously in a straight line.
P3 : To describe a circle with any center and distance.

To emphasize the connection to the axioms of neutral geometry, we interpreted the postulates as claims of existence (of lines and circles). Consider instead a more literal reading: they are not claiming the existence of objects, but telling us that we can make them. This chapter is dedicated to constructing geometric objects using two classical tools, a compass and a straightedge. The compass makes circles and arcs, and the straightedge makes segments, rays, and lines. Together they make the kinds of shapes that Euclid promised in his postulates.

The straightedge

The straightedge is a simple tool: it is just something that can draw lines. In all likelihood, your straightedge will be a ruler, and if so, you need to be aware of the distinction between a ruler and a straightedge. Unlike a ruler, a straightedge has no markings (nor can you add any). Therefore, you cannot measure distance with it. But a straightedge can

- draw a segment between two points,
- draw a ray from a point through another point,
- draw a line through two points,
- extend a segment to either a ray or the line containing it,
- extend a ray to the line containing it.

The compass

Not to be confused with the ever-northward-pointing navigational compass, the compass of geometry is a tool for creating a circle. More precisely, a compass can,

- given two distinct points P and Q, draw the circle centered at P that passes through Q,
- given points P and Q on a circle with center R, draw the arc $\overset{\frown}{PQ}$.

    A pencil tied to a piece of string makes a primitive compass, although perhaps not the most accurate one. Metal compasses (such as the one pictured) are more precise instruments. Nowadays, plastic compasses are common, but I have found them to be fragile and unreliable.

    There is one word of caution regarding compasses. It is tempting to try to use the compass to transfer distance. That is, to draw a circle of a certain radius, lift up the compass and move it to another location, then place it back down to draw another circle with the same radius. That process effectively transfers a distance (the radius) from one location to another, and so is a convenient way to construct a congruent copy of one segment in another location. It is a simple maneuver. The problem is that, according to the classical rules of the game, a compass does not have this transfer ability. The classical compass is "collapsing", meaning that as soon as it is used to create a circle, it falls apart. We will soon see that the two types of compasses are not fundamentally different, and therefore that the non-collapsing feature is only a convenience. Once we have shown that, we should have no qualms about using a non-collapsing compass when it will streamline the construction process. Until then, distance transfer using a compass is not allowed.

The digital compass and straightedge

There are several good computer programs that will allow you to build constructions digitally. There are both advantages and disadvantages to the digital approach. Drawing lines and circles

Euclidean Constructions

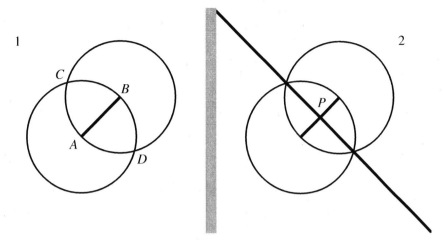

Figure 2. Constructing a perpendicular bisector.

on a real piece of paper with a real pencil is a tactile experience in a way that computer drawings are not. For more complicated constructions, paper and pencil can get messy. In addition, a construction on paper is static, while computer constructions can be dynamic, so you can drag points around and watch the rest of the construction adjust accordingly. Often the dynamism really reveals the power of the theorems in a way that static images cannot.

The perpendicular bisector

We begin with the construction of the perpendicular bisector of a segment. In figure 2

(1) Begin with a segment AB. With the compass construct two circles: one centered at A that passes through B and one centered at B that passes through A. The circles intersect twice, at C and D, once on each side of AB.
(2) Use the straightedge to draw the line $\overline{CD}$. It is perpendicular to AB, and its intersection P with AB is the midpoint of AB.

Exercise 16.1 Verify that the line constructed above is the perpendicular bisector of AB.

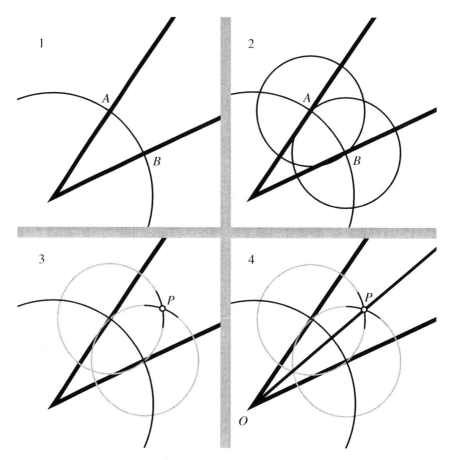

Figure 3. Constructing an angle bisector.

The bisector of an angle

In figure 3

(1) Given an angle whose vertex is O, draw a circle centered at O, and mark where it intersects the rays that form the angle as A and B.
(2) Draw two circles, one centered at A passing through B, and one centered at B passing through A.
(3) Label their intersection as P.
(4) Draw the ray $\overrightarrow{OP}$. It is the bisector of $\angle AOB$.

Exercise 16.2 Prove that $\overrightarrow{OP}$ is the bisector of $\angle AOB$.

Euclidean Constructions

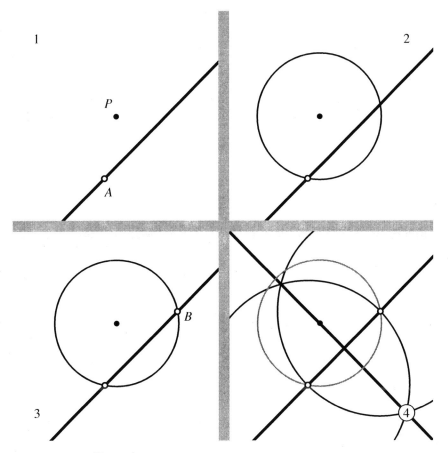

Figure 4. Constructing a perpendicular line, steps 1–4.

The perpendicular to a line ℓ through a point P

Case 1: if P is not on ℓ   In figure 4

(1) Mark a point A on ℓ.
(2) Draw the circle centered at P and passing through A.
(3) If the circle intersects ℓ only once (at P), then ℓ is tangent to the circle and AP is the perpendicular to ℓ through P (highly unlikely). Otherwise, label the second intersection B.
(4) Use the previous construction to find the perpendicular bisector to AB. This is the line we want.

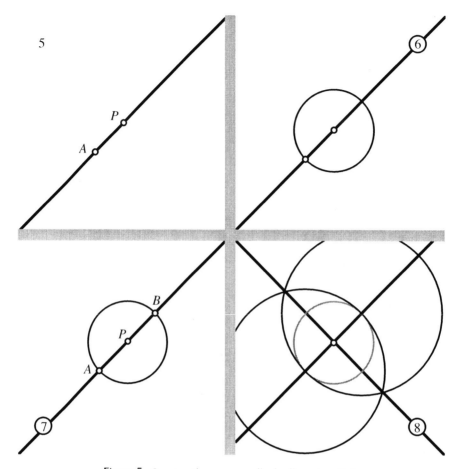

Figure 5. Constructing a perpendicular line, steps 5–8.

Case 2: if P is on ℓ In figure 5

(5) Mark a point A on ℓ other than P.
(6) Draw the circle centered at P passing through A.
(7) Mark the second intersection of this circle with ℓ as B.
(8) Use the previous construction to find the perpendicular bisector to AB. This is the line we want.

Exercise 16.3 Verify that the constructions create the desired perpendicular lines.

Given the perpendicular line construction, it is easy to describe a procedure to draw parallel lines: starting from a line, construct a perpendicular, and then a perpendicular to that. According to the alternate interior angle theorem, the result will be parallel to the initial line. That construction requires a fair number of steps, though. There is a more streamlined construction, but it requires a non-collapsing compass. So it is now time to look into the issue of collapsing versus non-collapsing compasses.

Collapsing v. non-collapsing

The apparent difference between a collapsing and a non-collapsing compass is that with a non-collapsing compass, we can draw a circle, move the compass to another location, and draw another circle of the same size. In effect, the non-collapsing compass becomes a mechanism for relaying information about size from one location in the plane to another. As we discussed at the start of this chapter, the compass is traditionally not permitted to retain and transfer that kind of information. The good news is that, in spite of this added feature, a non-collapsing compass is not any more powerful than a collapsing one. Everything that can be constructed with a non-collapsing compass can also be constructed with a collapsing one. The reason is simple: a collapsing compass can also transfer a circle from one location to another, though it requires a few more steps.

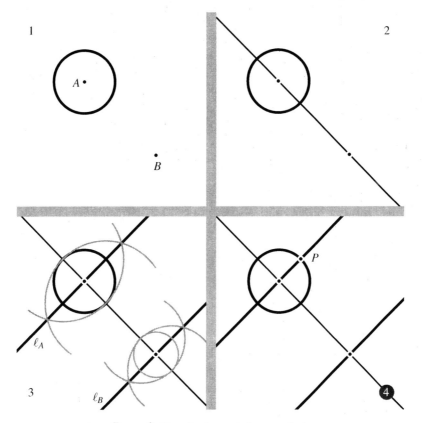

Figure 6. Transferring a circle, steps 1–4.

In figure 6

(1) Begin with a circle $\mathcal{C}$ with center A. Suppose we wish to draw another circle of the same size, this time centered at a point B.
(2) Construct the line $\overline{AB}$.
(3) Construct two lines perpendicular to $\overline{AB}$: ℓ_A through A and ℓ_B through B.
(4) Now ℓ_A intersects $\mathcal{C}$ twice: identify one point of intersection as P.

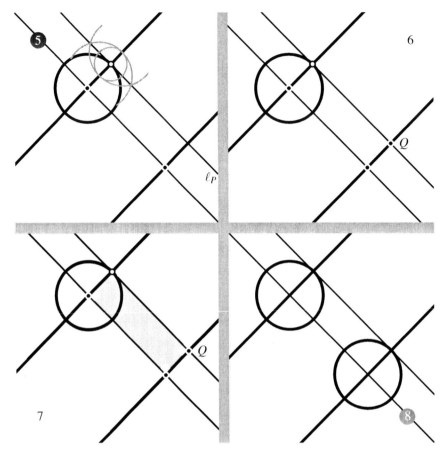

Figure 7. Transferring a circle, steps 5–8.

In figure 7

(5) Construct the line ℓ_P that passes through P and is perpendicular to ℓ_A.
(6) This line intersects ℓ_B. Identify the intersection of ℓ_P and ℓ_B as Q.
(7) Now A, B, P, and Q are the corners of a rectangle. The opposite sides AP and BQ must be congruent. So finally,
(8) Construct the circle with center B that passes through Q. This circle has the same radius as $\mathcal{C}$.

This means that a collapsing compass can do the same things a non-collapsing compass can. From now on, let's assume that our compass has the non-collapsing capability.

Transferring segments

Given a segment AB and a ray r whose endpoint is C, it is easy to find the point D on r so that $CD \simeq AB$. Just construct the circle centered at A with radius AB, and then (since the compass is non-collapsing) move the compass to construct a circle centered at C with the same radius. The intersection of this circle and r is D.

Transferring angles

Transferring an angle to a new location is a little more complicated. Suppose that we are given an angle with vertex P and a ray r with endpoint Q, and that we want to build congruent copies of $\angle P$ off of r (there is one on each side of r).

In figure 8

(1) Draw a circle with center P, and label its intersections with the two rays of $\angle P$ as A and B.
(2) Using the non-collapsing compass, transfer this circle to one that is centered at Q. Call it $\mathcal{C}$ and label its intersection with r as C.
(3) Draw another circle, this time one centered at A that passes through B. Then transfer it to one centered at C. The resulting circle will intersect $\mathcal{C}$ twice, once on each side of r. Label the intersection points as D_1 and D_2.
(4) By SSS, $\triangle PAB$, $\triangle QD_1C$, and $\triangle QD_2C$ are congruent. Therefore

$$\angle D_1 QC \simeq \angle P \simeq \angle D_2 QC.$$

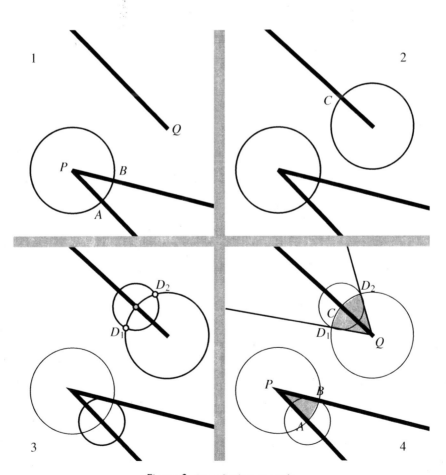

Figure 8. Transferring an angle.

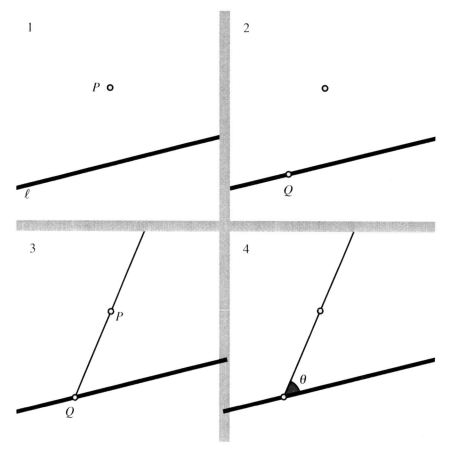

Figure 9. Constructing parallel lines, steps 1–4.

The parallel to a line through a point

In figure 9

(1) With a non-collapsing compass and angle transfer, we can now draw parallels more easily. Start with a line ℓ, and a point P that is not on that line.
(2) Mark a point Q on ℓ.
(3) Construct the ray $\overrightarrow{QP}$.
(4) This ray and ℓ form two angles, one on each side of $\overrightarrow{QP}$. Choose one of the angles and call it θ.

In figure 10

(5) Transfer it to another congruent angle θ' that comes off of the ray $\overrightarrow{PQ}$. There are two such angles, one on each side of the ray, but for the purposes of this construction, we want the one on the opposite side of $\overrightarrow{PQ}$ from θ.
(6) Now $\overrightarrow{PQ}$ is one of the rays defining θ'. Extend the other ray to the line containing it: call this line ℓ'. By the alternate interior angle theorem, ℓ' is parallel to ℓ.

Euclidean Constructions

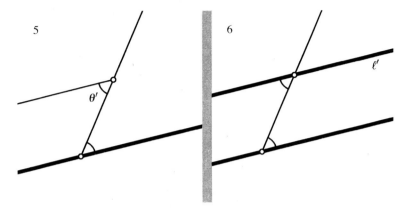

Figure 10. Constructing parallel lines, steps 5 and 6.

A rational multiple of a segment

Given a segment OP, we can construct a segment whose length is any rational multiple m/n of $|OP|$. In figure 11

(1) Along $\overrightarrow{OP}$, lay down m congruent copies of OP, end-to-end, to create a segment of length $m|OP|$. Label the endpoint of this segment as P_m.
(2) Draw another ray with endpoint O (other than $\overrightarrow{OP}$ or $\overrightarrow{OP^{op}}$), and label a point on it Q.

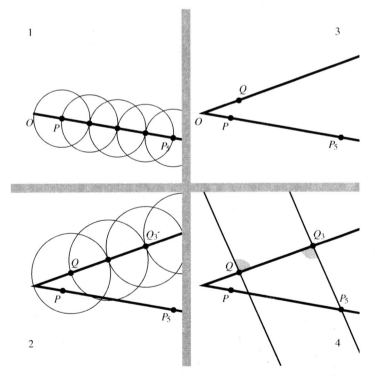

Figure 11. Constructing rational multiples, steps 1–4. In this example, we are given a segment OP, and construct a segment of length $5/3 \cdot |OP|$.

(3) Along $\overrightarrow{OQ}$, lay down n congruent copies of OQ, end-to-end, to create a segment of length $n|OQ|$. Label the endpoint of this segment as Q_n.
(4) Draw $\overline{P_m Q_n}$ and construct the line through Q that is parallel to $\overline{P_m Q_n}$.

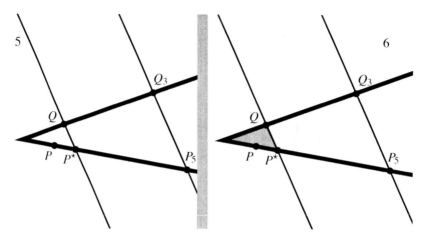

Figure 12. Constructing rational multiples, steps 5 and 6.

In figure 12

(5) It intersects $\overrightarrow{OP}$. Label the intersection as $P^\star$.
(6) The segment $OP^\star$ is the one we want: that is,

$$|OP|^\star = m/n \cdot |OP|.$$

To see why, observe that O, $P^\star$, and P_n are all parallel projections from O, Q, and Q_m, respectively. Therefore,

$$\frac{|OP^\star|}{|OP_m|} = \frac{|OQ|}{|OQ_n|} \implies \frac{|OP^\star|}{m \cdot |OP|} = \frac{1}{n} \implies |OP^\star| = \frac{m}{n}|OP|.$$

To round out this chapter we will look briefly at one of the central questions in the classical theory of constructions: given a circle, is it possible to construct a regular n-gon inscribed in it? This question has now been answered: it turns out that the answer is yes for some values of n, but no for others. In fact, a regular n-gon can be constructed if and only if n is a power of 2, or a product of a power of 2 and distinct Fermat primes (a Fermat prime is a prime of the form $2^{2^n} + 1$, and the only known Fermat primes are 3, 5, 17, 257, and 65537). A proof of this result requires Galois theory and as such falls well outside the scope of this book. We will look at a few of the small values of n where the construction is possible. In all cases, the key is to construct a central angle at O that measures $2\pi/n$.

An equilateral triangle inscribed in a circle

In this case, we need to construct a central angle of $2\pi/3$, and this can be done by constructing the supplementary angle of $\pi/3$. In figure 13

(1) Given a circle $\mathcal{C}$ with center O, mark a point A on it.
(2) Draw the diameter through A, and mark the other endpoint of it as B.
(3) Construct the perpendicular bisector to OB. Mark its intersections with $\mathcal{C}$ as C and D.
(4) The triangles $\triangle BOC$ and $\triangle BOD$ are equilateral, so $(\angle BOC) = (\angle BOD) = \pi/3$ and so the two supplementary angles $\angle AOC$ and $\angle AOD$ each measure $2\pi/3$. Construct the segments AC and AD to complete the equilateral triangle $\triangle ACD$.

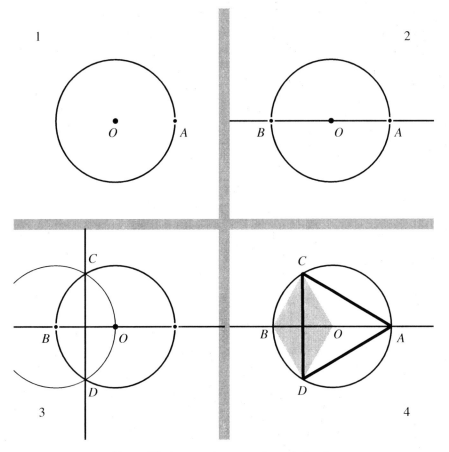

Figure 13. Constructing an equilateral triangle.

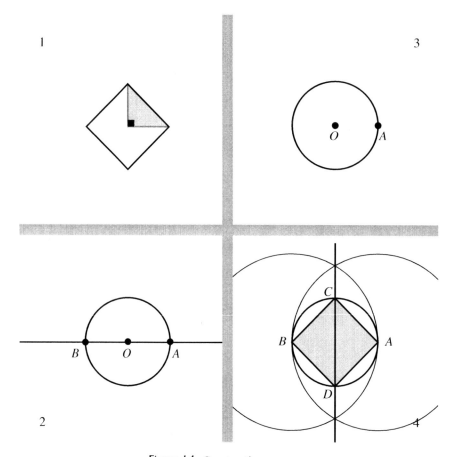

Figure 14. Constructing a square.

A square inscribed in a circle

In figure 14

(1) This is even easier, since the central angle needs to measure $\pi/2$, a right angle.
(2) Given a circle $\mathcal{C}$ with center O, mark a point A on it.
(3) Draw the diameter through A and mark the other endpoint as B.
(4) Construct the perpendicular bisector to AB and mark its intersections with $\mathcal{C}$ as C and D. The points A, B, C, and D are the vertices of the square. Connect them to get the square itself.

A regular pentagon inscribed in a circle

This one is considerably trickier. The central angle we are going to need is $2\pi/5$, which is 72°, an angle that you see a lot less frequently than the $2\pi/3$ and the $\pi/2$ of the previous constructions.

Euclidean Constructions

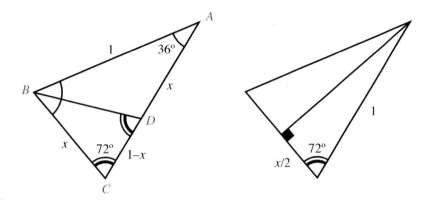

Figure 15. The golden triangle.

Before starting the construction, then, let's investigate the geometry of an angle measuring $2\pi/5$. There is a configuration of isosceles triangles that answers a lot of questions, shown in figure 15.

Exercise 16.4 Figure 15 shows a $36° - 72° - 72°$ triangle with sides of length 1, 1, and x. The angle bisector of $\angle B$ intersects AC at a point D. Show that $\triangle ABC$ and $\triangle BCD$ are similar. Show that $|CD| = 1 - x$.

Since $\triangle ABC$ and $\triangle BCD$ are similar, we can set up some ratios to solve for x:

$$\frac{1-x}{x} = \frac{x}{1} \implies 1-x = x^2 \implies x^2 + x - 1 = 0,$$

and with the quadratic formula, $x = (-1 \pm \sqrt{5})/2$. Of these solutions, x has to be the positive value since it represents a distance. The line from A to the midpoint of BC divides $\triangle ABC$ into two right triangles, and from them we can read off that

$$\cos(2\pi/5) = \frac{x/2}{1} = \frac{-1+\sqrt{5}}{4}.$$

This cosine value is the key to the construction of the regular pentagon.

The construction is straightforward, but it is not the most efficient construction. To simplify matters, we will construct the inscribed pentagon in a circle of radius one, though the same construction works in a circle of any radius.

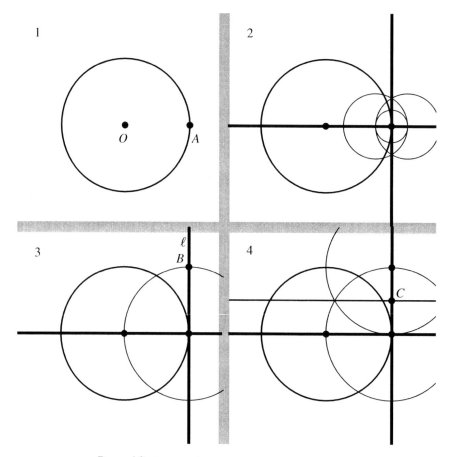

Figure 16. Constructing a regular pentagon, steps 1–4.

In figure 16

(1) Given a circle $\mathcal{C}$ with center O and radius 1. Mark a point A on $\mathcal{C}$.

Objective I. Construct a segment of length $\sqrt{5}/4$.

(2) Construct the line that passes through A and is perpendicular to $\overline{OA}$. Call it ℓ.
(3) Use the compass to mark a point B on ℓ that is a distance $|OA|$ from A.
(4) Construct the midpoint of AB, and call it C.

Euclidean Constructions

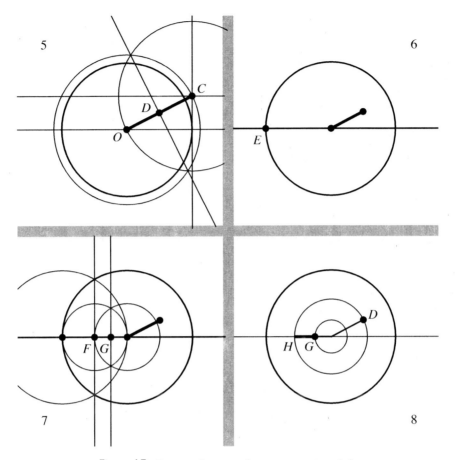

Figure 17. Constructing a regular pentagon, steps 5–8.

In figure 17

(5) Draw the segment OC. By the Pythagorean theorem,

$$|OC| = \sqrt{|OA|^2 + |AC|^2}$$
$$= \sqrt{1 + (1/2)^2}$$
$$= \sqrt{5}/2.$$

Locate the midpoint of OC (which is at a distance $\sqrt{5}/4$ from O). Call it D.

Objective II. Construct a segment of length $1/4$.

(6) Extend OA until it reaches the other side of $\mathcal{C}$ (the other endpoint of the diameter). Label this point E.

(7) Find the midpoint F of OE, and then find the midpoint G of OF. Then $|OE| = 1$, $|OF| = 1/2$, and $|OG| = 1/4$.

(8) Draw the circle centered at O that passes through D. Mark its intersection with OE as H. Then GH is a segment whose length is $(\sqrt{5}/4) - (1/4)$, as needed.

176 Euclidean Constructions

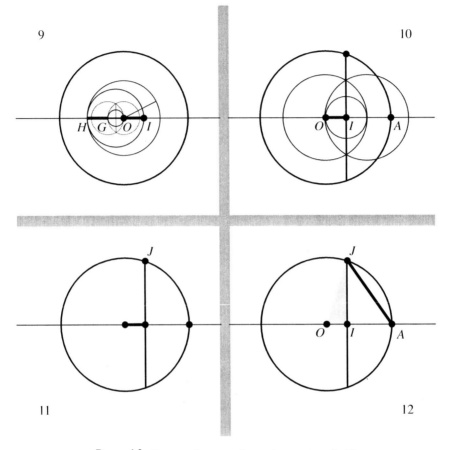

Figure 18. Constructing a regular pentagon, steps 9–12.

Objective III. Construct a segment of length $(-1 + \sqrt{5})/4$. In figure 18

(9) Use segment transfer to place a congruent copy of GH along the ray $\overrightarrow{OA}$, with one endpoint at O. Label the other endpoint I.

Objective IV. Mark a vertex of the pentagon.

(10) We will use A as one vertex of the pentagon. For the next, construct the line perpendicular to OA that passes through I.
(11) Mark one of the intersections of the perpendicular with $\mathcal{C}$ as J.
(12) Look at $\angle O$ in the right triangle $\triangle OIJ$:

$$\cos(\angle O) = \frac{|OI|}{|OJ|} = \frac{(-1+\sqrt{5})/4}{1}.$$

According to our previous calculation, $(\angle OIJ) = 2\pi/5$.

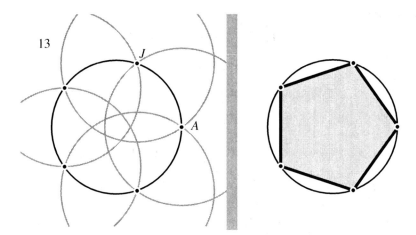

Figure 19. Constructing a regular pentagon, step 13 and the completed pentagon.

Objective V. The pentagon itself. In figure 19

(13) Segment AJ is one of the sides of the pentagon. Transfer congruent copies of the segment around the circle to get the other four sides of the pentagon.
(14) The completed pentagon.

16.1 Exercises

16.5. Given a segment AB, construct a segment of length $(7/3)|AB|$.

16.6. Given a circular arc, construct the center of the circle of which it is a part.

16.7. Given two segments AB and CD with $|CD| > |AB|$, construct a right triangle with a leg of length $|AB|$ and a hypotenuse of length $|CD|$.

16.8. Given a circle $\mathcal{C}$ tangent to a segment AB, construct a triangle $\triangle ABC$ so that $\mathcal{C}$ is also tangent to $\overline{AC}$ and $\overline{BC}$. Depending on the sizes of AB and $\mathcal{C}$, the circle may or may not lie inside the triangle.

16.9. In a circle, construct a regular (i) octagon, (ii) dodecagon, (iii) decagon.

16.10. Given a circle $\mathcal{C}$ and a point A outside it, construct the lines through A that are tangent to $\mathcal{C}$.

16.11. Given a circle $\mathcal{C}_A$ centered at A and a point B outside of $\mathcal{C}_A$, construct a circle $\mathcal{C}_B$ centered at B that intersects $\mathcal{C}_A$ at two points so that the tangent lines to $\mathcal{C}_A$ and $\mathcal{C}_B$ at each intersection are perpendicular. The two circles are said to be *orthogonal*.

16.12. (A foreshadowing.) (i) Given a triangle, construct the perpendicular bisectors to its three sides. (ii) Given a triangle, construct bisectors of its three interior angles.

16.13. Given three non-collinear points, construct a triangle that has them as the midpoints of its sides.

We haven't discussed area yet, but if you are willing to do some things out of order, here are a few area-based constructions.

16.14. Given a square whose area is A, construct a square whose area is $2A$.

16.15. Given a rectangle, construct a square with the same area.

16.16. Given a triangle, construct a rectangle with the same area.

16.2 References

Famously, it is impossible to trisect an angle with compass and straightedge. The proof of this impossibility requires a little Galois theory, but for the reader who has seen abstract algebra, is quite accessible. Proofs are often given in abstract algebra books (see, for instance, Durbin's *Modern Algebra* book [Dur92]).

17
Concurrence I

Start with three (or more) points. There is a small chance that they are collinear. In all likelihood, though, they are not. Should we find a configuration of points that are consistently collinear, that could be a sign of something interesting. Likewise, with three (or more) lines, the greatest likelihood is that each pair of lines intersect, but that none of the intersections coincide. It is unusual for two lines to be parallel, and it is unusual for three or more lines to intersect at the same point.

Definition 17.1 When three (or more) lines intersect at the same point, the lines are said to be *concurrent*. The intersection is called the *point of concurrence*.

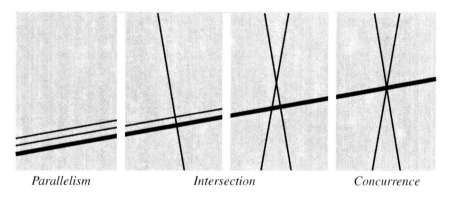

Parallelism *Intersection* *Concurrence*

Figure 1. Possible intersections of three lines.

In this chapter we are going to look at four concurrences of lines associated with a triangle. Geometers have catalogued thousands of these concurrences, so this is just the tip of a very substantial iceberg (see [Kim]).

17.1 The circumcenter

We begin by considering the perpendicular bisectors of the sides of a triangle. Recall the definition of a perpendicular bisector.

Definition 17.2 The *perpendicular bisector* of a segment AB is the line that is perpendicular to AB and passes through its midpoint.

We first need another characterization of the points of the perpendicular bisector.

Lemma 17.3 *A point X is on the perpendicular bisector of the segment AB if and only if $AX \simeq BX$.*

The proof of this is a simple application of triangle congruence theorems; it is left as an exercise.

Exercise 17.1 Prove that if X is on the perpendicular bisector to AB, then $AX \simeq BX$.

Exercise 17.2 Prove that if $AX \simeq BX$, then X is on the perpendicular bisector to AB.

Exercises 17.1 and 17.2 prepare the way for the first concurrence.

Theorem 17.4 *(Circumcenter) The perpendicular bisectors to the three sides of a triangle $\triangle ABC$ intersect at a point. The point of concurrence is called the* circumcenter *of the triangle.*

Proof The first thing to notice is that no two sides of the triangle can be parallel. Therefore, none of the perpendicular bisectors can be parallel so they all intersect each other. Let P be the intersection point of the perpendicular bisectors to AB and BC. Since P is on the perpendicular bisector to AB, $PA \simeq PB$. Since P is on the perpendicular bisector of BC, $PB \simeq PC$. Therefore, $PA \simeq PC$, and so P is on the perpendicular bisector to AC. □

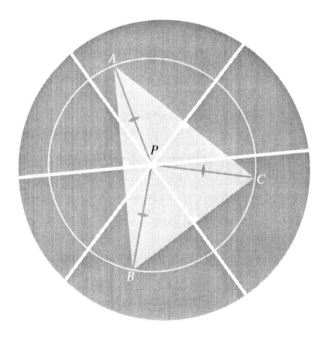

Figure 2. The circumcenter of a triangle.

17.2 The orthocenter

Because P is equidistant from A, B, and C, there is a circle centered at P that passes through A, B, and C. It is called the *circumcircle* of $\triangle ABC$.

Exercise 17.3 Let A, B, and C be three non-collinear points. Prove that the circumcircle of $\triangle ABC$ is the only circle that passes through them.

17.2 The orthocenter

Most people will be familiar with the altitudes of a triangle from area calculations in elementary geometry.

Definition 17.5 An *altitude* of a triangle is a line that passes through a vertex and is perpendicular to the opposite side.

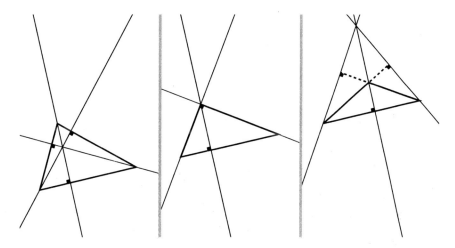

Figure 3. Altitudes for acute, right, and obtuse triangles.

An altitude of a triangle does not have to pass through the interior of the triangle at all. If the triangle is acute then all three altitudes will cross the triangle interior, but if the triangle is right, two of the altitudes will lie along the legs, and if the triangle is obtuse, two of the altitudes will only touch the triangle at their respective vertices. In any case, the altitude from the largest angle will cross through the interior of the triangle.

Theorem 17.6 *(Orthocenter) The altitudes of a triangle $\triangle ABC$ intersect at a point. This point of concurrence is called the* orthocenter *of the triangle.*

The key to the proof is that the altitudes of $\triangle ABC$ serve as the perpendicular bisectors of another (larger) triangle $\triangle abc$. That takes us back to what we have just shown, that the perpendicular bisectors of a triangle are concurrent.

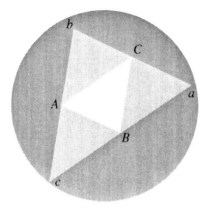

Figure 4. The orthocenter: constructing the larger triangle.

Proof First, we have to build $\triangle abc$. Draw three lines

- ℓ_1 that passes through A and is parallel to BC,
- ℓ_2 that passes through B and is parallel to AC,
- ℓ_3 that passes through C and is parallel to AB.

Each pair of lines intersect (they cannot be parallel since the sides of $\triangle ABC$ are not parallel), for a total of three intersections:

$$\ell_1 \cap \ell_2 = c, \quad \ell_2 \cap \ell_3 = a, \quad \text{and} \quad \ell_3 \cap \ell_1 = b.$$

Now we need to show that an altitude of $\triangle ABC$ is a perpendicular bisector of $\triangle abc$. The argument is the same for each altitude (other than letter shuffling), so let's focus on the altitude through A: call it α_A. I claim that α_A is the perpendicular bisector to bc. There are two conditions: (1) $\alpha_A \perp bc$ and (2) their intersection, A, is the midpoint of bc.

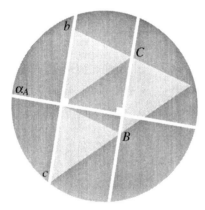

Figure 5. The orthocenter: the altitude α_A is perpendicular to bc.

(1) The first is easy thanks to the interplay between parallel and perpendicular lines in Euclidean geometry.

$$bc \parallel BC \text{ and } BC \perp \alpha_A \implies bc \perp \alpha_A.$$

17.2 The orthocenter

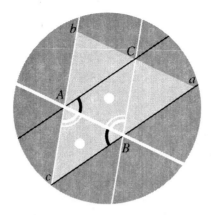

Figure 6. The orthocenter: establishing that $\triangle ABC \simeq \triangle BAc$.

(2) To get the second, we need to establish some congruent triangles that we have created. First,

A: $AC \parallel ac \implies \angle cBA \simeq \angle BAC$
S: $AB = AB$
A: $BC \parallel bc \implies \angle cAB \simeq \angle ABC$.

By ASA, $\triangle ABC \simeq \triangle BAc$. Next,

A: $AB \parallel ab \implies \angle BAC \simeq \angle bCA$
S: $AC = AC$
A: $BC \parallel bc \implies \angle BCA \simeq \angle bAC$.

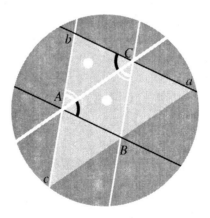

Figure 7. The orthocenter: establishing that $\triangle ABC \simeq \triangle CbA$.

Again by ASA, $\triangle ABC \simeq \triangle CbA$.

Matching the sides across the triangles, $Ac \simeq BC \simeq Ab$. This places A at the midpoint of bc and so α_A the perpendicular bisector to bc. Likewise, the altitude through B is the perpendicular bisector to ac and the altitude through C is the perpendicular bisector to ab. As perpendicular bisectors of $\triangle abc$, the lines must intersect at a point. □

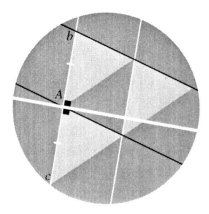

Figure 8. a_A is the perpendicular bisector of bc.

17.3 The centroid

Definition 17.7 A *median* of a triangle is a line segment from a vertex of the triangle to the midpoint of the opposite side.

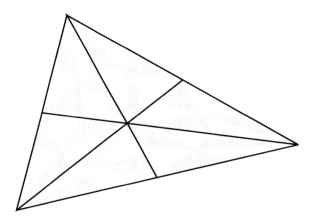

Figure 9. The three medians of a triangle.

Theorem 17.8 *(Centroid) The three medians of a triangle intersect at a point. This point of concurrence is called the* centroid *of the triangle.*

The key to this proof is that we can find the location of the intersection of any two medians—it is found two-thirds of the way down the median from the vertex. To understand why this is so, we will need to look at a sequence of three parallel projections.

17.3 The centroid

Proof   On $\triangle ABC$, label the midpoints of the three edges,

  a: the midpoint of BC,
  b: the midpoint of AC,
  c: the midpoint of AB,

so that Aa, Bb, and Cc are the medians. Now consider the following parallel projections.

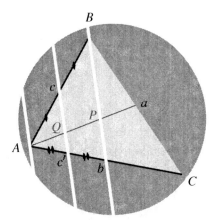

Figure 10. Centroid, step 1.

(1) Label the intersection of Aa and Bb as P. See figure 10. Extend a line from c that is parallel to Bb. Label its intersection with Aa as Q and its intersection with AC as c'. The first parallel projection, from AB to AC, associates the points

$$A \mapsto A, \quad B \mapsto b, \quad \text{and } c \mapsto c'.$$

Since $Ac \simeq cB$, this implies $Ac' \simeq c'b$.

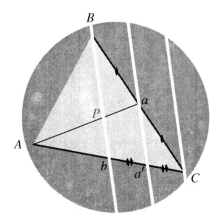

Figure 11. Centroid, step 2.

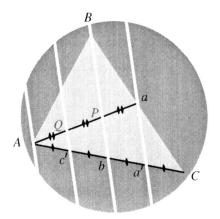

Figure 12. Centroid, step 3.

(2) Extend a line from a that is parallel to Bb. See figure 11. Label its intersection with AC as a'. The second parallel projection, from BC to AC, associates the points

$$C \mapsto C \quad B \mapsto b \quad a \mapsto a'.$$

Since $Ca \simeq aB$, we have $Ca' \simeq a'b$.

(3) Now b divides AC into two congruent segments, and a' and c' evenly subdivide them. See figure 12. In all, a', b, and c' split AC into four congruent segments. The third parallel projection is from AC back onto Aa:

$$A \mapsto A, \quad c' \mapsto Q, \quad b \mapsto P \text{ and } a' \mapsto a.$$

Since $Ac' \simeq c'b \simeq ba'$, this implies $AQ \simeq QP \simeq Pa$.

Therefore P, the intersection of Bb and Aa, will be found on Aa two-thirds of the way down from the vertex A. The letters in this argument are arbitrary, so with a permutation of letters, we could show that any pair of medians intersect at the two-thirds mark. Therefore, Cc will also intersect Aa at P, and so the three medians concur. □

Students who have taken calculus may already be familiar with the centroid. In calculus, the centroid of a planar shape D can be thought of as its balancing point, and its coordinates can be calculated as

$$\frac{1}{\iint_D 1\, dxdy} \left(\iint_D x\, dxdy, \iint_D y\, dxdy \right).$$

Exercise 17.4 Show that for triangles, the calculus description of the centroid matches the geometric description.

17.4 The incenter

This chapter began with bisectors of the sides of a triangle. It ends with the bisectors of the interior angles of a triangle.

17.4 The incenter

Theorem 17.9 *(Incenter) The bisectors of the three interior angles of a triangle intersect at a point, called the* **incenter** *of the triangle.*

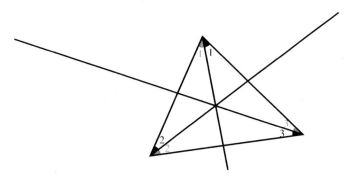

Figure 13. The three angle bisectors of a triangle.

The key to the proof is that the incenter is equidistant from the three sides of $\triangle ABC$.

Proof Take two of the angle bisectors, say the bisectors of $\angle A$ and $\angle B$, and label their intersection as P. We need to show that $\overrightarrow{CP}$ bisects $\angle C$. From P, drop perpendiculars to the sides of $\triangle ABC$. Label the feet of the perpendiculars a on BC, b on AC, and c on AB.

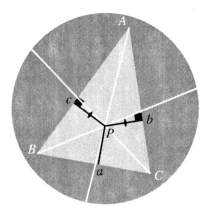   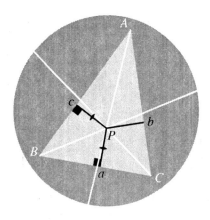

Figure 14. The incenter: showing $bP \simeq cP$.   Figure 15. The incenter: showing $cP \simeq aP$.

Then
$$\angle PbA \simeq \angle PcA, \quad \angle bAP \simeq \angle cAP, \quad \text{and} \quad AP = AP,$$
so by AAS, $\triangle AcP$ is congruent to $\triangle AbP$. In particular $bP \simeq cP$. Also,
$$\angle PaB \simeq \angle PcB, \quad \angle aBP \simeq \angle cBP, \quad \text{and} \quad BP = BP$$
and so by AAS, $\triangle BaP$ is congruent to $\triangle BcP$. In particular $cP \simeq aP$. Thus $aP \simeq bP$, and so the two right triangles $\triangle PaC$ and $\triangle PbC$ have congruent legs and share the same hypotenuse PC. By the HL congruence theorem for right triangles, they are congruent. Thus, $\angle aCP \simeq \angle bCP$, and so $\overrightarrow{CP}$ is the bisector of $\angle C$. □

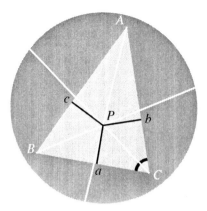

Figure 16. The incenter: establishing that $\overrightarrow{CP}$ bisects $\angle C$.

Exercise 17.5 Prove that the incenter is the only point in the interior of $\triangle ABC$ that is equidistant from $\overline{AB}$, $\overline{AC}$, and $\overline{BC}$.

Exercise 17.6 Demonstrate (by example) that there are points outside of $\triangle ABC$ that are equidistant from $\overline{AB}$, $\overline{AC}$, and $\overline{BC}$.

Since P is the same distance from the feet a, b, and c, there is a circle centered at P that is tangent to the three sides of the triangle. It is called the *inscribed circle* or *incircle* of the triangle.

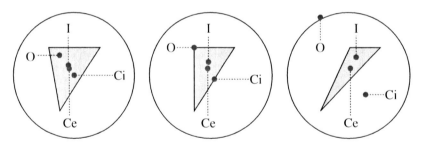

Figure 17. Positions of the four centers in an acute, right, and obtuse triangle.

In figure 17 the points of concurrence are labeled as

Ci	circumcenter	concurrence of perpendicular bisectors
O	orthocenter	concurrence of altitudes
Ce	centroid	concurrence of medians
I	incenter	concurrence of angle bisectors

17.5 Exercises

17.7. Using only compass and straightedge, construct the circumcenter, orthocenter, centroid, and incenter of a triangle.

17.5 Exercises

17.8. Using only compass and straightedge, construct the circumcircle and incircle of a triangle.

17.9. Let A, B, and C be three non-collinear points. Show that the incircle is the unique circle that is contained in $\triangle ABC$ and is tangent to its sides.

17.10. Under what circumstances does the circumcenter of a triangle lie outside the triangle? What about the orthocenter?

17.11. Under what circumstances do the orthocenter and circumcenter coincide? What about the orthocenter and centroid? What about the circumcenter and centroid?

17.12. Consider $\square ABCD$ with
$$(\angle A) + (\angle C) = \pi.$$
Prove that $\square ABCD$ is cyclic (this is the converse of exercise 14.8).

17.13. For a triangle $\triangle ABC$, there is an associated triangle called the orthic triangle whose vertices are the feet of the altitudes of $\triangle ABC$. Prove that the orthocenter of $\triangle ABC$ is the incenter of its orthic triangle. Hint: look for cyclic quadrilaterals and recall that the opposite angles of a cyclic quadrilateral are supplementary.

17.14. Suppose that $\triangle ABC$ and $\triangle abc$ are similar triangles with a scaling constant k, so that $|AB|/|ab| = k$. Let P be a center of $\triangle ABC$ (circumcenter, orthocenter, centroid, or incenter) and let p be the corresponding center of $\triangle abc$. (1) Show that $|AP|/|ap| = k$. (2) Let D denote the distance from P to AB and let d denote the distance from p to ab. Show that $D/d = k$.

17.15. In the proof of the orthocenter concurrence, we built a triangle larger by a factor of 2 around the given triangle. The process could be continued, building larger and larger triangles. Show that all the triangles would have the same point as their centroid.

18

Concurrence II

18.1 The Euler line

At the end of the last chapter, figure 17 showed three triangles and their centers. If you looked closely, you might have noticed something interesting: in each one, it certainly appears that the circumcenter, orthocenter, and centroid are collinear. This is no coincidence.

Theorem 18.1 *(The Euler line) The circumcenter, orthocenter, and centroid of a triangle are collinear, on a line called the* Euler *line.*

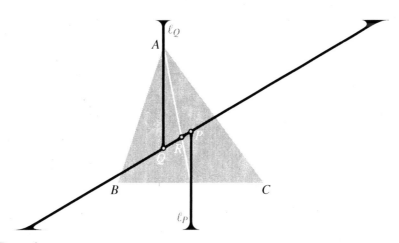

Figure 1. The circumcenter P, orthcenter Q, and centroid R all lie on the Euler line.

The proof of the theorem relies on two similar triangles. There are, however, some degenerate cases where the triangles are not properly formed. We will consider those cases before the formal proof. First, let's introduce some notation. On $\triangle ABC$, label

P: the circumcenter
Q: the orthocenter
R: the centroid
M: the midpoint of BC
ℓ_P: the perpendicular bisector to BC

ℓ_Q: the altitude through A

ℓ_R: the line containing the median AM.

A computer-generated dynamic sketch of the points and lines will give you a better sense of how they interact. One of the most readily apparent features of the construction is that ℓ_P and ℓ_Q are perpendicular to BC, so they cannot intersect unless they coincide. With a dynamic sketch, you can see that they can coincide.

Figure 2. Aligning an altitude and a perpendicular bisector.

Exercise 18.1 Show that if $\ell_P = \ell_Q$, then $\triangle ABC$ is isosceles and P, Q, and R are concurrent.

It is possible to line up P, Q, and R along the median AM without having ℓ_P, ℓ_Q, and ℓ_R coincide. That's because ℓ_P intersects AM at M and ℓ_Q intersects AM at A, and it is possible to place P at M and Q at A.

Exercise 18.2 Show that $P = M$ if and only if $Q = A$, that in that case $\triangle ABC$ is right, and that P, Q, and R are then concurrent.

In all other cases, P and Q will not be found on the median, and this is where things get interesting. Having addressed the degenerate cases, we can now look at the general case. At the heart of the argument are $\triangle AQR$ and $\triangle MPR$.

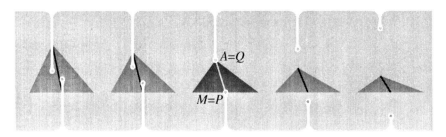

Figure 3. Putting the circumcenter and orthocenter on a median.

Proof We have

- **S:** We saw in the last chapter that the centroid is located two-thirds of the way down the median AM from A, so $|AR| = 2|MR|$.
- **A:** $\angle QAR \simeq \angle PMR$, since they are alternate interior angles between the parallel lines ℓ_P and ℓ_Q.

18.2 The nine point circle

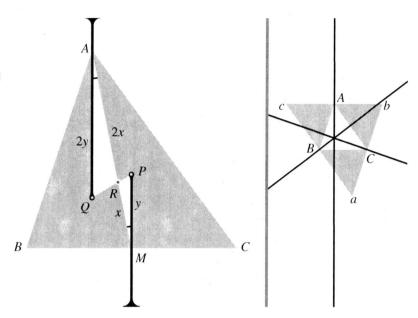

Figure 4. The general case. To show $|AQ| = 2|MP|$, we use the fact that the altitudes of $\triangle ABC$ are the perpendicular bisectors of $\triangle abc$, as shown on the right.

> **S:** Q, the orthocenter of $\triangle ABC$, is also the circumcenter of another triangle $\triangle abc$, similar to $\triangle ABC$, but twice as big. Since similarity scales distances from centers proportionally (see exercise 17.14), the distance from Q, the circumcenter of $\triangle abc$, to the side bc is double the distance from P, the circumcenter of $\triangle ABC$, to the side BC. In short, $|AQ| = 2|MP|$.

By SAS similarity, $\triangle AQR \sim \triangle MPR$. Thus $\angle PRM$ is congruent to $\angle QRA$. Since $\angle PRA$ is the supplement of $\angle PRM$, $\angle PRA$ must also be the supplement of $\angle QRA$. Therefore P, Q, and R are collinear. □

18.2 The nine point circle

While only three points are needed to define a unique circle, the next result lists nine points associated with a triangle that are always on one circle. Six of the points were identified by Feuerbach (and for this reason the circle sometimes bears his name). Several more beyond the traditional nine have been found since. If you are interested in the development of the theorem, there is a brief history in *Geometry Revisited* by Coxeter and Greitzer [CG67].

Theorem 18.2 *(The nine point circle) For a triangle, the following nine points all lie on the same circle: the feet of the altitudes, the midpoints of the sides, and the midpoints of the segments connecting the orthocenter to each vertex. This circle is the* nine point circle *associated with the triangle.*

This is a relatively long proof. We will use the following notation. Given $\triangle A_1 A_2 A_3$ with orthocenter R, label the following nine points:

L_i: the foot of the altitude that passes through A_i,
M_i: the midpoint of the side that is opposite A_i,
N_i: the midpoint of the segment $A_i R$.

The proof is based upon a key fact that is not mentioned in the statement of the theorem, that the segments $M_i N_i$ are diameters of the nine point circle. We will consider the circle $\mathcal{C}$ with diameter $M_1 N_1$ and show that the remaining seven points are on it. There are two key facts that will play pivotal roles.

- Thales' theorem (theorem 14.13): $\triangle ABC$ has a right angle at C if and only if C is on the circle with diameter AB.
- The diagonals of a parallelogram bisect one another (exercise 11.12).

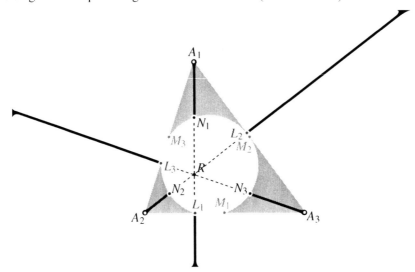

Figure 5. Labels for the nine points of the nine point circle.

Proof Lines that are parallel. First, we need to prove that several sets of lines are parallel. The key in each case is SAS triangle similarity, and the argument for the similarity is the same each time. Let's concentrate on getting the first one right, and then skip the details on all that follow. In $\triangle A_3 M_1 M_2$ and $\triangle A_3 A_2 A_1$

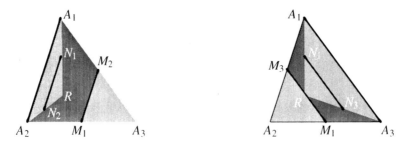

Figure 6. Parallel lines in the nine point circle.

18.2 The nine point circle

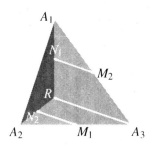 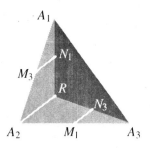

Figure 7. More parallel lines.

$$|A_3M_2| = \tfrac{1}{2}|A_3A_1|, \quad \angle A_3 = \angle A_3, \quad \text{and} \quad |A_3M_1| = \tfrac{1}{2}|A_3A_2|.$$

By the SAS similarity theorem, they are similar. In particular, the corresponding angles $\angle M_2$ and $\angle A_1$ are congruent. According to the alternate interior angle theorem, M_1M_2 and A_1A_2 are parallel. Let's employ the same argument many more times.

$$\triangle A_3M_1M_2 \sim \triangle A_3A_2A_1 \implies M_1M_2 \parallel A_1A_2$$
$$\triangle RN_1N_2 \sim \triangle RA_1A_2 \implies N_1N_2 \parallel A_1A_2$$
$$\triangle A_1N_1M_2 \sim \triangle A_1RA_3 \implies N_1M_2 \parallel A_3R$$
$$\triangle A_2M_1N_2 \sim \triangle A_2A_3R \implies M_1N_2 \parallel A_3R$$

$$\triangle A_2M_1M_3 \sim \triangle A_2A_3A_1 \implies M_1M_3 \parallel A_1A_3$$
$$\triangle RN_1N_3 \sim \triangle RA_1A_3 \implies N_1N_3 \parallel A_1A_3$$
$$\triangle A_1M_3N_1 \sim \triangle A_1A_2R \implies M_3N_1 \parallel A_2R$$
$$\triangle A_3M_1N_3 \sim \triangle A_3A_2R \implies M_1N_3 \parallel A_2R.$$

Angles that are right. A_3R is a portion of the altitude perpendicular to A_1A_2, so the first set of parallel lines are perpendicular to the second set of parallel lines. Therefore M_1M_2 and M_2N_1 are perpendicular, so $\angle M_1M_2N_1$ is a right angle; and N_1N_2 and N_2M_1 are perpendicular, so $\angle M_1N_2N_1$ is a right angle. By Thales' theorem, M_2 and N_2 are on $\mathcal{C}$.

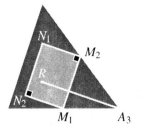 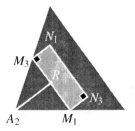

Figure 8. Right angles in the nine point circle.

Similarly, A_2R is perpendicular to A_1A_3 (an altitude and a base), so M_1M_3 and M_3N_1 are perpendicular, and so $\angle M_1M_3N_1$ is a right angle. Likewise, N_1N_3 and N_3M_1 are perpendicular, so $\angle M_1N_3N_1$ is a right angle. Thales' theorem tells us that M_3 and N_3 are on $\mathcal{C}$.

Figure 9. Diameters of the circle.

Segments that are diameters. Remember that M_1N_1 is a diameter of $\mathcal{C}$. From that, it is just a quick hop to show that L_1 is also on $\mathcal{C}$. It would be nice to do the same for L_2 and L_3, but to do that we will have to know that M_2N_2 and M_3N_3 are also diameters. Based on our work above,

$$M_1M_2 \parallel N_1N_2 \text{ and } M_1N_2 \parallel M_2N_1.$$

That makes $\square M_1M_2N_1N_2$ a parallelogram (in fact it is a rectangle). Its two diagonals, M_1N_1 and M_2N_2, must bisect each other. In other words, M_2N_2 crosses M_1N_1 at its midpoint. The midpoint of M_1N_1 is the center of $\mathcal{C}$. Therefore M_2N_2 passes through the center of $\mathcal{C}$ and is a diameter. The same argument works for M_3N_3, using the parallelogram $\square M_1M_3N_1N_3$ with bisecting diagonals M_1N_1 and M_3N_3.

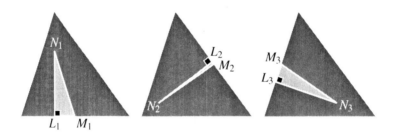

Figure 10. More angles that are right.

More angles that are right. M_1N_1, M_2N_2, and M_3N_3 are diameters of $\mathcal{C}$. $\angle M_1L_1N_1$, $\angle M_2L_2N_2$, and $\angle M_3L_3N_3$ are formed by the intersection of an altitude and a base, and so are right angles. Therefore, by Thales' theorem, L_1, L_2, and L_3 are on $\mathcal{C}$. □

18.3 The center of the nine point circle

The third result of this chapter ties together the previous two. Its proof weaves together a lot of what we have developed over the last two chapters.

18.3 The center of the nine point circle

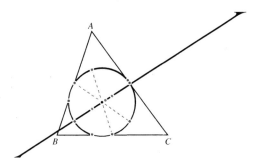

Figure 11. Locating the center of the nine point circle.

Theorem 18.3 *The center of the nine point circle is on the Euler line.*

Proof In $\triangle ABC$, label the circumcenter P and the orthocenter Q. Then $\overline{PQ}$ is the Euler line. Label the center of the nine point circle as O. Our last proof hinged upon a diameter of the nine point circle. Let's recycle some of that. If M is the midpoint of BC and N is the midpoint of QA, then MN is a diameter of the nine point circle. Now this proof really boils down to a single triangle congruence: we need to show that $\triangle ONQ$ and $\triangle OMP$ are congruent.

- **S:** $ON \simeq OM$. The center O of the nine point circle bisects the diameter MN.
- **A:** $\angle M \simeq \angle N$. These are alternate interior angles between two parallel lines, the altitude and bisector perpendicular to BC.
- **S:** $NQ \simeq MP$. In the Euler line proof we saw that $|AQ| = 2|MP|$. Well, $|NQ| = \frac{1}{2}|AQ|$, so $|NQ| = |MP|$.

By SAS, $\triangle ONQ$ and $\triangle OMP$ are congruent, and in particular $\angle QON \simeq \angle POM$. Since $\angle NOP$ is supplementary to $\angle POM$, it is also supplementary to $\angle QON$. Therefore Q, O, and P are collinear, and so O is on the Euler line. □

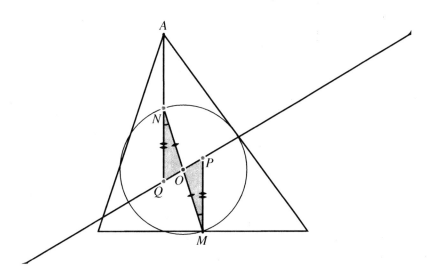

Figure 12. The triangles $\triangle OMP$ and $\triangle ONQ$ are congruent.

18.4 Exercises

18.3. Given $\triangle ABC$, let D and E be the feet of the altitudes on the sides AC and BC. Prove that there is a circle that passes through the points A, B, D, and E.

18.4. Under what conditions does the incenter lie on the Euler line?

18.5. Given an isosceles triangle $\triangle ABC$ with $AB \simeq AC$, let D be a point on the arc between B and C of the circumscribing circle. Show that DA bisects the angle $\angle BDC$.

18.6. Given $\triangle ABC$ let a be a point on BC, b be a point on AC, and c be a point on AB. Let

$\mathcal{C}_A$: the circle through points A, b, and c,
$\mathcal{C}_B$: the circle through points B, a, and c, and
$\mathcal{C}_C$: the circle through points C, a, and b.

Prove that $\mathcal{C}_A$, $\mathcal{C}_B$, and $\mathcal{C}_C$ intersect at a point. It is called the *Miguel point*. Hint: a quadrilateral is cyclic if and only if its opposite angles are supplementary—see exercise 14.8 in chapter 14 and exercise 17.12 in chapter 17.

18.7. Let P be a point on the circumcircle of $\triangle ABC$. Let L be the foot of the perpendicular from P to AB, M be the foot of the perpendicular from P to AC, and N be the foot of the perpendicular from P to BC. Show that L, M, and N are collinear. This line is called a *Simpson line*. Hint: as in the last exercise, look for cyclic quadrilaterals.

18.8. Given $\triangle ABC$, let

$\mathcal{C}_A$: the circle through B and C which is tangent to AC at C,
$\mathcal{C}_B$: the circle through A and C which is tangent to AB at A,
$\mathcal{C}_C$: the circle through A and B which is tangent to BC at B.

The circles exist and are unique according to exercise 14.7 in chapter 14. Prove that $\mathcal{C}_A$, $\mathcal{C}_B$, and $\mathcal{C}_C$ intersect at a point. It is called a *Brocard point*. Hint: use the inscribed angle theorem, noting that if, for instance, O_C is the center of $\mathcal{C}_C$, then $\triangle ABO_C$ is isosceles.

19

Concurrence III

19.1 Excenters and excircles

In the first chapter on concurrence, we saw that the bisectors of the interior angles of a triangle concur at the incenter. If you did exercise 17.13 (the orthic triangle) at the end of the chapter then you may have noticed that the sides of the original triangle are the bisectors of the exterior angles of the orthic triangle. We begin this last chapter on concurrence with another result that connects interior and exterior angle bisectors. First, a preliminary exercise.

Exercise 19.1 Given $\triangle ABC$, let ℓ_A be the line bisecting the exterior angle at A, and let ℓ_B be the line bisecting the exterior angle at B. Show that ℓ_A and ℓ_B intersect.

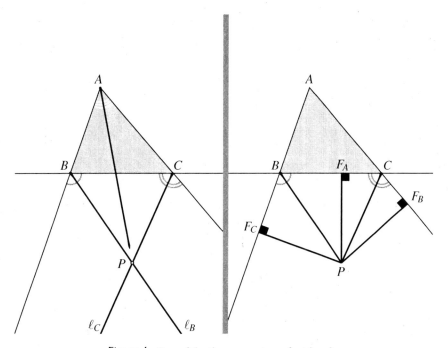

Figure 1. One of the three excenters of a triangle.

Theorem 19.1 *(Excenters) The exterior angle bisectors at two vertices of a triangle and the interior angle bisector at its third vertex intersect at a point called an* excenter *of the triangle.*

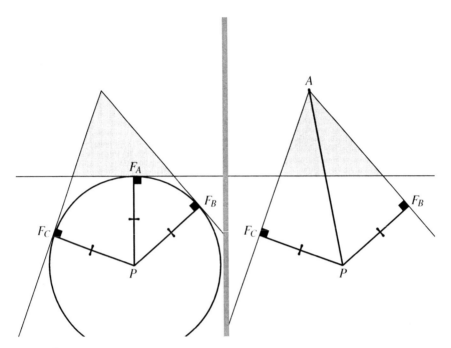

Figure 2. An excenter is equidistant from the three lines containing a side of the triangle.

Proof Let ℓ_B and ℓ_C be the lines bisecting the exterior angles at vertices B and C of $\triangle ABC$. They must intersect. Label the point of intersection as P. See figure 1. Now we need to show that the interior angle bisector at A also crosses through P. Label some more points: let F_A, F_B, and F_C be the feet of the perpendiculars through P to the sides BC, AC, and AB, respectively. Then, by AAS,

$$\triangle PF_AC \simeq \triangle PF_BC \text{ and } \triangle PF_AB \simeq \triangle PF_CB.$$

Therefore $PF_A \simeq PF_B \simeq PF_C$. By HL right triangle congruence, $\triangle PF_CA \simeq \triangle PF_BA$. In particular, $\angle PAF_C \simeq \angle PAF_B$ and so P is on the bisector of angle A. □

There are three excenters of a triangle. Each of the excircles (like the incenter) is equidistant from the three lines containing the sides of the triangle. Because of that, an excenter is the center of a circle tangent to them. The circles are called the *excircles* of the triangle.

19.2 Ceva's theorem

By now, we have seen enough concurrence theorems and enough of their proofs to have some sense of how they work. Most of them ultimately turn on a few hidden triangles that are congruent or similar. Take, for example, the concurrence of the medians. The proof required a 2 : 1 ratio of triangles. The hidden triangles can be quite difficult to find though. It would be helpful to have

19.2 Ceva's theorem

a computation that could determine when triples of segments concur. As motivation, consider the following special case.

Exercise 19.2 Suppose that $\triangle ABC$ is isosceles, with $AC \simeq BC$, that a is on BC, b is on AC, and c is on AB, so that Aa, Bb, and Cc are concurrent at a point P. Suppose as well that $Ab \simeq Ba$. Prove that $Ac \simeq Bc$.

This does not hold generally—if aC and bC are not congruent, or if $\triangle ABC$ is not isosceles, then some of the congruences break down. Ceva's theorem addresses the question more generally, using a sequence of similar triangles.

Theorem 19.2 *(Ceva's theorem) Three segments Aa, Bb, and Cc that connect the vertices of $\triangle ABC$ to their opposite sides are concurrent if and only if*

$$\frac{|Ab|}{|bC|} \cdot \frac{|Ca|}{|aB|} \cdot \frac{|Bc|}{|cA|} = 1.$$

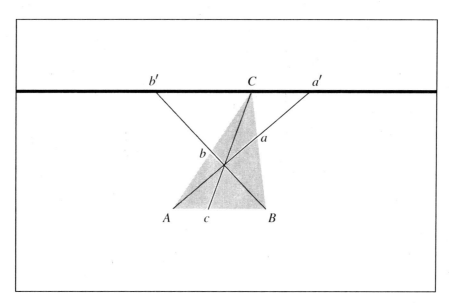

Figure 3. Setting up similar triangles to prove Ceva's theorem.

Proof $\implies$ Assume that Aa, Bb, and Cc concur at a point P. We will use similar triangles in the proof, but to get them, we need to extend the illustration. Draw the line that passes through C and is parallel to AB and then extend Aa and Bb so that they intersect it. Mark the intersection points as a' and b'. In all, we need to look at four pairs of similar triangles, and the ratios we can derive from them. They are shown in figure 4.

1. $\triangle AcP \sim \triangle a'CP$: $\quad \dfrac{|CP|}{|cP|} = \dfrac{|a'C|}{|Ac|}$.

2. $\triangle ABa \sim \triangle a'Ca$: $\quad \dfrac{|a'C|}{|AB|} = \dfrac{|aC|}{|aB|} \implies |a'C| = \dfrac{|AB| \cdot |aC|}{|aB|}$.

Figure 4. Ratios of sides in similar triangles.

Substituting from (2) into (1),

$$\frac{|CP|}{|cP|} = \frac{|AB| \cdot |aC|}{|aB| \cdot |Ac|}.$$

3. $\triangle BcP \sim \triangle b'CP$: $\quad \dfrac{|CP|}{|cP|} = \dfrac{|b'C|}{|Bc|}.$

4. $\triangle ABb \sim \triangle Cb'b$: $\quad \dfrac{|b'C|}{|AB|} = \dfrac{|bC|}{|Ab|} \implies |b'C| = \dfrac{|AB| \cdot |bC|}{|Ab|}.$

Substituting from (4) into (3),

$$\frac{|CP|}{|cP|} = \frac{|AB| \cdot |bC|}{|Ab| \cdot |BC|}.$$

Set the two ways of writing $|CP|/|cP|$ equal to each other and simplify:

$$\frac{|AB| \cdot |aC|}{|aB| \cdot |Ac|} = \frac{|AB| \cdot |bC|}{|Ab| \cdot |BC|} \implies \frac{|Ab|}{|bC|} \cdot \frac{|Ca|}{|aB|} \cdot \frac{|Bc|}{|cA|} = 1.$$

19.2 Ceva's theorem

⟸ A similar tactic works for the other direction. For this part, we are going to assume that

$$\frac{|Ab|}{|bC|} \cdot \frac{|Ca|}{|aB|} \cdot \frac{|Bc|}{|cA|} = 1,$$

and show that Aa, Bb, and Cc are concurrent. Label

  P: the intersection of Aa and Cc
  Q: the intersection of Bb and Cc.

For all the segments to concur, P and Q will have to be the same point. We can show this by computing the ratios $|CP|/|cP|$ and $|CQ|/|cQ|$ and seeing that they are equal. That shows that P and Q have to be the same distance down the segment Cc from C, and thus guarantee that they are the same. Again we can use ratios from similar triangles, shown in Figure 5.

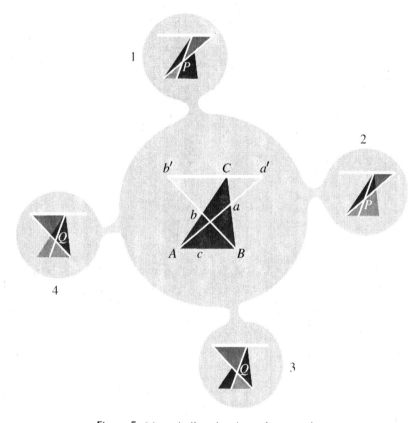

Figure 5. More similar triangles and more ratios.

1. $\triangle A \cdot P \sim \triangle a'CP$ $\dfrac{|C\cdot|}{|c\cdot|} = \dfrac{|a'\cdot|}{|A\cdot|}$

2. $\triangle \cdot \cdot \sim \triangle a'Cc$ $\dfrac{|a'\cdot|}{|A\cdot|}$ $\dfrac{|aC|}{|a\cdot|}$ $\cdot C|$ $\dfrac{|AB| \cdot C|}{|a\cdot|}$

Substituting from (2) into (1),

$$\frac{|CP|}{|cP|} = \frac{|aC| \cdot |AB|}{|aB| \cdot |Ac|}.$$

3. $\triangle BcQ \sim \triangle b'CQ$: $\quad \dfrac{|CQ|}{|cQ|} = \dfrac{|b'C|}{|Bc|}.$

4. $\triangle ABb \sim \triangle Cb'b$: $\quad \dfrac{|b'C|}{|AB|} = \dfrac{|bC|}{|Ab|} \implies |b'C| = \dfrac{|AB| \cdot |bC|}{|Ab|}.$

Substituting from (4) into (3),

$$\frac{|CQ|}{|cQ|} = \frac{|AB| \cdot |bC|}{|Ab| \cdot |Bc|}.$$

Divide and simplify:

$$\frac{|CP|}{|cP|} \bigg/ \frac{|CQ|}{|cQ|} = \frac{|aC| \cdot |AB| \cdot |Ab| \cdot |Bc|}{|aB| \cdot |Ac| \cdot |AB| \cdot |bC|} = \frac{|Ab|}{|bC|} \cdot \frac{|Ca|}{|aB|} \cdot \frac{|Bc|}{|cA|} = 1.$$

Therefore $|CP|/|cP| = |CQ|/|cQ|$, so $P = Q$. $\qquad\square$

Exercise 19.3 In the proof, it was stated that if P and Q are two points on a segment Cc and

$$\frac{|CP|}{|cP|} = \frac{|CQ|}{|cQ|},$$

then $P = Q$. Prove this.

Ceva's theorem is great for concurrences inside the triangle, but we have seen that concurrences can happen outside the triangle as well (such as the orthocenter of an obtuse triangle). Will this calculation still tell us about those concurrences? Unfortunately, no, at least as we have stated the theorem. The next exercise indicates where the previous proof runs into trouble.

Exercise 19.4 Show that there are distinct points P and Q on $\overline{Cc}$ so that

$$\frac{|CP|}{|cP|} = \frac{|CQ|}{|cQ|}.$$

There is a way to repair this, though, by introducing the notion of "signed distance". We assign a direction to the three lines containing a side of the triangle (saying one way is positive, the other way is negative). For two points A and B on a line, the signed distance from A to B is defined as

$$[AB] = \begin{cases} |AB| & \text{if } \overrightarrow{AB} \text{ points in the positive direction} \\ -|AB| & \text{if } \overrightarrow{AB} \text{ points in the negative direction} \end{cases}$$

This simple modification is all that is needed to extend Ceva's theorem.

19.3 Menelaus's theorem

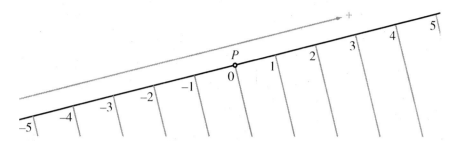

Figure 6. Signed distance from P. The sign is determined by a choice of direction.

Theorem 19.3 *(Ceva's theorem, extended version) Three lines Aa, Bb, and Cc that connect the vertices of $\triangle ABC$ to the lines containing their respective opposite sides are concurrent if and only if*

$$\frac{[Ab]}{[bC]} \cdot \frac{[Ca]}{[aB]} \cdot \frac{[Bc]}{[cA]} = 1.$$

Exercise 19.5 Prove the signed distance version of Ceva's theorem.

19.3 Menelaus's theorem

Ceva's theorem is one of a pair. The other half is its projective dual, Menelaus's theorem. We will not look at projective geometry in this book, but one of its underlying concepts is that there is a duality between points and lines. For some fundamental results, the duality allows the roles of the two to be interchanged.

Theorem 19.4 *(Menelaus's theorem) For $\triangle ABC$ and distinct points a on $\overline{BC}$, b on $\overline{AC}$, and c on $\overline{AB}$, a, b, and c are collinear if and only if*

$$\frac{[Ab]}{[bC]} \cdot \frac{[Ca]}{[aB]} \cdot \frac{[Bc]}{[cA]} = -1.$$

Proof We will prove that if a, b, and c are collinear, then the product of ratios is -1. The converse is left as an exercise. Suppose that a, b, and c lie along a line ℓ. The requirement that a, b, and c are distinct prohibits any of the intersections from occurring at a vertex. According to Pasch's lemma, then, ℓ will intersect two sides of the triangle, or it will miss all three sides entirely. Either way, it has to miss one of the sides. Let's say that the missed side is BC. There are two ways this can happen:

- ℓ intersects line BC on the opposite side of B from C
- ℓ intersects line BC on the opposite side of C from B.

The two cases are proved similarly, so let's just look at the second one. Draw the line through C parallel to ℓ. Label its intersection with $\overline{AB}$ as P. That sets up a pair of useful parallel

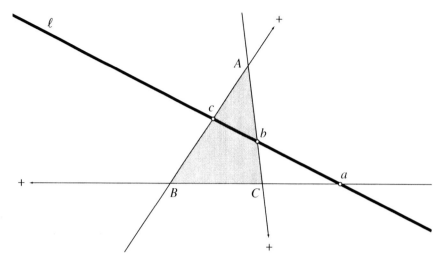

Figure 7. Menelaus's theorem gives a condition for three points on the sides of a triangle to be collinear.

projections, as shown in figure 8. The first parallel projection is from AB to AC:

$$A \mapsto A, \quad c \mapsto b, \quad P \mapsto C.$$

Comparing ratios,

$$\frac{|cP|}{|bC|} = \frac{|Ac|}{|Ab|} \implies |cP| = \frac{|Ac|}{|Ab|} \cdot |bC|.$$

The second parallel projection is from AB to BC:

$$B \mapsto B, \quad c \mapsto a, \quad P \mapsto C.$$

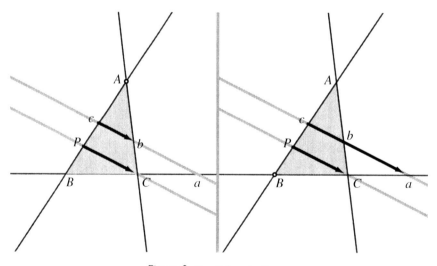

Figure 8. Parallel projections.

19.4 The Nagel point

Comparing ratios,

$$\frac{|cP|}{|aC|} = \frac{|Bc|}{|Ba|} \implies |cP| = \frac{|Bc|}{|Ba|} \cdot |aC|.$$

Divide the second $|cP|$ by the first $|cP|$ to get

$$1 = \frac{|cP|}{|cP|} = \frac{|Ab| \cdot |aC| \cdot |Bc|}{|Ac| \cdot |bC| \cdot |Ba|} = \frac{|Ab|}{|bC|} \cdot \frac{|Ca|}{|aB|} \cdot \frac{|Bc|}{|cA|}.$$

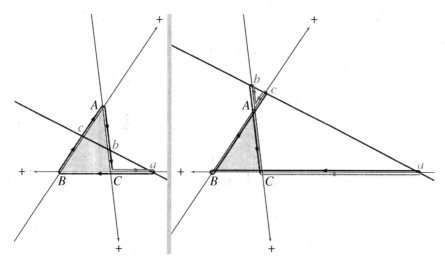

Figure 9. Keeping track of signs around the triangle.

That's close, but we are after an equation that calls for *signed* distance. So orient the lines of the triangle so that $\overrightarrow{AC}$, $\overrightarrow{CB}$, and $\overrightarrow{BA}$ point in the positive direction (any other orientation will flip pairs of signs that will cancel each other out). With this orientation, if ℓ intersects two sides of the triangle, then all the signed distances involved are positive except $[Ca] = -|Ca|$. If ℓ misses all three sides of the triangle, then three of the signed distances are positive, but three are not:

$$[Ab] = -|Ab|, \quad [Ca] = -|Ca|, \quad \text{and} \quad [cA] = -|cA|.$$

Either way, an odd number of signs are changed, so

$$\frac{[Ab]}{[bC]} \frac{[Ca]}{[aB]} \frac{[Bc]}{[cA]} = -1. \qquad \square$$

Exercise 19.6 Finish the proof of Menelaus's theorem by showing that if the product of ratios is -1, then the points are collinear.

19.4 The Nagel point

Let's return to excircles for one more concurrence, and this time we will use Ceva's theorem to prove it.

Theorem 19.5 *(Nagel point) If $\mathcal{C}_A$, $\mathcal{C}_B$, and $\mathcal{C}_C$ are the excircles of $\triangle ABC$ so that $\mathcal{C}_A$ is in the interior of $\angle A$, $\mathcal{C}_B$ is in the interior of $\angle B$, and $\mathcal{C}_C$ is in the interior of $\angle C$; and if F_A is the intersection of $\mathcal{C}_A$ with BC, F_B is the intersection of $\mathcal{C}_B$ with AC, and F_C is the intersection of $\mathcal{C}_C$ with AB; then AF_A, BF_B, and CF_C are concurrent. The point of concurrence is called the* Nagel point.

Proof This is a direct application of Ceva's theorem. Fortunately, the Nagel point must occur inside the triangle (why?), so that we may use regular distance rather than signed distance in the calculations. Thus, Ceva's theorem promises concurrence if we can show that

$$\frac{|AF_C|}{|F_CB|} \cdot \frac{|BF_A|}{|F_AC|} \cdot \frac{|CF_B|}{|F_BA|} = 1.$$

Label P_A, the center of excircle $\mathcal{C}_A$; P_B, the center of excircle $\mathcal{C}_B$; and P_C, the center of excircle $\mathcal{C}_C$.

By AA triangle similarity,

$$\triangle P_A F_A C \sim \triangle P_B F_B C$$
$$\triangle P_B F_B A \sim \triangle P_C F_C A$$
$$\triangle P_C F_C B \sim \triangle P_A F_A B.$$

The triangle similarities give some useful ratios:

$$\frac{|AF_C|}{|AF_B|} = \frac{|P_C F_C|}{|P_B F_B|} \quad \frac{|BF_A|}{|BF_C|} = \frac{|P_A F_A|}{|P_C F_C|} \quad \frac{|CF_B|}{|CF_A|} = \frac{|P_B F_B|}{|P_A F_A|}.$$

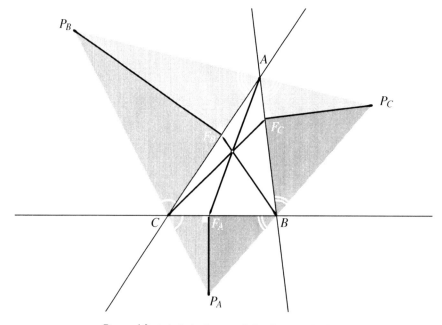

Figure 10. Labels in the search for the Nagel point.

19.5 Exercises

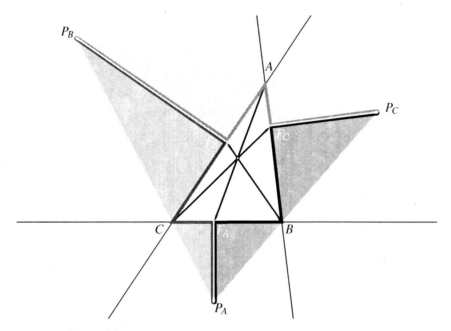

Figure 11. Use similar triangles to match ratios in Ceva's theorem.

Therefore

$$\frac{|AF_C|}{|F_CB|}\frac{|BF_A|}{|F_AC|}\frac{|CF_B|}{|F_BA|} = \frac{|AF_C|}{|AF_B|}\frac{|BF_A|}{|BF_C|}\frac{|CF_B|}{|CF_A|}$$
$$= \frac{|P_CF_C|}{|P_BF_B|}\frac{|P_AF_A|}{|P_CF_C|}\frac{|P_BF_B|}{|P_AF_A|}$$
$$= 1.$$

By Ceva's theorem, the segments are concurrent. □

19.5 Exercises

19.7. Use Ceva's theorem to prove that the medians of a triangle are concurrent.

19.8. Use Ceva's theorem to prove that the altitudes of a triangle are concurrent.

19.9. Given non-opposite rays r_1 and r_2 from O, let A and B be points on r_1 and A' and B' be points on r_2 so that $\overline{AA'}$ and $\overline{BB'}$ intersect at a point P. Use Menelaus's theorem to prove that

$$\frac{[AP]}{[BP]} \cdot \frac{[OB]}{[OA]} = \frac{[A'P]}{[B'P]} \cdot \frac{[OB']}{[OA']}.$$

19.10. Give a compass and straightedge construction of the three excircles and the nine point circle of a given triangle. If your construction is accurate enough, you should notice that each excircle is tangent to the nine point circle (a result commonly called Feuerbach's theorem).

19.11. Given $\triangle ABC$, label

> a: the point of tangency of the incircle with side BC,
> b: the point of tangency of the incircle with side AC, and
> c: the point of tangency of the incircle with side AB.

Prove that Aa, Bb, and Cc are concurrent. The point of concurrence is called the *Gergonne point*.

The remaining exercises lead up to Pappus's theorem, an important result in projective geometry.

19.12. Given two distinct lines ℓ_1 and ℓ_2 that may be parallel or intersecting, let A, B, and C be points on ℓ_1, and A', B', and C' be points on ℓ_2. Prove that if $AB' \parallel BC'$ and $BA' \parallel CB'$, then $AA' \parallel CC'$.

19.13. Assuming the configuration of points as described in the last exercise, label three intersections (provided they exist):

> P: the intersection of $\overline{AB'}$ and $\overline{BA'}$,
> Q: the intersection of $\overline{AC'}$ and $\overline{CA'}$, and
> R: the intersection of $\overline{BC'}$ and $\overline{CB'}$.

Prove that P, Q, and R are collinear.

19.14. Let r_1 and r_2 be rays that share an endpoint O. Suppose that A, B, and C are on r_1 and A', B', and C' are on r_2, and let X be the intersection of $\overline{AA'}$ and $\overline{CC'}$. Prove that $\overline{BB'}$ passes through X if and only if

$$\frac{[OA]}{[OC]} \cdot \frac{[BC]}{[BA]} = \frac{[OA']}{[OC']} \cdot \frac{[B'C']}{[B'A']}.$$

19.15. The result of the last exercise does not specify whether X is to be in the interior of $\angle O$ or not. For visualization purposes, draw a sketch of both possibilities.

19.16. (Pappus's theorem) Let ℓ_1 and ℓ_2 be two distinct lines, and suppose that A, B, and C are on ℓ_1 and A', B', and C' are on ℓ_2. Label

> P: the intersection of $\overline{AB'}$ and $\overline{BA'}$,
> Q: the intersection of $\overline{AC'}$ and $\overline{CA'}$, and
> R: the intersection of $\overline{BC'}$ and $\overline{CB'}$.

Prove that P, Q, and R are collinear. Hint: if both $AB' \parallel BC'$ and $BA' \parallel CB'$, then this follows from exercise 19.13. Hence we may assume that one pair of lines intersect, say $\overline{AB'}$ intersects $\overline{BC'}$ at a point S. The key, then, is to use result of exercise 19.14. In $\angle S$, you will need to use the equation once with A' serving the role of X, and once with C serving the role of X. Some algebra gives the equation that guarantees that P, Q, and R are collinear.

20
Trilinear Coordinates

This is the last chapter under the heading of Euclidean geometry. Looking back to the start, we have built an impressive structure from modest beginnings. Throughout it all, we have taken a synthetic approach to the subject, which is to say that we have avoided attaching a coordinate system to the plane, with all the powerful analytic techniques that come by doing so. I feel that it is in the classical spirit of the subject to try to maintain the synthetic stance for as long as possible. But as we now move into the more modern development of the subject, it is time to shift positions. As a result, much of the rest of this work will take on a decidedly different flavor. In this chapter, we begin to shift our stance, from the synthetic to the analytic.

20.1 Trilinear coordinates

In this chapter, we will look at trilinear coordinates, a coordinate system that is closely tied to the concurrence results of the last few chapters. Essentially, trilinear coordinates are defined by measuring signed distances from the sides of a triangle.

Definition 20.1 Given a side s of $\triangle ABC$ and a point P, let $|P, s|$ denote the (minimum) distance from P to the line containing s. Then define the signed distance from P to s as

$$[P, s] = \begin{cases} |P, s| & \text{if } P \text{ is on the same side of } s \text{ as the triangle} \\ -|P, s| & \text{if } P \text{ is on the opposite side of } s \text{ from the triangle.} \end{cases}$$

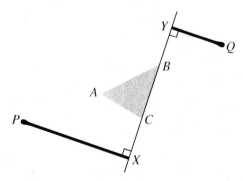

Figure 1. The signed distance to the side BC: $[P, BC] = |PX|$ while $[Q, BC] = -|QY|$.

211

From these signed distances, every triangle creates a kind of coordinate system in which a point P in the plane is assigned three coordinates

$$\alpha = [P, BC], \quad \beta = [P, AC], \quad \text{and} \quad \gamma = [P, AB].$$

This information is consolidated into the notation $P = [\alpha : \beta : \gamma]$. It is important to notice that while every point corresponds to a triple of real numbers, not every triple of real numbers corresponds to a point. For instance, when $\triangle ABC$ is equilateral with sides of length one, there is no point with coordinates $[2 : 2 : 2]$. Fortunately, there is a way around this limitation, using an equivalence relation.

Definition 20.2 Two sets of trilinear coordinates $[a : b : c]$ and $[a' : b' : c']$ are equivalent, written $[a : b : c] \sim [a' : b' : c']$, if there is a real number $k \neq 0$ so that

$$a' = ka, \quad b' = kb, \quad \text{and} \quad c' = kc.$$

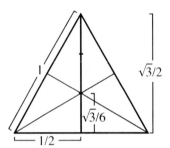

Figure 2. Measurements in an equilateral triangle.

Consider again the equilateral triangle $\triangle ABC$ with sides of length one. Yes, there is no point that is a distance of two from each side. But $[2 : 2 : 2]$ is equivalent to $[\sqrt{3}/6 : \sqrt{3}/6 : \sqrt{3}/6]$, and there is a point which is a distance of $\sqrt{3}/6$ from each side, the center of the triangle. That brings us to the definition of trilinear coordinates.

Definition 20.3 The *trilinear coordinates* of a point P with respect to $\triangle ABC$ is the equivalence class of triples $[k\alpha : k\beta : k\gamma]$ (with $k \neq 0$) where

$$\alpha = [P, BC], \quad \beta = [P, AC], \quad \text{and} \quad \gamma = [P, AB].$$

The coordinates corresponding to the actual signed distances, when $k = 1$, are called the exact trilinear coordinates of P. Because each coordinate is actually an equivalence class, there is an immediately useful relationship between trilinear coordinates in similar triangles.

Exercise 20.1 Suppose that $\triangle ABC$ and $\triangle A'B'C'$ are similar, with a scaling constant k so that

$$|A'B'| = k|AB|, \quad |B'C'| = k|BC|, \quad \text{and} \quad |C'A'| = k|CA|.$$

Given a point P, show that there is a unique point P' so that

$$[P', A'B'] = k[P, AB], \quad [P', B'C'] = k[P, BC], \quad \text{and} \quad [P', C'A'] = k[P, AC].$$

We call P and P' *similarly positioned* in their respective triangles $\triangle ABC$ and $\triangle A'B'C'$.

20.1 Trilinear coordinates

This idea of similar positioning will make more sense when we have studied similarity in the context of dilations. For now, note that the coordinates of P as determined by $\triangle ABC$ will be equivalent to the coordinates of P' as determined by $\triangle A'B'C'$. With that in mind, let's get back to the question of whether every equivalence class of triples of real numbers corresponds to a point.

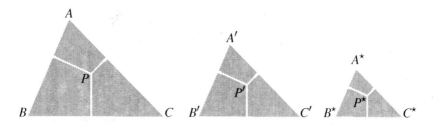

Figure 3. The exact trilinear coordinates of similarly positioned points in similar triangles: $P = [2 : 1 : 2]$, $P' = [1.5 : 0.75 : 1.5]$, and $P^\star = [1 : 0.5 : 1]$.

Theorem 20.4 *Given a triangle $\triangle ABC$ and real numbers x, y, and z, not all zero, there is a point whose trilinear coordinates with respect to $\triangle ABC$ are $[x : y : z]$.*

Proof There are two cases: one where x, y, and z have the same sign, and one where they do not. The proof will look at the first case in detail. The second differs at just one step, so its proof is only sketched. In both cases, the approach is constructive, but an indirect construction: instead of trying to find a point inside $\triangle ABC$ with the correct coordinates, we will start with a point P, and then build a new triangle $\triangle abc$ around it. That new triangle will

- be similar to the original $\triangle ABC$, and
- be positioned so that the trilinear coordinates of P with respect to $\triangle abc$ are $[x : y : z]$.

Then the similarly positioned point in $\triangle ABC$ will have the same coordinates relative to $\triangle ABC$.
    Case 1. $[+ : + : +] \sim [- : - : -]$
Consider the situation where x, y, and z are greater than or equal to zero (they cannot all be zero). This also handles the case where the coordinates are negative, since $[x : y : z] \sim [-x : -y : -z]$. Mark a point F_x that is a distance x away from P. On opposite sides of $\overrightarrow{PF_x}$, draw two more rays to form angles measuring $\pi - (\angle B)$ and $\pi - (\angle C)$. On the first ray, mark the point F_z that is a distance z from P. On the second, mark the point F_y that is a distance y from P. Label

- ℓ_x: the line through F_x that is perpendicular to PF_x,
- ℓ_y: the line through F_y that is perpendicular to PF_y,
- ℓ_z: the line through F_z that is perpendicular to PF_z.

Label their points of intersection as

$$a = \ell_y \cap \ell_z, \quad b = \ell_x \cap \ell_z, \quad \text{and} \quad c = \ell_x \cap \ell_y.$$

Clearly, the trilinear coordinates of P relative to $\triangle abc$ are $[x : y : z]$. To see that $\triangle abc$ and $\triangle ABC$ are similar, let's compare their interior angles. The quadrilateral $\square PF_xbF_z$ has right angles at vertices F_x and F_z and an angle measuring $\pi - (\angle B)$ at vertex P. Since the angle sum of a quadrilateral is 2π, that implies $(\angle b) = (\angle B)$, so they are congruent. By a similar argument, $\angle c$ and $\angle C$ are congruent. By AA similarity, then, $\triangle ABC$ and $\triangle abc$ are similar. According to exercise 20.1, there is a point that is similarly positioned relative to $\triangle ABC$ with the same trilinear coordinates.

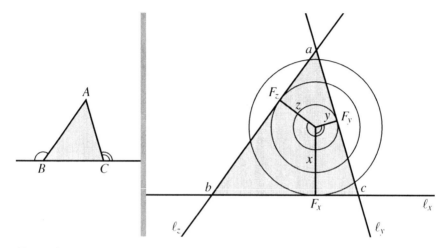

Figure 4. Constructing a triangle around P so that its trilinear coordinates are $[x : y : z]$.

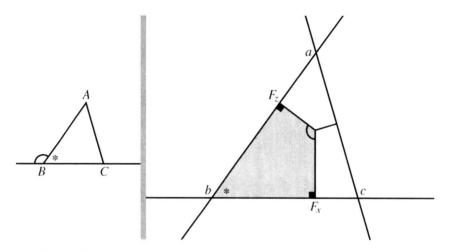

Figure 5. By AA similarity, the constructed triangle is similar to the original one.

Case 2. $[+ : - : -] \sim [- : + : +]$

With a little letter shuffling, this also handles cases of the form $[- : + : -]$, $[+ : - : +]$, $[- : - : +]$, and $[+ : + : -]$. The construction is the same as in the previous case, with one important change: in the previous construction, we needed

$$(\angle F_z P F_x) = \pi - (\angle B) \text{ and } (\angle F_y P F_x) = \pi - (\angle C).$$

20.2 Trilinears of the classical centers

This time we will instead want

$$(\angle F_z P F_x) = (\angle B) \quad \text{and} \quad (\angle F_y P F_x) = (\angle C).$$

The construction still forms a triangle $\triangle abc$ that is similar to $\triangle ABC$, but now P lies outside of it. Depending upon the location of a relative to the line ℓ_x, the signed distances from P to BC, AC, and AB, respectively are either x, y, and z, or $-x$, $-y$, and $-z$. Either way, since $[x:y:z]$ and $[-x:-y:-z]$ are equivalent, P has the correct coordinates. Then the similarly positioned point with respect to $\triangle ABC$ has the same trilinear coordinates.

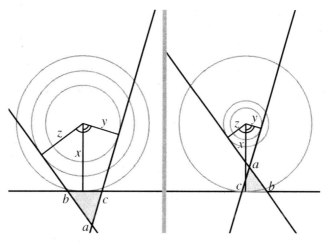

Figure 6. The second case, in which the desired coordinates do not all have the same sign. □

Exercise 20.2 Fill in the details of case 2 in the previous proof.

20.2 Trilinears of the classical centers

The classical triangle centers that we have studied in the last few chapters tend to have elegant trilinear coordinates. The rest of the chapter is dedicated to finding a few of them. The easiest is the incenter. Since it is equidistant from the sides of the triangle, its trilinear coordinates are $[1:1:1]$. The others will require a little bit more work. The formulas are valid for all triangles. However, we will need to use trigonometry to locate these centers, and we have only discussed the trigonometry of acute angles. For that reason, in the proofs we will restrict our attention to acute triangles. You have surely seen the unit circle extension of the trigonometric functions to all angle measures, so you are encouraged to complete the proofs by considering triangles that are not acute.

Theorem 20.5 *The trilinear coordinates of the circumcenter of $\triangle ABC$ are*

$$[\cos A : \cos B : \cos C].$$

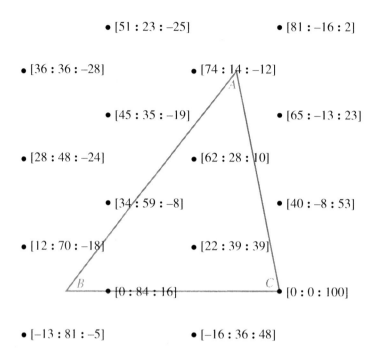

Figure 7. Trilinear coordinates of a few points, normalized so that the sum of the magnitudes of the coordinates is 100, and rounded to the nearest integer.

Proof First the labels. Label the circumcenter P. The circumcenter is the intersection of the perpendicular bisectors of the three sides of the triangle. Take one of them: the perpendicular bisector to BC. It intersects BC at its midpoint—call that point X. Now we can calculate the first exact trilinear coordinate in just a few steps.

As shown in figure 8, the minimum distance from P to BC is along the perpendicular—so $|P, BC| = |P, X|$. We have assumed that $\triangle ABC$ is acute. That places P inside the triangle, on

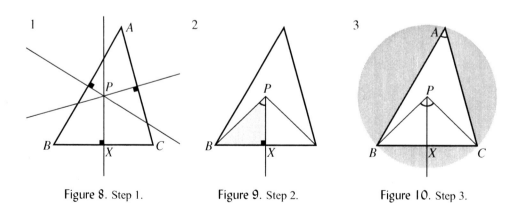

Figure 8. Step 1.   Figure 9. Step 2.   Figure 10. Step 3.

20.2 Trilinears of the classical centers 217

the same side of BC as A, which means that the signed distance $[P, BC]$ is positive. Therefore

$$[P, BC] = |P, BC| = |PX|.$$

Now examine $\angle BPX$ in $\triangle BPX$, shown in figure 9:

$$\cos(\angle BPX) = \frac{|PX|}{|PB|} \implies |PX| = |PB|\cos(\angle BPX).$$

As shown in figure 10, PX splits $\triangle BPC$ into two pieces, $\triangle BPX$ and $\triangle CPX$ that are congruent by SAS. Thus PX evenly divides $\angle BPC$ into two congruent pieces, and so

$$(\angle BPX) = \tfrac{1}{2}(\angle BPC).$$

The circumcenter is the center of the circle that passes through vertices A, B, and C. With respect to that circle, $\angle BAC$ is an inscribed angle, and $\angle BPC$ is the corresponding central angle. According to the inscribed angle theorem, $(\angle BAC) = (\angle BPC)/2$, so $(\angle BPX) = (\angle BAC)$. Combining these pieces,

$$[P, BC] = |PX| = |PB|\cos(\angle BPX) = |PB|\cos(\angle BAC).$$

With the same argument we can find the signed distances to the other two sides as well:

$$[P, AC] = |PC|\cos(\angle ABC) \text{ and } [P, AB] = |PA|\cos(\angle BCA).$$

Gather the information together to get the exact trilinear coordinates of the circumcenter:

$$P = [|PB|\cos(\angle A) : |PC|\cos(\angle B) : |PA|\cos(\angle C)].$$

Finally, observe that PA, PB, and PC are all the same length because they are radii of the circumcircle. Therefore, we can get the equivalent coordinates

$$P = [\cos(\angle A) : \cos(\angle B) : \cos(\angle C)]. \qquad \square$$

Theorem 20.6 *The trilinear coordinates of the orthocenter of $\triangle ABC$ are*

$$[\cos(\angle B)\cos(\angle C) : \cos(\angle A)\cos(\angle C) : \cos(\angle A)\cos(\angle B)].$$

Proof Label the orthocenter Q. It is the intersection of the three altitudes of the triangle. Label the feet of the altitudes

F_A : the foot of the altitude through A,
F_B : the foot of the altitude through B, and
F_C : the foot of the altitude through C.

Now think back to the way we proved that the altitudes concur in chapter 17, by showing that they are the perpendicular bisectors of a larger triangle $\triangle abc$, where

bc : passed through A and was parallel to BC,
ac : passed through B and was parallel to AC, and
ab : passed through C and was parallel to AB.

We will need that triangle again.

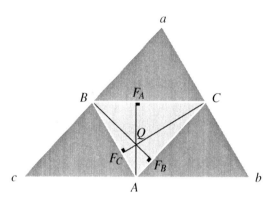

Figure 11. The orthocenter of $\triangle ABC$ is the circumcenter of $\triangle abc$.

Figure 12. We want to measure QF_A. Figure 13. The angles at Q and C are congruent.

In figure 12, the distance from Q to BC is measured along the perpendicular, so $|Q, BC| = |QF_A|$, but since we assumed our triangle is acute, Q will be inside $\triangle ABC$ and so the signed distance $[Q, BC]$ is positive. So

$$[Q, BC] = |Q, BC| = |QF_A|.$$

As shown in figure 13, in the right triangle $\triangle F_A QB$,

$$\cos(\angle F_A QB) = \frac{|QF_A|}{|QB|} \implies |QF_A| = |QB|\cos(\angle F_A QB).$$

By AA, $\triangle F_A QB \sim \triangle F_B CB$ (they share the angle at B and both have a right angle). Therefore $\angle F_A QB \simeq \angle F_B CB$.

20.2 Trilinears of the classical centers

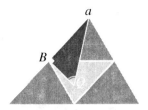 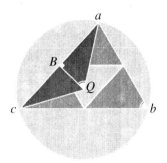

Figure 14. BQ is adjacent to $\angle Q$ in $\triangle QBa$.   Figure 15. $\angle Q \simeq \angle b$.

In the right triangle $\triangle aQB$ (shown in figure 14),

$$\cos(\angle aQB) = \frac{|QB|}{|Qa|} \implies |QB| = |Qa|\cos(\angle aQB).$$

In figure 15, the orthocenter Q of $\triangle ABC$ is the circumcenter of the larger triangle $\triangle abc$. Then $\angle abc$ is an inscribed angle in the circumcircle whose corresponding central angle is $\angle aQc$. By the inscribed angle theorem,

$$(\angle abc) = \tfrac{1}{2}(\angle aQc).$$

The segment QB bisects $\angle aQc$, so

$$(\angle aQB) = \tfrac{1}{2}(\angle aQc).$$

That means $\angle aQB \simeq \angle abc$, which is congruent to $\angle B$ in the original triangle. We can put this all together:

$$[Q, BC] = |QF_A| = |QB|\cos(\angle F_A QB) = |QB|\cos(\angle C)$$
$$= |Qa|\cos(\angle aQB)\cos(\angle C) = |Qa|\cos(\angle B)\cos(\angle C).$$

Through similar calculations,

$$[Q, AC] = |Qb|\cos(\angle A)\cos(\angle C)$$
$$[Q, AB] = |Qc|\cos(\angle A)\cos(\angle B).$$

That gives the exact trilinear coordinates for the orthocenter as

$$Q = [|Qa|\cos(\angle B)\cos(\angle C) : |Qb|\cos(\angle A)\cos(\angle C) : |Qc|\cos(\angle A)\cos(\angle B)]$$

Qa, Qb, and Qc are all the same length, though, since they are radii of the circumcircle of $\triangle abc$. Factoring gives an equivalent set of coordinates

$$Q = [\cos(\angle B)\cos(\angle C) : \cos(\angle A)\cos(\angle C) : \cos(\angle A)\cos(\angle B)]. \qquad \square$$

The last few results relied upon some essential property of the center in question. For the circumcenter it was the fact that it is equidistant from the three vertices and for the orthocenter, that it is the circumcenter of a larger triangle. To find the trilinear coordinates of the centroid, we will again draw on such a property, that the centroid is located 2/3 of the way down a median from the vertex.

Theorem 20.7 *The trilinear coordinates of the centroid of* $\triangle ABC$ *are*

$$[|AB| \cdot |AC| : |BA| \cdot |BC| : |CA| \cdot |CB|].$$

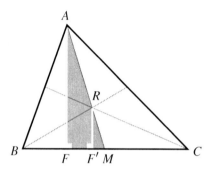

Figure 16. Labels for the centroid.

Proof First the labels:

F: the foot of the altitude through A,
M: the midpoint of the side BC,
R: the centroid of $\triangle ABC$ (the intersection of the medians),
F': the foot of the perpendicular through R to the side BC.

For convenience write $a = |BC|$, $b = |AC|$, and $c = |AB|$. Let's look at $[R, BC]$, which is one of the signed distances needed for the trilinear coordinates.

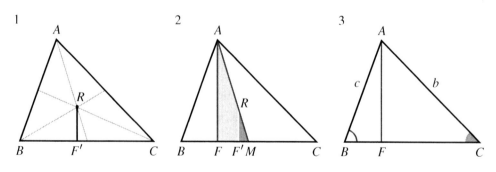

Figure 17. We need $|RF'|$. Figure 18. $|RF'| = |AF|/3$. Figure 19. The law of sines.

Unlike the circumcenter and orthocenter, the median is always in the interior of the triangle, even when the triangle is right or obtuse. Therefore the signed distance $[R, BC]$ is the positive distance $|R, BC|$ shown in figure 17. Since RF' is the perpendicular to BC that passes through R, $|RF'|$ measures the distance.

20.2 Trilinears of the classical centers

As shown in figure 18, between the median AM and the parallel lines AF and RF' there are two triangles, $\triangle AFM$ and $\triangle RF'M$. They are similar by AA (they share the angle at M and the right angles at F and F' are congruent). Furthermore, because R is located 2/3 of the way down the median from the vertex, $|RM| = \frac{1}{3}|AM|$. The legs of the triangles must be in the same ratio, so $|RF'| = \frac{1}{3}|AF|$.

The goal is to relate $|AF|$ to the sides and angles of the original triangle, and we can now easily do that in two ways. In figure 19, in the right triangle $\triangle AFB$,

$$\sin(\angle B) = \frac{|AF|}{c} \implies |AF| = c\sin(\angle B),$$

and in the right triangle $\triangle AFC$,

$$\sin(\angle C) = \frac{|AF|}{b} \implies |AF| = b\sin(\angle C).$$

Therefore

$$[R, BC] = |RF'| = \tfrac{1}{3}|AF| = \tfrac{1}{3}c\sin(\angle B) = \tfrac{1}{3}b\sin(\angle C)$$

Similarly, we can calculate the distances to the other two sides as

$$[R, AC] = \tfrac{1}{3}a\sin(\angle C) = \tfrac{1}{3}c\sin(\angle A)$$
$$[R, AB] = \tfrac{1}{3}b\sin(\angle A) = \tfrac{1}{3}a\sin(\angle B)$$

and so the exact trilinear coordinates of the centroid can be written as

$$R = \left[\tfrac{1}{3}c\sin(\angle B) : \tfrac{1}{3}a\sin(\angle C) : \tfrac{1}{3}b\sin(\angle A)\right].$$

There is still a little more work to get to the more symmetric form presented in the theorem. From the calculation in step (3),

$$c\sin(\angle B) = b\sin(\angle C) \implies \frac{\sin(\angle B)}{b} = \frac{\sin(\angle C)}{c}.$$

Likewise, the ratio $\sin(\angle A)/a$ has the same value (this is the "law of sines"). Therefore

$$\frac{3a}{\sin(\angle A)} = \frac{3b}{\sin(\angle B)} = \frac{3c}{\sin(\angle C)}.$$

Multiply the first coordinate of R by $3b/\sin(\angle B)$, the second coordinate of R by $3c/\sin(\angle C)$, and the third coordinate of R by $3a/\sin(\angle A)$. Since they are equal, the result is an equivalent set of trilinear coordinates for the centroid $R = [bc : ca : ab]$. □

At this point, it is worth mentioning that there is another important triangular coordinate system called *barycentric coordinates*. The premise of barycentric coordinates is this: the vertices of a triangle are assigned masses m_1, m_2, m_3, and then the point that is the center of mass of the three-point system is given barycentric coordinates $[m_1 : m_2 : m_3]$. Hence, while $[1 : 1 : 1]$ are the trilinear coordinates of the incenter, $[1 : 1 : 1]$ are the barycentric coordinates of the centroid. Barycentric coordinates are usually presented in conjunction with trilinear coordinates as the two are closely related. However, barycentric coordinates are also connected with issues of area, which we have not yet discussed. We will return to barycentric coordinates eventually, in chapter 32.

20.3 Exercises

20.3. (On the uniqueness of trilinear coordinate representations) For $\triangle ABC$, is it possible for two distinct points P and Q to have the same trilinear coordinates?

20.4. What are the trilinear coordinates of the three excenters of a triangle?

20.5. Show that the trilinear coordinates of the center of the nine point circle of $\triangle ABC$ are

$$[\cos((\angle B) - (\angle C)) : \cos((\angle C) - (\angle A)) : \cos((\angle A) - (\angle B))].$$

This one is a little tricky, so here is a hint if you are not sure where to start. Suppose that $\angle B$ is larger than $\angle C$. Label

O : the center of the nine point circle,
P : the circumcenter,
M : the midpoint of BC, and
X : the foot of the perpendicular from O to BC.

The key is to show that $\angle POX$ is congruent to $\angle B$ and that $\angle POM$ is congruent to $\angle C$. That shows that $(\angle MOX) = (\angle B) - (\angle C)$.

Part III

Euclidean Transformations

In the third part of this book, we will look at Euclidean geometry from a different perspective, that of Euclidean transformations. It is a point of view that has been most closely associated with Felix Klein, that the way to study some property (such as congruence) is to study the maps that preserve it. The first chapter sets the scene with a quick development of analytic geometry. Then it is on to Euclidean isometries, bijections of the Euclidean plane that preserve distance. Over several chapters we will study the isometries, and ultimately we will classify all Euclidean isometries into four types: reflections, rotations, translations, and glide reflections. Then it is time to loosen the restriction a bit to consider bijections that preserve congruence, but not necessarily distance. Finally, we will look at inversion, a type of bijection of the punctured plane (the Euclidean plane minus a point). As luck would have it, inversion provides a convenient bridge into non-Euclidean geometry.

21
Analytic Geometry

This chapter is a quick development of analytic geometry and trigonometry in the language of Euclidean geometry. I feel an obligation to provide this bridge between traditional Euclidean geometry (as we have developed it) and more contemporary analytic geometry, but you should already be comfortable with this material, so feel free to skim through it.

21.1 Analytic geometry

At the heart of analytic geometry, there is a correspondence between points and coordinates, ordered pairs of real numbers. The Cartesian approach to the correspondence is a familiar one that goes something like this. Begin with two perpendicular lines (the choice is arbitrary). These are the x- and y-axes. Their intersection is the origin O. We will want to measure signed distances from O along the axes, which means we need to assign a positive direction to them. Once directions have been chosen, each axis will be divided into two rays that share O as their common vertex: a positive axis consisting of points whose signed distance from O is positive, and a negative axis consisting of points whose signed distance from O is negative. From a geometric point of view, the choice of directions is arbitrary, but there is an established convention: the axes are assigned positive directions so that the positive y-axis is a 90° counterclockwise turn from the positive x-axis. Geometry provides no way to distinguish which direction is the counterclockwise direction, so it is a convention that must be communicated by way of illustrations (and clocks).

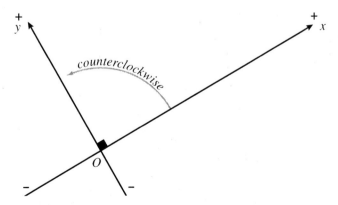

Figure 1. A counterclockwise coordinate frame.

A point P on the x-axis is assigned the coordinates $(p, 0)$, where p is the signed distance from O to P. A point Q on the y-axis is assigned the coordinates $(0, q)$ where q is the signed distance from O to Q. Most points will not lie on either axis. For their coordinates, we must consider their projections onto the axes. If R is such a point, then we draw the lines that pass through R and are perpendicular to the axes. If the points where the perpendiculars cross the axes have coordinates $(a, 0)$ and $(0, b)$, then the coordinates of R are (a, b). With this correspondence, every point corresponds to a unique coordinate pair, and every coordinate pair corresponds to a unique point.

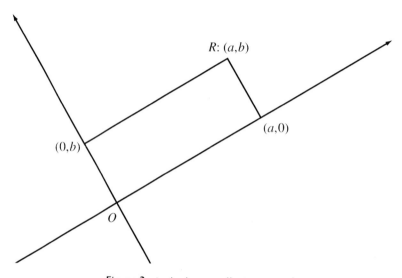

Figure 2. Assigning coordinates to a point.

The next step is to figure out how to calculate the distance between points in terms of their coordinates. This is essential for everything that we are going to do. Let's begin with two special cases.

Lemma 21.1 *(Horizontal and vertical distance)* For points that share an x-coordinate, $P_1 = (x, y_1)$ and $P_2 = (x, y_2)$, $|P_1 P_2| = |y_1 - y_2|$. For points that share a y-coordinate, $P_3 = (x_3, y)$ and $P_4 = (x_4, y)$, $|P_3 P_4| = |x_3 - x_4|$.

Proof We will just prove the second statement. Label two more points, $Q_1 = (x_1, 0)$ and $Q_2 = (x_2, 0)$. Then $\square P_1 P_2 Q_2 Q_1$ is a rectangle, so its opposite sides $P_1 P_2$ and $Q_1 Q_2$ have to be the same length. This is where we make the direct connection between coordinates and distance: the coordinates along each axis were chosen to reflect their signed distance from the origin O. To be thorough, there are several cases to consider:

$$O * Q_1 * Q_2 : \quad |Q_1 Q_2| = |O Q_2| - |O Q_1| = x_2 - x_1 = |x_1 - x_2|$$
$$O * Q_2 * Q_1 : \quad |Q_1 Q_2| = |O Q_1| - |O Q_2| = x_1 - x_2 = |x_1 - x_2|$$
$$Q_1 * O * Q_2 : \quad |Q_1 Q_2| = |O Q_1| + |O Q_2| = -x_1 + x_2 = |x_1 - x_2|$$

21.1 Analytic geometry

$Q_2 * O * Q_1: \quad |Q_1 Q_2| = |OQ_2| + |OQ_1| = -x_2 + x_1 = |x_1 - x_2|$

$Q_1 * Q_2 * O: \quad |Q_1 Q_2| = |OQ_1| - |OQ_2| = -x_1 - (-x_2) = |x_1 - x_2|$

$Q_2 * Q_1 * O: \quad |Q_1 Q_2| = |OQ_2| - |OQ_1| = -x_2 - (-x_1) = |x_1 - x_2|.$

In all cases, $|P_1 P_2| = |Q_1 Q_2| = |x_1 - x_2|$. □

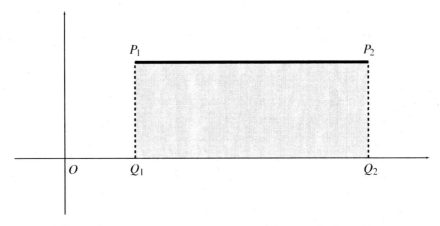

Figure 3. Measuring the distance between points on a horizontal line.

The general distance formula is now an easy consequence of the Pythagorean theorem.

Theorem 21.2 *(The distance formula) For two points $P = (x_1, y_1)$ and $Q = (x_2, y_2)$,*

$$|PQ| = \sqrt{(x_1 - x_2)^2 + (y_1 - y_2)^2}.$$

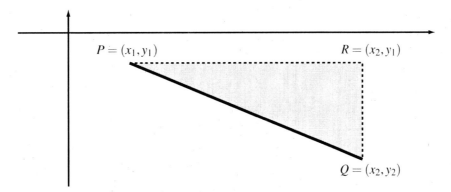

Figure 4. The distance formula is a consequence of the Pythagorean theorem.

Proof If P and Q share either x-coordinates or y-coordinates, then the formula reduces to the special case in the previous lemma (because $\sqrt{a^2} = |a|$). If not, mark one more point: $R = (x_2, y_1)$, the intersection of the horizontal line through P and the vertical line through Q.

Then $|PR| = |x_1 - x_2|$, $|RQ| = |y_1 - y_2|$, and $\triangle PRQ$ is a right triangle. By the Pythagorean theorem,

$$|PQ|^2 = |PR|^2 + |QR|^2 = (x_1 - x_2)^2 + (y_1 - y_2)^2.$$

Take the square root to get the formula. □

Corollary 21.3 *(Equation of a circle) The equation of a circle $\mathcal{C}$ with center at $P = (h, k)$ and radius r is*

$$(x - h)^2 + (y - k)^2 = r^2.$$

Proof By definition, the points of $\mathcal{C}$ are those points that are a distance of r from P. Therefore (x, y) is on $\mathcal{C}$ if and only if

$$\sqrt{(x - h)^2 + (y - k)^2} = r.$$

Square both sides of the equation to get the standard form. □

Now let's try to find equations that will describe lines. The key is the idea that a line describes the shortest path between points. That is captured more formally in the triangle inequality, which you should recall states that $|AB| + |BC| \geq |AC|$, but that the equality only happens when $A * B * C$.

Theorem 21.4 *Given two distinct points $P_1 = (x_1, y_1)$ and $P_2 = (x_2, y_2)$ on a line ℓ, a third point $P = (x, y)$ lies on ℓ if and only if its coordinates can be written in the form*

$$x = x_1 + t(x_2 - x_1) \text{ and } y = y_1 + t(y_2 - y_1)$$

for some $t \in \mathbb{R}$.

Proof The orderings of P, P_1, and P_2 on the line create several cases. Let's consider only the case where t is between 0 and 1 and P is between P_1 and P_2. It is representative of the other two cases.

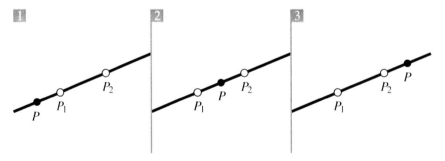

Figure 5. Finding the equation for an arbitrary point P on a line through P_1 and P_2. There are three cases depending upon the order of the points P, P_1, and P_2 on that line.

21.1 Analytic geometry

$\Longrightarrow$ First, let us show that if $P = (x_1 + t(x_2 - x_1), y_1 + t(y_2 - y_1))$ for some value of t between 0 and 1, then P is between P_1 and P_2. We can directly calculate $|P_1 P|$ and $|P P_2|$:

$$|P_1 P| = [(x - x_1)^2 + (y - y_1)^2]^{1/2}$$
$$= [(x_1 + t(x_2 - x_1) - x_1)^2 + (y_1 + t(y_2 - y_1) - y_1)^2]^{1/2}$$
$$= [(tx_2 - tx_1)^2 + (ty_2 - ty_1)^2]^{1/2}$$
$$= t[(x_2 - x_1)^2 + (y_2 - y_1)^2]^{1/2}$$
$$= t|P_1 P_2|.$$

$$|P P_2| = [(x_2 - x)^2 + (y_2 - y)^2]^{1/2}$$
$$= [(x_2 - (x_1 + t(x_2 - x_1)))^2 + (y_2 - (y_1 + t(y_2 - y_1)))^2]^{1/2}$$
$$= [((1 - t)x_2 - (1 - t)x_1)^2 + ((1 - t)y_2 - (1 - t)y_1)^2]^{1/2}$$
$$= (1 - t)[(x_2 - x_1)^2 + (y_2 - y_1)^2]^{1/2}$$
$$= (1 - t)|P_1 P_2|.$$

According to the triangle inequality, P is between P_1 and P_2, since

$$|P_1 P| + |P P_2| = t|P_1 P_2| + (1 - t)|P_1 P_2| = |P_1 P_2|.$$

$\Longleftarrow$ Conversely, let us show that if P is between P_1 and P_2, then the coordinates of P can be written in the parametric form $(x_1 + t(x_2 - x_1), y_1 + t(y_2 - y_1))$ for some value of t between 0 and 1. P is the only point in the plane that is a distance $d_1 = |P_1 P|$ from P_1 and a distance $d_2 = |P P_2|$ from P_2. Because of the uniqueness, we just need to find a point in parametric form that is also those distances from P_1 and P_2. The point that we are looking for is the one where $t = d_1/(d_1 + d_2)$. The two calculations, that the distance from this point to P_1 is d_1, and that the distance from this point to P_2 is d_2, are both straightforward, so are left as exercises.

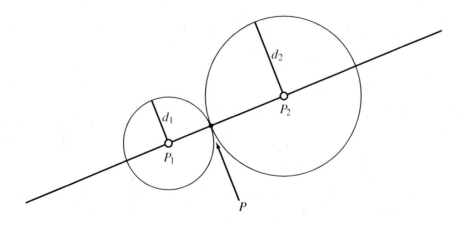

Figure 6. A point P is between P_1 and P_2 if the triangle inequality is an equality. Furthermore, P is the only point that is a distance d_1 from P_1 and a distance d_2 from P_2. $\square$

Exercise 21.1 Complete the calculation in the last proof: using the parametrized form for $r(t)$ given there, show that $r(d_1/(d_1 + d_2))$ gives the coordinates of a point that is a distance d_1 from P_1 and d_2 from P_2.

From the parametric form it is easy to get to standard form, and from there to point-slope form, slope-intercept form, and so on. The latter steps are usually done in a pre-calculus course, so I will only go one step further.

Theorem 21.5 *The coordinates (x, y) of the points of a line satisfy an equation of the form $Ax + By = C$ where A, B, and C are real numbers.*

Proof Suppose that (x_1, y_1) and (x_2, y_2) are distinct points on the line. As we saw in the last theorem, the other points on the line have coordinates (x, y) that satisfy the equations

$$x = x_1 + t(x_2 - x_1)$$
$$y = y_1 + t(y_2 - y_1).$$

Now it is just a matter of combining the equations to eliminate the parameter t.

$$x - x_1 = t(x_2 - x_1)$$
$$y - y_1 = t(y_2 - y_1).$$

It is tempting to divide the second equation by the first. That eliminates t and also serves as a definition of the slope of a line (in particular, it shows that the slope is constant). But we cannot divide by zero. Instead, multiply:

$$(x - x_1)(y_2 - y_1) = t(x_2 - x_1)(y_2 - y_1)$$
$$(y - y_1)(x_2 - x_1) = t(y_2 - y_1)(x_2 - x_1).$$

Set the equations equal and simplify:

$$(x - x_1)(y_2 - y_1) = (y - y_1)(x_2 - x_1)$$
$$x(y_2 - y_1) - x_1(y_2 - y_1) = y(x_2 - x_1) - y_1(x_2 - x_1)$$
$$x(y_2 - y_1) - y(x_2 - x_1) = x_1(y_2 - y_1) - y_1(x_2 - x_1).$$

This equation has the proper form, with

$$A = (y_2 - y_1), \quad B = -(x_2 - x_1), \quad \text{and} \quad C = x_1(y_2 - y_1) - y_1(x_2 - x_1). \qquad \square$$

Exercise 21.2 Suppose that A, B, and C are real numbers. Show that if A and B are not both zero, then the equation $Ax + By = C$ describes a line.

21.2 The unit circle approach to trigonometry

At the end of chapter 13 on similarity, we defined the six trigonometric functions in terms of the angles of a right triangle, which means that they were restricted to values in the interval $(0, \pi/2)$. There is also a unit circle approach that extends the definitions beyond that narrow window. This

21.2 The unit circle approach to trigonometry

is standard fare, so what follows is a brief recap. A point with two positive coordinates (x, y) on the unit circle corresponds to a right triangle whose vertices are $(0, 0)$, $(x, 0)$, and (x, y). If θ is the measure of the angle at the origin, then $\cos\theta = x$ and $\sin\theta = y$ (because the hypotenuse has length one). Now just continue that: any ray from the origin forms an angle θ measured in the counterclockwise direction from the x-axis. The ray intersects the unit circle at a point (x, y) and we define

$$\cos(\theta) = x, \quad \sin(\theta) = y.$$

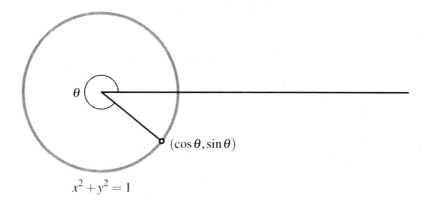

Figure 7. Coordinates on the unit circle correspond to the cosine and sine functions.

Allowing for proper and reflex angles, that extends the domains of sine and cosine to $[0, 2\pi)$, but we can go farther. We can allow the ray to spin around the circle more than once (for θ values greater than 2π) or in the clockwise direction (for negative θ). By imposing periodicity:

$$\cos(\theta + 2n\pi) = \cos(\theta), \quad \sin(\theta + 2n\pi) = \sin(\theta), \quad n \text{ an integer.}$$

The other four trigonometric functions (tangent, cotangent, secant, cosecant) are defined similarly as the ratios

$$\tan(\theta) = y/x, \quad \cot(\theta) = x/y, \quad \sec(\theta) = 1/x, \quad \csc(\theta) = 1/y.$$

There are many relationships among the trigonometric functions, some obvious and some subtle. Let's get the obvious ones out of the way. From the definitions of the functions, we get the reciprocal identities

$$\sec\theta = \frac{1}{\cos\theta}, \quad \csc\theta = \frac{1}{\sin\theta}, \quad \cot\theta = \frac{1}{\tan\theta},$$

and identities that relate tangent and cotangent to sine and cosine:

$$\tan\theta = \frac{\sin\theta}{\cos\theta}, \quad \cot\theta = \frac{\cos\theta}{\sin\theta}.$$

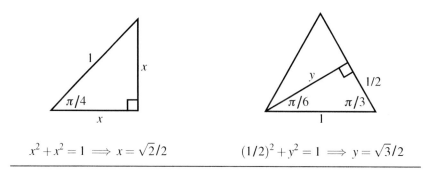

$$x^2 + x^2 = 1 \implies x = \sqrt{2}/2 \qquad (1/2)^2 + y^2 = 1 \implies y = \sqrt{3}/2$$

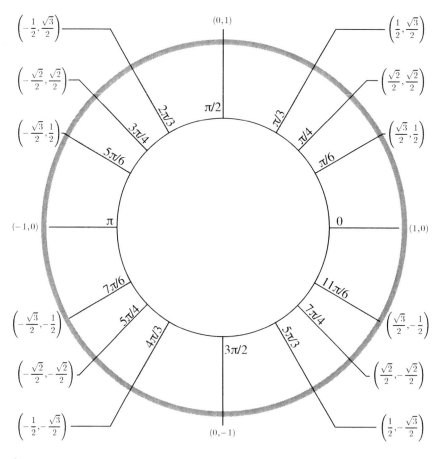

Figure 8. We can use an equilateral triangle and an isosceles right triangle to find the sine and cosine values for the angles $\pi/6$, $\pi/3$, and $\pi/4$. Then use the symmetry of the circle to extend beyond the first quadrant.

21.2 The unit circle approach to trigonometry

From the equation of the circle $x^2 + y^2 = 1$, we get the Pythagorean identities

$$\sin^2 \theta + \cos^2 \theta = 1, \quad \tan^2 \theta + 1 = \sec^2 \theta, \quad 1 + \cot^2 \theta = \csc^2 \theta.$$

By comparing angles taken in the counterclockwise and clockwise directions, we see that cosine and secant are even functions (where $f(-x) = f(x)$) and that the other four are odd functions (where $f(-x) = -f(x)$).

Beyond these, there is a second tier of identities—double angle, half angle, power reduction, etc.—that are not so immediately clear. They can be derived from two big identities, the addition formulas for sine and cosine, but the proofs of those formulas require a more careful look at the geometry of the unit circle. To close this chapter, we will prove them.

Theorem 21.6 *(Addition rule for cosine)* For angles measures α and β,

$$\cos(\alpha + \beta) = \cos \alpha \cos \beta - \sin \alpha \sin \beta.$$

The key to the proof is to compare two distances that we know to be the same, one distance expressed in terms of the angle $\alpha + \beta$, the other in terms of the individual angles α and β. The trick is to make the right choice of distances.

Proof On the unit circle, label:

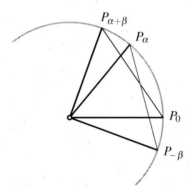

Figure 9. Marking points to derive the cosine addition formula.

$$P_0 = (1, 0)$$
$$P_\alpha = (\cos \alpha, \sin \alpha)$$
$$P_{-\beta} = (\cos(-\beta), \sin(-\beta)) = (\cos \beta, -\sin \beta)$$
$$P_{\alpha + \beta} = (\cos(\alpha + \beta), \sin(\alpha + \beta)).$$

If O is the origin, then $\triangle O P_0 P_{\alpha+\beta}$ and $\triangle O P_{-\beta} P_\alpha$ are congruent (SAS: in each triangle, two of the sides are radii, and the angle between them measures $\alpha + \beta$). So the two segments $P_0 P_{\alpha+\beta}$ and $P_{-\beta} P_\alpha$ are congruent, and we can compare their lengths (it is actually easier to work with the squares of those lengths). Throughout these calculations, we make repeated use of the

Pythagorean identity $\sin^2 x + \cos^2 x = 1$.

$$|P_0 P_{\alpha+\beta}|^2 = (\cos(\alpha+\beta) - 1)^2 + (\sin(\alpha+\beta) - 0)^2$$
$$= \cos(\alpha+\beta)^2 - 2\cos(\alpha+\beta) + 1 + \sin^2(\alpha+\beta)$$
$$= 2 - 2\cos(\alpha+\beta).$$
$$|P_{-\beta} P_\alpha|^2 = (\cos\alpha - \cos\beta)^2 + (\sin\alpha + \sin\beta)^2$$
$$= \cos^2\alpha - 2\cos\alpha\cos\beta + \cos^2\beta + \sin^2\alpha + 2\sin\alpha\sin\beta + \sin^2\beta$$
$$= 2 - 2\cos\alpha\cos\beta + 2\sin\alpha\sin\beta.$$

Set the expressions equal and simplify:

$$2 - 2\cos(\alpha+\beta) = 2 - 2\cos\alpha\cos\beta + 2\sin\alpha\sin\beta$$
$$-2\cos(\alpha+\beta) = -2\cos\alpha\cos\beta + 2\sin\alpha\sin\beta$$
$$\cos(\alpha+\beta) = \cos\alpha\cos\beta - \sin\alpha\sin\beta. \qquad \square$$

Now let's look at the addition rule for the sine function. For this proof, one approach would be to use the cofunction identity $\sin(x) = \cos(\pi/2 - x)$ followed by the addition rule for cosine that we just derived (although we have not yet derived this cofunction identity). Instead, let's contrive something in the spirit of the last proof by comparing some distances and then doing a little algebra.

Theorem 21.7 *(Addition rule for sine)*

$$\sin(\alpha + \beta) = \sin\alpha\cos\beta + \cos\alpha\sin\beta.$$

Proof On the unit circle, label:

$$P_0 = (1, 0)$$
$$P_\alpha = (\cos\alpha, \sin\alpha)$$
$$P_\beta = (\cos\beta, \sin\beta)$$
$$P_{\alpha+\beta} = (\cos(\alpha+\beta), \sin(\alpha+\beta)).$$

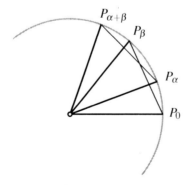

Figure 10. Marking points to derive the sine addition formula.

By SAS, $P_\alpha P_{\alpha+\beta}$ and $P_0 P_\beta$ are congruent. Let's compare the lengths (we use the addition rule for cosine midway through the first distance calculation).

$$\begin{aligned}
|P_\alpha P_{\alpha+\beta}|^2 &= (\cos(\alpha+\beta) - \cos(\alpha))^2 + (\sin(\alpha+\beta) - \sin(\alpha))^2 \\
&= \cos^2(\alpha+\beta) - 2\cos\alpha\cos(\alpha+\beta) + \cos^2\alpha \\
&\quad + \sin^2(\alpha+\beta) - 2\sin\alpha\sin(\alpha+\beta) + \sin^2\alpha \\
&= 2 - 2\cos\alpha\cos(\alpha+\beta) - 2\sin\alpha\sin(\alpha+\beta) \\
&= 2 - 2\cos\alpha(\cos\alpha\cos\beta - \sin\alpha\sin\beta) - 2\sin\alpha\sin(\alpha+\beta) \\
&= 2 - 2\cos^2\alpha\cos\beta + 2\sin\alpha\cos\alpha\sin\beta - 2\sin\alpha\sin(\alpha+\beta)
\end{aligned}$$

and

$$\begin{aligned}
|P_0 P_\beta|^2 &= (\cos\beta - 1)^2 + (\sin\beta - 0)^2 \\
&= \cos^2\beta - 2\cos\beta + 1 + \sin^2\beta \\
&= 2 - 2\cos\beta.
\end{aligned}$$

Set the expressions equal and simplify:

$$\begin{aligned}
2 - 2\cos^2\alpha\cos\beta + 2\sin\alpha\cos\alpha\sin\beta - 2\sin\alpha\sin(\alpha+\beta) &= 2 - 2\cos\beta \\
-2\cos^2\alpha\cos\beta + 2\sin\alpha\cos\alpha\sin\beta - 2\sin\alpha\sin(\alpha+\beta) &= -2\cos\beta \\
\cos^2\alpha\cos\beta - \sin\alpha\cos\alpha\sin\beta + \sin\alpha\sin(\alpha+\beta) &= \cos\beta.
\end{aligned}$$

Solve for $\sin(\alpha+\beta)$:

$$\begin{aligned}
\sin\alpha\sin(\alpha+\beta) &= \cos\beta - \cos^2\alpha\cos\beta + \sin\alpha\cos\alpha\sin\beta \\
&= \cos\beta(1 - \cos^2\alpha) + \sin\alpha\cos\alpha\sin\beta \\
&= \cos\beta\sin^2\alpha + \sin\alpha\cos\alpha\sin\beta \\
&= \sin\alpha(\sin\alpha\cos\beta + \cos\alpha\sin\beta).
\end{aligned}$$

As long as $\sin\alpha$ is not zero, we can divide both sides by it, and what's left is what we want. That leaves one loose thread: what if $\sin\alpha$ is zero? That is left as an exercise. $\square$

Exercise 21.3 Suppose that $\sin(\alpha) = 0$, and hence that α is an integer multiple of π. Prove that the addition formula for sine holds in this case as well.

21.3 Exercises

21.4. (The midpoint formula) Let $P = (a, b)$ and $Q = (c, d)$. Verify that the coordinates of the midpoint of PQ are

$$\left(\frac{a+c}{2}, \frac{b+d}{2}\right).$$

21.5. Show that the points on the circle with center (h, k) and radius r can be described by the parametric equations
$$x(\theta) = h + r\cos\theta$$
$$y(\theta) = k + r\sin\theta.$$

21.6. Let ℓ_1 and ℓ_2 be perpendicular lines, neither vertical. Show that the slopes of ℓ_1 and ℓ_2 are negative reciprocals of one another.

21.7. Verify that the triangle with vertices at $(0, 0)$, $(2a, 0)$, and $(a, a\sqrt{3})$ is equilateral.

21.8. Find the equation of the circle that passes through $(0, 0)$, $(4, 2)$, and $(2, 6)$.

21.9. Let $\triangle ABC$ be the triangle with vertices $A = (0, 0)$, $B = (1, 0)$, and $C = (a, b)$. Find the coordinates of its circumcenter, orthocenter, and centroid in terms of a and b.

21.10. All the angle values identified in figure 8 can be written in the form $n\pi/12$, but not all values of that form are represented. Find the coordinates on the unit circle for the angles $\theta = \pi/12, 5\pi/12, 7\pi/12$, and $11\pi/12$.

The next few exercises verify some common trigonometric identities that we will need to for later calculations. You don't need to do them all—I just want to have all of these identities together in one place.

21.11. Use the addition formulas to derive the cofunction identities
$$\sin\left(\frac{\pi}{2} - \theta\right) = \cos\theta \qquad \cos\left(\frac{\pi}{2} - \theta\right) = \sin\theta$$
$$\tan\left(\frac{\pi}{2} - \theta\right) = \cot\theta \qquad \cot\left(\frac{\pi}{2} - \theta\right) = \tan\theta$$
$$\sec\left(\frac{\pi}{2} - \theta\right) = \csc\theta \qquad \csc\left(\frac{\pi}{2} - \theta\right) = \sec\theta.$$

21.12. Use the addition formulas to derive the double angle formulas
$$\sin(2\theta) = 2\sin\theta\cos\theta$$
$$\cos(2\theta) = \cos^2\theta - \sin^2\theta$$
$$= 2\cos^2\theta - 1$$
$$= 1 - 2\sin^2\theta$$
$$\tan(2\theta) = \frac{2\tan\theta}{1 - \tan^2\theta}.$$

21.13. Use the double angle formulas for cosine to derive the power-reduction formulas
$$\sin^2\theta = \frac{1 - \cos(2\theta)}{2}$$
$$\cos^2\theta = \frac{1 + \cos(2\theta)}{2}$$
$$\tan^2\theta = \frac{1 - \cos(2\theta)}{1 + \cos(2\theta)}.$$

21.3 Exercises

21.14. Use the power-reduction formulas to derive the half-angle formulas

$$\sin \frac{\theta}{2} = \pm \sqrt{\frac{1 - \cos \theta}{2}}$$

$$\cos \frac{\theta}{2} = \pm \sqrt{\frac{1 + \cos \theta}{2}}$$

$$\tan \frac{\theta}{2} = \frac{1 - \cos \theta}{\sin \theta} = \frac{\sin \theta}{1 + \cos \theta}.$$

21.15. Verify the product-to-sum formulas

$$\sin \alpha \sin \beta = \frac{1}{2}[\cos(\alpha - \beta) - \cos(\alpha + \beta)]$$

$$\cos \alpha \cos \beta = \frac{1}{2}[\cos(\alpha + \beta) + \cos(\alpha - \beta)]$$

$$\sin \alpha \cos \beta = \frac{1}{2}[\sin(\alpha + \beta) + \sin(\alpha - \beta)].$$

21.16. Verify the sum-to-product formulas

$$\sin \alpha + \sin \beta = 2 \sin\left(\frac{\alpha + \beta}{2}\right) \cos\left(\frac{\alpha - \beta}{2}\right)$$

$$\sin \alpha - \sin \beta = 2 \cos\left(\frac{\alpha + \beta}{2}\right) \sin\left(\frac{\alpha - \beta}{2}\right)$$

$$\cos \alpha + \cos \beta = 2 \cos\left(\frac{\alpha + \beta}{2}\right) \cos\left(\frac{\alpha - \beta}{2}\right)$$

$$\cos \alpha - \cos \beta = -2 \sin\left(\frac{\alpha + \beta}{2}\right) \sin\left(\frac{\alpha - \beta}{2}\right).$$

21.17. (A parabola) Let $Q = (0, a)$ and ℓ be the line $y = -a$, where $a \neq 0$. Show that the points $P = (x, y)$ so that $d(P, Q) = d(P, \ell)$ satisfy the equation $y = x^2/(4a)$. The point Q is called the focus, and the line $y = -a$ is called the directrix.

21.18. (An ellipse) Let $Q_1 = (0, a/2)$, $Q_2 = (0, -a/2)$, and let $c > |a|$, where $a \neq 0$. Show that the points $P = (x, y)$ so that

$$d(P, Q_1) + d(P, Q_2) = c$$

satisfy the equation

$$\frac{x^2}{b^2} + \frac{y^2}{c^2} = \frac{1}{4}$$

where $b^2 = c^2 - a^2$. The points Q_1 and Q_2 are called the foci of the ellipse.

21.19. (A hyperbola) Let $Q_1 = (0, a/2)$, $Q_2 = (0, -a/2)$, and let $|c| < |a|$, where $a \neq 0$. Show that the points $P = (x, y)$ so that
$$d(P, Q_1) - d(P, Q_2) = c$$
satisfy the equation
$$-\frac{x^2}{b^2} + \frac{y^2}{c^2} = \frac{1}{4}$$
where $b^2 = a^2 - c^2$. The points Q_1 and Q_2 are called foci of the hyperbola.

22

Isometries

One way to study a mathematical object is to study the types of maps that preserve it, that is, that leave it invariant. For instance, in group theory we study group homomorphisms because they preserve the group operation (in the sense that $f(a \cdot b) = f(a) \cdot f(b)$). In Euclidean geometry there are several structures that might be worth preserving—incidence, order, congruence—but in the next few chapters our focus will be on mappings that preserve distance.

22.1 Definitions

Let's start with a review of some basic terminology associated with maps from one set to another.

Definition 22.1 A map $f : X \to Y$ is:

- *one-to-one* if for all x_1, x_2 in X, $f(x_1) = f(x_2) \implies x_1 = x_2$,
- *onto* if for every $y \in Y$ there is an $x \in X$ such that $f(x) = y$,
- *bijective* if it is one-to-one and onto.

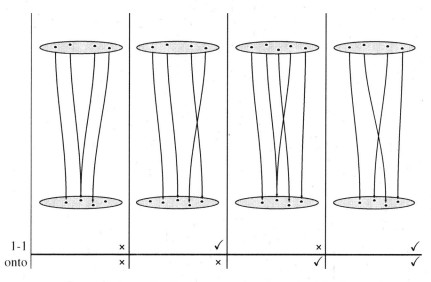

Figure 1. A mapping that is one-to-one and onto is a bijection.

Under the right circumstances, two mappings may be chained together: the composition of $f : X \to Y$ and $g : Y \to Z$ is

$$g \circ f : X \to Z : g \circ f(x) = g(f(x)).$$

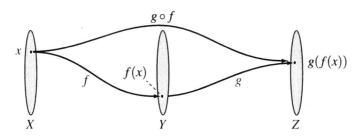

Figure 2. The composition of f and g.

This type of composition is usually not commutative and in fact, $f \circ g$ may not even be defined. It is associative, though, and that is an essential property. For any space X the map $id : X \to X$ defined by $id(x) = x$ is called the identity map. Two maps $f : X \to Y$ and $g : Y \to X$ are inverses of one another if $f \circ g$ is the identity map on Y and $g \circ f$ is the identity map on X. For a map to have an inverse, it must be bijective (and conversely, any bijection is invertible).

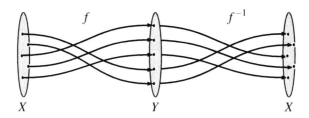

Figure 3. A function and its inverse.

Definition 22.2 An *automorphism* is a bijective mapping from a set to itself.

We are interested in automorphisms of the Euclidean plane, but not just any automorphisms. We want those that do not distort the distances between points. They are called Euclidean isometries.

Definition 22.3 Let $\mathbb{E}$ denote the set of points of the Euclidean plane. A Euclidean *isometry* is an automorphism $f : \mathbb{E} \to \mathbb{E}$ that preserves the distance between points: for all A, B in $\mathbb{E}$, $|f(A)f(B)| = |AB|$.

At this point, there are a few fairly fundamental and obvious points to make that, taken together, show that the set of Euclidean isometries is a group under the composition operation.

Lemma 22.4 *The composition of two isometries is an isometry. The identity map is an isometry. The inverse of an isometry is an isometry.*

22.1 Definitions

Exercise 22.1 Prove the previous lemma.

Everything we have done in Euclidean geometry floats on five undefined terms: point, line, on, between, and congruence. An isometry is defined in terms of its behavior on points, but the distance preservation condition has implications for the remaining undefined terms as well.

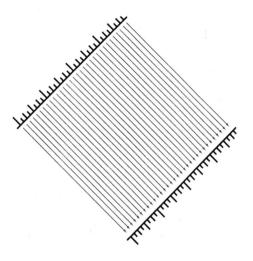

Figure 4. An isometry is a bijection that preserves distances between points.

Lemma 22.5 *An isometry preserves segment and angle congruence. That is,*

$$AB \simeq A'B' \implies f(A)f(B) \simeq f(A')f(B')$$
$$\angle ABC \simeq \angle A'B'C' \implies \angle f(A)f(B)f(C) \simeq \angle f(A')f(B')f(C').$$

Proof Isometries preserve distance and hence segment length, and it is those lengths that determine whether or not segments are congruent. That is, if $AB \simeq A'B'$, then

$$|f(A)f(B)| = |AB| = |A'B'| = |f(A')f(B')|,$$

and so $f(A)f(B) \simeq f(A')f(B')$.

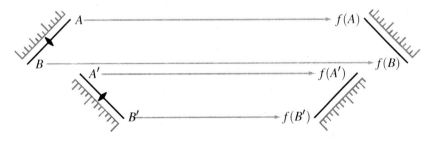

Figure 5. Isometries preserve the congruence relation between segments.

Now consider the case of angle congruence. Relocate, if necessary, A' and C' on their rays so that $BA \simeq B'A'$ and $BC \simeq B'C'$. By SAS, $\triangle ABC$ and $\triangle A'B'C'$ are congruent. The corresponding sides of the triangles are congruent, and from the first part of the proof, the congruences are transferred by f:

$$AB \simeq A'B' \implies f(A)f(B) \simeq f(A')f(B'),$$
$$BC \simeq B'C' \implies f(B)f(C) \simeq f(B')f(C'),$$
$$CA \simeq C'A' \implies f(C)f(A) \simeq f(C')f(A').$$

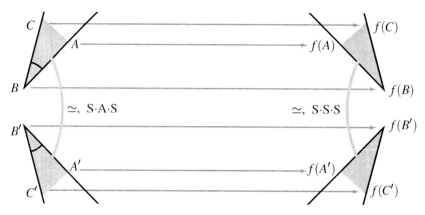

Figure 6. Isometries preserve the congruence relation between angles.

By SSS, $\triangle f(A)f(B)f(C)$ and $\triangle f(A')f(B')f(C')$ are congruent, and so $\angle f(A)f(B)f(C)$ and $\angle f(A')f(B')f(C')$ are congruent. $\square$

Exercise 22.2 Prove that if f is an isometry, then for an angle $\angle ABC$, $(\angle ABC) = (\angle f(A)f(B)f(C))$.

Therefore an isometry f preserves segment length and angle measure. This is useful and we will use it in the last proof of this chapter.

Lemma 22.6 *If A, B, and C are collinear in the order $A * B * C$ and f is an isometry, then $f(A)$, $f(B)$, and $f(C)$ are collinear in the order $f(A) * f(B) * f(C)$.*

Proof Suppose $A * B * C$. Then, by segment addition,

$$|AC| = |AB| + |BC|.$$

Distance is invariant under f, so we can make the substitutions

$$|f(A)f(B)| = |AB|, \quad |f(B)f(C)| = |BC|, \quad \text{and} \quad |f(A)f(C)| = |AC|,$$

to get

$$|f(A)f(C)| = |f(A)f(B)| + |f(B)f(C)|.$$

22.2 Fixed points

This is the degenerate case of the triangle inequality: the only way the equation can be true is if $f(A)$, $f(B)$, and $f(C)$ are collinear with $f(B)$ between $f(A)$ and $f(C)$. □

In the last result we were talking about three points, but by extension, it implies that all the points on a line are mapped to collinear points. In other words, an isometry, which is defined as a bijection of points, is also a bijection of the lines of the geometry. Further, an isometry maps segments to segments, rays to rays, angles to angles, and circles to circles. This suggests an opportunity to simplify notation. When applying an isometry f to a segment AB, for example, instead of writing $f(A)f(B)$, we can write the more streamlined $f(AB)$. For $\angle ABC$, we can write $f(\angle ABC)$ instead of the rather inelegant $\angle f(A)f(B)f(C)$, and so on.

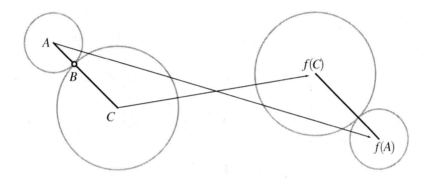

Figure 7. An isometry preserves the relations of incidence and order: there is only one possible location for $f(B)$.

22.2 Fixed points

The goal of the next few chapters is to classify Euclidean isometries. It turns out that one of the keys to this is *fixed points*.

Definition 22.7 A point P is a *fixed point* of an isometry f if $f(P) = P$.

The first step towards a classification is to answer the question

Given isometries f_1 and f_2, which may be described in different ways, how do we figure out if they are really the same?

Showing that they are not the same is usually easy: we just need to find one point P where $f_1(P) \neq f_2(P)$. Showing that they *are* the same seems like a more difficult task. Isometries are functions of the Euclidean plane. Without any additional structure, the only way to show two functions are equal is to show that they agree on the value of all points. Fortunately, the bijection and distance-preserving properties of an isometry impose constraints on its behavior that allow us to determine whether two isometries are the same by looking at just a few points.

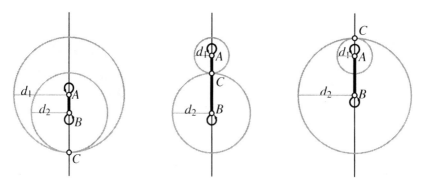

Figure 8. If an isometry fixes two points (A and B) on a line, then it fixes all points on the line. The image of a third point C must be the same distances from A and B as C is.

Theorem 22.8 *If an isometry f fixes two distinct points A and B, then it fixes all the points on $\overline{AB}$.*

Proof Let C be a third point on $\overline{AB}$. Label its distances from A as d_1 and from B as d_2. The key here is that C is the only point that is a distance d_1 from A and a distance d_2 from B (this is intuitively clear, but for a more formal point of view, look back at our investigation of the intersections of circles in chapter 14). Now apply the isometry f to the points. This will not alter distances, so $f(C)$ is still a distance d_1 from $f(A) = A$, and $f(C)$ is still a distance d_2 from $f(B) = B$, so $f(C)$ must be C. □

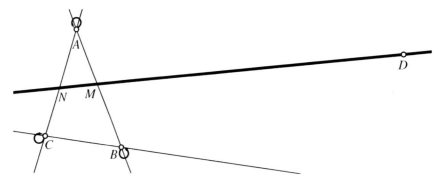

Figure 9. The argument for why an isometry that fixes three non-collinear points (A, B, and C) fixes all the points of the plane.

Theorem 22.9 *If an isometry f fixes three non-collinear points A, B, and C, then it fixes all points (it is the identity isometry).*

Proof By the last result, f fixes the points on $\overline{AB}$, $\overline{AC}$, and $\overline{BC}$. Suppose that D is a point that is not on any of the lines. We need to show that D is a fixed point as well. Choose a point M that is between A and B. It is fixed by f. According to Pasch's lemma, $\overline{DM}$ must intersect at least one other side of $\triangle ABC$. Call the intersection N. It too is fixed by f. Therefore D is on a line $\overline{MN}$ with two fixed points. According to the previous result, it is a fixed point. □

22.3 The analytic viewpoint

Now we can answer the question posed at the start of this section: how much do we need to know about two isometries before we can say they are the same?

Theorem 22.10 *If two isometries f_1 and f_2 agree on three non-collinear points, then they are equal.*

Proof Suppose that A, B, and C are three non-collinear points, and that

$$f_1(A) = f_2(A), \quad f_1(B) = f_2(B), \quad \text{and} \quad f_1(C) = f_2(C).$$

Applying f_2^{-1} to the equations,

$$f_2^{-1} \circ f_1(A) = f_2^{-1} \circ f_2(A) = id(A) = A,$$
$$f_2^{-1} \circ f_1(B) = f_2^{-1} \circ f_2(B) = id(B) = B,$$
$$f_2^{-1} \circ f_1(C) = f_2^{-1} \circ f_2(C) = id(C) = C.$$

Therefore $f_2^{-1} \circ f_1$ has three non-collinear fixed points. It must be the identity, and so

$$f_2^{-1} \circ f_1 = id$$
$$f_2 \circ f_2^{-1} \circ f_1 = f_2 \circ id$$
$$id \circ f_1 = f_2$$
$$f_1 = f_2. \qquad \square$$

22.3 The analytic viewpoint

We now look at isometries from the analytic point of view. Any isometry defines a function on coordinate pairs. As we have seen, isometries themselves are structured, so it makes sense, then, that the functions they define on the coordinate pairs would have to be similarly inflexible. That is indeed the case.

Theorem 22.11 *(General form for an isometry) A Euclidean isometry T has analytic equations that can be written in one of two matrix forms:*

$$T \begin{pmatrix} x \\ y \end{pmatrix} = \begin{pmatrix} h \\ k \end{pmatrix} + \begin{pmatrix} \cos\theta & -\sin\theta \\ \sin\theta & \cos\theta \end{pmatrix} \begin{pmatrix} x \\ y \end{pmatrix}$$

or

$$T \begin{pmatrix} x \\ y \end{pmatrix} = \begin{pmatrix} h \\ k \end{pmatrix} + \begin{pmatrix} \cos\theta & \sin\theta \\ \sin\theta & -\cos\theta \end{pmatrix} \begin{pmatrix} x \\ y \end{pmatrix}$$

where h, k, and θ are real numbers.

Proof Let T be an isometry. We want to know the effect of T on an arbitrary point (x, y), but it will take a few steps to get there. We will start with the origin, move to the point $(x, 0)$, and then to (x, y).

The origin $(0, 0)$. This is the easy one. Since the origin is our first point of consideration, there are no limitations on where it goes (we don't know it yet, but there are isometries that take any point to any other point of the plane). Set h and k by looking at what happens to the origin: set $(h, k) = T(0, 0)$.

Figure 10. Tracking points under an isometry: the origin.

The point $(x, 0)$. An isometry preserves distances, and the distance from $(x, 0)$ to the origin is $|x|$. Applying the isometry to both points, the distance from $T(x, 0)$ to (h, k) also has to be $|x|$. In other words, $T(x, 0)$ is on the circle with center (h, k) and radius $|x|$. As shown in Exercise 21.5, in which you were to derive the standard parametrization of a circle (commonly used in calculus), $T(x, 0)$ has the form

$$(h + |x| \cos \theta, \ k + |x| \sin \theta)$$

for some value of θ. You can verify that the absolute value signs are not needed and we can choose θ so that $T(x, 0) = (h + x \cos \theta, k + x \sin \theta)$.

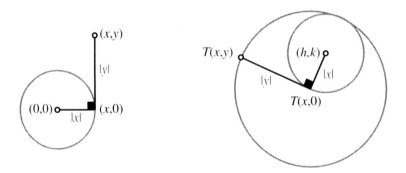

Figure 11. Tracking points under an isometry: the points $(x, 0)$ and (x, y).

The point (x, y). Since the distance from $(x, 0)$ to (x, y) is $|y|$, $T(x, y)$ has to be on the circle centered at $T(x, 0)$ with radius $|y|$. Therefore its coordinates can be written in the form

$$(h + x \cos \theta + |y| \cos \phi, k + x \sin \theta + |y| \sin \phi)$$

for some value of ϕ. The possibilities are more limited than that, though: the rays from $(x, 0)$ through $(0, 0)$ and (x, y) form a right angle at $(x, 0)$. Since an isometry preserves angle measures, the images of the rays must also form a right angle. This can only happen if $\phi = \theta + \pi/2$ or $\phi = \theta - \pi/2$. As before, the absolute value signs around the y can be dropped so $T(x, y)$ has coordinates

$$\left(h + x\cos\theta + y\cos\left(\theta \pm \frac{\pi}{2}\right), k + x\sin\theta + y\sin\left(\theta \pm \frac{\pi}{2}\right)\right).$$

Now use the addition formulas for sine and cosine

$$\cos(\theta \pm \pi/2) = \cos\theta\cos(\pm\pi/2) - \sin\theta\sin(\pm\pi/2) = \mp\sin\theta$$
$$\sin(\theta \pm \pi/2) = \sin\theta\cos(\pm\pi/2) + \cos\theta\sin(\pm\pi/2) = \pm\cos\theta$$

and the coordinates for $T(x, y)$ take on one of two forms:

$$T(x, y) = (h + x\cos\theta - y\sin\theta, k + x\sin\theta + y\cos\theta)$$

or

$$T(x, y) = (h + x\cos\theta + y\sin\theta, k + x\sin\theta - y\cos\theta).$$

Written in matrix form, these are

$$T\begin{pmatrix}x\\y\end{pmatrix} = \begin{pmatrix}h\\k\end{pmatrix} + \begin{pmatrix}\cos\theta & -\sin\theta\\ \sin\theta & \cos\theta\end{pmatrix}\begin{pmatrix}x\\y\end{pmatrix}$$

or

$$T\begin{pmatrix}x\\y\end{pmatrix} = \begin{pmatrix}h\\k\end{pmatrix} + \begin{pmatrix}\cos\theta & \sin\theta\\ \sin\theta & -\cos\theta\end{pmatrix}\begin{pmatrix}x\\y\end{pmatrix}.$$ □

22.4 Exercises

22.3. Let T be an isometry and let r be a ray with endpoint O. Prove that $T(r)$ is also a ray, with endpoint $T(O)$.

22.4. Verify that if ℓ_1 and ℓ_2 are parallel lines and T is an isometry, then $T(\ell_1)$ and $T(\ell_2)$ are parallel.

22.5. Let T be an isometry and let A and B be two points that are on the same side of a line ℓ. Prove that $T(A)$ and $T(B)$ are on the same side of $T(\ell)$.

22.6. Let T be an isometry and let D be a point in the interior of $\angle ABC$. Prove that $T(D)$ is a point in the interior of $T(\angle ABC)$.

22.7. Suppose that T is an isometry that fixes two distinct points P and Q, and that there exist two distinct points R_1 and R_2 so that $T(R_1) = R_2$ and $T(R_2) = R_1$. Prove that $T = T^{-1}$.

22.8. Suppose that a circle $\mathcal{C}$ is invariant under an isometry T. That is, if P is on $\mathcal{C}$, then so is $T(P)$. Prove that the center of $\mathcal{C}$ is a fixed point of T.

22.9. Let M be the midpoint of a segment AB, and let T be an isometry so that $T(A) = B$ and $T(B) = A$. Prove that M is a fixed point of the isometry.

22.10. Given a proper angle $\angle ABC$ and an isometry T such that $T(\overrightarrow{BA}) = \overrightarrow{BC}$ and $T(\overrightarrow{BC}) = \overrightarrow{BA}$, show that T fixes all the points of the angle bisector of $\angle ABC$.

22.11. In the final theorem of this chapter we saw that every isometry can be written in one of two forms. Prove the converse, that a mapping of that form is an isometry.

23
Reflections

This chapter introduces the first type of isometry, reflection across a line. As it turns out, reflections are the building blocks for all isometries. In this chapter we will see why, in the three reflections theorem. This theorem provides the strategy that we will use over the next few chapters to classify all isometries. First, we must define a reflection.

Definition 23.1 Given a line ℓ, the *reflection* across ℓ is the mapping $s : \mathbb{E} \to \mathbb{E}$ so that

- if P is on ℓ, then $s(P) = P$, and
- if P is not on ℓ, then $s(P)$ is the unique point on the line through P perpendicular to ℓ that is the same distance from ℓ as P, but that is on the opposite side of ℓ from P.

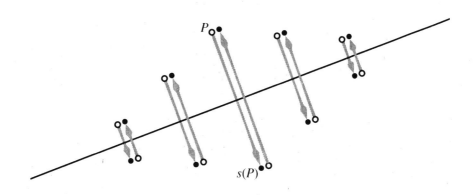

Figure 1. Reflecting points across a line.

We first verify that a reflection is an isometry.

Theorem 23.2 *A reflection is an isometry.*

Proof It is easy to see that a reflection s is a bijection. Look at the composition $s \circ s$: the swap of points done by the first application of s is immediately undone by the second application of s, so that $s^2 = id$. Therefore s is its own inverse, and for a mapping to have an inverse, it must be a bijection.

249

The other step is to show that s preserves distances, that $|s(PQ)| = |PQ|$ for any points P and Q. The only thing that makes this part difficult is that there are many possible positions of P and Q relative to each other and to ℓ, the line of reflection:

I: P and Q are both on ℓ.
II: One of P and Q is on ℓ, while the other is not.
  [1] $\overline{PQ}$ is perpendicular to ℓ
  [2] $\overline{PQ}$ is not perpendicular to ℓ
III: Neither P nor Q is on ℓ.
  [1] $\overline{PQ}$ is perpendicular to ℓ
    [i] P and Q are on the same side of ℓ
    [ii] P and Q are on opposite sides of ℓ
  [2] $\overline{PQ}$ is not perpendicular to ℓ
    [i] P and Q are on the same side of ℓ
    [ii] P and Q are on opposite sides of ℓ

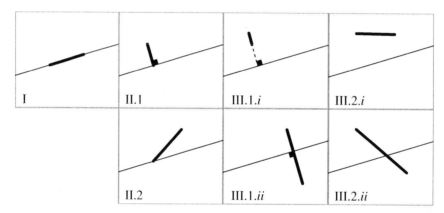

Figure 2. Verifying that a reflection is an isometry. There are many cases, depending on the position of the segment PQ relative to the reflecting line ℓ.

At this point, none of these cases should cause any trouble. Consider case III.2.*i*, which is the archetypal case in this proof. To verify the claim in this case, first label two more points (both of which are fixed by s):

F_P: the foot of the perpendicular to ℓ through P, and
F_Q: the foot of the perpendicular to ℓ through Q.

From the definition of a reflection,

$$PF_P \simeq s(PF_P) \quad \text{and} \quad QF_Q \simeq s(QF_Q)$$

and the angles at F_P and F_Q are right angles. Because F_PF_Q is congruent to itself, by SASAS, $\square PF_PF_QQ$ and $s(\square PF_PF_QQ)$ are congruent, and therefore PQ and $s(PQ)$ are the same length. □

We saw in the last chapter that if an isometry fixes two points, it must fix all the points on the line through them. Of course, every reflection fixes all the points of a line. A good question

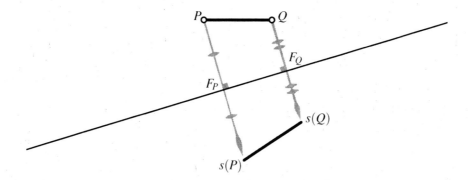

Figure 3. Use SASAS to verify that reflections preserve isometries in case III.2.*i*.

to ask, then, is how common is this "line-fixing" behavior? Not that common, as it turns out, and so this is a useful characterization of a reflection.

Theorem 23.3 *If an isometry fixes all the points of a line but is not the identity, then it must be a reflection.*

Proof Let f be an isometry that fixes all the points on a line ℓ but is not the identity. Let s be the reflection across ℓ. We already know that f and s agree on ℓ, so we just need to show that they agree on one point that isn't on ℓ. Take two points A and B on ℓ and a third point C that is not on ℓ. Since an isometry preserves distance, and since A and B are fixed,

$$AC \simeq f(AC) \simeq Af(C) \implies f(C) \text{ is on the circle with center } A \text{ and radius } |AC|, \text{ and}$$
$$BC \simeq f(BC) \simeq Bf(C) \implies f(C) \text{ is on the circle with center } B \text{ and radius } |BC|.$$

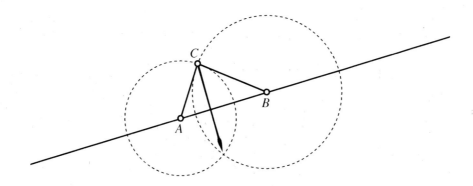

Figure 4. If a non-identity isometry fixes a line, then it is a reflection.

Therefore $f(C)$ is at an intersection of these two circles, and there are only two (distinct circles intersect at most twice). Furthermore, one of them is C, and if $f(C) = C$, then f would fix three non-collinear points and would be the identity. We excluded that possibility, so $f(C)$ has to be the other point of intersection. For the same reasons, $s(C)$ must be the second intersection. Therefore $f(C) = s(C)$; the two isometries agree on three non-collinear points, A, B, and C and so they must be equal. □

Theorem 23.4 *(Three reflections theorem) Any isometry can be written as a reflection, a composition of two reflections, or a composition of three reflections.*

We saw in the last chapter that when isometries agree on three non-collinear points they have to be the same. That is how we will proceed. We just need to find a composition of up to three reflections $s_3 \circ s_2 \circ s_1$ that agrees with T on three non-collinear points. There are three steps to this, as we build up each of the three reflections. At each step we will make the composition of reflections match T on another point without moving any of the previously set points.

Proof Let A, B, and C be three non-collinear points and let T be an isometry.

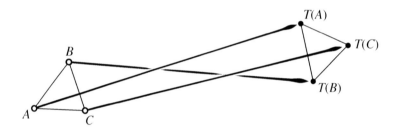

Figure 5. The image of a triangle under an isometry T.

Step One. With the first isometry, s_1, we put A into position. If $A = T(A)$, let s_1 be the identity isometry. If $A \neq T(A)$, let s_1 be the reflection across the perpendicular bisector of $AT(A)$. Either way, $s_1(A) = T(A)$.

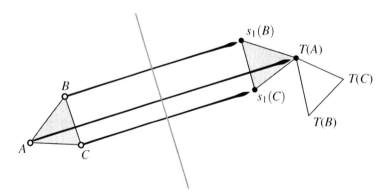

Figure 6. The first reflection.

Step Two. With the second isometry, s_2, we put B into position. To do this, we look at where $s_1(B)$ ended up after step one. It is possible (but unlikely) that $s_1(B)$ ended up on the line $T(\overline{AB})$. If that is the case, then because

$$|s_1(AB)| = |AB| = |T(AB)|,$$

there are only two possible spots for $s_1(B)$, one on either side of $T(A)$. If $s_1(B)$ is on the same side of $T(A)$ as $T(B)$, then $s_1(B) = T(B)$ already, so we can let s_2 be the identity isometry. If $s_1(B)$ is on the opposite side of $T(A)$ from $T(B)$, then let s_2 be the reflection across the line that passes through $T(A)$ and is perpendicular to $s_1(B)T(B)$. That reflection fixes $T(A)$ and maps $s_1(B)$ to $T(B)$.

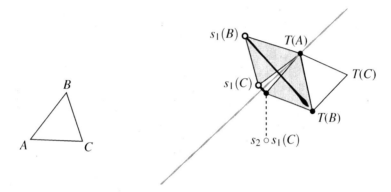

Figure 7. The second reflection.

The more likely possibility is that $s_1(B)$ is not on $T(\overline{AB})$. In that case, let s_2 be the reflection across the bisector of $\angle s_1(B)T(A)T(B)$. Then $T(A)$ is on the line of reflection, so it will be fixed by s_2. Furthermore, the reflecting line cuts $\triangle s_1(B)T(A)T(B)$ in two pieces, that, by SAS, are congruent. Therefore the reflecting line is the perpendicular bisector to $s_1(B)T(B)$, and that means s_2 will map $s_1(B)$ to $T(B)$. Here is where we stand after step two:

$$s_2 \circ s_1(A) = s_2 \circ T(A) = T(A),$$
$$s_2 \circ s_1(B) = T(B).$$

Step Three. That leaves point C. As in the previous step, what we do next depends upon where $s_2 \circ s_1(C)$ is. There aren't many possibilities. We know that $s_2 \circ s_1(AB) = T(AB)$, and we know that $s_2 \circ s_1(\triangle ABC)$ is congruent to $\triangle ABC$, which is congruent to $T(\triangle ABC)$. There are only two ways to build that triangle on a given side of $T(AB)$, one on either side of it. If $s_2 \circ s_1(C)$ is on the same side of $T(AB)$ as $T(C)$, then $s_2 \circ s_1(C) = T(C)$ already, so let s_3 be the identity map. If $s_2 \circ s_1(C)$ is on the opposite side of $T(AB)$ from $T(C)$, then let s_3 be the reflection across the line $T(AB)$. That fixes $T(A)$ and $T(B)$, but maps $s_2 \circ s_1(C)$ to $T(C)$. Putting it all together,

$$s_3 \circ s_2 \circ s_1(A) = T(A)$$
$$s_3 \circ s_2 \circ s_1(B) = T(B)$$
$$s_3 \circ s_2 \circ s_1(C) = T(C).$$

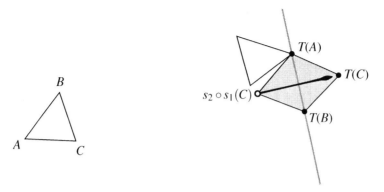

Figure 8. The third reflection.

Since the two isometries agree on three non-collinear points, they must be the same. As long as at least one of s_1, s_2, and s_3 is a reflection, we have met the requirements of the theorem. What if all of them are the identity map? In that case, T is the identity map, and the identity can be written as the composition of any reflection s with itself: $T = s \circ s$. □

Over the next few chapters, we will use this result to classify all isometries. In the next chapter, we will look at what happens when we compose two reflections. Then, after a brief discussion of orientation, we will look at what happens when we tack on a third reflection.

23.1 The analytic viewpoint

It is a little messy to try to work out an equation for an arbitrary reflection at this point. We can, however, find an equation for a reflection across a line that passes through the origin. Let's close out this chapter by doing so.

Theorem 23.5 *Let ℓ be a line through the origin, and let (a, b) be the coordinates of an intersection of ℓ with the unit circle. Then the reflection s across ℓ is given by*

$$s\begin{pmatrix} x \\ y \end{pmatrix} = \begin{pmatrix} a^2 - b^2 & 2ab \\ 2ab & b^2 - a^2 \end{pmatrix} \begin{pmatrix} x \\ y \end{pmatrix}.$$

Proof Since (a, b) is on the unit circle, it can be written as $(\cos \theta, \sin \theta)$ for some real number θ. Let D be the distance from the point (x, y) to the origin and let ϕ be its angle measure as measured from the x-axis, in the counterclockwise direction, so that

$$\begin{matrix} \cos \phi = x/D \\ \sin \phi = y/D \end{matrix} \implies \begin{pmatrix} x \\ y \end{pmatrix} = \begin{pmatrix} D \cos \phi \\ D \sin \phi \end{pmatrix}.$$

If α is the angle between ϕ and θ, $\alpha = \phi - \theta$, then $s(x, y)$ will still be at a distance D from the origin, but at an angle

$$\phi - 2\alpha = \phi - 2(\phi - \theta) = 2\theta - \phi.$$

23.1 The analytic viewpoint

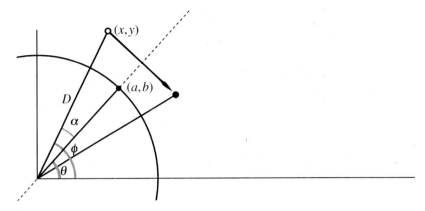

Figure 9. A reflection across a line that passes through the origin. In this case $\phi > \theta$ so that $\alpha > 0$, but the calculations are the same if $\alpha < 0$.

Therefore

$$s\begin{pmatrix} x \\ y \end{pmatrix} = \begin{pmatrix} D\cos(2\theta - \phi) \\ D\sin(2\theta - \phi) \end{pmatrix}$$

and we can use the addition rules for sine and cosine:

$$s\begin{pmatrix} x \\ y \end{pmatrix} = \begin{pmatrix} D\cos(2\theta)\cos(-\phi) - D\sin(2\theta)\sin(-\phi) \\ D\sin(2\theta)\cos(-\phi) + D\cos(2\theta)\sin(-\phi) \end{pmatrix}$$

$$= \begin{pmatrix} D\cos(2\theta)\cos\phi + D\sin(2\theta)\sin\phi \\ D\sin(2\theta)\cos\phi - D\cos(2\theta)\sin\phi \end{pmatrix}.$$

This can factored into a matrix form, and from there, the double angle formulas will take us the rest of the way.

$$s\begin{pmatrix} x \\ y \end{pmatrix} = \begin{pmatrix} \cos(2\theta) & \sin(2\theta) \\ \sin(2\theta) & -\cos(\theta) \end{pmatrix} \begin{pmatrix} D\cos\phi \\ D\sin\phi \end{pmatrix}$$

$$= \begin{pmatrix} \cos^2\theta - \sin^2\theta & 2\sin\theta\cos\theta \\ 2\sin\theta\cos\theta & \sin^2\theta - \cos^2\theta \end{pmatrix} \begin{pmatrix} D\cos\phi \\ D\sin\phi \end{pmatrix}$$

$$= \begin{pmatrix} a^2 - b^2 & 2ab \\ 2ab & b^2 - a^2 \end{pmatrix} \begin{pmatrix} x \\ y \end{pmatrix}. \qquad \square$$

There are two special cases worth noting. The equation for reflecting across the x-axis is

$$s\begin{pmatrix} x \\ y \end{pmatrix} = \begin{pmatrix} 1 & 0 \\ 0 & -1 \end{pmatrix} \begin{pmatrix} x \\ y \end{pmatrix} = \begin{pmatrix} x \\ -y \end{pmatrix},$$

and the equation for reflecting across the y-axis is

$$s\begin{pmatrix} x \\ y \end{pmatrix} = \begin{pmatrix} -1 & 0 \\ 0 & 1 \end{pmatrix} \begin{pmatrix} x \\ y \end{pmatrix} = \begin{pmatrix} -x \\ y \end{pmatrix}.$$

23.2 Exercises

23.1. Given $\triangle ABC$, let s be the reflection across the perpendicular bisector of BC. Prove that if $\triangle ABC$ is invariant under s (that is, $s(\triangle ABC) = \triangle ABC$), then $\triangle ABC$ is isosceles.

23.2. Given $\square ABCD$, let E and F be midpoints of AB and CD respectively. Let s be the reflection across the line EF. Prove that if $s(A) = B$ and $s(C) = D$, then $\square AEFD \simeq \square BEFC$.

23.3. Given a convex quadrilateral $\square ABCD$, let s_1 be the reflection across the diagonal AC, and let s_2 be the reflection across the diagonal BD. If $s_1(B) = D$ and $s_2(A) = C$, what can you say about $\square ABCD$?

23.4. What is the matrix equation for a reflection across the line $y = x$?

23.5. What is the matrix equation for a reflection across the horizontal line $y = k$?

23.6. Suppose that s_1 and s_2 are reflections, and that for some point P, $s_1(P) = s_2(P) \neq P$. Is it necessarily true that $s_1 = s_2$?

23.7. Suppose that s_1 and s_2 are reflections, and that for two distinct points P and Q, $s_1(P) = s_2(P)$ and $s_1(Q) = s_2(Q)$. Is it necessarily true that $s_1 = s_2$?

23.8. Given lines ℓ_1 and ℓ_2 is there necessarily a reflection s so that $s(\ell_1) = \ell_2$? If so, is it unique?

23.9. Let s_1 and s_2 be reflections across perpendicular lines ℓ_1 and ℓ_2 that intersect at a point P. Show that if Q is any other point, then P is the midpoint of the segment connecting Q to $s_2 \circ s_1(Q)$.

23.10. Let s_1 and s_2 be reflections across perpendicular lines ℓ_1 and ℓ_2. For any point P, prove that

$$P = s_2 \circ s_1 \circ s_2 \circ s_1(P),$$

and that there is a circle which passes through the points P, $s_1(P)$, $s_2 \circ s_1(P)$, and $s_1 \circ s_2 \circ s_1(P)$.

24
Translations and Rotations

The big result of the last chapter was that every isometry can be written as a reflection or as a composition of two or three reflections. In this chapter we will look at the types of isometries that we can get by composing two reflections. Any reflection composed with itself results in the identity, so we are interested in compositions of two distinct reflections. Then there are two cases: either the reflecting lines are parallel or the reflecting lines are intersecting.

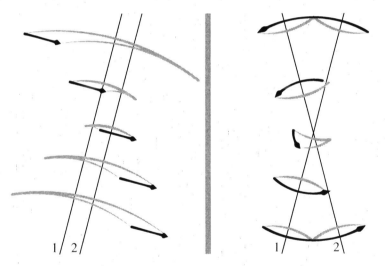

Figure 1. Pairs of reflections: (l) across parallel lines and (r) across intersecting lines.

The two cases describe fundamentally different types of isometries. In the second, the intersection point of the two lines is fixed by the composition of isometries. This doesn't happen in the first, since there is no intersection point, and in fact, this composition does not have any fixed points.

Definition 24.1 A *translation* is a composition of reflections across parallel lines. A *rotation* is a composition of reflections across intersecting lines.

By defining translations and rotations as compositions of isometries, it is automatically true that they will be isometries as well. But the definitions do not do a good job of revealing what a translation or rotation actually looks like. Doing that is the purpose of this chapter.

257

24.1 Translation

First, let's tackle the case of the translation. To do that, it is helpful to take a more measured look at the behavior of a single reflection. Consider a reflection s across a line ℓ. Let P be a point that is not on ℓ and let $\ell_\perp$ be the line through P that is perpendicular to ℓ. Now let's set up $\ell_\perp$ as a number line. That is, choose an arbitrary point O to be the origin, and a ray from O that points in the positive direction; then every point on $\ell_\perp$ has a coordinate, its signed distance from O. Suppose that P is at coordinate x and that ℓ and $\ell_\perp$ intersect at the point Q with coordinate y. Given the definition of a reflection, $s(P)$ has to be somewhere on $\ell_\perp$, and so it too must correspond to some coordinate. What coordinate? The distance from P to Q is $|y - x|$. Since s is an isometry and Q is a fixed point, the distance from $s(P)$ to Q is $|y - x|$ too. That limits the coordinate for $s(P)$ to one of two possibilities:

$$y + |y - x| = \begin{cases} y + (y - x) = 2y - x & \text{if } y - x \geq 0 \\ y - (y - x) = x & \text{if } y - x < 0 \end{cases}$$

or

$$y - |y - x| = \begin{cases} y - (y - x) = x & \text{if } y - x \geq 0 \\ y - (-(y - x)) = 2y - x & \text{if } y - x < 0. \end{cases}$$

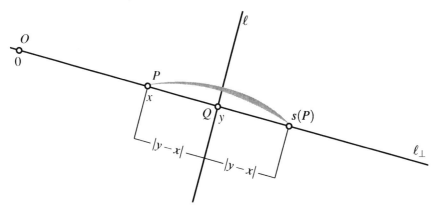

Figure 2. Measuring reflections along a line.

Since P is not on ℓ it is not a fixed point, so $s(P)$ is not at the coordinate x. The only other possibility, then, is that $s(P)$ is at the coordinate $2y - x$. The formula still works even if P is on ℓ. For in that case P is fixed, so $s(P)$ is also at coordinate x. That is what the formula reveals: if P is on ℓ, $y = x$, and so $2y - x = x$. Having this formula in hand will make it a little easier to compose parallel reflections.

Theorem 24.2 *Suppose that t is the translation $s_2 \circ s_1$ where s_1 and s_2 are reflections across parallel lines ℓ_1 and ℓ_2 that are separated by a distance d. Then for any point P, $t(P)$ is located*

- *on the line through P that is perpendicular to both ℓ_1 and ℓ_2,*
- *in the direction of the ray that points from ℓ_1 to ℓ_2,*
- *at a distance $2d$ from P.*

24.1 Translation

Proof Take a point P, and let $\ell_\perp$ be the line through P that is perpendicular to ℓ_1 and ℓ_2. By definition, $s_1(P)$ will still be on $\ell_\perp$, and then so will $s_2(s_1(P))$. Let's just look along this line, and, as in the preceding discussion, lay out a number line along it. It does not matter where we put the origin on the line, but it does help the discussion to choose the positive direction so that going from ℓ_1 to ℓ_2 moves in the positive direction. Then mark the coordinates

x : coordinate of P,
y_1 : coordinate for the intersection of $\ell_\perp$ and ℓ_1,
y_2 : coordinate for the intersection of $\ell_\perp$ and ℓ_2.

According to our previous calculations, $s_1(x)$ will be at the coordinate $2y_1 - x$ and $s_2 \circ s_1(x)$ will be at the coordinate

$$2y_2 - (2y_1 - x) = x + 2(y_2 - y_1) = x + 2d.$$

Therefore $s_2 \circ s_1(P)$ will be $2d$ farther along the line $\ell_\perp$ than P in the direction pointing from ℓ_1 to ℓ_2.

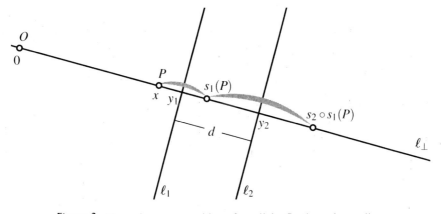

Figure 3. Measuring a composition of parallel reflections along a line. □

To recap, a translation moves all points along lines that are perpendicular to ℓ_1 and ℓ_2. They move in parallel, in the same direction, over the same distance. All of that—the parallel lines, the direction, and the distance—can be determined by looking at the effect of the translation on a single point. Therefore a translation is completely determined by its behavior on a single point. Because of that, we can get a precise idea of how many translations there are.

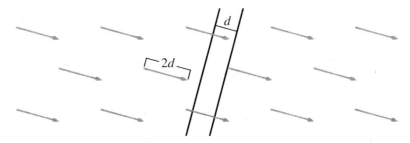

Figure 4. Transations shift all points the same amount in the same direction.

Theorem 24.3 *Given two distinct points P and Q, there is exactly one translation t so that $t(P) = Q$.*

Proof To prove existence, we will describe (as a composition of reflections) a translation that maps P to Q. The two reflections, s_1 and s_2, will be across lines that are perpendicular to PQ (and hence are parallel to one another). Let s_1 be the reflection across the line through P. Let s_2 be the reflection across the line through the midpoint of PQ. Then $s_2 \circ s_1$ is a translation and

$$s_2 \circ s_1(P) = s_2(P) = Q.$$

Proving uniqueness is easy: since a translation is completely determined by its behavior on one point, there can be only one translation taking P to Q. □

It is cumbersome to try to think of a translation as a composition of reflections. The properties we derived give a much better sense of the effects of a translation, and they can be formalized as follows. A *directed segment* is a line segment that distinguishes between its ends: one is called the initial endpoint, the other the terminal endpoint. We can define an equivalence relation $\sim$ on the set of all directed segments as follows: two directed segments σ_1 and σ_2 are equivalent if there is a translation t mapping σ_1 to σ_2 so that initial point is mapped to initial point and terminal point is mapped to terminal point.

Exercise 24.1 Verify that the relation $\sim$ is an equivalence relation on the set of all directed segments.

Definition 24.4 A *vector* is an equivalence class of directed segments under the $\sim$ equivalence relation.

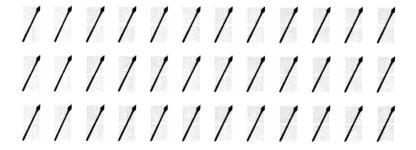

Figure 5. Some equivalence class representatives of a vector (one over, two up).

Associated to any transformation t is the vector that is represented by directed segments of the form $Pt(P)$ with initial point P and terminal point $t(P)$. The vector defines and is defined by t. It is called the *translation vector* of t. It is almost always more convenient and natural to think about a translation in terms of its translation vector than as a composition of reflections. For instance, if you think of a translation t as a composition of reflections, it might not be that clear that t has no fixed points. If you think of that translation in terms of its translation vector,

24.2 Rotations

The illustrations at the start of this chapter suggest that when ℓ_1 and ℓ_2 intersect, the rotation $r = s_2 \circ s_1$ acts by turning points around the intersection point O. To measure the effect of the turning, we establish an angular coordinate system around O (just as we established a linear coordinate system on $\ell_\perp$ when ℓ_1 and ℓ_2 were parallel). Choose a ray $\overrightarrow{OR}$ with endpoint O. This marks the zero angle and a direction around that point (this amounts to a choice of clockwise or counterclockwise, which is discussed in the next chapter). After choosing, every ray from O will form an angle with $\overrightarrow{OR}$ and we can then associate each point on the ray with the angle measure. Before attempting two reflections, let's try to understand how the angular coordinate of a point behaves when we apply one reflection s across a line ℓ to it. Pick a point O on ℓ and set up an angular coordinate system as described. Let P be an arbitrary point that is not on ℓ. Then label

θ : the angular coordinate at P
ϕ : the angular coordinate of one of the rays from O that make up ℓ.

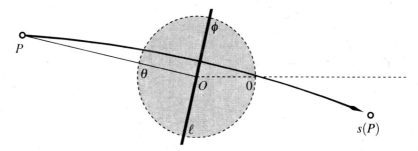

Figure 6. Measuring reflections around a fixed point.

There are two choices for ϕ that differ by π, but for the calculation, it makes no difference which one we pick. The angle between ℓ and OP has a measure of $|\phi - \theta|$. Since isometries preserve angle measure and ℓ is fixed by s, the angle between ℓ and $Os(P)$ also has a measure of $|\phi - \theta|$. That limits the possibilities for the angular coordinates of $s(P)$ to

$$\phi + |\phi - \theta| = \begin{cases} 2\phi - \theta & \text{if } \phi - \theta \geq 0 \\ \theta & \text{if } \phi - \theta < 0 \end{cases}$$

or

$$\phi - |\phi - \theta| = \begin{cases} \theta & \text{if } \phi - \theta \geq 0 \\ 2\phi - \theta & \text{if } \phi - \theta < 0. \end{cases}$$

Since P is not on ℓ, it is not fixed, and therefore $s(P)$ will not be at angle θ. The only other possibility is that $s(P)$ is at angle $2\phi - \theta$. Furthermore, the formula still holds when P is on ℓ. In that case, P is fixed, so $s(P)$ is also at angle θ. That is what the formula indicates: if P is on ℓ, then $\phi = \theta$ and so $2\phi - \theta = \theta$. Now let's take the formula and use it to see what happens when we compose two intersecting reflections.

Theorem 24.5 *Suppose that r is the rotation $s_2 \circ s_1$ where s_1 and s_2 are reflections across lines ℓ_1 and ℓ_2 intersecting at a point O and at an angle of θ to one another. For any point P, $r(P)$ is located*

- *on the circle centered at O that passes through P*
- *so that OP and $Os(P)$ form an angle with measure 2θ,*
- *in the direction indicated by the arc from ℓ_1 to ℓ_2.*

Proof Since $s_2 \circ s_1$ preserves distances and O is a fixed point, the distance from O to $s(P)$ is the same as the distance from O to P. That places $s(P)$ on the circle centered at O passing through P. Where precisely? As in the discussion above, set up an angular coordinate system centered at O. Mark the coordinates

α : the angular coordinate for P,
ϕ_1 : the angular coordinate for ℓ_1,
ϕ_2 : the angular coordinate for ℓ_2.

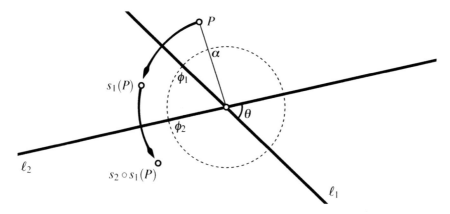

Figure 7. Measuring a composition of reflections around the fixed point.

The intersection of ℓ_1 and ℓ_2 forms two vertical angle pairs. It is helpful to measure angles in the direction so that the directed angle from ℓ_1 to ℓ_2 is the smaller of them (if ℓ_1 and ℓ_2 intersect at right angles, it doesn't matter which orientation you choose). According to the previous discussion, $s_1(P)$ will have the coordinate $2\phi_1 - \alpha$. Then $s_2(s_1(P))$ will have the angular coordinate

$$2\phi_2 - (2\phi_1 - \alpha) = \alpha + 2(\phi_2 - \phi_1) = \alpha + 2\theta.$$

Therefore OP and $Os(P)$ form an angle of 2θ, measured in the direction from ℓ_1 to ℓ_2. □

It is generally more convenient to think of a rotation in terms of the angle 2θ, the *rotation angle*, and the fixed point, the *center of rotation*, rather than as a composition of reflections. For instance, by thinking of a rotation in terms of its rotation angle and center, it is clear that a rotation only has one fixed point, the center of rotation. This viewpoint also gives a good perspective on just how common rotations are. The proof of the following result is left to the reader.

Theorem 24.6 *For a point O and angle measure $0 < \theta < 2\pi$, there is exactly one rotation in each of the two directions around O that has rotation center O and rotation angle θ. When $\theta = \pi$, the rotations in those two directions coincide (a half-turn).*

Exercise 24.2 Prove theorem 24.6.

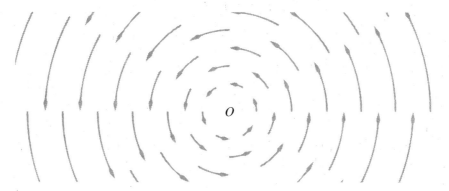

Figure 8. A counterclockwise rotation by $\pi/6$ centered at O.

24.3 The analytic viewpoint

Analytically, translations are the simplest of the isometries. If we break the translation vector of a translation T down into a horizontal component h and a vertical component k, then

$$T\begin{pmatrix} x \\ y \end{pmatrix} = \begin{pmatrix} x + h \\ y + k \end{pmatrix}.$$

The equations for rotations are more challenging. For now, let's restrict our attention to rotations that are centered at the origin.

Theorem 24.7 *The analytic equation for a rotation r around the origin by an angle θ in the counterclockwise direction is*

$$r\begin{pmatrix} x \\ y \end{pmatrix} = \begin{pmatrix} \cos\theta & -\sin\theta \\ \sin\theta & \cos\theta \end{pmatrix} \begin{pmatrix} x \\ y \end{pmatrix}.$$

Proof We can realize this rotation as a composition of two reflections across lines through the origin. For convenience, let's choose the reflections s_1 across ℓ_1 and s_2 across ℓ_2, where ℓ_1 is the x-axis and ℓ_2 forms an angle of $\theta/2$ (counterclockwise) with the x-axis.

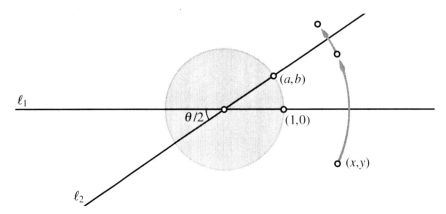

Figure 9. A rotation by θ around the origin can be written as a composition of reflections across two lines which form a $\theta/2$ angle at the origin.

Then $s_2 \circ s_1$ will be a rotation by an angle of $2 \cdot \theta/2 = \theta$. In the last chapter, we found out that equations for these reflections take the form

$$s\begin{pmatrix} x \\ y \end{pmatrix} = \begin{pmatrix} a^2 - b^2 & 2ab \\ 2ab & b^2 - a^2 \end{pmatrix}$$

where (a, b) marks the intersection of the line and the unit circle. We can use that equation now. The first line intersects the unit circle at $(1, 0)$, so

$$s_1\begin{pmatrix} x \\ y \end{pmatrix} = \begin{pmatrix} 1 & 0 \\ 0 & -1 \end{pmatrix}\begin{pmatrix} x \\ y \end{pmatrix}.$$

The second line intersects the unit circle at $(\cos\theta/2, \sin\theta/2)$, so

$$s_2\begin{pmatrix} x \\ y \end{pmatrix} = \begin{pmatrix} \cos^2(\theta/2) - \sin^2(\theta/2) & 2\cos(\theta/2)\sin(\theta/2) \\ 2\cos(\theta/2)\sin(\theta/2) & \sin^2(\theta/2) - \cos^2(\theta/2) \end{pmatrix}\begin{pmatrix} x \\ y \end{pmatrix}.$$

We can use the double angle formulas to write

$$\cos^2(\theta/2) - \sin^2(\theta/2) = \cos(\theta),$$
$$2\cos(\theta/2)\sin(\theta/2) = \sin(\theta),$$

which simplifies the matrix considerably to

$$s_2\begin{pmatrix} x \\ y \end{pmatrix} = \begin{pmatrix} \cos\theta & \sin\theta \\ \sin\theta & -\cos\theta \end{pmatrix}\begin{pmatrix} x \\ y \end{pmatrix}.$$

To compute the composition of the transformations, multiply the matrices:

$$r\begin{pmatrix} x \\ y \end{pmatrix} = \begin{pmatrix} \cos\theta & \sin\theta \\ \sin\theta & -\cos\theta \end{pmatrix}\begin{pmatrix} 1 & 0 \\ 0 & -1 \end{pmatrix}\begin{pmatrix} x \\ y \end{pmatrix} = \begin{pmatrix} \cos\theta & -\sin\theta \\ \sin\theta & \cos\theta \end{pmatrix}\begin{pmatrix} x \\ y \end{pmatrix}. \qquad \square$$

24.4 Exercises

24.3. Prove that every isometry T can be written as a composition $t_1 \circ t_2$ where t_1 is a translation (or the identity) and t_2 is an isometry that fixes the origin.

24.4 Exercises

24.4. Find the analytic equations for reflections across the lines $x = a$ and $x = b$. Then verify that their composition is a translation.

24.5. Suppose that r_1 and r_2 are counterclockwise rotations about the origin by angles of θ_1 and θ_2 respectively. From the matrix equations for r_2 and r_1, show that the matrix equations for $r_2 \circ r_1$ have the form of a rotation or the identity.

24.6. Suppose that ℓ is an invariant line of a rotation r. That is, if P is a point on ℓ, then $r(P)$ is also on ℓ. Show that ℓ passes through the center of rotation and the angle of rotation is π.

24.7. Let r be a rotation that maps the point $(0, 0)$ to the point $(1, 0)$ and the point $(1, 0)$ to the point (a, b). What conditions are required of a and b for this to be possible?

24.8. In exercise 21.17 of chapter 21, we found the equation of a parabola with focus $(0, a)$ and directrix line $y = -a$, where $a \neq 0$. Find the equation of a parabola with focus $(a, 0)$ and directrix line $x = -a$ in two ways. First, do a direct calculation, as in chapter 21. Second, write a point on the parabola from exercise 21.17 in the parametrized form $(t, t^2/(4a))$ and apply a rotation.

24.9. Working from the previous exercise, find the equation for a parabola with focus $(3, 5)$ and directrix line $x = -1$.

24.10. Show that the parabola with focus at $(1, 1)$ and directrix line $y = -2 - x$ has the equation
$$(x - y)^2 - 8(x + y) = 0.$$

24.11. Find the equation of the ellipse with foci $(2, 0)$ and $(6, 0)$ that passes through the point $(0, 0)$.

24.12. Find the equation of the ellipse with foci $Q_1 = (1, \sqrt{3})$ and $Q_2 = (-1, -\sqrt{3})$ so that for a point $P = (x, y)$ on it
$$d(P, Q_1) + d(P, Q_2) = 8.$$

25
Orientation

The remaining step in the classification of isometries is to tackle compositions of three reflections. We would like to know whether compositions of three reflections can devolve into any of the previously identified isometries (reflection, translation, rotation). It turns out that composing three reflections can yield another reflection, but it will never result in a translation or rotation. This is helpful. It comes down to the issue of orientation, which we will investigate in this chapter.

An *orthonormal frame* is an ordered pair of perpendicular unit length segments that share a common endpoint. One such frame, $\mathscr{F}_+ = \{OP_x, OP_y\}$, centered at the origin with

$$O = (0,0), \quad P_x = (1,0), \quad \text{and} \quad P_y = (0,1),$$

is at the heart of the coordinate system. There is another such frame, $\mathscr{F}_- = \{OP_x, OP'_y\}$, that shares the same first segment as $\mathscr{F}_+$, but that has $P'_y = (0,-1)$. In general, a frame can be viewed as a way to represent information about orientation, whether it is clockwise or counterclockwise. Until now, we have only made that choice at the origin: in $\mathscr{F}_+$, the directed minor arc from P_x to P_y points in the counterclockwise direction; in $\mathscr{F}_-$, the directed minor arc from P_x to P_y points in the clockwise direction. By using translations and rotations, we can now propagate that choice across the rest of the plane. Consider a frame $\mathscr{F} = \{QQ_x, QQ_y\}$. Let r be the rotation centered at O so that the vectors $r(OP_x)$ and QQ_x are parallel and pointed in the same direction. Let t be the translation which maps the point O to Q. Then

$$t \circ r(O) = Q \quad \text{and} \quad t \circ r(P_x) = Q_x.$$

What about P_y and P'_y? Both $t \circ r(OP_y)$ and $t \circ r(OP'_y)$ are perpendicular to QQ_x, and both $t \circ r(P_y)$ and $t \circ r(P'_y)$ are located a distance of 1 from Q along their segments. Because of the segment and angle construction axioms, there are only two points that meet both requirements. One of them is Q_y. Therefore, exactly one of two things may happen: either $t \circ r(P_y) = Q_y$ or $t \circ r(P'_y) = Q_y$. In the first case, $t \circ r$ maps $\mathscr{F}_+$ to $\mathscr{F}$; then we call $\mathscr{F}$ a counterclockwise frame, and the directed minor arc from Q_x to Q_y is in the counterclockwise direction. In the second case, $t \circ r$ maps $\mathscr{F}_-$ to $\mathscr{F}$; then we call $\mathscr{F}$ a clockwise frame, and the directed minor arc from Q_x to Q_y is in the clockwise direction.

This is a two-step process, in the order $t \circ r$, where r is a rotation around the origin, followed by a translation t. As long as we stick to that way of doing it, there is one and only one way to map one of the two original frames to any other. Hence there is no issue about making a consistent choice because there is no choice to make. This is not a trivial issue because not all

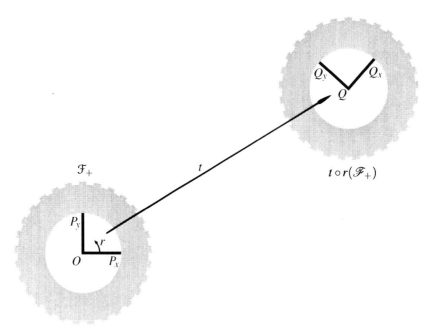

Figure 1. A rotation followed by a translation maps the frame $\mathscr{F}_+$ to the frame $\mathscr{F} = \{QQ_x, QQ_y\}$. Therefore $\mathscr{F}$ is a counterclockwise frame.

surfaces can be oriented consistently like this. The most famous non-orientable surface is the Möbius strip. It is formed by taking a strip, giving it a half-twist, and joining the ends. A frame $\mathscr{F}$ on the Möbius strip can be translated from one point to another in two different ways, t_1 and t_2, and the resulting frames $t_1(\mathscr{F})$ and $t_2(\mathscr{F})$ are not oriented the same way. The unfortunate side effect of the two-step process is that it doesn't determine whether other combinations of rotations and translations might actually reverse orientation, even though we instinctively know that to not be true. We will show that soon as an easy consequence of the fact that a reflection reverses the orientation of a frame, which we will tackle next. That argument will require the following lemma.

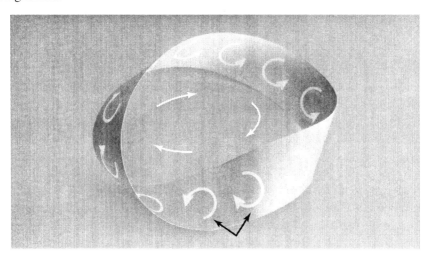

Figure 2. One lap around the Möbius strip flips orientation.

Orientation

Lemma 25.1 *Let r be a rotation by an angle θ (not a half-turn), and let ℓ be a line. Then the angle between ℓ and $r(\ell)$ is θ.*

Exercise 25.1 Prove the previous lemma.

Theorem 25.2 *Let s be a reflection and $\mathscr{F}$ be a frame. Then $\mathscr{F}$ and $s(\mathscr{F})$ have the opposite orientation.*

Proof Let $\mathscr{F}$ be a counterclockwise frame (the clockwise case is essentially the same). Now let's suppose that $s(\mathscr{F})$ is also counterclockwise and try to arrive at a contradiction. Because both are counterclockwise, there are rotations r_1 and r_2 around the origin and translations t_1 and t_2 so that

$$t_1 \circ r_1(\mathscr{F}_+) = \mathscr{F} \quad \text{and} \quad t_2 \circ r_2(\mathscr{F}_+) = s(\mathscr{F})$$

so

$$s \circ t_1 \circ r_1(\mathscr{F}_+) = s(\mathscr{F}) = t_2 \circ r_2(\mathscr{F}_+).$$

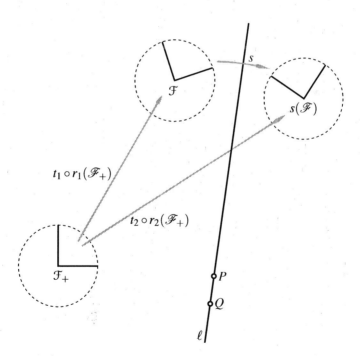

Figure 3. A reflection will always reverse the orientation of a frame because the composition of a rotation, translation, and reflection cannot have the same effect as the composition of just a rotation and a translation.

We have seen that if isometries agree on three non-collinear points, then they must be the same. The frame $\mathscr{F}_+$ contains three non-collinear points, and thus, $s \circ t_1 \circ r_1$ and $t_2 \circ r_2$ are the same. Therefore

$$s \circ t_1 \circ r_1 = t_2 \circ r_2$$
$$s \circ t_1 = t_2 \circ r_2 \circ r_1^{-1}$$
$$s = t_2 \circ r_2 \circ r_1^{-1} \circ t_1^{-1}.$$

On the right hand side, $r_2 \circ r_1^{-1}$ is a composition of two rotations around the origin. Either the two rotations cancel, leaving the identity map, or they compose to form another rotation around the origin. Let ρ denote this composition. On the left, the map s is a reflection, so it fixes a line of points. Choose two distinct points on it, P and Q. Then

$$t_2 \circ \rho \circ t_1^{-1}(P) = s(P) = P \implies \rho \circ t_1^{-1}(P) = t_2^{-1}(P)$$
$$t_2 \circ \rho \circ t_1^{-1}(Q) = s(Q) = Q \implies \rho \circ t_1^{-1}(Q) = t_2^{-1}(Q).$$

According to the last lemma, the angle between $\rho \circ t_1^{-1}(PQ)$ and $t_1^{-1}(PQ)$ is the angle of rotation of ρ, say θ. Both t_1^{-1} and t_2^{-1} are translations, however, so both $t_1^{-1}(PQ)$ and $t_2^{-1}(PQ)$ are parallel to PQ, and hence to each other. Therefore $\rho \circ t_1^{-1}(PQ)$ also forms an angle of θ with $t_2^{-1}(PQ)$. The only way the two segments can coincide, then, is if ρ is either a half-turn or the identity. We may rule out the half-turn since a half-turn would reverse the direction of the vector PQ. Hence ρ must be the identity. That means that $s = t_2 \circ t_1^{-1}$ and we arrive at a contradiction: the composition $t_2 \circ t_1^{-1}$ of two translations is another translation, not a reflection. □

This shows that a reflection is a wholesale swap of orientations. All the clockwise frames reflect to counterclockwise ones and all the counterclockwise frames reflect to clockwise ones.

Definition 25.3 An isometry is *orientation-preserving* if it maps clockwise frames to clockwise frames and counterclockwise frames to counterclockwise frames. A isometry is *orientation-reversing* if it swaps clockwise and counterclockwise frames.

We have just proved that reflections are orientation-reversing. Because every isometry can be written as a composition of reflections, determining whether an isometry is orientation preserving or reversing is a matter of counting flips.

Corollary 25.4 *A composition of two orientation-preserving maps is orientation-preserving; a composition of two orientation-reversing maps is orientation-preserving; a composition of one orientation-preserving map and one orientation-reversing map is orientation-reversing.*

Exercise 25.2 Prove the corollary.

Corollary 25.5 *Translations, rotations, and the identity map are orientation-preserving. They are the only orientation-preserving isometries.*

Exercise 25.3 Prove the corollary.

Let's now recap our progress in the classification of isometries.

# of reflections	isometry	orientation	fixed points
1	reflection	reversing	line
2	identity	preserving	all
	translation	"	none
	rotation	"	point
3	?	reversing	?

In the next chapter we find what goes in place of the question mark.

25.1 Exercises

25.4. Show that if τ is an orientation-preserving isometry that fixes two points, then it is the identity. Show that if τ is an orientation-preserving isometry that has at least one fixed point and at least one non-fixed point, then it is a rotation.

25.5. Let τ_1 be a counterclockwise rotation by $\pi/2$ about the origin. Let τ_2 be a counterclockwise rotation by $\pi/2$ about the point $(1, 0)$. Show that $\tau_1 \circ \tau_2$ is a rotation.

25.6. Another way to think about orientation is with normal vectors formed by cross products. Label the three points $O = (0, 0)$, $P = (1, 0)$, and $Q = (0, 1)$ and let T be an isometry. Form the vectors

v_x: the vector from O to P,
v_y: the vector from O to Q,
w_x: the vector from $T(O)$ to $T(P)$, and
w_y: the vector from $T(O)$ to $T(Q)$.

Compare the cross products $v_1 \times v_2$ and $T(v_1) \times T(v_2)$, where T is

- a reflection across a line through O,
- a rotation around O,
- a translation.

25.7. Let T be an orientation-preserving isometry. Prove that T is a rotation if and only if there exist points P and Q so that $|PT(P)| \neq |QT(Q)|$.

25.8. Let t be a translation and r be a rotation. Prove that $t \circ r$ and $r \circ t$ are rotations.

25.9. Consider the matrix equation

$$T \begin{pmatrix} x \\ y \end{pmatrix} = M \begin{pmatrix} x \\ y \end{pmatrix}$$

(where M is a 2×2 matrix) for (a) a reflection across a line through the origin and (b) a rotation around the origin. Show that for a reflection, $\det(M) = -1$ and that for a rotation, $\det(M) = 1$. In this way, the determinant of M captures information about orientation.

The remaining exercises look at surfaces that can be formed by bending and warping a flat sheet and gluing its sides together. More complete investigations can be found in many places. For instance, Massey has a nice write-up in the first chapter of *A Basic Course in Algebraic Topology* [Mas91].

25.10. Let S be the square with vertices $(0, 0)$, $(1, 0)$, $(0, 1)$, and $(1, 1)$. Orient each of those edges with vectors

$\vec{a}$: pointing from $(0, 0)$ to $(1, 0)$
$\vec{b}$: pointing from $(0, 0)$ to $(0, 1)$
$\vec{A}$: pointing from $(0, 1)$ to $(1, 1)$
$\vec{B}$: pointing from $(1, 0)$ to $(1, 1)$.

Imagine that a square S is made from a flexible material. Show that by bending and warping S, and gluing $\vec{a}$ to $\vec{A}$ and $\vec{b}$ to $\vec{B}$ so that they are oriented consistently, the resulting shape, a *torus*, can be made to look like an like an inner tube.

25.11. Let S be the square in the last exercise, with $\vec{a}$, $\vec{b}$, and $\vec{A}$ oriented as before. Orient $\vec{B}$ so that it points from $(1, 1)$ to $(1, 0)$. The resulting shape, a *Klein bottle* cannot be embedded in three-dimensional space without a self-intersection. Show that it is a non-orientable surface. Hint: show that it contains a Möbius strip.

25.12. Let S be the octagon $P_1 P_2 P_3 \cdots P_8$. Glue the following vectors along its edges, matching the orientation of those vectors:

- $\overrightarrow{P_1 P_2}$ with $\overrightarrow{P_4 P_3}$,
- $\overrightarrow{P_2 P_3}$ with $\overrightarrow{P_5 P_4}$,
- $\overrightarrow{P_5 P_6}$ with $\overrightarrow{P_8 P_7}$, and
- $\overrightarrow{P_6 P_7}$ with $\overrightarrow{P_1 P_8}$.

Show that the result can be deformed to form a inner-tube shape with two holes (a "two-holed torus").

26
Glide Reflections

Now let's look at a composition of three reflections. The first two reflections will get us to either the identity map, a translation, or a rotation. We are going to add another reflection to that. Composing a reflection with the identity will, of course, give a reflection. What about composing a reflection with a translation or a rotation? That is the subject of this chapter.

26.1 Glide reflections

Take a reflection s across a line ℓ followed by a translation t whose translation vector is parallel to ℓ. The composition $t \circ s$ swaps the two sides of ℓ and translates along ℓ. Therefore it has no fixed points. We have only seen one type of isometry that has no fixed points so far, a translation. But this isometry, a composition of three reflections, will be orientation-reversing, so it cannot be a translation.

Definition 26.1 A *glide reflection* is a composition of a translation t followed by a reflection s across a line that is parallel to the translation vector.

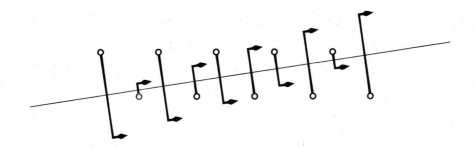

Figure 1. The path of a few points under a glide reflection.

In general, it matters in which order you compose isometries: $f_1 \circ f_2$ is typically not the same as $f_2 \circ f_1$. But the s and t that make up a glide reflection are interchangeable.

Lemma 26.2 *Let s be a reflection across a line ℓ and let t be a translation parallel to ℓ. Then $s \circ t = t \circ s$.*

Proof If P is a point on the reflecting line ℓ, then so is its translation $t(P)$, and in that case, the reflection has no effect on either one of them, so

$$s \circ t(P) = t(P) = t \circ s(P).$$

Now suppose that P is not on ℓ. In that case, compare the quadrilaterals

- with vertices $P, s(P), t(P)$, and $s \circ t(P)$,
- with vertices $P, s(P), t(P)$, and $t \circ s(P)$.

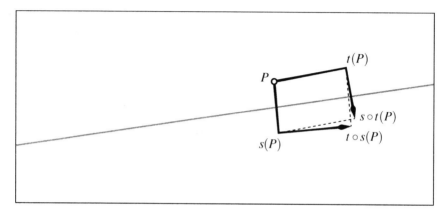

Figure 2. The two isometries that make up a glide reflection, a translation and a reflection, commute.

They share two sides, $Ps(P)$ and $Pt(P)$, and since the reflection and translation are perpendicular motions, both have right angles at three of the four vertices, at P, $s(P)$, and at $t(P)$. Then the fourth angle must also be a right angle, and so the two quadrilaterals are rectangles. There is only one way to build a rectangle given two of its adjacent sides. Therefore $s \circ t(P)$ and $t \circ s(P)$ must be the same. □

For what we are going to do, we need an easy way to recognize glide reflections. The key is that along the reflecting line, a glide reflection looks like a translation. Call the line of reflection the "glide line", and the distance of translation along the line the "glide distance". The next two lemmas investigate this.

Lemma 26.3 *Let τ be an isometry with an invariant line ℓ, and suppose that there is a translation t so that $\tau(P) = t(P)$ for all points P on ℓ. If $\tau \neq t$, then τ is a glide reflection.*

Proof Look at the effect of the composition of τ and t^{-1} on a point P of the line ℓ:

$$t^{-1} \circ \tau(P) = t^{-1} \circ t(P) = P.$$

It fixes all points on ℓ. Assuming $\tau \neq t$, $t^{-1} \circ \tau$ cannot be the identity map. The only other isometry that fixes an entire line is a reflection (this was theorem 23.3). Therefore $t^{-1} \circ \tau = s$ where s is the reflection across ℓ, and so $\tau = t \circ s = s \circ t$ is a glide reflection. □

26.2 Compositions of three reflections

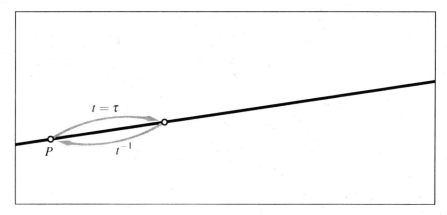

Figure 3. An isometry that acts like a translation along a line but is not a translation must be a glide reflection.

Lemma 26.4 *Let $\overline{PQ}$ be an invariant line of an isometry τ and suppose that $\tau(P) = t(P)$ and $\tau(Q) = t(Q)$ for some translation t. Then $\tau = t$ for all points on $\overline{PQ}$.*

Exercise 26.1 Prove the lemma.

By combining the two lemmas we see that an orientation-reversing isometry that agrees with a translation for two distinct points on an invariant line must be a glide reflection.

26.2 Compositions of three reflections

Let's start the hunt for other isometries by looking at what happens when we compose a translation and a reflection. If the translation is parallel to the line of reflection then that is the definition of a glide reflection. But what if the translation is not along the reflecting line?

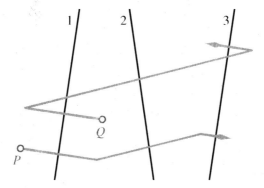

Figure 4. A composition of three reflections. Is the result another reflection? a glide reflection? or something new?

Exercise 26.2 Let t be a translation with translation vector v, and let s be a reflection across a line ℓ. Prove that if v is perpendicular to ℓ then $s \circ t$ is a reflection.

Theorem 26.5 *Let t be a translation with translation vector v and let s be a reflection across line ℓ. Suppose that ℓ is not perpendicular to v and let θ be the angle between v and ℓ. Then $s \circ t$ is a glide reflection whose glide line is parallel to ℓ, at a distance $(|v|\sin\theta)/2$ from ℓ, and whose glide distance is $|v|\cos\theta$.*

As the previous lemmas suggest, if we want to show that $s \circ t$ is a glide reflection, then we need to find its glide line. The best way to do that is to experiment with the translation-reflection combination. We are looking for a line along which $s \circ t$ acts as a translation: first t will move the points off the glide line, and then s will move them back, shifted from their original location.

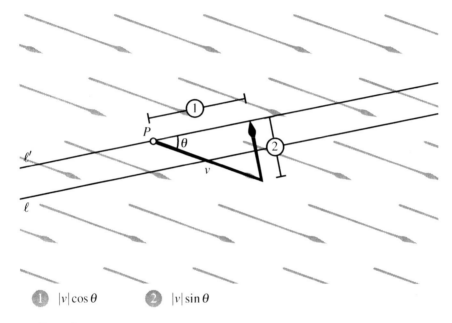

① $|v|\cos\theta$ ② $|v|\sin\theta$

Figure 5. Locating the glide line in a composition of a translation and a reflection.

Proof   Let ℓ' be the line that

- is parallel to ℓ,
- is on the opposite side of ℓ from the direction that v points, and
- is separated from ℓ by a distance of $(|v|\sin\theta)/2$.

Consider a point P on ℓ'. We can break the translation $t(P)$ down into two steps: first a translation by $|v|\cos\theta$ along ℓ', and then a translation by $|v|\sin\theta$ perpendicular to ℓ'. The second translation shows that $t(P)$ is located on the opposite side of ℓ from P, at a distance of $(|v|\sin\theta)/2$ from ℓ. Therefore, when we apply the reflection s to $t(P)$, the result $s \circ t(P)$ is back on ℓ', but shifted up from P a distance of $|v|\cos\theta$. All the points on ℓ' exhibit this behavior, so $s \circ t$ acts as a translation along ℓ'. Since $s \circ t$ is orientation-reversing, it cannot be a translation. According to lemma 26.3 above, it must be a glide reflection. □

We have taken care of combinations of a translation with a reflection. What happens when we combine a reflection and a rotation? There are two cases, depending on whether the reflecting

26.2 Compositions of three reflections

line passes through the center of rotation. The case where it does is easier, so let's start with that one.

Theorem 26.6 *Let r be a rotation by an angle of θ centered at a point O, and let s be a reflection across a line ℓ that passes through O. Then $s \circ r$ is a reflection across a line that passes through O and forms a (signed) angle of $-\theta/2$ with ℓ.*

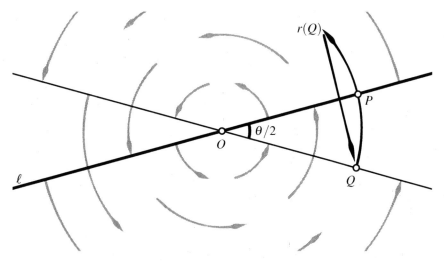

Figure 6. O and Q are fixed by the composition $s \circ r$, so $\overline{OQ}$ must be the fixed line of a reflection.

Proof O is a fixed point of $s \circ r$. We need to find one more fixed point. Consider a point P on ℓ other than O and rotate it by $-\theta/2$ about O (that is, rotate it in the opposite direction from r) to a point Q. This point Q is the one we want: $\overrightarrow{Or(Q)}$ will form an angle of $\theta/2$ with ℓ. Reflecting back across ℓ, $\overrightarrow{Os \circ r(Q)}$ will form an angle of $-\theta/2$ with ℓ. Since its distance from O remains unchanged, that means $s \circ r(Q) = Q$. Since $s \circ r$ fixes two points, O and Q, it must fix all the points on $\overline{OQ}$. As a result, $s \circ r$ will either be the identity or a reflection. But $s \circ r$ can't be the identity since it is orientation-reversing. □

If the reflecting line does not pass through the center of rotation, then the situation is more complicated.

Theorem 26.7 *Let r be a rotation by an angle of θ centered at a point O, let s be a reflection across a line ℓ that does not pass through O, and let Q be the closest point on ℓ to O. Then $s \circ r$ is a glide reflection along a line that passes through Q at an angle of $\theta/2$ to ℓ.*

Proof We again need to find the glide line. Use the labels

$P: (s \circ r)^{-1}(Q) = r^{-1} \circ s^{-1}(Q)$
$R: (s \circ r)(Q)$
$R': r(Q)$

F_P: the foot of perpendicular from P to ℓ
F_R: the foot of perpendicular from R to ℓ.

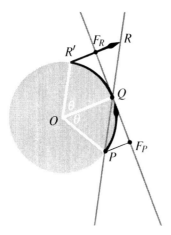

Figure 7. Finding the glide line in the composition of a rotation and a reflection.

The labels are set up so that

$$s \circ r(P) = Q \text{ and } s \circ r(Q) = R.$$

We will show that the glide line is the line through P, Q, and R. There are two things to do to show that. First, we need to show that P, Q, and R are collinear. Second, we need to show that $s \circ r$ moves P and Q in the same way that a translation does, that is, it moves P and Q in the same direction and $|PQ| = |QR|$. Consider the following triangle congruences, as shown in figure 8.

(1) The segments OP, OQ, and OR' are congruent because they are radii of the same circle. Then $\angle POQ$ and $\angle QOR'$ both measure θ, so they are congruent. By SAS, $\triangle OQP \simeq \triangle OQR'$.
(2) $\angle OQF_P$ and $\angle OQF_R$ are both right angles, so by angle subtraction, $\angle PQF_P$ and $\angle R'QF_R$ are congruent. The angles at F_P and F_R are right angles. From the previous triangle congruence, $PQ \simeq QR'$. By AAS, $\triangle PQF_P \simeq \triangle R'QF_R$.
(3) The segments $F_R R'$ and $F_R R$ are congruent since R is the reflection of R' across $\overline{F_R Q}$. The angles $\angle R'F_R Q$ and $\angle RF_R Q$ are right angles and hence congruent, and $F_R Q$ is congruent to itself. By SAS, $\triangle R'QF_R \simeq \triangle RQF_R$.

Therefore, $\triangle PQF_P$ and $\triangle RQF_R$ are congruent. Their corresponding angles $\angle PQF_P$ and $\angle RQF_R$ are congruent, and since F_P, Q, and F_R are collinear, P, Q, and R must be collinear too. By comparing the lengths of the hypotenuses of the congruent triangles, $|PQ| = |QR|$. Therefore $s \circ r$ acts like a translation for the two points P and Q. It follows that $s \circ r$ acts like a translation for all points on $\overline{PQ}$. Since $s \circ r$ is not a translation (it is orientation-reversing), it must be a glide reflection. □

26.2 Compositions of three reflections

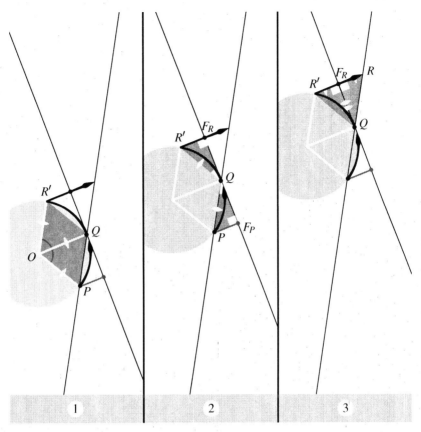

Figure 8. Use three triangle congruences to verify that $s \circ r$ acts like a translation on the line through P, Q, and R.

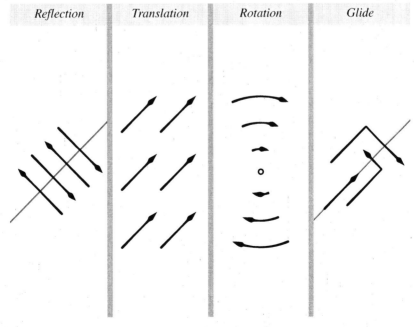

Figure 9. The four non-identity Euclidean isometries.

Table 1. Isometries of the Euclidean plane

# of reflections	isometry	orientation	fixed points
1	reflection	reversing	line
2	identity	preserving	all
	translation	"	none
	rotation	"	point
3	glide reflection	reversing	none

That's it! We have looked at all possible combinations of at most three reflections and seen the following types of isometries: the identity, reflections, translations, rotations, and glide reflections. The results are consolidated into table 1.

26.3 Exercises

26.3. Let g_1 and g_2 be glide reflections whose reflecting lines are parallel. Prove that $g_1 \circ g_2$ is a translation or the identity.

26.4. Let g_1 and g_2 be glide reflections whose reflecting lines intersect. Prove that $g_1 \circ g_2$ is a rotation.

26.5. Give analytic equations for the glide reflection formed by reflecting across the line $y = mx$ and then translating a distance d along this line. Assume $m > 0$ and choose the translation vector so that it points from the origin into the first quadrant.

26.6. We saw that the composition of a rotation and a reflection is a glide reflection if the center of rotation is not on the line of reflection. What is the glide distance in this case (in terms of the rotation center, the rotation angle, and the line of reflection)?

26.7. Let r be a counterclockwise rotation by $\pi/4$ about the origin. Let s be the reflection across the line $y = 1$. What is the equation of the glide line of the glide reflection $s \circ r$?

26.8. Let g be a glide reflection. What is the minimum number of points required to completely determine g (to find its glide line and glide distance)?

26.9. Describe the isometries τ that satisfy the condition $\tau^2 = \text{id}$. Describe the isometries that satisfy the condition $\tau^n = \text{id}$ for $n > 2$.

26.10. Show that the composition of a glide reflection and a reflection is either a rotation or a translation. Give examples in which each occurs.

26.11. Show that the composition of two glide reflections is either the identity, a rotation, or a translation. Give examples in which each occurs.

27
Change of Coordinates

27.1 Vector arithmetic

In chapter 24, we used translations to define vectors, but then did little with vectors beyond that. Let's take a more detailed look now. In general, a vector holds two pieces of information: a length and a direction. It is represented by a directed segment, and it is common to distinguish the endpoints of the segment with the names "tail" and "head", so that the segment points from the tail to the head. There is one exception: the zero vector is a vector with length zero and no direction. You can think of it as what happens when a segment shrinks to a point and the head and tail merge. It is common practice to conflate a vector with one of its representative directed segments, and there is generally no problem with that. In this introductory section, it is probably a good idea to maintain a little distance between the two so we will write $\vec{v}$ for a vector, and v for one of its representative directed segments. After this section, it will be safe to mix up the two notions.

One of the strengths of vectors is that they have an inherent arithmetic that points do not. Two vectors can be added using a "head-to-tail" procedure as follows. Given two vectors $\vec{u}$ and $\vec{v}$, their sum $\vec{u} + \vec{v}$ is the vector that is represented by a directed segment $u + v$ that is defined as follows. Let u be a representative of $\vec{u}$ and let v be the representative of $\vec{v}$ whose tail is located at the head of u. Then $u + v$ is the directed segment from the tail of u to the head of v. A vector $\vec{v}$ can be multiplied by a real number r. The resulting vector $r \cdot \vec{v}$ is represented by a directed segment that

- has the same tail as v and is on the same line as v,
- has length $|r| \cdot |v|$, and
- is in the same direction as v if $r > 0$ and in the opposite direction if $r < 0$.

The calculations require a choice of representatives. This raises the potential issue that these operations might not be well-defined: different choices for the representatives could lead to different answers. This does not happen, as indicated in the following exercise.

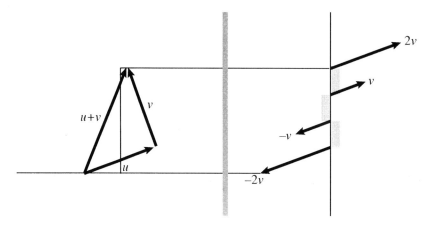

Figure 1. (l) Vector addition. (r) Scalar multiplication.

Exercise 27.1 Let u_1 and u_2 be directed segments representing $\vec{u}$, and let v_1 and v_2 be directed segments representing $\vec{v}$. Let r be a real number. Prove that $u_1 + v_1$ and $u_2 + v_2$ represent the same vector. Prove that $r \cdot u_1$ and $r \cdot u_2$ represent the same vector.

There is an analytic side of the story too. Let $\vec{v}$ be a vector represented by a directed segment v, and mark

(t_x, t_y) : the coordinates of the tail of v
(h_x, h_y) : the coordinates of the head of v.

Then $h_x - t_x$ is called the horizontal component or x-component of $\vec{v}$, and $h_y - t_y$ is called the vertical component or y-component of $\vec{v}$. The values do not depend on the choice of v. We write the vector $\vec{v}$ in terms of its components as $\vec{v} = \langle h_x - t_x, h_y - t_y \rangle$. Vectors and points are not the same thing, so point coordinates (x, y) should not be equated with vector components $\langle x, y \rangle$. There is, however, a connection between the two. If $\vec{v} = \langle x, y \rangle$, then the representative of $\vec{v}$ that has its tail at the origin will have its head at the point with coordinates (x, y).

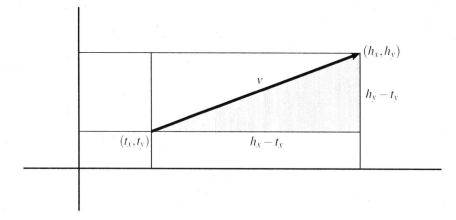

Figure 2. The horizontal and vertical components of a vector.

27.1 Vector arithmetic

Lemma 27.1 *Let $\vec{u} = \langle u_x, u_y \rangle$ and $\vec{v} = \langle v_x, v_y \rangle$. Then $\vec{u} + \vec{v} = \langle u_x + v_x, u_y + v_y \rangle$.*

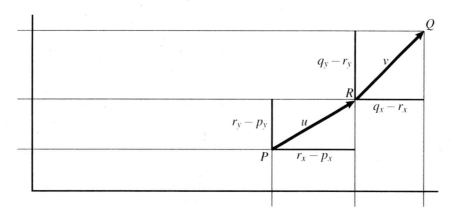

Figure 3. Addition of vectors is done component-wise.

Proof Position u and v head-to-tail. Label the coordinates of the tail of u as (p_x, p_y), of the head of v as (q_x, q_y), and of the head of u, which is the tail of v, as (r_x, r_y). Then the horizontal component of $\vec{u} + \vec{v}$ is

$$q_x - p_x = (q_x - r_x) + (r_x - p_x) = u_x + v_x,$$

and the vertical component of $\vec{u} + \vec{v}$ is

$$q_y - p_y = (q_y - r_y) + (r_y - p_y) = u_y + v_y. \qquad \square$$

Lemma 27.2 *Let $\vec{v} = \langle v_x, v_y \rangle$ and k be a real number. Then*

$$k \cdot \vec{v} = \langle k v_x, k v_y \rangle.$$

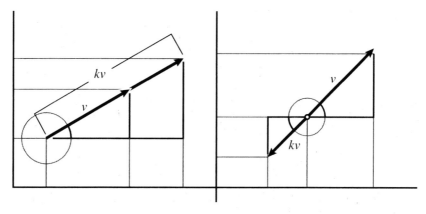

Figure 4. Scalar multiplication is done component-wise. On the left, $k > 0$; on the right, $k < 0$.

Proof From the previous part, we can break $\vec{v}$ down into two vectors, one containing the horizontal component, the other the vertical:

$$\vec{v} = \langle v_x, 0 \rangle + \langle 0, v_y \rangle.$$

The two vectors, together with $\vec{v}$, form a right triangle. In the same way, we can form a right triangle from $k \cdot \vec{v}$ and its horizontal and vertical components. The two triangles are similar. Comparing the hypotenuses, the (signed) scaling factor between those triangles is k. Scaling the legs by the same amount, $k \cdot \vec{v}$ has a horizontal component of kv_x and a vertical component of kv_y. □

The arithmetic of vectors has some familiar properties. They are more linear algebra than geometry, so their proofs are omitted.

Theorem 27.3 *The following are true for all vectors $\vec{u}$, $\vec{v}$, and $\vec{w}$ and for all real numbers k and l:*

- *Additive associativity:* $(\vec{u} + \vec{v}) + \vec{w} = \vec{u} + (\vec{v} + \vec{w})$
- *Additive commutativity:* $\vec{u} + \vec{v} = \vec{v} + \vec{u}$
- *Additive identity: the sum of the zero vector and $\vec{v}$ is $\vec{v}$*
- *Additive inverse: every vector $\vec{v}$ has an additive inverse $\vec{w}$ so that $\vec{v} + \vec{w}$ is the zero vector*
- *Distributive 1:* $k(\vec{u} + \vec{v}) = k\vec{u} + k\vec{v}$
- *Distributive 2:* $(k + l)\vec{v} = k\vec{v} + l\vec{v}$
- *Multiplicative associativity:* $kl(\vec{v}) = k(l\vec{v})$
- *Multiplicative identity:* $1(\vec{v}) = \vec{v}$.

There is one more term to define. The *norm* (or length, or size, or magnitude) of a vector $\vec{v}$, written $|\vec{v}|$, is the length of any of its representative segments. Using the distance formula, the norm of a vector may be calculated from its components to be

$$|\langle v_x, v_y \rangle| = \sqrt{(v_x)^2 + (v_y)^2}.$$

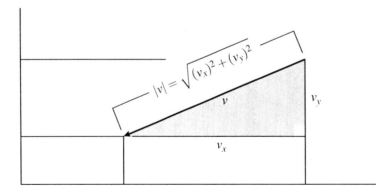

Figure 5. The norm of a vector may be calculated with the distance formula.

27.2 Change of coordinates

Our study of the analytic side of geometry began with choices about where to put the origin, and how to point the x- and y-axes. We introduced frames in chapter 25 as a a pair of perpendicular unit length segments $\{v_x, v_y\}$ from a common endpoint. Each frame provides the necessary pieces to set up a coordinate system: the vertex O of the frame is the origin, and the segments v_x and v_y point in the directions of the positive x- and y-axes. Each frame $\mathscr{F}$ determines a coordinate system $C_{\mathscr{F}}$. We can describe the coordinates (x, y) of a point P relative to this frame in the same fashion as in chapter 21. From the perspective of vectors,

$$\vec{OP} = x \cdot \vec{v}_x + y \cdot \vec{v}_y.$$

The matrix equation of an isometry depends on the coordinate system. In theory, any isometry can be described analytically in any coordinate system. In practice, however, some coordinate systems are more convenient than others. To switch from a coordinate system, $C_{\mathscr{F}}$, to the more convenient one, $C_{\mathscr{G}}$, we need to understand how a point's $C_{\mathscr{F}}$-coordinates are related to its $C_{\mathscr{G}}$-coordinates. As you might expect, the key to this is an isometry that maps the frame $\mathscr{F}$ to the frame $\mathscr{G}$. There are a few more things that we need to know before we can proceed. These are gathered into the next four exercises.

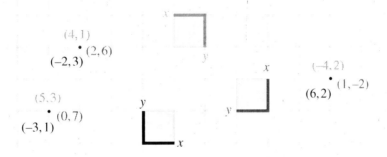

Figure 6. The coordinates of three points in three different systems.

Exercise 27.2 Prove that for any frames $\mathscr{F}$ and $\mathscr{G}$, there is a unique isometry T so that $T(\mathscr{F}) = \mathscr{G}$.

Exercise 27.3 Prove that if M is a matrix, v_1 and v_2 are vectors, and k is a constant, then

- $M(v_1 + v_2) = Mv_1 + Mv_2$
- $M(kv_1) = kM(v_1).$

Exercise 27.4 Suppose that the isometry T is given by a matrix equation of the form

$$T\begin{pmatrix} x \\ y \end{pmatrix} = M \begin{pmatrix} x \\ y \end{pmatrix} + \begin{pmatrix} e \\ f \end{pmatrix}$$

where M is a 2×2 matrix with a determinant of ± 1 (we have seen that every isometry can be written in this form). Show that

$$T^{-1}\begin{pmatrix} x \\ y \end{pmatrix} = M^{-1}\left(\begin{pmatrix} x \\ y \end{pmatrix} - \begin{pmatrix} e \\ f \end{pmatrix}\right).$$

Exercise 27.5 Suppose that the isometry

$$T\begin{pmatrix} x \\ y \end{pmatrix} = M\begin{pmatrix} x \\ y \end{pmatrix} + \begin{pmatrix} e \\ f \end{pmatrix}$$

maps the $\mathscr{F}$ frame $\{v_x, v_y\}$ to the $\mathscr{G}$ frame $\{w_x, w_y\}$. Prove that $M(\vec{v}_x) = \vec{w}_x$ and that $M(\vec{v}_y) = \vec{w}_y$ (these are not the segments that define the frame, but the associated vectors).

With those preliminaries done, we can now turn our attention to the change of coordinates. This would just be the standard change of basis argument of linear algebra except for the possibility of a translation, which complicates matters.

Theorem 27.4 *Let $C_\mathscr{F}$ and $C_\mathscr{G}$ be the coordinate systems determined by the frames $\mathscr{F}$ and $\mathscr{G}$ respectively, and let T be the isometry that maps $\mathscr{F}$ to $\mathscr{G}$. Then the $C_\mathscr{G}$-coordinates of a point P are the same as the $C_\mathscr{F}$-coordinates of $T^{-1}(P)$.*

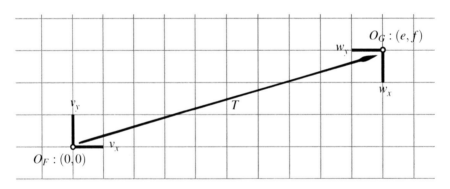

Figure 7. There is a unique isometry from any orthonormal frame to any other.

Proof Let $\mathscr{F} = \{v_x, v_y\}$, $\mathscr{G} = \{w_x, w_y\}$, and let O_f and O_g be the vertices of the frames $\mathscr{F}$ and $\mathscr{G}$, respectively. Throughout this proof, we will describe the isometry T and its inverse in terms of the $C_\mathscr{F}$-coordinate system. That is, write

$$T\begin{pmatrix} x \\ y \end{pmatrix} = M\begin{pmatrix} x \\ y \end{pmatrix} + \begin{pmatrix} e \\ f \end{pmatrix}$$

with the understanding that the matrix M and the coordinates (x, y) and (e, f) are written in terms of $C_\mathscr{F}$. Then

$$O_g = T(O_f) = M\begin{pmatrix} 0 \\ 0 \end{pmatrix} + \begin{pmatrix} e \\ f \end{pmatrix} = \begin{pmatrix} e \\ f \end{pmatrix},$$

27.2 Change of coordinates

and so the point O_G has coordinates (e, f) in the $C_\mathscr{F}$-coordinate system. Let P be a point whose coordinates are (x, y) in the $C_\mathscr{G}$-coordinate system. In terms of vectors, this means that $\vec{O_g P} = x\vec{w}_x + y\vec{w}_y$, and so

$$\vec{O_f P} = \vec{O_f O_g} + \vec{O_g P} = \begin{pmatrix} e \\ f \end{pmatrix} + x\vec{w}_x + y\vec{w}_y.$$

Now compute $T^{-1}(P)$, using the fact that the matrix multiplication acts linearly:

$$\begin{aligned} T^{-1}(P) &= T^{-1}\left(\begin{pmatrix} e \\ f \end{pmatrix} + x\vec{w}_x + y\vec{w}_y\right) \\ &= M^{-1}\left(\begin{pmatrix} e \\ f \end{pmatrix} + x\vec{w}_x + y\vec{w}_y - \begin{pmatrix} e \\ f \end{pmatrix}\right) \\ &= M^{-1}(x\vec{w}_x + y\vec{w}_y) \\ &= M^{-1}(x\vec{w}_x) + M^{-1}(y\vec{w}_y) \\ &= x \cdot M^{-1}(\vec{w}_x) + y \cdot M^{-1}(\vec{w}_y). \end{aligned}$$

According to exercise 27.5, $M(v_x) = w_x$ and $M(v_y) = w_y$. Therefore $M^{-1}(w_x) = v_x$ and $M^{-1}(w_y) = v_y$, and so

$$T^{-1}(P) = x \cdot v_x + y \cdot v_y.$$

That is, the coordinates of $T^{-1}(P)$ in the $C_\mathscr{F}$-coordinate system are (x, y). □

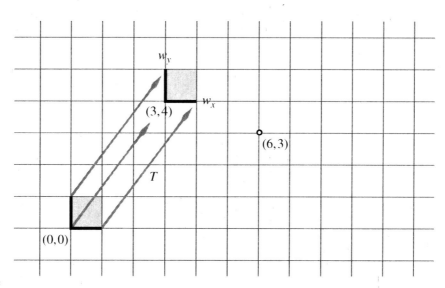

Figure 8. What are the coordinates of (6, 3) in the $\{w_x, w_y\}$ frame?

The value of this theorem is in situations where calculations are difficult in one coordinate system, but easy in another. We first look at examples where the coordinates of a point can be easily determined in both systems.

Example 27.5 Let $\mathcal{G}$ be the frame $\{w_x, w_y\}$ where in $C_{\mathcal{F}}$-coordinates,

- w_x has endpoints $(3, 4)$ and $(4, 4)$, and
- w_y has endpoints $(3, 4)$ and $(3, 5)$.

Consider a point P with $C_{\mathcal{F}}$-coordinates $(6, 3)$. It seems clear that its $C_{\mathcal{G}}$-coordinates are $(3, -1)$. The previous theorem should confirm that. The isometry T that maps $\mathcal{F}$ to $\mathcal{G}$ is a translation by $\langle 3, 4 \rangle$. Its inverse is the translation in the opposite direction:

$$T^{-1}\begin{pmatrix} x \\ y \end{pmatrix} = \begin{pmatrix} x \\ y \end{pmatrix} - \begin{pmatrix} 3 \\ 4 \end{pmatrix},$$

and so, as anticipated,

$$T^{-1}\begin{pmatrix} 6 \\ 3 \end{pmatrix} = \begin{pmatrix} 6 \\ 3 \end{pmatrix} - \begin{pmatrix} 3 \\ 4 \end{pmatrix} = \begin{pmatrix} 3 \\ -1 \end{pmatrix}.$$

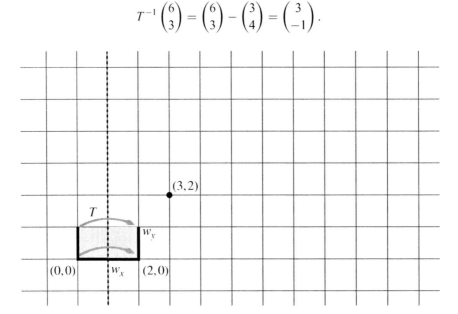

Figure 9. What are the coordinates of $(3, 2)$ in the $\{w_x, w_y\}$ frame?

Example 27.6 Let $\mathcal{G}$ be the frame $\{w_x, w_y\}$ where in $C_{\mathcal{F}}$-coordinates,

- w_x has endpoint $(2, 0)$ and $(1, 0)$, and
- w_y has endpoint $(2, 0)$ and $(2, 1)$.

Consider a point P with $C_{\mathcal{F}}$-coordinates $(3, 2)$. We can see that the $C_{\mathcal{G}}$-coordinates are $(-1, 2)$. This time, the isometry that maps $\mathcal{F}$ to $\mathcal{G}$ is a reflection that is given by

$$T\begin{pmatrix} x \\ y \end{pmatrix} = \begin{pmatrix} -1 & 0 \\ 0 & 1 \end{pmatrix}\begin{pmatrix} x \\ y \end{pmatrix} + \begin{pmatrix} 2 \\ 0 \end{pmatrix} = \begin{pmatrix} -x \\ y \end{pmatrix} + \begin{pmatrix} 2 \\ 0 \end{pmatrix} = \begin{pmatrix} 2-x \\ y \end{pmatrix}.$$

Since it is a reflection, it is its own inverse and we can calculate

$$T^{-1}\begin{pmatrix} 3 \\ 2 \end{pmatrix} = \begin{pmatrix} 2-3 \\ 2 \end{pmatrix} = \begin{pmatrix} -1 \\ 2 \end{pmatrix}.$$

This confirms what we expected.

27.2 Change of coordinates

In the last few chapters, we worked out the matrix equations for some, but not all, isometries. In particular, we gave equations only for rotations about the origin and reflections across lines through the origin. With the right change of coordinate frame, we can now move the origin around, and so get equations for any rotation or reflection. Let's consider an example.

Example 27.7 Suppose we want to find the matrix equation of a counterclockwise rotation by $\pi/2$ around the point $(3, 1)$. Begin with the coordinates (x, y) of a point P. The only formula we have for a rotation is one for rotation about the origin. To use it, we will have to switch to a coordinate system with $(3, 1)$ as its origin. We can do it with a translation. There are three steps:

- *Find the coordinates of P in the new coordinate system.* The translation

$$T\begin{pmatrix} x \\ y \end{pmatrix} = \begin{pmatrix} x \\ y \end{pmatrix} + \begin{pmatrix} 3 \\ 1 \end{pmatrix}$$

takes the current coordinate frame to one with the origin at $(3, 1)$. To find the coordinates of P in the new system, we just need its inverse:

$$T^{-1}\begin{pmatrix} x \\ y \end{pmatrix} = \begin{pmatrix} x \\ y \end{pmatrix} - \begin{pmatrix} 3 \\ 1 \end{pmatrix}.$$

- *Calculate the rotation of this point.* The matrix for the rotation is

$$\begin{pmatrix} \cos \pi/2 & -\sin \pi/2 \\ \sin \pi/2 & \cos \pi/2 \end{pmatrix} = \begin{pmatrix} 0 & -1 \\ 1 & 0 \end{pmatrix}.$$

- *Write the result in the original coordinate system.* Once we have computed the effect of the rotation in the new coordinate system, we can return to the original coordinate system to report our answer. We do that by applying T.

Combining the steps gives the equation of the rotation:

$$R\begin{pmatrix} x \\ y \end{pmatrix} = \begin{pmatrix} 0 & -1 \\ 1 & 0 \end{pmatrix} \left[\begin{pmatrix} x \\ y \end{pmatrix} - \begin{pmatrix} 3 \\ 1 \end{pmatrix} \right] + \begin{pmatrix} 3 \\ 1 \end{pmatrix}$$

$$= \begin{pmatrix} 0 & -1 \\ 1 & 0 \end{pmatrix} \begin{pmatrix} x - 3 \\ y - 1 \end{pmatrix} + \begin{pmatrix} 3 \\ 1 \end{pmatrix}$$

$$= \begin{pmatrix} 1 - y \\ x - 3 \end{pmatrix} + \begin{pmatrix} 3 \\ 1 \end{pmatrix}$$

$$= \begin{pmatrix} 4 - y \\ x - 2 \end{pmatrix}.$$

The example illustrates the general procedure. Let τ be an isometry. Suppose $\mathscr{F}$ and $\mathscr{G}$ are frames and that S is the matrix equation of the isometry that maps $\mathscr{F}$ to $\mathscr{G}$ (written in terms of the $\mathscr{F}$-coordinate system). Suppose that τ can be expressed as a matrix equation T in the $\mathscr{G}$-coordinate system. Then τ can be expressed as the matrix equation $S \circ T \circ S^{-1}$ in the $\mathscr{F}$ coordinate system.

27.3 Exercises

27.6. Verify that vector addition is commutative and associative.

27.7. Take a vector $\langle a, b \rangle$. Let S be the set consisting of the identity isometry and translations whose translation vectors have the form $\langle ma, nb \rangle$. Show that the composition of two elements of S is an element of S. Show that the inverse of an element of S is an element of S. So S is a subgroup of the group of isometries.

27.8. What is the image of the point $(3, 0)$ under the counterclockwise rotation by an angle $\pi/6$ about the point $(1, 1)$?

27.9. What is the matrix equation for the counterclockwise rotation by $\pi/4$ around the point $(1, 1)$?

27.10. What is the matrix equation for the counterclockwise rotation by π around the point $(5, 0)$?

27.11. What is the matrix equation for the reflection across the line $y = x + 2$?

27.12. What is the matrix equation for the reflection across the line $y = 2x - 1$?

27.13. What is the matrix equation for a glide reflection whose glide line is $y = 2x + 1$ and whose glide distance is 5 (and the glide vector points from the origin into the first quadrant)?

27.14. Use a change of coordinates to find the general form for the counterclockwise rotation by an angle θ about a point (h, k).

27.15. Use a change of coordinates to find the general form for the reflection across the line $y = mx + b$.

27.16. Show that the composition of two rotations is a translation or the identity if the sum of the rotation angles is a multiple of 2π, and a rotation otherwise.

27.17. Let r_1 be a rotation by θ_1 and let r_2 be a rotation by θ_2, where $\theta_1 + \theta_2$ is not a multiple of 2π. Prove that $r_1 \circ r_2$ is a rotation by $\theta_1 + \theta_2$.

28
Dilation

28.1 Similarity mappings

Throughout our study of Euclidean geometry, we have dealt with two fundamentally important equivalence relations for triangles, congruence and similarity. The isometries of the last few chapters are closely tied to the congruence relation: if T is a triangle and τ is an isometry, then $\tau(T)$ is congruent to T. In this chapter, we will look at mappings that are tied to the similarity relation.

Definition 28.1 A bijective mapping σ of the Euclidean plane is called a *similarity mapping* if for every triangle T, T and its image $\sigma(T)$ are similar.

The first and most important thing to do is to understand the effect that a similarity mapping has on distance. To do this, we need a lemma about positioning segments. We will say that two segments AB and CD are in general position if neither A nor B is on $\overline{CD}$ and neither C nor D is on $\overline{AB}$.

Lemma 28.2 *Given segments s_1 and s_2, there is a segment s_3 that is in general position with respect to both s_1 and s_2.*

Exercise 28.1 Prove the lemma.

If two segments are in general position, then any three of their four endpoints will form a triangle (restricting to general position eliminates the degenerate cases). We need these triangles to work with similarity mappings, so we will use general positioning in the next proof as a way to form them.

Theorem 28.3 *A bijection σ is a similarity mapping if and only if it scales all distances by a constant. That is, σ is a similarity mapping if and only if there is a positive real number k so that $|\sigma(AB)| = k|AB|$ for all segments AB.*

Proof $\Longleftarrow$ Suppose that σ scales all distances by a constant k. Then given $\triangle ABC$,

$$|\sigma(AB)| = k|AB|, \quad |\sigma(AC)| = k|AC|, \quad \text{and} \quad |\sigma(BC)| = k|BC|.$$

By the SSS similarity theorem, $\triangle ABC$ and $\sigma(\triangle ABC)$ are similar, and so σ meets the requirements of a similarity mapping.

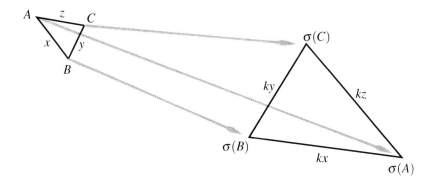

Figure 1. The effect of a similarity mapping σ (with scaling factor k) on a triangle.

$\implies$ Suppose that σ is a similarity mapping, and that $|\sigma(AB)| = k|AB|$ for some segment AB. Let CD be any other segment. According to the previous lemma, there is a segment EF that is in general position with respect to AB and CD. We can repeatedly use the fact that σ is a similarity mapping to track the scaling constant from AB to EF to CD:

$$\sigma(\triangle AEB) \sim \triangle AEB \implies |\sigma(EB)| = k|EB|$$
$$\sigma(\triangle BEF) \sim \triangle BEF \implies |\sigma(EF)| = k|EF|$$
$$\sigma(\triangle ECF) \sim \triangle ECF \implies |\sigma(CF)| = k|CF|$$
$$\sigma(\triangle CFD) \sim \triangle CFD \implies |\sigma(CD)| = k|CD|.$$

Therefore σ scales CD by the same amount as it does AB. □

Let's investigate some of the properties of a similarity mapping σ.

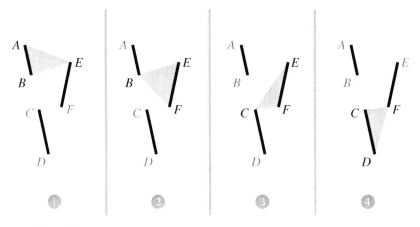

Figure 2. A series of triangles to connect the scaling of AB to that of CD.

Lemma 28.4 *If $A * B * C$, then $\sigma(A) * \sigma(B) * \sigma(C)$.*

28.2 Dilations

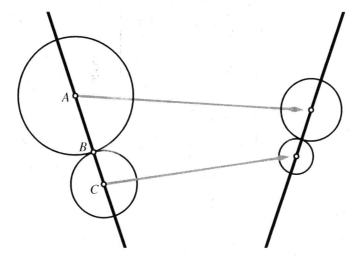

Figure 3. There is only one possible location for B. Therefore, a similarity mapping preserves the relationships of incidence and order.

Proof Since $A * B * C$, $|AC| = |AB| + |BC|$. Multiply by the scaling constant k to get

$$k|AC| = k|AB| + k|BC|$$
$$|\sigma(AC)| = |\sigma(AB)| + |\sigma(BC)|.$$

This is the degenerate case of the triangle inequality. The only way it can be true is if $\sigma(A)$, $\sigma(B)$, and $\sigma(C)$ are collinear, and $\sigma(B)$ is between $\sigma(A)$ and $\sigma(C)$. □

More generally, the images of any number of collinear points are collinear, and their order is retained.

Lemma 28.5 *If $AB \simeq A'B'$, then $\sigma(AB) \simeq \sigma(A'B')$.*

Exercise 28.2 Prove lemma 28.5.

Lemma 28.6 *For an angle $\angle A$, $\sigma(\angle A) \simeq \angle A$. If $\angle A \simeq \angle A'$, then $\sigma(\angle A) \simeq \sigma(\angle A')$.*

Exercise 28.3 Prove lemma 28.6.

This property together with the distance scaling property shows that a similarity mapping will map any polygon to a similar polygon. While similarity mappings distort distances, they do so in a relatively tame way, and the synthetic relations of incidence, order, and congruence are preserved.

28.2 Dilations

We have looked at some properties of similarity mappings without ever asking whether there are mappings (other than isometries) that have the condition. There are; we use them daily whenever we use a map, a blueprint, or a scale model.

Definition 28.7 Let O be a point and k be a positive real number. The *dilation* by a factor of k centered at O is the map d of the Euclidean plane so that

- $d(O) = O$, and
- for any other point P, $d(P)$ is the point on $\overrightarrow{OP}$ that is a distance $k|OP|$ from O.

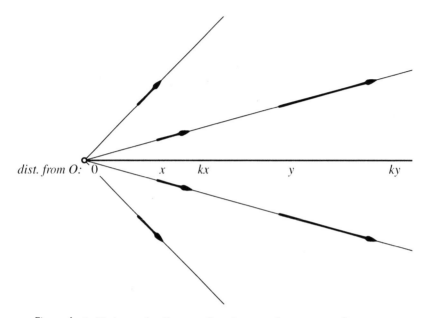

Figure 4. A dilation scales distances from its center by a constant factor.

Dilations are also called scalings, dilatations, and occasionally homotheties. It is clear that a dilation is a bijection. It is easy to describe its inverse: if d is the dilation by k centered at O, its inverse is another dilation centered at O, by a factor of $1/k$. When $k = 1$, d is the identity map. Otherwise, a dilation will not be an isometry and it will alter distance.

Theorem 28.8 *A dilation is a similarity mapping.*

Proof Let d be a dilation centered at O with a scaling factor of k. By definition, a segment with one endpoint on O will be scaled by k. To show that d is a similarity mapping, we need to show that any other segment AB is scaled by the same amount. There are a few cases to consider.

Case 1. Suppose that A and B are on the same ray from O, and for convenience, let's suppose that A is between O and B. Then $d(A)$ and $d(B)$ are still on the same ray from O, at respective distances of $k|OA|$ and $k|OB|$, and so $d(A)$ is still between O and $d(B)$. Therefore

$$|d(AB)| = |d(OB)| - |d(OA)| = k|OB| - k|OA| = k(|OB| - |OA|) = k|AB|.$$

Case 2. Suppose that A and B are on opposite rays from O. Then $d(A)$ and $d(B)$ are on the same opposite rays, and so

$$|d(AB)| = |d(OA)| + |d(OB)| = k|OA| + k|OB| = k(|OA| + |OB|) = k|AB|.$$

28.2 Dilations

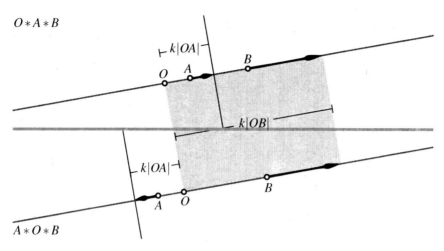

Figure 5. A dilation scales the distance between two points by a constant factor. Case 1: $O * A * B$. Case 2: $A * O * B$.

Case 3. The most common case is when A and B are neither on the same ray nor on opposite rays from O. Compare $\triangle AOB$ and $d(\triangle AOB)$. Since $d(O) = O$, and $d(A)$ and $d(B)$ are on the same rays from O as A and B, $\angle AOB = d(\angle AOB)$. In addition, $|d(OA)| = k|OA|$ and $|d(OB)| = k|OB|$. By the SAS similarity theorem, $\triangle AOB$ and $d(\triangle AOB)$ are similar. Comparing the third sides of the triangles, $|d(AB)| = k|AB|$. □

As with isometries, the effect of a dilation can be described with a matrix equation.

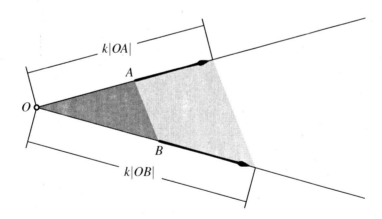

Figure 6. A dilation scales the distance between two points by a constant factor. Case 3: O, A, and B are not collinear.

Theorem 28.9 *The matrix equation for a dilation d by a factor of k centered at the point $(0, 0)$ is*

$$d\begin{pmatrix}x\\y\end{pmatrix} = \begin{pmatrix}kx\\ky\end{pmatrix}.$$

Proof We show that the mapping d that is given by the equation has the same effect on points as a dilation by k does. There are three things to show:

(1) that d fixes the origin O;
(2) that for any other point P, $d(P)$ is on $\overrightarrow{OP}$; and
(3) that the distance from O to $d(P)$ is $k|OP|$.

1. $d\begin{pmatrix} 0 \\ 0 \end{pmatrix} = \begin{pmatrix} k \cdot 0 \\ k \cdot 0 \end{pmatrix} = \begin{pmatrix} 0 \\ 0 \end{pmatrix}$.

2. The line that passes through the origin and the point (x, y) can be parametrized as $r(t) = (xt, yt)$ (it passes through the origin when $t = 0$ and the point (x, y) when $t = 1$). It also passes through the point (kx, ky) when $t = k$. Furthermore, since we have specified that k be a positive number, both (x, y) and (kx, ky) are on the same side of the origin and so they are on the same ray from O.

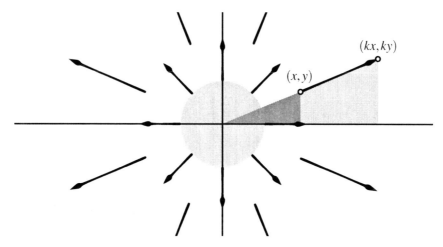

Figure 7. A dilation centered at the origin.

3. The distance from $(0, 0)$ to (kx, ky) is

$$\sqrt{(kx - 0)^2 + (ky - 0)^2} = \sqrt{k^2(x^2 + y^2)} = k\sqrt{(x - 0)^2 + (y - 0)^2}.$$

It is k times the distance from the origin to (x, y). □

As we did in the last chapter with isometries, we can use a change of coordinates to describe dilations about any point.

Theorem 28.10 *The matrix equation for a dilation d by a factor of k centered at the point (a, b) is*

$$d\begin{pmatrix} x \\ y \end{pmatrix} = \begin{pmatrix} kx + (1 - k)a \\ ky + (1 - k)b \end{pmatrix}.$$

Proof Let P be a point with coordinates (x, y). We can calculate the matrix equation for d with a change of coordinates:

(1) convert (x, y) to a coordinate system whose origin is at (a, b),
(2) perform the scaling by a factor of k, and then
(3) convert the result back to the original coordinate system.

1. The translation $t(x, y) = (x + a, y + b)$ shifts the standard coordinate frame centered at $(0, 0)$ to one that is centered at (a, b). To compute the coordinates of P in the new coordinate system, apply t^{-1}:

$$\begin{pmatrix} x \\ y \end{pmatrix} \mapsto \begin{pmatrix} x - a \\ y - b \end{pmatrix}.$$

2. Now scale by k, using the special formula from the previous theorem:

$$\begin{pmatrix} x - a \\ y - b \end{pmatrix} \mapsto \begin{pmatrix} k(x - a) \\ k(y - b) \end{pmatrix}.$$

3. Convert back to the original coordinate system by applying t:

$$\begin{pmatrix} k(x - a) \\ k(y - b) \end{pmatrix} \mapsto \begin{pmatrix} k(x - a) + a \\ k(y - b) + b \end{pmatrix} = \begin{pmatrix} kx + (1 - k)a \\ ky + (1 - k)b \end{pmatrix}.$$

□

28.3 Preserving incidence, order, and congruence

Dilations and isometries are similarity mappings. It is natural to wonder if there are others. Let's investigate a slightly more general question. Every similarity mapping preserves the relations of incidence, order, and congruence. We have seen two such types of mappings, dilations and isometries. What other types of bijections will preserve the structures? Our effort to answer this will seem familiar. It is closely connected to our work on parallel projection in chapter 12. The proper analogy is to think of a parallel projection as a one-dimensional similarity mapping. In the next few results, we will assume that a bijection f preserves incidence, order, and congruence. This is a quick way to say that if a collection of points are all on one line, then their images will be as well, in the same order; and that if segments or angles are congruent, then their images will be congruent as well.

Lemma 28.11 *Let f be a bijection that preserves incidence, order, and congruence. Let s_1 and s_2 be segments. If $|s_1| = \frac{1}{2}|s_2|$ and f scales s_2 by k, then f scales s_1 by k as well.*

Proof. Label the two endpoints of s_2 as A and B and its midpoint as M. Then s_1, AM, and BM are congruent and so their images must be as well. Then

$$|f(s_1)| = \frac{|f(A)| + |f(s_1)|}{2} = \frac{|f(AM)| + |f(BM)|}{2} = \frac{|f(AB)|}{2} = \frac{k|AB|}{2} = k|s_1|.$$

□

Lemma 28.12 *Let f be a bijection that preserves incidence, order, and congruence. If $A * B * C$ and if f scales AB and BC by k, then f scales AC by k.*

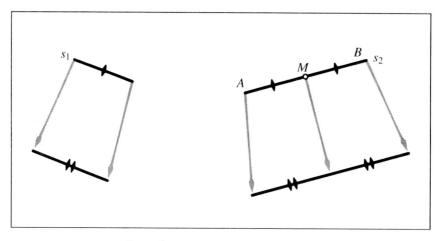

Figure 8. Halving and scaling segments.

Proof Since f preserves the order of points, $f(A) * f(B) * f(C)$, and so

$$|f(AC)| = |f(AB)| + |f(BC)| = k|AB| + k|BC| = k(|AB| + |BC|) = k|AC|. \qquad \square$$

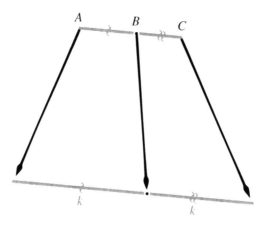

Figure 9. Chaining and scaling segments.

Lemma 28.13 *Let f be a bijection that preserves incidence, order, and congruence. If $|s_1| = (m/2^n) \cdot |s_2|$ where m and n are positive integers, and if f scales s_2 by k, then f scales s_1 by k as well.*

Proof The first lemma tells us that a segment of length $(1/2) \cdot |s_2|$ will be scaled by k. Applied again, it tells us that a segment of length $(1/4) \cdot |s_2|$ will be scaled by k. And so on, so that for all positive integers n, a segment of length $(1/2^n) \cdot |s_2|$ will be scaled by a factor of k. Then

28.3 Preserving incidence, order, and congruence

we can line up m segments of length $(1/2^n) \cdot |s_2|$, to get a segment of length $(m/2^n) \cdot |s_2|$. By repeatedly applying the second lemma, we can see that it too is scaled by k. □

Theorem 28.14 *A bijection of the Euclidean plane that preserves incidence, order, and congruence is a similarity mapping.*

Proof Let f be a bijection that preserves incidence, order, and congruence. Since f maps congruent segments to congruent segments, all segments of a given length will be scaled by the same amount. Let k be the scaling constant for a segment of length one. By subdividing and chaining (as described above), k is the scaling constant for all segments of length $m/2^n$. We need to show that k is the scaling constant for segments of all other lengths. Suppose that the segment OA has a length of x and that $|f(OA)| = k'x$. To get an idea of k', we can use dyadic approximations to pin OA between segments that are scaled by k. For each n, there is an m_n so that

$$\frac{m_n}{2^n} \leq x \leq \frac{m_n + 1}{2^n}.$$

Along $\overrightarrow{OA}$, mark points $M_n^<$ and $M_n^>$ bracketing A so that $|OM_n^<| = m_n/2^n$ and $|OM_n^>| = (m_n + 1)/2^n$. Reading off the points in order, $O * M_n^< * A * M_n^>$. The distance between $M_n^<$ and $M_n^>$ is $1/2^n$, so as n increases, the bracketing of A gets tighter and tighter. Since f preserves incidence and order, when we apply it to the points, we get a bracketing of $f(A)$ that can determine the scaling of OA:

$$f(O) * f(M_n^<) * f(A) * f(M_n^>)$$
$$|f(OM_n^<)| \leq |f(OA)| \leq |f(OM_n^>)|$$
$$k \cdot m_n/2^n \leq k' \cdot |OA| \leq k \cdot (m_n + 1)/2^n.$$

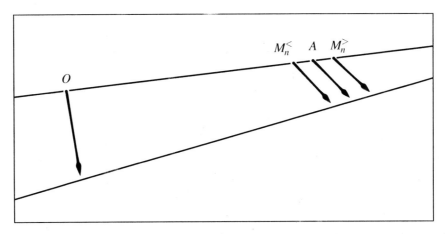

Figure 10. A mapping that preserves the relations of incidence, order, and congruence, must scale all distances by a constant k.

To find k', divide by $|OA|$:

$$k \cdot \frac{m_n/2^n}{|OA|} \leq k' \leq k \cdot \frac{(m_n+1)/2^n}{|OA|}.$$

The set of inequalities are true for all values of n. As n increases, the terms $(m_n/2^n)/|OA|$ and $((m_n + 1)/2^n)/|OA|$ approach 1. The only way that the inequalities can be satisfied for all n is for k' to be equal to k. Therefore f scales all distances by the same constant k. According to theorem 28.3, f must be a similarity mapping. □

One last theorem ties it all together.

Theorem 28.15 *A bijection that preserves incidence, order, and congruence can be written as a composition of an isometry and a dilation.*

Proof Let f be such a bijection. As we have just seen, f is a similarity mapping, so f scales distances by a constant k. Let d be the dilation centered at the origin that scales by a factor of k. Its inverse, d^{-1}, is a dilation by a factor of $1/k$, so for a segment s,

$$|d^{-1} \circ f(s)| = (1/k) \cdot |f(s)| = (1/k) \cdot k \cdot |s| = |s|.$$

Therefore $d^{-1} \circ f$ is an isometry. Writing τ for this isometry, $d^{-1} \circ f = \tau$. Apply the dilation d to both sides to get

$$d \circ d^{-1} \circ f = d \circ \tau \implies f = d \circ \tau,$$

and we have written f as a composition of an isometry and a dilation. □

28.4 Exercises

28.4. Let τ be a similarity mapping and $\mathcal{C}$ a circle. Prove that $\tau(\mathcal{C})$ is also a circle.

28.5. What is the image of the point $(2, 3)$ under the dilation by a factor of 4 centered at the point $(1, 5)$?

28.6. Show that if d_1 and d_2 are transformations with the same scaling factor, then there is an isometry τ so that $d_2 = \tau \circ d_1$.

28.7. Consider dilations

d_1: the dilation with center $(a, 0)$ and scaling factor k_a
d_2: the dilation with center $(b, 0)$ and scaling factor k_b.

Does $d_1 \circ d_2$ have a fixed point? If so, what are its coordinates?

28.8. Show that if d_1 is a dilation by a factor of k_1 and d_2 is a dilation by a factor of k_2, then $d_1 \circ d_2$ can be written as a composition of an isometry and a dilation by a factor of $k_1 \cdot k_2$.

28.9. Write an equation for the similarity mapping that is formed by first dilating by a factor of $1/2$ around the point $(1, 1)$, and then reflecting across the x-axis. Does the transformation have any fixed points?

28.4 Exercises

28.10. Prove that if $\triangle ABC \sim \triangle A'B'C'$, then there is a similarity mapping σ so that $\sigma(A) = A'$, $\sigma(B) = B'$, and $\sigma(C) = C'$.

28.11. Consider the similar triangles $\triangle ABC$ and $\triangle A'B'C'$ with vertices

$$A = (0,0), \quad B = (1,0), \quad \text{and} \quad C = (0,1)$$

and

$$A' = (2,0), \quad B' = (0,2), \quad \text{and} \quad C' = (0,-2).$$

Find the equation of the similarity mapping that maps $\triangle ABC$ to $\triangle A'B'C'$.

28.12. Let f be a similarity mapping and let ℓ be a line. Is it true that it is always possible to find an isometry τ and a parallel projection $\pi : \tau(\ell) \to f(\ell)$ so that $f(P) = \pi \circ \tau(P)$ for all points P on ℓ?

28.13. Let σ be a similarity mapping, and let P be a point whose trilinear coordinates with respect to $\triangle ABC$ are $[x : y : z]$. Prove that $\sigma(P)$ has the same trilinear coordinates with respect to $\sigma(\triangle ABC)$.

29

Applications of Transformations

We have spent the last several chapters building up a theory of Euclidean transformations, drawing on some of the Euclidean theory that we had previously developed. In this chapter we will turn that around by using the theory of transformations to prove three results of classical Euclidean geometry.

29.1 Varignon's theorem

The first result, Varignon's theorem, describes a construction of a parallelogram from a quadrilateral. Its proof uses half-turns. In chapter 24, we defined a half turn to be a rotation with a rotation angle of π, so a half-turn is its own inverse. Because of that, we don't have to specify whether the rotation is counterclockwise or clockwise: they are the same. In exercises 27.16 and 27.17 at the end of chapter 27, you were to investigate what happens when you compose two rotations. In particular, you were to verify that if the two angles of rotation sum to a multiple of 2π, then their composition is either the identity or a translation (it is a fairly straightforward, albeit messy, calculation using the matrix equations for a rotation). Because of that, when we compose two half-turns, their rotation angles add up to $\pi + \pi = 2\pi$, and the result must be either a translation or the identity.

Lemma 29.1 *Let r_A, r_B, r_C, and r_D be half-turns around four distinct points A, B, C, and D. If the composition $r_A \circ r_B \circ r_C \circ r_D$ is the identity map, then $\square ABCD$ is a parallelogram.*

Proof Let's break the composition into two pieces: $r_A \circ r_B$ and $r_C \circ r_D$. If we assume that their composition is the identity, then they are inverses of each other. That is,

$$r_C \circ r_D = (r_A \circ r_B)^{-1} = r_B^{-1} \circ r_A^{-1}.$$

Each of r_B and r_A is its own inverse, though, since they are half-turns. Thus $r_C \circ r_D = r_B \circ r_A$. In that case, we can apply both maps to the point A, sending it in two directions around the quadrilateral, and it will end up in the same place. Label the ending point P, and along the way label one more point, $Q = r_D(A)$. That is,

$$r_C \circ r_D(A) = r_C(Q) = P \quad \text{and} \quad r_B \circ r_A(A) = r_B(A) = P.$$

303

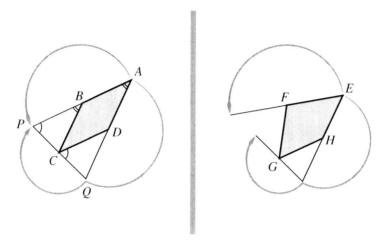

Figure 1. Four half-turns applied to points A and E. On the left, $\square ABCD$ is a parallelogram. On the right, $\square EFGH$ is not.

The points A, P, and Q form a triangle around the original quadrilateral. The triangle is well-balanced with respect to $\square ABCD$: because r_D is an isometry, $|AD| = |DQ|$, because r_C is an isometry, $|QC| = |CP|$, and because r_B is an isometry, $|PB| = |AB|$. Thus,

$$|AQ| = 2|DQ|, \quad |QP| = 2|CQ| = 2|CP|, \quad \text{and} \quad |PA| = 2|PB|.$$

By SAS similarity we have found three similar triangles: $\triangle AQP$ is similar to $\triangle DQC$ and $\triangle BCP$. Matching angles, $\angle DCQ \simeq \angle P$ and $\angle A \simeq \angle PBC$. The alternate interior angle theorem tells us that $CD \parallel AB$ and $AD \parallel BC$ and so $\square ABCD$ is, by definition, a parallelogram. $\square$

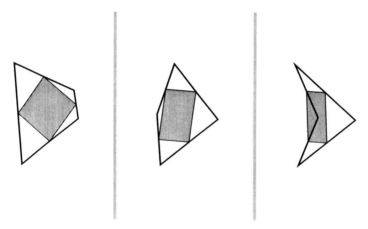

Figure 2. Varignon's theorem: the midpoints of the sides of a quadrilateral are the vertices of a parallelogram.

Theorem 29.2 *(Varignon's theorem) Let $\square A_1 A_2 A_3 A_4$ be a quadrilateral and label the midpoints of the sides B_1, B_2, B_3, and B_4, so that B_i is the midpoint of $A_i A_{i-1}$. Then $\square B_1 B_2 B_3 B_4$ is a parallelogram.*

29.2 Napoleon's theorem

The strategy should be obvious: use the last lemma. That means we need to look at the composition of the half-turns around the midpoints B_1, B_2, B_3, and B_4. We will show that the composition of the half-turns is the identity.

Proof For $1 \leq i \leq 4$, let r_i be the half-turn around the point B_i. The compositions $r_1 \circ r_2$ and $r_3 \circ r_4$ are both translations, so their composition is either a translation or the identity. However translations do not have any fixed points, and this composition does:

$$r_1 \circ r_2 \circ r_3 \circ r_4(A_4) = r_1 \circ r_2 \circ r_3(A_3) = r_1 \circ r_2(A_2) = r_1(A_1) = A_4.$$

Since $r_1 \circ r_2 \circ r_3 \circ r_4$ has a fixed point, it cannot be a translation and so it must be the identity. According to the previous lemma, $\square B_1 B_2 B_3 B_4$ is a parallelogram. □

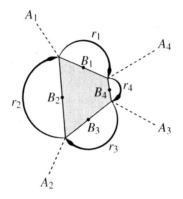

Figure 3. The proof of Varignon's theorem via half-turns.

29.2 Napoleon's theorem

Like Varignon's theorem, Napoleon's theorem reveals an unexpected symmetry. Yes, it is named after *that* Napoleon, although there is skepticism about whether he discovered it. Presumably, once you have conquered half of Europe, no one is going to raise a fuss if you claim a theorem or two as well.

Theorem 29.3 *(Napoleon's theorem) Given $\triangle ABC$, construct three equilateral triangles exterior to it, one on each of the sides AB, BC, and CA. The centers of the equilateral triangles are the vertices of another triangle, which is equilateral.*

This proof begins as Varignon's did, with a composition of rotations whose rotation angles add up to 2π. The fixed point is easy to find, meaning that the composition is the identity. It may not be clear how to use that fact, and so it is a bit of a scramble to the finish. The fundamental symmetry of the situation comes from the three equilateral triangles, and the rotations that capture the symmetry are 1/3-turns around their centers.

Proof First, we will enforce a consistency in the labeling of vertices: we will want the path that goes from A to B to C to A to make a clockwise loop around the triangle. If it instead

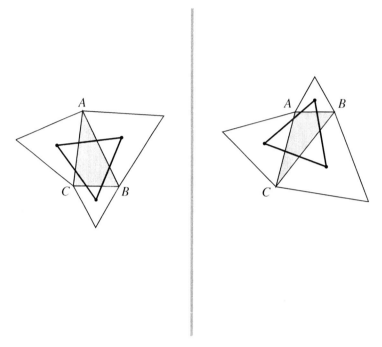

Figure 4. Napoleon's theorem: the centers of the equilateral triangles built on the sides of a triangle are themselves the vertices of an equilateral triangle.

makes a counterclockwise loop, you can just swap two of the labels. Label the centers of the equilateral triangles as a, b, and c, where

- a is the center of the triangle built on AB,
- b is the center of the triangle built on BC, and
- c is the center of the triangle built on CA.

Label the corresponding $2\pi/3$ counterclockwise rotations around the points as r_a, r_b, and r_c. When we compose the rotations, their rotation angles add up to $2\pi/3 + 2\pi/3 + 2\pi/3 = 2\pi$, so their composition $r_c \circ r_b \circ r_a$ is either a translation or the identity. Now look inside one of the equilateral triangles, say the one centered at a, and notice that in it $|aA| = |aB|$ and $(\angle AaB) = 2\pi/3$. Therefore r_a sends A to B. Likewise, r_b sends B to C and r_c sends C to A. In combination,

$$r_c \circ r_b \circ r_a(A) = r_c \circ r_b(B) = r_c(C) = A,$$

and so $r_c \circ r_b \circ r_a$ has a fixed point. It can't be a translation, so it must be the identity.

Let's see what happens when we substitute the point a into the composition (which we know is the identity, and so must map a to itself):

$$r_c \circ r_b \circ r_a(a) = a \implies r_c \circ r_b(a) = a \implies r_b(a) = r_c^{-1}(a).$$

This gives us one last point to label: $d = r_b(a)$. There are two triangles to look at. The first is $\triangle abd$. Since r_b maps the segment ba to the segment bd, ba and bd are congruent. Thus $\triangle abd$ is an isosceles triangle. Furthermore, at vertex b, we know the angle measure is $2\pi/3$. The

29.2 Napoleon's theorem

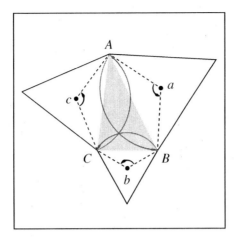

Figure 5. Rotations by $2\pi/3$ around the centers a, b, and c.

other two angles in this triangle add up to $\pi - 2\pi/3 = \pi/3$. According to the isosceles triangle theorem, they are congruent, so they each measure $\pi/6$. The second triangle is $\triangle acd$. The map r_c^{-1} is also a rotation by $2\pi/3$, but it is a clockwise rotation by that amount. It maps the segment ca to the segment cd, and so they must be congruent. Therefore, $\triangle acd$ is also isosceles. Its angle at vertex c has a measure of $2\pi/3$, so its other two angles each measure $\pi/6$.

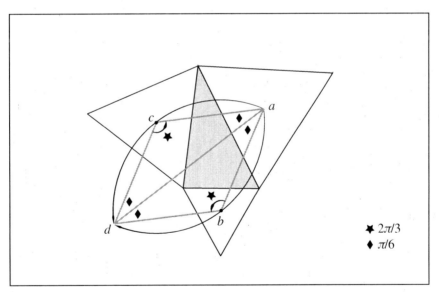

Figure 6. Triangles $\triangle abd$ and $\triangle acd$ are isosceles.

When we put the two pieces together, we get

$$(\angle bac) = (\angle bad) + (\angle cad) = \pi/6 + \pi/6 = \pi/3.$$

The angle at a is no different from those at b and c. A similar argument (in which the compositions of r_a, r_b, and r_c are taken in different orders) will show that the other two angles of $\triangle abc$ also measure $\pi/3$. Therefore $\triangle abc$ is equiangular and so it is equilateral. □

29.3 The nine point circle

For the last part of this chapter, let's look back at the nine point circle theorem. We proved this theorem in chapter 18 without using transformation methods: the key then was to find a diameter of the nine point circle. This time, the key is to find a transformation that maps the nine point circle to the circumcircle. In the chapter 18 proof, we also needed to know that the diagonals of a parallelogram bisect one another. In this proof, we will need the converse of that.

Lemma 29.4 *If segments AC and BD bisect each other, then □ABCD is a parallelogram.*

Exercise 29.1 Use isometries to prove this lemma. In particular, consider the half-turn around the intersection of AC and BD.

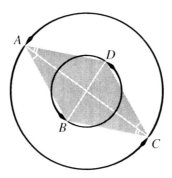

Figure 7. Diagonals of a parallelogram bisect one another.

Theorem 29.5 *(The nine point circle theorem, revisited) For any triangle, the following nine points all lie on the same circle: the feet of the three altitudes, the midpoints of the three sides, and the midpoints of the three segments connecting the orthocenter to each vertex. This circle is called the nine point circle associated with the triangle.*

Proof Given $\triangle A_1 A_2 A_3$ with orthocenter R, label

L_i : the foot of the altitude which passes through A_i,
M_i : the midpoint of the side that is opposite A_i, and
N_i : the midpoint of the segment $A_i R$.

Let d be the dilation by a factor of 2 centered at the orthocenter. We will show that $d(L_i)$, $d(M_i)$, and $d(N_i)$ are all on the circumcircle $\mathcal{C}$. (The proof does not handle a few degenerate cases, when $M_i = R$ and when $L_i = M_i$. Those cases are easily resolved though, so I have omitted them to keep the proof streamlined.)

29.3 The nine point circle

The points N_i. Since N_i is halfway from R to A_i, d maps N_i to the corresponding vertex A_i. The vertices are, of course, on $\mathcal{C}$.

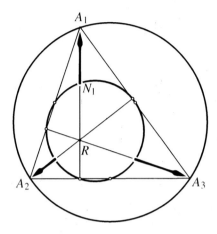

Figure 8. A dilation centered at R with scaling factor 2. The image of N_i is A_i.

The points M_i. This is the difficult part. Take for example M_1, the midpoint of A_2A_3. The dilation d maps M_1 to a point D that is twice as far away from R as M_1, and so M_1 is the midpoint of RD. Thus M_1 is the intersection of two bisecting diagonals, A_2A_3 and RD. As we just proved, this shows that $\square RA_2DA_3$ is a parallelogram. Therefore

- DA_3 is parallel to RA_2, the altitude perpendicular to A_1A_3. Hence DA_3 is perpendicular to the side A_1A_3.

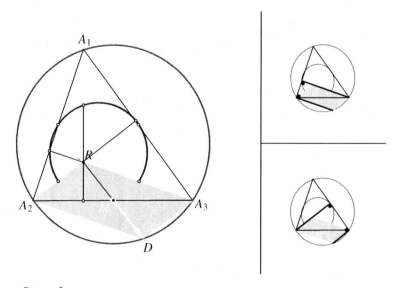

Figure 9. The image of M_i is the fourth vertex of a parallelogram.

- DA_2 is parallel to RA_3, the altitude perpendicular to A_1A_2. Hence DA_2 is perpendicular to the side A_1A_2.

In other words, both $\angle A_1A_2D$ and $\angle A_1A_3D$ are right angles. According to Thales' theorem, A_2 and A_3 have to be on the circle with diagonal A_1D. There is only one circle through the points A_1, A_2, and A_3, the circumcircle $\mathcal{C}$. Therefore $D = d(M_1)$ is on $\mathcal{C}$. It is just a matter of shuffling around the indices to show that d maps M_2 and M_3 to points of $\mathcal{C}$ as well. Furthermore, each of the segments $A_id(M_i)$ is a diameter of $\mathcal{C}$. (This is in keeping with the chapter 18 proof where we showed directly that N_iM_i is a diameter of the nine point circle. Here we see that its scaled image $d(N_iM_i) = A_id(M_i)$ is a diameter of the circumcircle.)

The points L_i. The intersection of each altitude with its corresponding side forms a right angle $\angle N_iL_iM_i$. Apply the dilation: the result, $d(\angle N_iL_iM_i)$, will still be a right angle. As we just saw, $d(N_iM_i)$ is a diameter of $\mathcal{C}$. By Thales' theorem, $d(L_i)$ is on $\mathcal{C}$ as well.

In conclusion, the dilation d maps the nine points L_i, M_i, and N_i to nine points of the circle $\mathcal{C}$. In reverse, d^{-1} will map nine points of $\mathcal{C}$ to L_i, M_i, and N_i. Since d^{-1} is a dilation, it will map the points of one circle to the points of another circle. Therefore L_i, M_i, and N_i are on the same circle. □

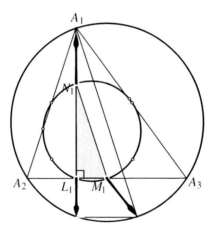

Figure 10. The image of L_i is on the circumcircle (Thales' theorem).

Transformations provide a fundamentally different perspective on the problems of geometry. I hope that these few examples give you a sense of that. Later, transformations will be a critical weapon in our arsenal.

29.4 References

The proof of Varignon's theorem (certainly not the most common) is from Wallace and West's *Roads to Geometry* [WW04]. The proof of Napoleon's theorem is from the geometry web site cuttheknot.org [Bog], where I. M. Yaglom's *Geometric Transformations I* is referred to. This proof of the nine point circle theorem is from Pedoe's book *A Course of Geometry for Colleges and Universities* [Ped70], which is now available under the title *Geometry, a Comprehensive Course* from Dover Publications.

29.5 Exercises

29.2. Prove that the composition of two half-turns around distinct points separated by a distance x is a translation by a distance $2x$.

29.3. Let r be a rotation around a point P. Prove that every line through P is invariant (that is, $r(\ell) = \ell$) if and only if r is a half-turn.

29.4. Given $\triangle ABC$, let

- r_A be the half-turn around the midpoint of BC,
- r_B be the half-turn around the midpoint of AC, and
- r_C be the half-turn around the midpoint of AB.

Then $r_A \circ r_B \circ r_C$ is a half-turn as well. What is its center of rotation?

29.5. Show that if $\square ABCD$ is a parallelogram, then the composition $r_A \circ r_B \circ r_C \circ r_D$ of half-turns around A, B, C, D is the identity (this is the converse of what we proved in the chapter).

29.6. Consider $\triangle ABC$ where $A = (0, 0)$, $B = (2, 0)$, and $C = (1, 3)$. Find the diameter of the circumcircle of $\triangle ABC$, and from that, the radius of the nine point circle.

29.7. Let P be a point on a circle $\mathcal{C}$. Let

$$S = \{P\} \cup \{X | P * Y * X \text{ where } Y \text{ is on } \mathcal{C} \text{ and } PY \simeq YX\}.$$

Prove that S is a circle.

29.8. Let S be the square with vertices $(1, 1), (1, -1)(-1, 1)$, and $(-1, -1)$. Identify the eight isometries that leave S invariant. Show that the composition of two of them is another of them.

29.9. Let $\triangle ABC$ be a right triangle with right angle at C. Let r be the half-turn around the midpoint of AB. Prove that $\square ACBr(C)$ is a rectangle.

30

Area I

30.1 The area function

We can now begin to talk about one of the more fundamental concepts of plane geometry: the *area* of a polygon. While a case could be made that it ought to have been introduced earlier, much of what needs to be said is best expressed in the language of isometries. When we talk about the area of a polygon, we are talking about a number, a positive real number. So you can think of area as a function from the set of polygons to the set of positive real numbers,

$$A : \{\text{polygons}\} \longrightarrow (0, \infty).$$

If the area function is going to live up to our expectations, it needs to meet a few other requirements.

- If two polygons are congruent, their areas should be the same. This statement can also be interpreted in terms of isometries. If P is a polygon and τ is an isometry, then $\tau(P)$ and P are congruent. Therefore area should be an invariant of any isometry.

Figure 1. Congruent polygons have the same area.

- If a polygon can be broken down into smaller pieces, then the area of the polygon should be the sum of the areas of the pieces. More precisely, let $\text{int}(P)$ denote the set of points in the interior of a polygon P, and let $\overline{P}$ denote the set of interior points together with the points on the edges of P. A set of polygons $\{P_i\}$ is a *decomposition* of P if

  $\cup \overline{P_i} = \overline{P}$ (the pieces cover P), and

  $\text{int}(P_i) \cap \text{int}(P_j) = \emptyset$ if $i \neq j$ (the pieces don't overlap).

  If $\{P_i\}$ is a decomposition of the polygon P, then $A(P)$ should equal $\sum A(P_i)$.

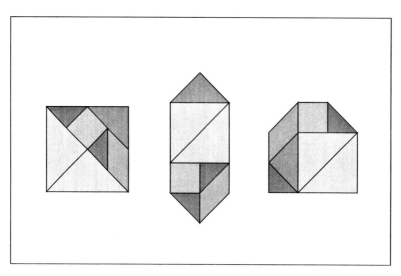

Figure 2. Three convex shapes. Since they can be decomposed into the same set of congruent pieces (the tangram tiles), they have the same areas.

- Finally, we need something to get us started, and it is this. In a rectangle, two adjacent sides (and interchangeably their respective lengths) are called its "base" and "height". The area of a rectangle with base b and height h is bh.

Figure 3. The area of a rectangle is $A = bh$.

The congruence and decomposition conditions allow us to cut apart and rearrange polygons, starting with rectangles, to find the areas of other familiar polygons. We will start that process in the next few results. Because these early results are just a few steps removed from the formula for the area of a rectangle, the formulas involve bases and heights, so we should clarify what is meant by the terms "base" and "height" in each shape.

Parallelogram: Any side of a parallelogram can serve as its base. The height is a segment that is perpendicular to the base; one of its endpoints is on the line containing the base and the other is on the line through the opposite side. It is often helpful to use a vertex of the parallelogram as one of the endpoints for the height.

30.1 The area function

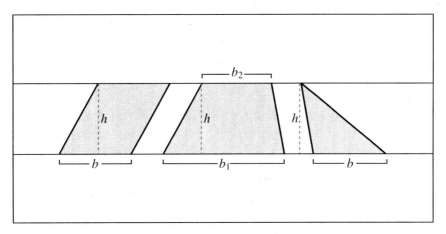

Figure 4. The base and height of a (l) parallelogram, (c) trapezoid, and (r) triangle.

Trapezoid: The two parallel sides are both considered bases (the area formula uses both). The height is as in the parallelogram, a segment perpendicular to, and connecting, the lines through the two parallel sides.

Triangle: Any of the sides of a triangle can serve as its base. The height is the segment from the opposite vertex to the line containing the base, perpendicular to it (the height runs along the altitude, but the altitude of a triangle was originally defined to be a line, not a segment).

Let's start cutting and gluing to find some area formulas.

Theorem 30.1 *A parallelogram with base b and height h has area $A = bh$.*

A child with scissors will tell you this: you can cut the corner off a parallelogram, shift it over to the other side, and make a rectangle. This proof uses that idea, but there is at least one complication. The line of the scissor cut should run along a height segment from one side to the opposite side. Often this is easily done, but in some tall and slanted parallelograms there is no height segment connecting the two sides. Nevertheless, the child's scissors cut approach feels right. Let's call the parallelograms where the approach works the "well-behaved" parallelograms. The proof will deal with them first, and then figure out how to deal with the unruly.

Proof Let $\square ABCD$ be a parallelogram. The result is clearly true if it is a rectangle. If it is not, then it has two acute angles and two obtuse angles. Let's suppose that $\angle D$ is one of the obtuse angles and consider the side AB to be the base (of length b) of the parallelogram. Now drop a perpendicular from D to $\overline{AB}$ (the distance from D to $\overline{AB}$ is the height h). The perpendicular clearly intersects $\overline{AB}$, but it very well might not intersect AB. If it intersects the segment, then the parallelogram is well-behaved in the sense described above. Let's deal with that first.

If the parallelogram is well-behaved, then a cut along the height line will separate the parallelogram into a triangle and a trapezoid. A translation that maps A to B will shift the triangle piece so that it fits perfectly with the trapezoid piece, forming a rectangle with base b and height h. The area of the rectangle is bh. Since it is formed from the same pieces as the parallelogram, the area of the parallelogram is also bh.

If the parallelogram is not well-behaved, the approach will not quite work. However, via a sequence of translations,

t_1: that maps A to B,
t_2: that maps $t_1(A)$ to $t_1(B)$,
t_3: that maps $t_2(A)$ to $t_2(B)$,

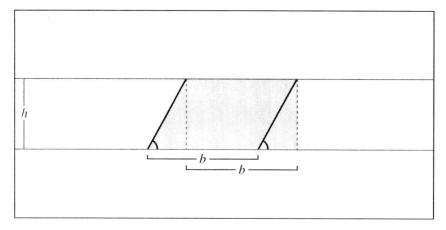

Figure 5. Cutting and rearranging a wide parallelogram into a rectangle.

and so on, we can lay out a strip of congruent copies of the parallelogram that will form one large parallelogram. Given enough iterations of this process (say n times), the resulting parallelogram will be well-behaved. It will be a parallelogram with a base of nb and a height of h, so its area will be $A = nbh$. It is made up of n congruent pieces, each with an area $A = (nbh)/n = bh$.

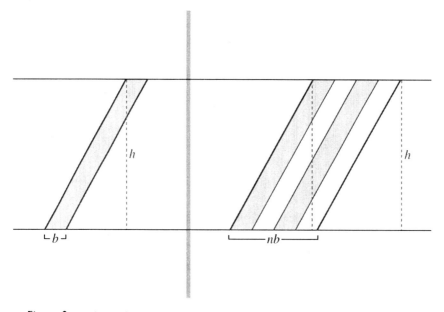

Figure 6. Cutting and rearranging a stack of narrow parallelograms into a rectangle.

30.1 The area function

Exercise 30.1 (a) Fill in the details, explaining why, in the first part of the problem, the triangle and trapezoid pieces will fit together to form a rectangle. (b) Fill in the details, explaining why, in the second part of the problem, the n copies of the parallelogram will fit together to form a larger parallelogram.

Exercise 30.2 In the second part of this proof, we claim that by stacking together sufficiently many parallelograms, the resulting parallelogram will eventually be well-behaved. Prove this.

There are two choices for what will be the base of the parallelogram (any of the four sides could be the bases, so there are really are four choices, but since the opposite sides of a parallelogram are congruent, there are two different choices). For the area of a parallelogram to be well-defined, it must not depend upon which choice we make.

Theorem 30.2 *The area of a parallelogram does not depend upon the choice of base.*

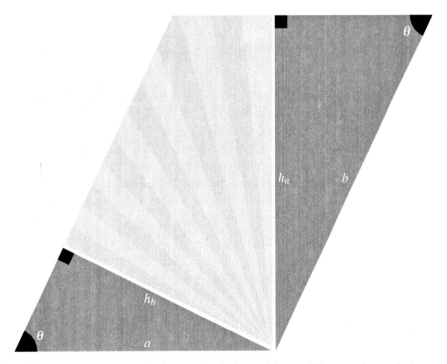

Figure 7. No matter which side is chosen to be the base of the parallelogram, the area is the same.

Proof Consider a parallelogram with sides of length a and b. Let h_a be the height corresponding to the base of length a, and let h_b be the height corresponding to the base of length b. Then we can write the area of the parallelogram as either $A = ah_a$ or $A = bh_b$. If θ is the angle between the sides of the parallelogram (take the acute angle for convenience), then $h_a = b\sin\theta$ and $h_b = a\sin\theta$, so either way $A = ab\sin\theta$. $\square$

Theorem 30.3 *A triangle with base b and height h has area* $A = \frac{1}{2}bh$.

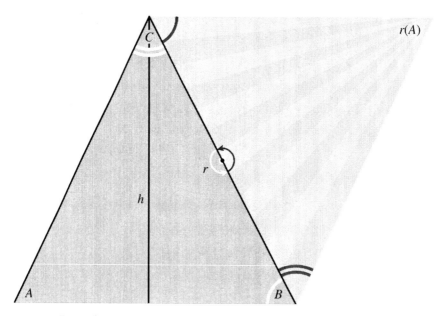

Figure 8. Two congruent triangles fit together to form a parallelogram.

Proof Begin with $\triangle ABC$. Identify its base b as the segment AB, and the corresponding height h. Consider a half-turn r through the midpoint of BC. The resulting triangle $r(\triangle ABC)$ is congruent to the original. The rotation r swaps the points B and C, which means the alternate interior angles at B and C are congruent. Thus the sides AB and $Cr(A)$ are parallel, as are the sides AC and $Br(A)$. We have created a parallelogram. It has a base b and a height h, so its area is bh. The areas of the triangles forming it, then, must be half of that so they have an area of $bh/2$. □

As with the parallelogram, there is an apparent choice of base. Does the choice effect the result?

Theorem 30.4 *The area of a triangle does not depend on the choice of base.*

Proof Start with $\triangle ABC$. There are three choices of base, and each can potentially lead to a different, non-congruent, parallelogram. Label the corresponding heights:

h_A : the height associated with BC,
h_B : the height associated with AC, and
h_C : the height associated with AB.

Look more closely, however, and see that the two parallelograms formed by turning across AB and AC both have base BC and height h_A, so they have the same area. The two parallelograms formed by turning across AB and BC both have base AC and height h_B, so they too have the same area. So yes, the parallelograms may not be congruent, but they have the same area. □

30.2 The laws of sines and cosines

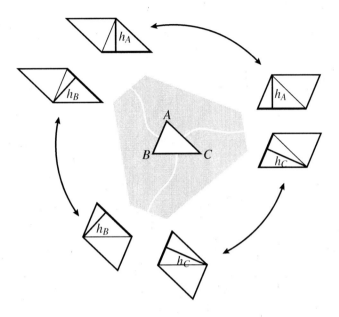

Figure 9. No matter which side is chosen to be the base of the triangle, the area is the same.

Theorem 30.5 *A trapezoid with bases b_1 and b_2 and height h has area*
$$A = \frac{b_1 + b_2}{2} \cdot h.$$

Exercise 30.3 Prove theorem 30.5.

30.2 The laws of sines and cosines

Standard trigonometry provides functions that describe the relationships between the sides and angles of a right triangle. The law of sines builds from them to describe some of the connections between the angles and the sides of an arbitrary triangle. While we could have derived the law of sines when we first looked at the trigonometric functions, it is nice to do it by thinking in terms of area.

Theorem 30.6 *(The law of sines) In $\triangle ABC$, let a denote the length of the side opposite $\angle A$, b denote the length of the side opposite $\angle B$, and c denote the length of the side opposite $\angle C$. Then*
$$\frac{\sin(\angle A)}{a} = \frac{\sin(\angle B)}{b} = \frac{\sin(\angle C)}{c}.$$

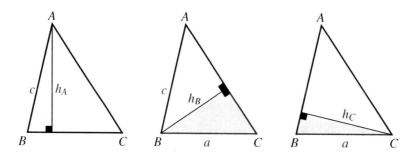

Figure 10. Writing the area of a triangle in terms of each base yields the law of sines.

Proof We know any side of the triangle can serve as the base in the calculation of its area, and that no matter which is chosen, the result is the same. Doing that calculation with each of the sides,

$$\tfrac{1}{2}ah_A = \tfrac{1}{2}bh_B = \tfrac{1}{2}ch_C$$

where h_A, h_B, h_C are the heights corresponding to the bases a, b, and c respectively. Work with the first equality and note that we can write $h_A = c\sin(\angle B)$ and $h_B = c\sin(\angle A)$. Therefore

$$\tfrac{1}{2}ac\sin(\angle B) = \tfrac{1}{2}bc\sin(\angle A)$$

$$a\sin(\angle B) = b\sin(\angle A)$$

$$\frac{\sin(\angle B)}{b} = \frac{\sin(\angle A)}{a}.$$

That gets the first half of the law of sines, and working with the second equality we have $h_B = a\sin(\angle C)$ and $h_C = a\sin(\angle B)$ so

$$\tfrac{1}{2}ba\sin(\angle C) = \tfrac{1}{2}ca\sin(\angle B)$$

$$b\sin(\angle C) = c\sin(\angle B)$$

$$\frac{\sin(\angle C)}{c} = \frac{\sin(\angle B)}{b}.$$

□

We have already seen one proof the Pythagorean theorem—it used ratios from similar triangles—but many of its traditional proofs are based on ideas of area. Let's look at another proof that is in that more traditional style (it also motivates the proof of the law of cosines).

Theorem 30.7 *(The Pythagorean theorem) In a right triangle with legs of length a and b and hypotenuse of length c,*

$$c^2 = a^2 + b^2.$$

30.2 The laws of sines and cosines

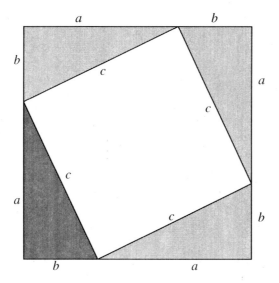

Figure 11. A proof of the Pythagorean theorem based on areas.

Proof   Position four congruent copies of the triangle around a square with sides of length c as shown in figure 11. Now look at how the angles come together at each corner of the square: the two acute angles of the right triangle, and then the right angle of the square. Taken together, the measures of the three angles add up to π, so the edges of the triangles join up in a straight line. The pieces fit to form a square with sides of length $a + b$. We can calculate the area of the big square in two ways.

- Directly in terms of its sides:
$$(a+b)^2 = a^2 + 2ab + b^2.$$

- By adding the areas of the center square and surrounding triangles:
$$c^2 + 4 \cdot \tfrac{1}{2}ab = c^2 + 2ab.$$

Set the two equal,
$$a^2 + 2ab + b^2 = c^2 + 2ab,$$
and subtract $2ab$ from both sides to get $a^2 + b^2 = c^2$, the Pythagorean theorem.   □

The Pythagorean theorem only applies to right triangles. The law of cosines is an extension of the Pythagorean theorem that can be used in any triangle. If $\angle\theta$ is a right angle, then the $2ab\cos\theta$ term in its statement is zero, and this gives the Pythagorean theorem.

Theorem 30.8 *(The law of cosines) Given a triangle with sides of length a, b, and c, and angle θ opposite side c,*

$$c^2 = a^2 + b^2 - 2ab\cos\theta.$$

In proving this, we can try borrow as much as we can from the preceding proof of the Pythagorean theorem. So we start by building four congruent copies of the triangle around a square with sides c. When the triangle is not right, there are complications, for it is when $\angle\theta$ is right that the sides of the two neighboring triangles line up with each other to form a square. If $\angle\theta$ is not a right angle, this does not happen. The special Pythagorean arrangement splits the more general problem into two cases, one when $\angle\theta$ is acute and one when $\angle\theta$ is obtuse. We will prove the acute case and leave the obtuse case as an exercise.

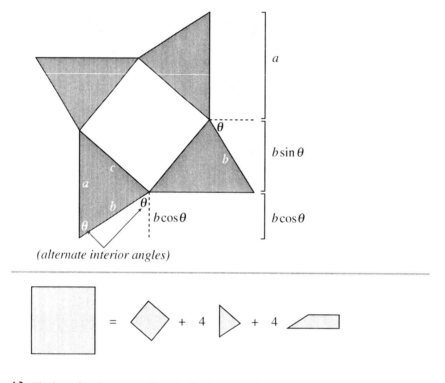

Figure 12. The law of cosines generalizes the Pythagorean theorem. For acute triangles, the arrangement of triangles in the Pythagorean proof does not form a square.

Proof (Acute case) Construct four congruent copies of the triangle around a square with sides of length c, as shown in figure 12. They form a pinwheel shape around the square. We can build a square that frames that pinwheel by drawing lines through each pinwheel tip parallel to the "a" sides of the triangle. Since adjacent triangles in the pinwheel are turned at right angles to each other, the new lines will also intersect at right angles. So we have a big square that is divided into four trapezoids, four triangles, and a smaller square. Now let's calculate the areas of the shapes.

Area of the big square:

$$(a+b(\sin\theta + \cos\theta))^2$$
$$= a^2 + 2ab(\sin\theta + \cos\theta) + b^2(\sin\theta + \cos\theta)^2$$
$$= a^2 + 2ab\sin\theta + 2ab\cos\theta + b^2\sin^2\theta + 2b^2\sin\theta\cos\theta + b^2\cos^2\theta$$
$$= a^2 + 2ab\sin\theta + 2ab\cos\theta + b^2 + 2b^2\sin\theta\cos\theta.$$

Area of the small square: c^2.

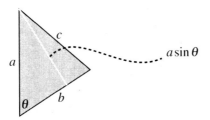

Figure 13. One of four triangles.

Area of one of the four triangles: $\frac{1}{2}ab\sin\theta$.

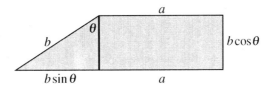

Figure 14. One of four trapezoids.

Area of one of the four trapezoids:

$$\tfrac{1}{2}(a + (a + b\sin\theta)) \cdot b\cos\theta = \tfrac{1}{2}(2ab\cos\theta + b^2\sin\theta\cos\theta)$$
$$= ab\cos\theta + \frac{1}{2}b^2\sin\theta\cos\theta.$$

Since the area of the whole is the sum of the areas of the parts,

$$a^2 + 2ab\sin\theta + 2ab\cos\theta + b^2 + 2b^2\sin\theta\cos\theta$$
$$= c^2 + 4\left(\tfrac{1}{2}ab\sin\theta\right) + 4\left(ab\cos\theta + \tfrac{1}{2}b^2\sin\theta\cos\theta\right).$$

Simplifying this,

$$a^2 + 2ab\sin\theta + 2ab\cos\theta + b^2 + 2b^2\sin\theta\cos\theta$$
$$= c^2 + 2ab\sin\theta + 4ab\cos\theta + 2b^2\sin\theta\cos\theta.$$

Cancel common terms to get the law of cosines,

$$a^2 + 2ab\cos\theta + b^2 = c^2 + 4ab\cos\theta$$
$$a^2 + b^2 - 2ab\cos\theta = c^2. \qquad \square$$

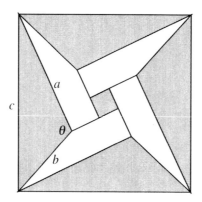

Figure 15. The law of cosines: the obtuse case.

Exercise 30.4 Use areas to prove the obtuse case of the law of cosines. Hint: it may be better to build the triangles inside the square with sides c, as shown in figure 15, rather than out around it.

30.3 Heron's formula

To close this chapter, we will use the law of cosines to derive another formula for the area of a triangle called *Heron's formula*. The SSS triangle congruence theorem says that a triangle is uniquely determined by the lengths of its sides so there should be a formula to calculate the area of a triangle using just the lengths of its sides. The formula $A = bh/2$ does not do that, since it also requires a height. But Heron's formula does. Heron's formula is not difficult from a theoretical point of view. The trouble is on the calculation side. To help with those calculations, it is helpful to introduce another term.

Definition 30.9 The *semiperimeter s* of a triangle is half its perimeter. If a triangle has sides of length a, b, and c, then its semiperimeter is

$$s = \tfrac{1}{2}(a+b+c).$$

30.3 Heron's formula

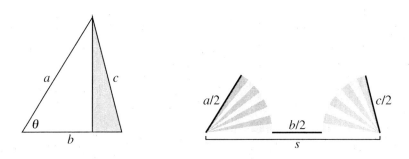

Figure 16. The semiperimeter of a triangle.

Theorem 30.10 *(Heron's formula) The area of a triangle with sides of length a, b, and c, and semiperimeter s is*

$$A = \sqrt{s(s-a)(s-b)(s-c)}.$$

It is always possible to relabel the side lengths so that a is the base and so that the angle θ between sides a and b is acute (at least two angles in a triangle have to be acute). The height of the triangle is then $b \sin \theta$, so its area is

$$A = \frac{1}{2} ab \sin \theta.$$

We want to eliminate θ. The law of sines might seem like the obvious choice, but it always relates an {angle and side} to another {angle and side}, so it doesn't eliminate angles. The law of cosines gives a way to relate an angle to the three sides. In this proof, we will write the area in terms of cosine, not sine, use the law of cosines, and then suffer the algebraic consequences.

Proof Label the sides of the triangle so that side a is the base and the angle θ between a and b is acute. Then the area of the triangle is

$$A = \frac{1}{2} ab \sin \theta.$$

By the Pythagorean identity,

$$\sin^2 \theta = 1 - \cos^2 \theta.$$

Normally, taking the square root of both sides would yield two solutions. In this case, since θ is an acute angle, $\sin \theta$ will be a positive number, and so

$$\sin \theta = \sqrt{1 - \cos^2 \theta},$$

and the area of the triangle is

$$A = \tfrac{1}{2} ab \sqrt{1 - \cos^2 \theta}.$$

Now use the law of cosines

$$c^2 = a^2 + b^2 - 2ab \cos \theta \implies \cos \theta = \frac{c^2 - a^2 - b^2}{2ab},$$

and substitute into the area formula to get a big algebra problem:

$$A = \tfrac{1}{2}ab\sqrt{1 - \left[\frac{c^2 - a^2 - b^2}{2ab}\right]^2}$$

$$= \tfrac{1}{2}ab\sqrt{\frac{4a^2b^2 - (c^2 - a^2 - b^2)^2}{4a^2b^2}}$$

$$= \tfrac{1}{2}ab \cdot \frac{1}{2ab}\sqrt{4a^2b^2 - (c^4 - 2a^2c^2 - 2b^2c^2 + a^4 + 2a^2b^2 + b^4)}$$

$$= \tfrac{1}{4}\sqrt{-(a^4 - 2a^2b^2 + b^4) + 2(b^2c^2 + c^2a^2) - c^4}$$

$$= \tfrac{1}{4}\sqrt{-(a^2 - b^2)^2 + 2c^2(a^2 + b^2) - c^4}.$$

Complete the square on the second and third terms inside the radical.

$$A = \tfrac{1}{4}\sqrt{-(a^2 - b^2)^2 + (a^2 + b^2)^2 - (a^2 + b^2)^2 + 2c^2(a^2 + b^2) - c^4}$$

$$= \tfrac{1}{4}\sqrt{(-a^4 + 2a^2b^2 - b^4 + a^4 + 2a^2b^2 + b^4) - ((a^2 + b^2) - c^2)^2}$$

$$= \tfrac{1}{4}\sqrt{4a^2b^2 - (a^2 + b^2 - c^2)^2}.$$

We have a difference of perfect squares inside the radical, so

$$A = \tfrac{1}{4}\sqrt{(2ab - (a^2 + b^2 - c^2))(2ab + (a^2 + b^2 - c^2))}$$

$$= \tfrac{1}{4}\sqrt{((-a^2 + 2ab - b^2) + c^2)((a^2 + 2ab + b^2) - c^2)}$$

$$= \tfrac{1}{4}\sqrt{(c^2 - (a - b)^2)((a + b)^2 - c^2)}.$$

Because of the differences of squares,

$$A = \tfrac{1}{4}\sqrt{(c + (a - b))(c - (a - b))((a + b) + c)((a + b) - c)}$$

$$= \sqrt{\frac{(a - b + c)(-a + b + c)(a + b + c)(a + b - c)}{16}}$$

$$= \sqrt{\frac{a - b + c}{2} \cdot \frac{-a + b + c}{2} \cdot \frac{a + b + c}{2} \cdot \frac{a + b - c}{2}}$$

$$= \sqrt{\left[\frac{a + b + c}{2} - b\right] \cdot \left[\frac{a + b + c}{2} - a\right] \cdot \left[\frac{a + b + c}{2}\right] \cdot \left[\frac{a + b + c}{2} - c\right]}$$

$$= \sqrt{(s - b)(s - a)s(s - c)}. \qquad \square$$

In this chapter, we started from the area of a rectangle and worked our way down to the area of a triangle. In the next chapter, we will build up from the area of a triangle to the area of polygons in general.

30.4 References

The idea behind the proof of Heron's Formula is simple enough, but without Coxeter's *Introduction to Geometry*[Cox64] (pages 12-13) I may have given up somewhere in the calculation. The exercise (below) relating the law of sines to the circumradius is also from that book.

30.5 Exercises

30.5. In this chapter, we started with a formula for the area of a rectangle; from that we derived the formula for the area of a parallelogram, and then the area of a triangle. It is possible to reverse the order. Starting from the area of a rectangle, derive the formula for the area of a triangle. Then, from that, derive the formula for the area of a parallelogram.

30.6. The "well-behaved" condition of a parallelogram could be described as a requirement that the parallelogram not be too thin, too tall, and too slanted. Make this precise: consider a parallelogram with base b, height h, and an acute interior angle θ. What are the requirements for the angle θ (in terms of b and h) for the parallelogram to be well-behaved?

30.7. In a parallelogram with adjacent sides of length a and b and acute interior angle θ, describe the number of strips n required to cut and form a rectangle (as described in the proof of the parallelogram area formula) in terms of a, b, and θ.

30.8. The Penrose tiles are a pair of rhombuses that tile the Euclidean plane in an aperiodic manner (they are named after Roger Penrose, who discovered their aperiodicity). One of the tiles has interior angles measuring $2\pi/5$ and $3\pi/5$. The other has interior angles measuring $\pi/5$ and $4\pi/5$. Find the areas of the Penrose tiles assuming the length of each side of a tile is 1.

30.9. Consider $\triangle ABC$ with vertices $A = (0, 0)$, $B = (3, 1)$, and $C = (4, 2)$. Calculate the area of $\triangle ABC$, first using the formula $A = bh/2$, then using Heron's formula.

30.10. Find a formula for the area of an equilateral triangle in terms of the length s of one of its sides.

30.11. (A more symmetric connection between area and the ratio that occurs in the law of sines) Given $\triangle ABC$, let a be the length of the side opposite $\angle A$, b be the length of the side opposite $\angle B$, and c be the length of the side opposite $\angle C$. If $\mathcal{A}$ is the area of the triangle prove that

$$\frac{\sin(\angle A)}{a} = \frac{2\mathcal{A}}{abc}.$$

30.12. (A connection between the law of sines and the circumcircle) Given $\triangle ABC$, let a be the length of the side opposite $\angle A$, b be the length of the side opposite $\angle B$, and c be the length of the side opposite $\angle C$. Let r be the radius of the circumcircle of $\triangle ABC$. Prove that

$$\frac{\sin(\angle A)}{a} = \frac{1}{2r}.$$

30.13. Consider a regular n-sided polygon inscribed in a circle with radius r. By dividing the polygon into triangles, find a formula for its area in terms of n and r. (We will take a different approach to this problem in the next chapter).

30.14. Let $\mathcal{C}_1$ and $\mathcal{C}_2$ be circles with centers O_1 and O_2 and radii r_1 and r_2, respectively. Suppose that the center of $\mathcal{C}_2$ lies on $\mathcal{C}_1$. Let P be one of the points of intersection of $\mathcal{C}_1$ and $\mathcal{C}_2$. Use the law of cosines to give an equation relating $(\angle O_1 P O_2)$ to r_1 and r_2.

30.15. Given $\triangle ABC$, label the midpoints of the sides

α the midpoint of BC,

β the midpoint of AC, and

γ the midpoint of AB

and rays

r_A the bisector of $\angle A$,

r_B the bisector of $\angle B$, and

r_C the bisector of $\angle C$.

The segment $A\alpha$ forms an angle with the ray r_A. On the opposite side of r_A from α, draw another ray to form an angle congruent to it. This ray intersects BC at a point a (so, in other words, r_A bisects $\angle a A\alpha$). Likewise label b and c on AC and AB, respectively. The segments Aa, Bb, and Cc are called *symmedians*. Prove that the symmedians are concurrent. The point of concurrence is called the *Lemoine point*. Hint: use Ceva's theorem and the law of sines, remembering that $\sin(\pi - x) = \sin(x)$.

30.16. Use the law of sines and trigonometric identities to derive the law of tangents: in $\triangle ABC$, let a be the length of the side opposite A and b be the length of the side opposite B. Show that

$$\frac{\tan\left(\frac{(\angle A) - (\angle B)}{2}\right)}{\tan\left(\frac{(\angle A) + (\angle B)}{2}\right)} = \frac{a-b}{a+b}.$$

30.17. We took as definition that the area of a rectangle is given by the formula $A = bh$. That can be derived from a much more minimal condition: that the area of a 1×1 square is one. Derive the general formula for the area of a rectangle from this. Hint: build up to this formula gradually. First show the result for a $b \times 1$ rectangle, by stacking and subdividing squares in one direction. Then stack and subdivide $b \times 1$ rectangles in the other direction.

31
Area II

31.1 Areas of polygons

The goal of this chapter is to establish a formula for the area of a general simple polygon. We will use a proof by induction to prove it. We need two things to make the proof work. First, we need a way to decompose a polygon into smaller pieces. This is handled by the next result, which states that any simple polygon has a diagonal that cuts it into two smaller pieces. Second, we need a working formula for the "base" case, the area of a triangle. We found a few formulas for the area of a triangle in the last chapter, but none is appropriate for this problem, so we will derive another one, this time in terms of the coordinates of its vertices. Those two steps are the hard work of this section. Once they are done, it is easy to slot those pieces into the induction proof.

Theorem 31.1 *Every simple polygon $\mathcal{P}$ has a diagonal that lies entirely in its interior.*

If $\mathcal{P}$ is convex, then any diagonal will work. If $\mathcal{P}$ is not convex, some of the diagonals will not be contained entirely in $\mathcal{P}$. We need to show, then, that even the most contorted polygon has at least one diagonal that lies entirely inside it.

Proof Let's consider the coordinates of the vertices of $\mathcal{P}$. We look for the "lowest" point on the polygon, the vertex with the smallest y-coordinate. Call it P_i. Now consider the segment that connects P_i's two neighbors, P_{i-1} and P_{i+1}. If $P_{i-1}P_{i+1}$ lies entirely inside of $\mathcal{P}$, then we have found our diagonal, easy enough.

What if it doesn't? In that case, it is because at least some of the remaining vertices of $\mathcal{P}$ lie inside $\triangle P_{i-1}P_iP_{i+1}$. From this subset of vertices, let P_j be the lowest one, the one with the smallest y-coordinate. I claim that the segment P_iP_j lies entirely inside $\mathcal{P}$, so that it can serve as our diagonal. To see why, you need to remember that a point Q is inside a polygon $\mathcal{P}$ if any ray from Q crosses the polygon an odd number of times (counting multiplicities). In this case, if Q is a point on P_iP_j, it is lower than any of the vertices of $\mathcal{P}$ except for P_i, and possibly P_{i-1} or P_{i+1}. Therefore $\overrightarrow{QP_i}$ intersects the sides $P_{i-1}P_i$ and P_iP_{i+1} only once at the shared endpoint P_i, and it does not intersect any of the other sides of $\mathcal{P}$ at all. Since P_j is inside $\triangle P_{i-1}P_iP_{i+1}$, $\overrightarrow{QP_i}$ splits the polygon at P_i (the adjacent vertices P_{i-1} and P_{i+1} are separated by the line P_iP_j). Therefore, there is one intersection of $\overrightarrow{QP_i}$ with $\mathcal{P}$ and it has multiplicity one. That's an odd

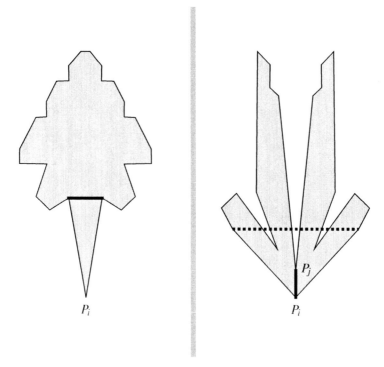

Figure 1. Finding a diagonal: on the left, $P_{i-1}P_{i+1}$ is a diagonal. On the right, it is not, but P_iP_j is.

number of intersections, so Q is inside $\mathcal{P}$. That is true for all points on the segment P_iP_j, so P_iP_j is a diagonal that lies entirely inside $\mathcal{P}$. □

Exercise 31.1 In the previous proof, we tacitly assumed that $\mathcal{P}$ has a unique lowest point. This may not be the case. Prove that there is an isometry f so that $f(\mathcal{P})$ has a unique lowest point.

Now let's go back to the question of the area of a triangle. Let's get some motivation by looking at the problem from the perspective of vector calculus. For two three-dimensional vectors $\vec{v} = \langle v_x, v_y, v_z \rangle$ and $\vec{w} = \langle w_x, w_y, w_z \rangle$, the cross product $\vec{v} \times \vec{w}$ is given by the determinant

$$\vec{v} \times \vec{w} = \begin{vmatrix} i & j & k \\ v_x & v_y & v_z \\ w_x & w_y & w_z \end{vmatrix}.$$

It is a well-known fact from calculus that the length of $\vec{v} \times \vec{w}$ is the area of the parallelogram formed by $\vec{v}$ and $\vec{w}$, so half of it would be the area of the triangle with sides $\vec{v}$ and $\vec{w}$. Let's use that idea to calculate the area of the triangle with vertices at (x_1, y_1), (x_2, y_2), and (x_3, y_3). We can make vectors out of two of the sides and embed them in three-dimensional space by setting the last coordinate equal to zero:

$$\vec{v} = \langle x_2 - x_1, y_2 - y_1, 0 \rangle \text{ and } \vec{w} = \langle x_3 - x_1, y_3 - y_1, 0 \rangle.$$

31.1 Areas of polygons

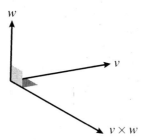

Figure 2. Given vectors v and w, the length of $v \times w$ is the area of the parallelogram that has v and w as sides.

Now compute

$$\vec{v} \times \vec{w} = \begin{vmatrix} i & j & k \\ x_2 - x_1 & y_2 - y_1 & 0 \\ x_3 - x_1 & y_3 - y_1 & 0 \end{vmatrix}$$

$$= [(x_2 - x_1)(y_3 - y_1) - (x_3 - x_1)(y_2 - y_1)]k$$

$$= [(x_2 y_3 - x_2 y_1 - x_1 y_3 + x_1 y_1) - (x_3 y_2 - x_3 y_1 - x_1 y_2 + x_1 y_1)]k$$

$$= [(x_1 y_2 - x_2 y_1) + (x_2 y_3 - x_3 y_2) + (x_3 y_1 - x_1 y_3)]k$$

$$= \left(\begin{vmatrix} x_1 & y_1 \\ x_2 & y_2 \end{vmatrix} + \begin{vmatrix} x_2 & y_2 \\ x_3 & y_3 \end{vmatrix} + \begin{vmatrix} x_3 & y_3 \\ x_1 & y_1 \end{vmatrix} \right) k.$$

It is easy to read off the length of $\vec{v} \times \vec{w}$, and half that amount gets you a formula for the area of a triangle. While all of this may be familiar, it takes us out of the plane, and it draws on some facts about vectors that we have not developed. Instead, let's look for a more elementary proof of the formula.

Theorem 31.2 *Label the three vertices of a triangle in counterclockwise order: $P_1 = (x_1, y_1)$, $P_2 = (x_2, y_2)$, and $P_3 = (x_3, y_3)$. The area of $\triangle P_1 P_2 P_3$ is*

$$A = \frac{1}{2} \left(\begin{vmatrix} x_1 & y_1 \\ x_2 & y_2 \end{vmatrix} + \begin{vmatrix} x_2 & y_2 \\ x_3 & y_3 \end{vmatrix} + \begin{vmatrix} x_3 & y_3 \\ x_1 & y_1 \end{vmatrix} \right).$$

Proof Designate the side $P_1 P_2$ to be the base of the triangle, and put $b = |P_1 P_2|$. With the right isometry (isometries do not alter areas of shapes), we can reposition the triangle so that its base lies along the x-axis. Then it is easy to read off the height. The necessary isometry is composed of two pieces.

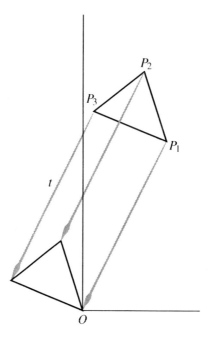

Figure 3. First translate the triangle to move one vertex to the origin.

The first piece is a translation t to move P_1 to the origin:

$$t\begin{pmatrix} x \\ y \end{pmatrix} = \begin{pmatrix} x - x_1 \\ y - y_1 \end{pmatrix}$$

$$t(P_1) = (0, 0)$$
$$t(P_2) = (x_2 - x_1, y_2 - y_1)$$
$$t(P_3) = (x_3 - x_1, y_3 - y_1).$$

The second piece is a rotation r about the origin to move $t(P_2)$ onto the x-axis. To find the angle for the rotation, look at the angle θ between the x-axis and the line from the origin through $t(P_2)$. The sine and cosine values of this angle are

$$\cos\theta = \frac{x_2 - x_1}{b} \quad \text{and} \quad \sin\theta = \frac{y_2 - y_1}{b}.$$

To put the base of the triangle along the x-axis, then, we need to rotate by $-\theta$. The matrix equation for the rotation is

$$r\begin{pmatrix} x \\ y \end{pmatrix} = \begin{pmatrix} \cos(-\theta) & -\sin(-\theta) \\ \sin(-\theta) & \cos(-\theta) \end{pmatrix} \begin{pmatrix} x \\ y \end{pmatrix}$$

$$= \begin{pmatrix} \cos\theta & \sin\theta \\ -\sin\theta & \cos\theta \end{pmatrix} \begin{pmatrix} x \\ y \end{pmatrix}$$

$$= \frac{1}{b} \begin{pmatrix} x_2 - x_1 & y_2 - y_1 \\ y_1 - y_2 & x_2 - x_1 \end{pmatrix} \begin{pmatrix} x \\ y \end{pmatrix}.$$

31.1 Areas of polygons

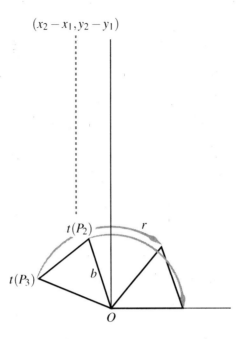

Figure 4. Then rotate about the origin to place a second vertex on the positive x-axis.

The point $t(P_1)$ stays at the origin, while the point $t(P_2)$ rotates around to $(b, 0)$. The key to finding the height of the triangle, though, lies with the third point:

$$r \circ t(P_3) = \frac{1}{b} \begin{pmatrix} x_2 - x_1 & y_2 - y_1 \\ y_1 - y_2 & x_2 - x_1 \end{pmatrix} \begin{pmatrix} x_3 - x_1 \\ y_3 - y_1 \end{pmatrix}$$

$$= \frac{1}{b} \begin{pmatrix} [x_2 - x_1][x_3 - x_1] + [y_2 - y_1][y_3 - y_1] \\ [x_3 - x_1][y_1 - y_2] + [x_2 - x_1][y_3 - y_1] \end{pmatrix}.$$

Since $r \circ t$ is a composition of a rotation and a translation, it is orientation-preserving. Since the points P_1, P_2, P_3 are listed in counterclockwise order, their images under $r \circ t$ are also in counterclockwise order. That means $r \circ t(P_3)$ must lie above the x-axis, and so the height of the triangle is just the y-coordinate of $r \circ t(P_3)$:

$$h = \frac{1}{b}[(x_3 - x_1)(y_1 - y_2) + (x_2 - x_1)(y_3 - y_1)].$$

The rest is algebra:

$$A = \frac{1}{2}bh$$
$$= \frac{1}{2}b \cdot \frac{1}{b}[(x_3 - x_1)(y_1 - y_2) + (x_2 - x_1)(y_3 - y_1)]$$
$$= \frac{1}{2}(x_3 y_1 - x_1 y_1 - x_3 y_2 + x_1 y_2 + x_2 y_3 - x_2 y_1 - x_1 y_3 + x_1 y_1)$$
$$= \frac{1}{2}((x_1 y_2 - x_2 y_1) + (x_2 y_3 - x_3 y_2) + (x_3 y_1 - x_1 y_3))$$
$$= \frac{1}{2}\left(\begin{vmatrix} x_1 & y_1 \\ x_2 & y_2 \end{vmatrix} + \begin{vmatrix} x_2 & y_2 \\ x_3 & y_3 \end{vmatrix} + \begin{vmatrix} x_3 & y_3 \\ x_1 & y_1 \end{vmatrix} \right).$$

□

With this area formula in hand, we can now turn back to the bigger question of polygon area.

Theorem 31.3 *Let $P_1 = (x_1, y_1)$, $P_2 = (x_2, y_2)$, $P_3 = (x_3, y_3)$, ..., $P_n = (x_n, y_n)$ be the coordinates of the vertices of a simple polygon listed in counterclockwise order. For notational convenience, put $x_{n+1} = x_1$ and $y_{n+1} = y_1$. Then the area of the polygon is*

$$A = \frac{1}{2} \sum_{k=1}^{n} \begin{vmatrix} x_k & y_k \\ x_{k+1} & y_{k+1} \end{vmatrix}.$$

Proof The proof uses induction on n, the number of sides of the polygon. In the base case, when $n = 3$, the polygons are triangles, and the area formula is just the coordinate formula for triangular area that we proved above. Now move to the inductive step: suppose that the formula gives the proper area for every polygon with fewer than n sides, and let $\mathcal{P}$ be an polygon with n sides. As we saw at the start of the chapter, $\mathcal{P}$ can be cut in two along a diagonal Δ.

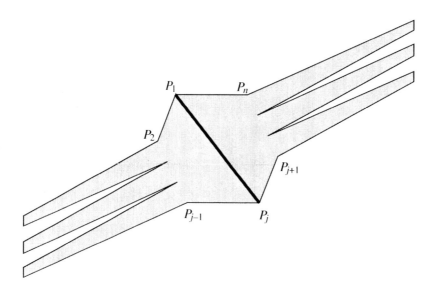

Figure 5. Induction: use a diagonal to cut the polygon into two pieces with fewer sides.

To keep indices as simple as possible, let's relabel the points P_i and the corresponding coordinates (x_i, y_i) so that one end of Δ is the vertex $P_1 = (x_1, y_1)$. Continue around $\mathcal{P}$ in the counterclockwise direction, labeling the remaining vertices $P_2 = (x_2, y_2)$, $P_3 = (x_3, y_3)$, ..., $P_n = (x_n, y_n)$ and then loop back around to the start by setting $x_{n+1} = x_1$ and $y_{n+1} = y_1$. At some point along the way, we come to the other end of Δ. Identify that point as $P_j = (x_j, y_j)$. With those labels, Δ cuts $\mathcal{P}$ into two smaller polygons with fewer than n sides, $P_1 P_2 \ldots P_j$ and $P_j P_{j+1} \ldots P_n P_1$. The area of $\mathcal{P}$ is the sum of the areas of the two pieces, and by the induction

31.1 Areas of polygons

hypothesis we know the area formula works for them. Therefore

$$A(\mathcal{P}) = A(P_1 P_2 \ldots P_j) + A(P_j P_{j+1} \ldots P_n P_1)$$

$$= \frac{1}{2} \sum_{k=1}^{j-1} \begin{vmatrix} x_k & y_k \\ x_{k+1} & y_{k+1} \end{vmatrix} + \frac{1}{2} \begin{vmatrix} x_j & y_j \\ x_1 & y_1 \end{vmatrix} + \frac{1}{2} \sum_{k=j}^{n} \begin{vmatrix} x_k & y_k \\ x_{k+1} & y_{k+1} \end{vmatrix} + \frac{1}{2} \begin{vmatrix} x_1 & y_1 \\ x_j & y_j \end{vmatrix}$$

$$= \frac{1}{2} \sum_{k=1}^{n} \begin{vmatrix} x_k & y_k \\ x_{k+1} & y_{k+1} \end{vmatrix} + \frac{1}{2}(x_j y_1 - x_1 y_j) + \frac{1}{2}(x_1 y_j - x_j y_1)$$

$$= \frac{1}{2} \sum_{k=1}^{n} \begin{vmatrix} x_k & y_k \\ x_{k+1} & y_{k+1} \end{vmatrix}.$$

By induction, the formula holds for all polygons. □

What's really going on here is that by repeatedly cutting $\mathcal{P}$ along diagonals, we can eventually break $\mathcal{P}$ down into triangles: we can triangulate $\mathcal{P}$. The area of a triangle in the triangulation is calculated by three determinants, one for each edge. Different triangulations lead to different edges, but (and this is key) each internal edge is an edge of two triangles, and if the counterclockwise orientation of one triangle points it in the direction from v_i to v_j, then the counterclockwise orientation of the other triangle points it in the direction from v_j to v_i.

Figure 6. Each internal edge is shared by two neighboring triangles, but is oriented oppositely in the triangles. In the overall area calculation, the components cancel one another.

In the end, the internal edge makes a contribution to the overall area computation of

$$\begin{vmatrix} x_i & y_i \\ x_j & y_j \end{vmatrix} + \begin{vmatrix} x_j & y_j \\ x_i & y_i \end{vmatrix} = (x_i y_j - x_j y_i) + (x_j y_i - x_i y_j) = 0.$$

The contributions of all the internal edges cancel out, leaving just the contributions from the edges of the polygon. This is what happens along the internal edge Δ in the proof above. If you have studied multivariable calculus, the internal cancellation may seem familiar. This same thing happens in Green's theorem, where a double integral across a region is converted to a line integral around the region. In fact, this area formula is a special case of Green's theorem. The connection is explored more thoroughly in the exercises.

31.2 The area of a circle

We now are in position to find a formula for the area of a circle. There are several ways to derive this famous formula, but I want to incorporate the coordinate formula we just derived. First we will use the formula to find the area of a regular polygon. Then we will trap the circle between circumscribed and circumscribing regular polygons, and use their areas as upper and lower bounds for the area of the circle (as we did in the derivation of the circumference formula in chapter 17).

Theorem 31.4 *Let $\mathcal{P}$ be a regular polygon with n sides and a radius of r (this is the radius of the circumscribing circle). Then the area of $\mathcal{P}$ is*

$$A = \frac{1}{2} n r^2 \sin\left(\frac{2\pi}{n}\right).$$

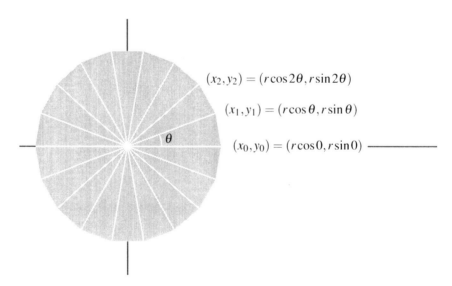

Figure 7. The determinant formula provides an easy calculation of the area of a regular n-gon with radius r.

31.2 The area of a circle

Proof All regular polygons with the same radius and the same number of sides are congruent, so we will just use the one that is easiest, and that is the one where $\mathcal{P}$ is centered at the origin with its n vertices at the coordinates

$$\left(r\cos\left(\frac{2\pi k}{n}\right), r\sin\left(\frac{2\pi k}{n}\right)\right), \quad 1 \le k \le n.$$

It makes for easier reading if we put $\theta_x = 2\pi x/n$. Then the area of $\mathcal{P}$ is

$$A = \frac{1}{2}\sum_{k=1}^{n} \begin{vmatrix} r\cos\theta_k & r\sin\theta_k \\ r\cos\theta_{k+1} & r\sin\theta_{k+1} \end{vmatrix}$$

$$= \frac{1}{2}\sum_{k=1}^{n} \left(r^2 \cos\theta_k \sin\theta_{k+1} - r^2 \sin\theta_k \cos\theta_{k+1}\right).$$

Factor out the r^2. Using the fact that sine is an odd function and that cosine is an even one, we can get this into the right form to use the addition formula for sine. Then the rest is easy:

$$A = \frac{1}{2}r^2 \sum_{k=1}^{n} [\cos(-\theta_k)\sin(\theta_{k+1}) + \sin(-\theta_k)\cos(\theta_{k+1})].$$

$$= \frac{1}{2}r^2 \sum_{k=1}^{n} \sin(\theta_{k+1} - \theta_k)$$

$$= \frac{1}{2}r^2 \sum_{k=1}^{n} \sin(2\pi/n)$$

$$= \frac{1}{2}r^2 \cdot n\sin(2\pi/n). \qquad \square$$

By trapping a circle between circumscribed and circumscribing regular polygons, it is possible to pin down its area.

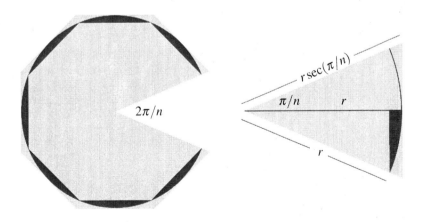

Figure 8. The area of a circle as a limit of areas of regular polygons.

Theorem 31.5 *The area of a circle with radius r is πr^2.*

Proof The radius of the inscribed regular polygon is r. The radius of the circumscribed regular polygon is $r \sec(\pi/n)$ (as shown in figure 8). By substituting the radii into the area formula we just derived, we get upper and lower bounds on the area of the circle:

$$\frac{1}{2}nr^2 \sin\left(\frac{2\pi}{n}\right) \leq A \leq \frac{1}{2}n(r \sec \pi/n)^2 \sin\left(\frac{2\pi}{n}\right).$$

The inequalities are true for all n, so let's see what happens when we take the limit as n approaches ∞:

- π/n approaches 0, so $\sec(\pi/n)$ approaches 1.
- for the term $n \sin(2\pi/n)$, make the substitution $m = n/2$. As n approaches ∞, so does m, and so

$$\lim_{n \to \infty} n \sin(2\pi/n) = \lim_{m \to \infty} 2m \sin(\pi/m) = 2 \lim_{m \to \infty} m \sin(\pi/m).$$

Back in chapter 15, this limit was the definition of π (although we were using degrees instead of radians at the time), so this term is approaching 2π.

Now let's put it back together:

$$\lim_{n \to \infty} \frac{1}{2}nr^2 \sin\left(\frac{2\pi}{n}\right) \leq A \leq \lim_{n \to \infty} \frac{1}{2}n(r \sec \pi/n)^2 \sin\left(\frac{2\pi}{n}\right).$$

$$\frac{1}{2}2\pi r^2 \leq A \leq \frac{1}{2}2\pi r^2.$$

Because A is trapped between two values that are both closing in upon πr^2, $A = \pi r^2$. □

31.3 Exercises

31.2. Use the determinant formula for area from this chapter to find the area of the triangle with vertices at $(0, 0)$, $(3, 1)$, and $(4, 2)$.

31.3. Let $\mathcal{P} = P_1 P_2 \cdots P_n$ be a polygon and suppose that τ is a similarity mapping with a scaling factor of k. What is the relationship between the area of $\mathcal{P}$ and the area of $\tau(\mathcal{P})$?

31.4. Find the area of a regular five pointed star (Scläfli symbol $\{5/2\}$) inscribed in a circle of radius one.

31.5. Consider $\triangle ABC$, where $A = (1, 0)$, $B = (0, 1)$, and $C = (0, 0)$. What is the area of the incircle of $\triangle ABC$? What is the area of the circumcircle of $\triangle ABC$?

31.6. Give an inductive proof that any simple polygon $\mathcal{P}$ can be triangulated.

31.7. An alternate proof of the formula for the area of the circle involves cutting n congruent pie pieces, and then rearranging them into an approximate parallelogram. Work out the details of this approach.

31.3 Exercises

31.8. Let O be the center of a circle $\mathcal{C}$, and let A and B be points on $\mathcal{C}$. Prove that the area of the sector bounded by the segments OA and OB and the arc $\overparen{AB}$ is

$$\mathcal{A} = \frac{1}{2}r^2\theta,$$

where r is the radius of $\mathcal{C}$ and $\theta = (\angle AOB)$.

31.9. Let $\mathcal{C}$ be a circle with radius r and center O, and let P be a point outside $\mathcal{C}$. There are two lines through P that are tangent to $\mathcal{C}$. Let Q_1 and Q_2 be the points of tangency, and let $\theta = (\angle Q_1 P Q_2)$. What is the area of the region inside $\square P Q_1 O Q_2$ but outside $\mathcal{C}$ in terms of r and θ?

31.10. The line $y = 1$ cuts off a piece of the parabola $y = x^2$ with finite area. Calculate its area by using a limit of approximating polygon areas and the determinant formula for them. You may need to use the summation formulas

$$\sum_{i=1}^{n} i = \frac{n(n+1)}{2} \quad \text{and} \quad \sum_{i=1}^{n} i^2 = \frac{n(n+1)(2n+1)}{6}.$$

31.11. (For those who have studied Green's theorem in calculus) Let $\mathcal{P}$ be a simple n-sided polygon with vertices (taken in the counterclockwise direction) at coordinates (x_i, y_i), for $1 \le i \le n$. Use Green's theorem to show that the area of $\mathcal{P}$ is given by the integral

$$\frac{1}{2} \oint_{\mathcal{P}} -y\,dx + x\,dy.$$

Parametrize the edges of $\mathcal{P}$ and compute the integral to get the area formula given in the chapter.

32

Barycentric Coordinates

In chapter 20 we studied the trilinear coordinate system. We did not discuss the closely related barycentric coordinate system at that time, because we hadn't yet dealt with area. Now that we have looked at area, we can properly study the topic. Barycentric coordinates are connected to the idea of the center of mass, the balance point of a set of weights. Archimedes has the first word on this topic that is near and dear to heart of every kindergarten kid.

(The principle of the lever) Place two masses m_1 and m_2 on a seesaw at distances d_1 and d_2 from the fulcrum. The seesaw balances if

$$m_1 d_1 = m_2 d_2.$$

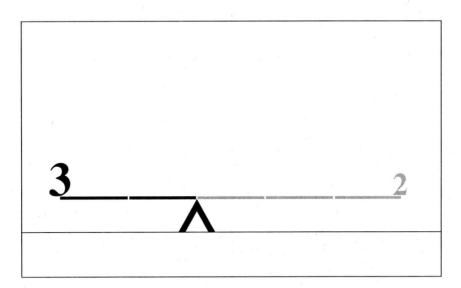

Figure 1. A simple lever.

Archimedes' lever is a one-dimensional construct because the points and the fulcrum are all on one line. For the two-dimensional case, with points lying in the plane, the better picture is one of mobiles, such as those of Alexander Calder. Let's think about how he (or some other mobile maker) would make a simple mobile, one that consists of just three equal weights located at points A, B, and C, and wired together as shown in Figure 2.

341

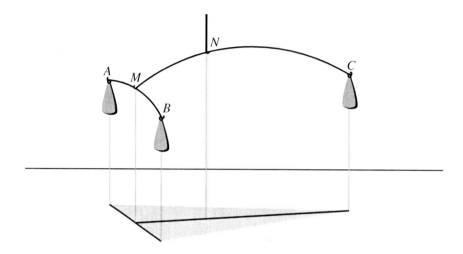

Figure 2. A mobile.

From a mathematical point of view, the interesting questions are: where should he put the hook M so that A and B balance?, and where should he put the hook N so that everything balances when the mobile is hung from the string? The answer to the first is easy: since the two weights are the same, the principle of the lever says that M needs to be at the midpoint of AB. The answer to the second is just a bit more involved: since M is now supporting twice the weight of C, the principle of the lever says that the distance from N to C must be twice the distance from N to M. In other words, N must be located two-thirds of the way down CM from C so it is at the centroid of $\triangle ABC$. This was a simple system since all three weights were the same, but imagine if we changed the weights so that they were not all the same. The corresponding balance point of the system (N) would move as well. This is the key to barycentric coordinates: by putting different weights at A, B, and C, we get different balance points. The barycentric coordinates of a point P are the weights that make P the balance point.

32.1 The vector approach

There is a vector approach to this problem. Start with the two person seesaw, with masses m_A and m_B at points A and B, respectively. Then the principle of the lever takes on the following form: the balance point occurs at the center of mass M, when

$$m_a \cdot \overrightarrow{MA} + m_b \cdot \overrightarrow{MB} = 0.$$

More generally, we can consider when a sum of terms of the form $m_i \cdot \overrightarrow{MP_i}$ add up to zero. The quantity $m_i \cdot \overrightarrow{MP_i}$ is a measure of the the tendency of the system to turn in the direction of P_i. The balancing point, the center of mass M, is where they all cancel out:

$$\sum_i m_i \cdot \overrightarrow{MP_i} = 0.$$

32.1 The vector approach

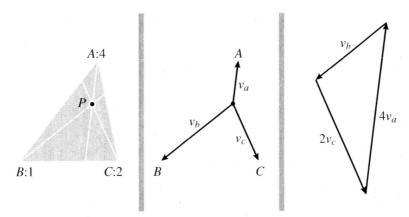

Figure 3. A vector approach to the center of mass. If P is the center of mass of points A with mass 4, B with mass 1, and C with mass 2, and if v_a, v_b, and v_c are vectors from P to A, B, and C, respectively, then $4v_a + v_b + 2v_c = 0$.

It is not really necessary to think of the coefficients m_i as masses at all. To avoid physics entirely, you can think of them as arbitrary scalar coefficients in a vector equation. Whether you think of them as masses or not, it is the coefficients that form the basis for barycentric coordinates. Let's start by investigating some properties of centers of mass, beginning with a two mass system. The center of mass of two objects will lie between them, as long as the masses are positive. If you are willing to allow negative mass, then the center of mass may not be between them. The calculations then require signed distances and signed areas but everything else still works. As this is meant to be a brief introduction to the topic, we will stick to positive masses and positive distances.

Theorem 32.1 *If M is the center of mass of a two mass system, with mass m_A at point A and mass m_B at point B, then*

$$|MA| = \frac{m_B}{m_A + m_B} \cdot |AB| \quad \text{and} \quad |MB| = \frac{m_A}{m_A + m_B} \cdot |AB|.$$

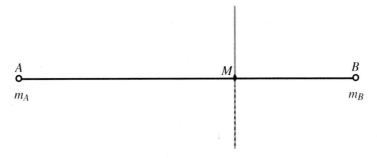

Figure 4. The center of mass of a two mass system.

Proof Since M is the center of mass, by definition

$$m_A \cdot \overrightarrow{MA} + m_B \cdot \overrightarrow{MB} = 0.$$

For $m_A \cdot \overrightarrow{MA}$ and $m_B \cdot \overrightarrow{MB}$ to cancel, they have to be the same length, so $m_A|MA| = m_B|MB|$, and thus $|MA|/|MB| = m_B/m_A$. Then

$$\frac{|MA|}{|AB|} = \frac{|MA|}{|MA|+|MB|} = \frac{1}{1+(|MB|/|MA|)} = \frac{1}{1+(m_A/m_B)} = \frac{m_B}{m_A+m_B}.$$

Therefore

$$|MA| = \frac{m_B}{m_A+m_B} \cdot |AB|.$$

The calculation of $|MB|$ is similar. $\square$

Theorem 32.2 *In $\triangle ABC$ with masses m_A at A, m_B at B, and m_C at C, label*

M_{AB} : *the center of mass of A and B,*
M_{AC} : *the center of mass of A and C, and*
M_{BC} : *the center of mass of B and C.*

Then the segments AM_{BC}, BM_{AC}, and CM_{AB} are concurrent.

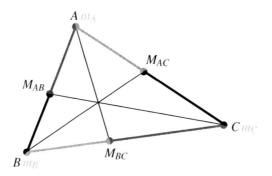

Figure 5. The center of mass of a three mass system.

Proof Ceva's theorem (theorem 19.2) guarantees a point of concurrence if a product of ratios around the edges of the triangle equals 1. In this case, the product is

$$\frac{|AM_{AB}|}{|M_{AB}B|} \cdot \frac{|BM_{BC}|}{|M_{BC}C|} \cdot \frac{|CM_{AC}|}{|M_{AC}A|}.$$

The first ratio using the previous proposition is

$$\frac{|AM_{AB}|}{|M_{AB}B|} = \frac{m_B/(m_A+m_B) \cdot |AB|}{m_A/(m_A+m_B) \cdot |AB|} = \frac{m_B}{m_A}.$$

32.1 The vector approach

Likewise,

$$\frac{|BM_{BC}|}{|M_{BC}C|} = \frac{m_C}{m_B} \quad \text{and} \quad \frac{|CM_{AC}|}{|M_{AC}A|} = \frac{m_A}{m_C},$$

and so

$$\frac{|AM_{AB}|}{|M_{AB}B|} \cdot \frac{|BM_{BC}|}{|M_{BC}C|} \cdot \frac{|CM_{AC}|}{|M_{AC}A|} = \frac{m_B}{m_A} \cdot \frac{m_C}{m_B} \cdot \frac{m_A}{m_C} = 1.$$

By Ceva's theorem, the segments are concurrent. □

Theorem 32.3 *The center of mass M of masses m_A at A, m_B at B, and m_C at C, is the point of concurrence of the segments AM_{BC}, BM_{AC}, and CM_{AB}.*

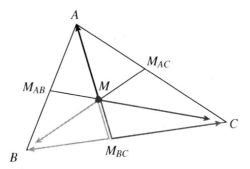

Figure 6. The center of mass, the connection to concurrence.

Proof Let's show that M is on the segment AM_{BC}. A similar argument will work to show it is on the other two segments, and therefore that it is at their mutual intersection. Since M is the center of mass of the three mass system, we may write

$$m_A \overrightarrow{MA} + m_B \overrightarrow{MB} + m_C \overrightarrow{MC} = 0.$$

Vector arithmetic gets us

$$m_A \overrightarrow{MA} + m_B(\overrightarrow{MM_{BC}} + \overrightarrow{M_{BC}B}) + m_C(\overrightarrow{MM_{BC}} + \overrightarrow{M_{BC}C}) = 0,$$

$$m_A \overrightarrow{MA} + (m_B + m_C)\overrightarrow{MM_{BC}} + (m_B \overrightarrow{M_{BC}B} + m_C \overrightarrow{M_{BC}C}) = 0.$$

The last term in parentheses is zero since M_{BC} is the center of mass of the system with masses m_B at B and m_C at C. Therefore

$$m_A \overrightarrow{MA} + (m_B + m_C)\overrightarrow{MM_{BC}} = 0.$$

For the vectors to cancel, they must be oppositely directed. That is, A, M, and M_{BC} must be collinear. □

Definition 32.4 Given $\triangle ABC$ and a point M, a set of *barycentric coordinates* of M (relative to $\triangle ABC$) is a triple $[m_a : m_b : m_C]$ (with not all of m_A, m_B, and m_C equal to zero) so that

$$m_a \overrightarrow{MA} + m_b \overrightarrow{MB} + m_c \overrightarrow{MC} = 0.$$

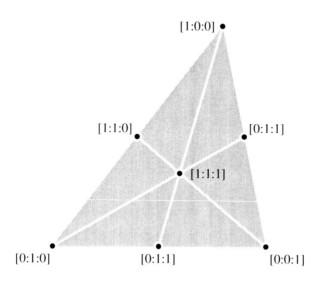

Figure 7. Barycentric coordinates of vertices, midpoints, and the centroid of a triangle.

One observation is that barycentric coordinates are defined only up to a constant multiple: if

$$m_a \overrightarrow{MA} + m_b \overrightarrow{MB} + m_c \overrightarrow{MC} = 0$$

then

$$k \cdot m_a \overrightarrow{MA} + k \cdot m_b \overrightarrow{MB} + k \cdot m_c \overrightarrow{MC} = 0$$

as well. Therefore, just as with trilinear coordinates, the barycentric coordinates of a point are not really a triple $[m_a : m_b : m_c]$, but an equivalence class of triples. That is, the coordinates $[m_a : m_b : m_c]$ and $[n_a : n_b : n_c]$ are equivalent if there is a nonzero constant k so that $m_a = kn_a$, $m_b = kn_b$, and $m_c = kn_c$.

32.2 The connection to area and trilinears

Barycentric coordinates can be calculated, quite directly, using either areas of triangles or trilinear coordinates. The key to doing so is the following theorem that relates the masses m_A, m_B, and m_C to the areas of certain triangles. Throughout the rest of this chapter, the notation $(\triangle ABC)$ will denote the area of $\triangle ABC$.

32.2 The connection to area and trilinears

Theorem 32.5 *Given $\triangle ABC$ with masses m_A at A, m_B at B, and m_C at C, and a center of mass M, then*

$$\frac{m_A}{m_B} = \frac{(\triangle BCM)}{(\triangle ACM)}, \quad \frac{m_B}{m_C} = \frac{(\triangle ACM)}{(\triangle ABM)}, \quad \text{and} \quad \frac{m_C}{m_A} = \frac{(\triangle ABM)}{(\triangle BCM)}.$$

Proof Let's look at the first (the other two are just a shuffling of labels). Label

F_C: the foot of the altitude from C
F_M: the foot of the perpendicular from M to AB
M_{AB}: the center of mass of AB.

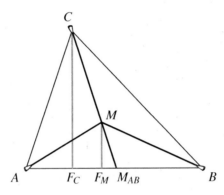

Figure 8. Relating ratios of masses to ratios of areas.

Then

$$(\triangle CAM) = (\triangle CAM_{AB}) - (\triangle MAM_{AB})$$
$$= \tfrac{1}{2}|CF_C| \cdot |AM_{AB}| - \tfrac{1}{2}|MF_M| \cdot |AM_{AB}|$$
$$= \tfrac{1}{2}|AM_{AB}|(|CF_C| - |MF_M|)$$

and

$$(\triangle CBM) = (\triangle CBM_{AB}) - (\triangle MBM_{AB})$$
$$= \tfrac{1}{2}|CF_C| \cdot |BM_{AB}| - \tfrac{1}{2}|MF_M| \cdot |BM_{AB}|$$
$$= \tfrac{1}{2}|BM_{AB}|(|CF_C| - |MF_M|)$$

so

$$\frac{(\triangle CAM)}{(\triangle CBM)} = \frac{\tfrac{1}{2}|AM_{AB}|(|CF_C| - |MF_M|)}{\tfrac{1}{2}|BM_{AB}|(|CF_C| - |MF_M|)} = \frac{|AM_{AB}|}{|BM_{AB}|} = \frac{m_B}{m_A}.$$

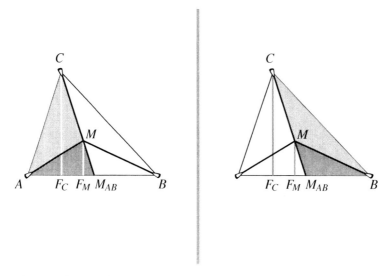

Figure 9. (l)The area of $\triangle CAM$ is the difference of the areas of $\triangle CAM_{AB}$ and $\triangle MAM_{AB}$. (r) Likewise for the area of $\triangle CBM$.

Likewise, with the proper interchange of letters,

$$\frac{m_B}{m_C} = \frac{(\triangle ACM)}{(\triangle ABM)} \quad \text{and} \quad \frac{m_C}{m_A} = \frac{(\triangle ABM)}{(\triangle BCM)}. \qquad \square$$

As an immediate consequence, we get a way to use triangle areas to calculate barycentric coordinates.

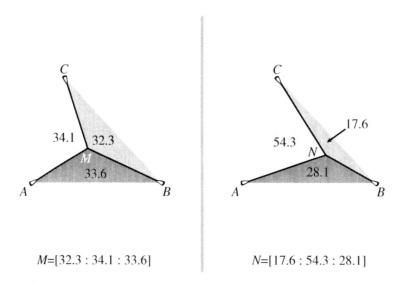

$M = [32.3 : 34.1 : 33.6]$ \qquad $N = [17.6 : 54.3 : 28.1]$

Figure 10. The barycentric coordinates of two points as determined by the areas of triangles.

32.2 The connection to area and trilinears

Corollary 32.6 *A point M subdivides $\triangle ABC$ into three pieces, $\triangle ABM$, $\triangle ACM$, and $\triangle BCM$. The barycentric coordinates of M can be computed from the the areas of the triangles as*

$$[(\triangle BCM) : (\triangle ACM) : (\triangle ABM)].$$

Proof Let $[m_a : m_b : m_c]$ be the barycentric coordinates of M. At least one of the three coordinates must be nonzero. Let's assume it is m_c. We will divide through by m_c, use the previous theorem to make the connection to area, and multiply through by $(\triangle ABM)$.

$$[m_a : m_b : m_c] = [m_a/m_c : m_b/m_c : 1]$$
$$= [(\triangle BCM)/(\triangle ABM) : (\triangle ACM)/(\triangle ABM) : 1]$$
$$= [(\triangle BCM) : (\triangle ACM) : (\triangle ABM)]. \qquad \square$$

So just how closely related are barycentric and trilinear coordinates?

Theorem 32.7 *If the trilinear coordinates of a point M relative to $\triangle ABC$ are $[a : b : c]$, then the barycentric coordinates of M (relative to the same triangle) are*

$$[a \cdot |BC| : b \cdot |AC| : c \cdot |AB|].$$

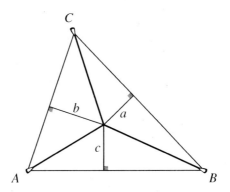

Figure 11. The distances that determine the trilinear coordinates are the heights of the triangles whose areas determine the barycentric coordinates.

Proof The barycentric coordinates of M can be computed from the areas of triangles as

$$[(\triangle BCM) : (\triangle ACM) : (\triangle ABM)] = \left[\tfrac{1}{2} h_a |BC| : \tfrac{1}{2} h_b |AC| : \tfrac{1}{2} h_c |AB|\right]$$
$$\sim \left[h_a |BC| : h_b |AC| : h_C |AB|\right],$$

where h_a, h_b, and h_c are the lengths of the altitudes from M in the triangles. But the trilinear coordinates of M can be normalized to measure exactly those lengths. Therefore $a = h_a$, $b = h_b$, and $c = h_c$. □

32.3 Barycentric coordinates of triangle centers

Barycentric coordinates are set up so that the coordinates of the centroid are $[1 : 1 : 1]$. What about some of the other triangle centers we have encountered? We now know an easy way to convert from trilinear coordinates to barycentric coordinates, and we have already found the trilinear coordinates of the centers. This is a good review of that material, though, and in fact the calculations of the barycentric coordinates closely follow the earlier trilinear calculations. The key to finding the barycentric coordinates of the circumcenter and incenter is the fact that they are the centers of circles: the circumcircle and the incircle.

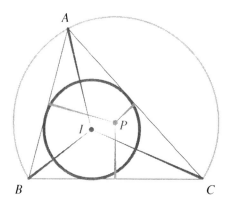

Figure 12. The circumcenter and incenter are both the centers of circles.

Theorem 32.8 *(Circumcenter) In $\triangle ABC$, the barycentric coordinates of the circumcenter are*

$$[|BC|\cos(\angle A) : |AC|\cos(\angle B) : |AB|\cos(\angle C)].$$

Proof Let P denote the circumcenter and remember that it is the center of the circumcircle, a circle that passes through each of A, B, and C, so that $|PA| = |PB| = |PC|$. Let r be the radius of the circumcircle. If F is the foot of the perpendicular through P to the line BC, then $\angle BPF = \frac{1}{2}\angle BPC = \angle A$ (by the inscribed angle theorem). Therefore

$$|PF| = r\cos(\angle A)$$

and so

$$(\triangle PBC) = \tfrac{1}{2}r\cos(\angle A)|BC|.$$

32.3 Barycentric coordinates of triangle centers

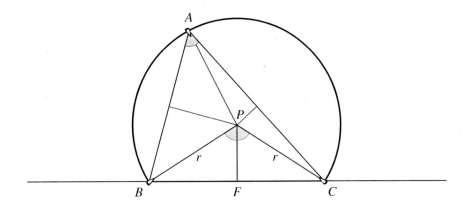

Figure 13. The circumcenter P.

Similarly

$$(\triangle PAC) = \tfrac{1}{2}r\cos(\angle B)|AC| \text{ and } (\triangle PAB) = \tfrac{1}{2}r\cos(\angle C)|AB|,$$

and we have seen that the areas of the triangles determine the barycentric coordinates of P:

$$\left[\tfrac{1}{2}r|BC|\cos(\angle A) : \tfrac{1}{2}r|AC|\cos(\angle B) : \tfrac{1}{2}r|AB|\cos(\angle C)\right]$$
$$\sim [|BC|\cos(\angle A) : |AC|\cos(\angle B) : |AB|\cos(\angle C)]. \qquad \square$$

Theorem 32.9 *(Incenter) In $\triangle ABC$, the barycentric coordinates of the incenter are*

$$[|BC| : |AC| : |AB|].$$

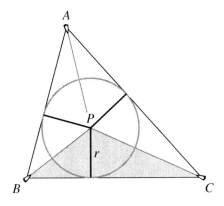

Figure 14. The incenter P.

Proof Let P be the incenter of $\triangle ABC$, which is equidistant from the sides of the triangle. Let r be the radius of the incircle. Then

$$(\triangle PBC) = \tfrac{1}{2}r|BC|, \quad (\triangle PAC) = \tfrac{1}{2}r|AC|, \quad \text{and} \quad (\triangle PAB) = \tfrac{1}{2}r|AB|,$$

so the barycentric coordinates of P are

$$\left[\tfrac{1}{2}r|BC| : \tfrac{1}{2}r|AC| : \tfrac{1}{2}r|AB|\right] \sim [|BC| : |AC| : |AB|]. \qquad \square$$

Theorem 32.10 *(Orthocenter) In $\triangle ABC$, the barycentric coordinates of the orthocenter are*

$$[|BC|\cos(B)\cos(C) : |AC|\cos(A)\cos(C) : |AB|\cos(A)\cos(B)].$$

Exercise 32.1 Prove theorem 32.10. The proof can be made to follow the argument for the trilinear coordinates of the orthocenter.

32.4 References

I referenced Clark Kimberling's web site [Kim] in an earlier chapter, but it also includes barycentric coordinates for many triangle centers. Once again, Coxeter's *Introduction to Geometry* [Cox64] provides a good perspective on this topic. There is also a letter from John Conway to Steve Sigur [Con] on the web that extolls the virtues of barycentric coordinates.

32.5 Exercises

32.2. Consider $\triangle ABC$ whose vertices are $A = (0,0)$, $B = (2,0)$, and $C = (0,4)$. Find the barycentric coordinates for the point $(1,1)$.

32.3. Show that the barycentric coordinates of the excenters of $\triangle ABC$ are $[-|BC| : |AC| : |AB|]$, $[|BC| : -|AC| : |AB|]$, and $[|BC| : |AC| : -|AB|]$.

32.4. Consider three points P, Q, and R, whose barycentric coordinates with respect to $\triangle ABC$ are $P = [x_p : y_p : z_p]$, $Q = [x_q : y_q : z_q]$, and $R = [x_r : y_r : z_r]$. Prove that P, Q, and R are collinear if and only if

$$\begin{vmatrix} x_p & y_p & z_p \\ x_q & y_q & z_q \\ x_r & y_r & z_r \end{vmatrix} = 0.$$

32.5. Use the result of the previous exercise to give another proof that the circumcenter, orthocenter, and centroid of $\triangle ABC$ are collinear. Hint: first use trigonometric identities to show that

$$\cos(A) = \sin(B)\sin(C) - \cos(B)\cos(C).$$

Then use that to show that the coordinates of the centroid can be written as a linear combination of the coordinates of the orthocenter and circumcenter.

33
Inversion

In the last few chapters we classified all the bijective mappings of the Euclidean plane that respect incidence, order, and congruence. Now we are going to look for mappings that fall short of that stringent list of conditions, but that still preserve enough of the Euclidean structure to tell us something interesting. The particular mapping that we will investigate in the next few chapters is called inversion. Inversions provide interesting insight into some of the classical problems of Euclidean geometry, particularly those that involve circles. Inversions also play an important role in the study of non-Euclidean geometry. I think that the most natural path into the topic of inversion is via stereographic projection. This means that we will have to momentarily step outside of the plane. Don't worry. By the time we get around to formally defining inversions, we will be comfortably back in the plane.

33.1 Stereographic projection

Ever since map-makers realized that the earth is round, they have sought ways to project a spherical surface on a flat plane. One approach which is nice mathematically (although maybe not so nice cartographically) is called *stereographic projection*. It works as follows. First put the center of the sphere (say of radius r) at the origin of the plane. Then draw rays from the north pole through points of the sphere. They will intersect the plane, establishing a bijection between the points of the sphere (except the north pole itself) and the points of the plane. With a few symbols, we can describe the process more precisely. Label

$\mathbb{E}$: the plane $z = 0$,
$\mathbb{S}$: the sphere of radius r, centered at the origin,
N : the north pole, the point with coordinates $(0, 0, r)$,
Φ : the stereographic projection from $\mathbb{S}$ to $\mathbb{E}$,
P : a point of $\mathbb{S}$ other than N.

Then $\overrightarrow{NP}$ will intersect $\mathbb{E}$ exactly once, and $\Phi(P)$ is defined to be the intersection point. Since Φ is a bijection, it has an inverse, Φ^{-1}, that is called *inverse stereographic projection*. We can do the same kind of projection from the south pole, and to define inversion, we will need to work from both poles, an inverse stereographic projection using the north pole followed by a stereographic projection using the south pole. It is straightforward to work out analytic equations to describe the mappings, and that is the first task of this chapter.

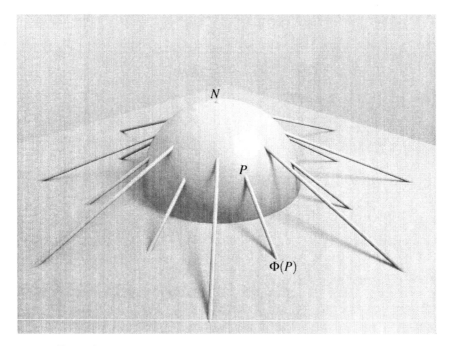

Figure 1. Stereographic projection of the points of a sphere onto a plane.

Theorem 33.1 *The inverse stereographic projection Φ_N^{-1} from the north pole $(0, 0, r)$ is given by the equation*

$$\Phi_N^{-1}(x, y) = \left(\frac{2xr^2}{d^2 + r^2}, \frac{2yr^2}{d^2 + r^2}, \frac{rd^2 - r^3}{d^2 + r^2} \right)$$

where $d = \sqrt{x^2 + y^2}$ is the distance from O to the point (x, y). The stereographic projection Φ_S from the south pole $(0, 0, -r)$ is given by the equation

$$\Phi_S(x, y, z) = \left(\frac{rx}{r + z}, \frac{ry}{r + z} \right).$$

Proof We will prove the first statement and leave the second as an exercise. The point (x, y) in the plane corresponds to the point $(x, y, 0)$ in three-dimensional space. Start with a parametrized equation for the line through $(x, y, 0)$ and the north pole, $(0, 0, r)$:

$$s(t) = \langle 0, 0, r \rangle + t \langle x - 0, y - 0, 0 - r \rangle = \langle tx, ty, r - rt \rangle.$$

We need to find when the line intersects the sphere. All the points on the sphere are a distance r from the origin, so $|s(t)|^2 = r^2$. That is,

$$(tx)^2 + (ty)^2 + (r - rt)^2 = r^2$$
$$t^2 x^2 + t^2 y^2 + r^2 - 2r^2 t + r^2 t^2 = r^2.$$

33.1 Stereographic projection

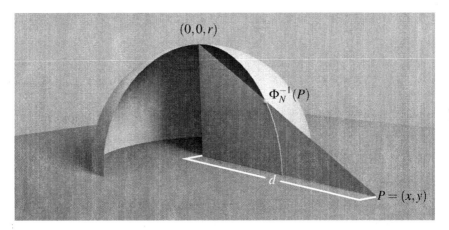

Figure 2. Inverse stereographic projection: the line from the north pole $(0, 0, r)$ to the point $P = (x, y, 0)$ in the plane intersects the sphere at $\Phi_N^{-1}(P)$.

Cancel r^2, substitute d^2 for $x^2 + y^2$, and factor to solve for t:

$$t^2(x^2 + y^2) - 2r^2 t + r^2 t^2 = 0$$
$$t^2 d^2 + r^2 t^2 - 2r^2 t = 0$$
$$t((d^2 + r^2)t - 2r^2) = 0.$$

The first solution, when $t = 0$, is at the north pole and is not the one we want. The other intersection occurs when

$$(d^2 + r^2)t - 2r^2 = 0 \implies t = \frac{2r^2}{d^2 + r^2}.$$

Substituting into $s(t)$ gives the vector that points to $\Phi_N^{-1}(x, y)$:

$$\left\langle x \cdot \frac{2r^2}{d^2 + r^2}, y \cdot \frac{2r^2}{d^2 + r^2}, r - r \cdot \frac{2r^2}{d^2 + r^2} \right\rangle = \left\langle \frac{2xr^2}{d^2 + r^2}, \frac{2yr^2}{d^2 + r^2}, \frac{rd^2 - r^3}{d^2 + r^2} \right\rangle.$$

A similar approach works for the second part. Find the equation of the line through the south pole and a point on the sphere, then locate its intersection with the plane $z = 0$. □

Exercise 33.1 Verify the formula given in the last theorem for stereographic projection from the south pole.

This is a book on plane geometry, so we should be looking for maps from the plane to itself. We can get such a map by composing Φ_N^{-1} and Φ_S. The first step in the composition takes the plane to the sphere and the second step brings it back. When we do this, there is a problem at the origin O, since $\Phi_N^{-1}(O) = S$ and $\Phi_S(S)$ is undefined. If we omit that one bad point, though, what's left is a bijection from $\mathbb{E} - O$ to itself. Let's call it σ. Then

$$\sigma(x, y) = \Phi_S \circ \Phi_N^{-1}(x, y) = \Phi_S\left(\frac{2xr^2}{d^2 + r^2}, \frac{2yr^2}{d^2 + r^2}, r - \frac{2r^3}{d^2 + r^2}\right).$$

This composition is going to get messy. Fortunately, its x and y coordinates are similar enough that we can just look at only one of them. The x-coordinate of Φ_S is $rx/(r + z)$. Substituting

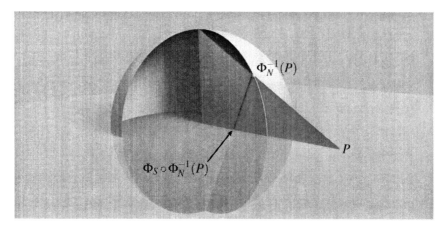

Figure 3. Composing inverse stereographic projection from the north pole and stereographic projection from the south pole defines a planar map.

the result of $\Phi_N^{-1}(x, y)$ into the expression gives

$$\frac{r\left(\dfrac{2xr^2}{d^2 + r^2}\right)}{r + \left(r - \dfrac{2r^3}{d^2 + r^2}\right)}.$$

Multiply top and bottom by $d^2 + r^2$ to get

$$\frac{2xr^3}{2r(d^2 + r^2) - 2r^3} = \frac{2xr^3}{2rd^2 + 2r^3 - 2r^3} = \frac{2xr^3}{2rd^2} = \frac{xr^2}{d^2}.$$

The second coordinate works similarly and simplifies to yr^2/d^2, so

$$\sigma(x, y) = \left(x \cdot \frac{r^2}{d^2},\; y \cdot \frac{r^2}{d^2}\right).$$

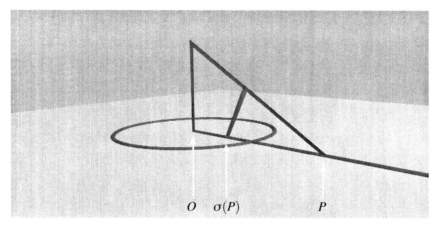

Figure 4. Inversion as a composition of projections.

33.1 Stereographic projection

So $\sigma(x, y)$ is on the same ray from the origin as (x, y), but its distance from the origin has been altered. It is now

$$\sqrt{\left(\frac{xr^2}{d^2}\right)^2 + \left(\frac{yr^2}{d^2}\right)^2} = \sqrt{\frac{x^2r^4 + y^2r^4}{d^4}} = \sqrt{\frac{d^2r^4}{d^4}} = \frac{r^2}{d}.$$

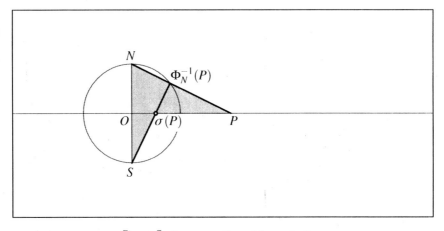

Figure 5. A cross section of the projection.

There is another view that may be more appealing than the previous calculations. Take a cross section of the sphere and plane as illustrated in figure 5.

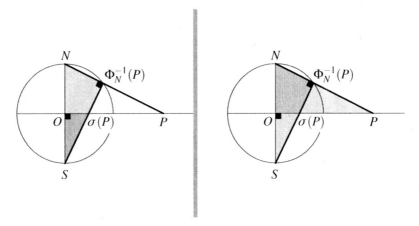

Figure 6. Thales' theorem provides a more synthetic interpretation of the relationship between P and $\sigma(P)$.

By Thales' theorem, $\overline{NP}$ and $\overline{S\sigma(P)}$ intersect at right angles at $\Phi_N^{-1}(P)$. Then by AA similarity, $\triangle S\sigma(P)O \sim \triangle SN\Phi_N^{-1}(P)$ (since they both have a right angle and they share the angle at S). Also by AA similarity, $\triangle SN\Phi_N^{-1}(P) \sim \triangle PNO$ (using the right angles and the shared angle at N). Therefore $\triangle S\sigma(P)O$ and $\triangle PNO$ are similar. Matching the corresponding ratios of the two legs of the triangles,

$$\frac{|O\sigma(P)|}{r} = \frac{r}{|OP|} \implies |O\sigma(P)| = \frac{r^2}{|OP|} = r^2/d.$$

33.2 Inversion

This map σ that we constructed in the previous section is an inversion. Using the above properties, we can now give a definition of inversion that does not stray from the plane. The sphere of radius r is replaced by its intersection with the plane, a circle of radius r. Furthermore, there is no longer any advantage to centering the circle at the origin.

Definition 33.2 Let $\mathcal{C}$ be a circle with center O and radius r. The inversion σ across $\mathcal{C}$ is the bijection of the points of $\mathbb{E} - O$ defined as follows. For any point $P \in \mathbb{E} - O$, $\sigma(P)$ is the point on $\overrightarrow{OP}$ that is a distance $r^2/|OP|$ from O.

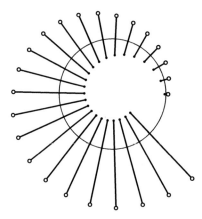

Figure 7. Inversion of a collection of points across a circle.

An inversion turns a circle inside out:

- If P is inside $\mathcal{C}$, then $|OP|$ is less than r, so $r^2/|OP|$ is greater than r, so $\sigma(P)$ is outside $\mathcal{C}$.
- If P is outside $\mathcal{C}$, then $|OP|$ is greater than r, so $r^2/|OP|$ is less than r, so $\sigma(P)$ is inside $\mathcal{C}$.
- If P is on $\mathcal{C}$, then $|OP| = r$, so $r^2/|OP| = r$, so $\sigma(P)$ is again on $\mathcal{C}$. Since $\overrightarrow{OP}$ intersects $\mathcal{C}$ only once, $P = \sigma(P)$.

The observation that σ fixes points of $\mathcal{C}$ is important. An inversion is a little like a reflection. Whereas a reflection fixes a line and swaps the two sides of it, an inversion fixes a circle and swaps the interior and exterior of it. Furthermore, it is easy to see that, like a reflection, an inversion is its own inverse. But it is important to note how an inversion differs from a reflection. For one thing, an inversion does not preserve distances. Also, an inversion does not scale distances by a constant. Points that are close to O may be thrown far apart from one another, while points that are far from O will be squeezed into a tiny region right around O.

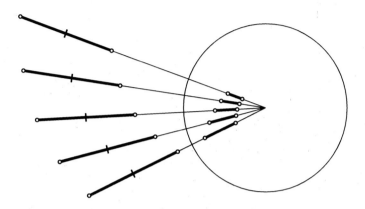

Figure 8. Distances are not all scaled by the same amount under an inversion.

Theorem 33.3 *Let σ be the inversion across a circle $\mathcal{C}$ with radius r and center O. Then for distinct points P and Q in $\mathbb{E} - O$,*

$$\triangle POQ \sim \triangle \sigma(Q)O\sigma(P).$$

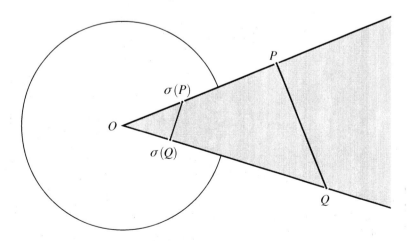

Figure 9. A pair of similar triangles, $\triangle OPQ$ and $\triangle O\sigma(Q)\sigma(P)$, created by inversion.

Proof The triangles share an angle at O. Compare the lengths of the sides:

$$|O\sigma(P)| = r^2/|OP| \quad \text{and} \quad |O\sigma(Q)| = r^2/|OQ|,$$

so

$$\frac{|O\sigma(P)|}{|O\sigma(Q)|} = \frac{r^2/|OP|}{r^2/|OQ|} = \frac{|OQ|}{|OP|}.$$

By the SAS similarity theorem, the two triangles are similar. $\square$

While the last theorem provides similar triangles, the sides OP and OQ are crossed up in the similarity so that, for instance, $\angle P$ is congruent to $\angle \sigma(Q)$. Now let's look at some larger structures. We have seen that all Euclidean transformations map lines to lines, but what happens when we invert a line? One situation is easy: a line that passes through O is mapped to itself. (Technically, it isn't quite mapped to itself, because σ is undefined at O and σ does not map any point to O. However, for the rest of the section, it is more convenient to ignore the problems that arise at O.) For a line that does not pass through O, the situation gets more interesting.

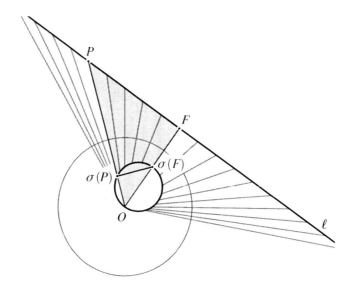

Figure 10. A line that does not pass through O is inverted to a circle that does.

Theorem 33.4 *Let σ be the inversion across a circle $\mathcal{C}$ with radius r and center O. Let ℓ be a line that does not pass through O. Then $\sigma(\ell)$ is a circle that passes through O.*

Proof Let F be the foot of the perpendicular from O to ℓ. We will show that $O\sigma(F)$ is the diameter of the circle $\sigma(\ell)$. To see why, take any other point P on ℓ. Then $\triangle OFP$ is a right triangle with right angle at F. As we have just seen, $\triangle OFP$ is similar to $\triangle O\sigma(P)\sigma(F)$ which

33.2 Inversion

means that $\triangle O\sigma(P)\sigma(F)$ is a right triangle whose right angle is at $\sigma(P)$. By Thales' theorem, $\sigma(P)$ is on the circle with diameter $O\sigma(P)$. □

It is easy to reverse that argument: any circle that passes through O inverts to a line (that does not pass through O). But that obviously leads to another question, what about circles that do not pass through O? We will again need to rely upon Thales' theorem. The setup is more complicated.

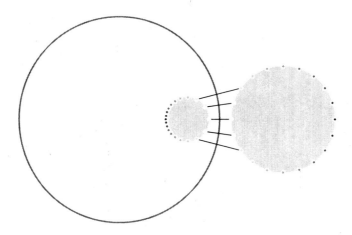

Figure 11. A circle that does not pass through O is inverted to another circle.

Theorem 33.5 *Let σ be the inversion across a circle $\mathcal{C}$ with radius r and center O. Let c be a circle that does not pass through O. Then $\sigma(c)$ is again a circle (that does not pass through O).*

Proof The ray from O through the center of c will intersect c twice. Label the intersections P and Q, with P the intersection that is closer to O. Then PQ is a diameter of c and I claim that $\sigma(P)\sigma(Q)$ is a diameter of $\sigma(c)$. Let R be another point on c. Then

$$\triangle OPR \sim \triangle O\sigma(R)\sigma(P) \implies \angle OPR \simeq \angle O\sigma(R)\sigma(P)$$
$$\triangle OQR \sim \triangle O\sigma(R)\sigma(Q) \implies \angle OQR \simeq \angle O\sigma(R)\sigma(Q).$$

Angle arithmetic gives

$$(\angle \sigma(P)\sigma(R)\sigma(Q)) = (\angle O\sigma(R)\sigma(P)) - (\angle O\sigma(R)\sigma(Q))$$
$$= (\angle OPR) - (\angle OQR).$$

Because $\angle OPR$ is an exterior angle of $\triangle PQR$,

$$(\angle OPR) = (\angle PQR) + (\angle PRQ).$$

Substituting,

$$(\angle \sigma(P)\sigma(R)\sigma(Q)) = ((\angle PQR) + (\angle PRQ)) - (\angle OQR) = (\angle PRQ).$$

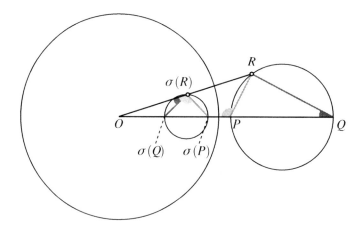

Figure 12. The image of a circle: some similar triangles.

Since R is on the circle with diameter PQ, $\angle PRQ$ is a right angle. Therefore $\angle \sigma(P)\sigma(R)\sigma(Q)$ is a right angle as well, and so $\sigma(R)$ lies on the circle with diameter $\sigma(P)\sigma(Q)$. □

A word of warning regarding the last theorem: while the image $\sigma(c)$ is a circle, σ does not map the center of c to the center of $\sigma(c)$.

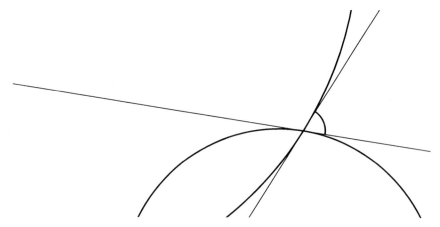

Figure 13. We measure the angle between intersecting curves (in particular circles) by the angle between their tangent lines at the point of intersection.

Since an inversion σ doesn't map lines to lines, the question of whether $\sigma(\angle ABC) \simeq (\angle ABC)$ is a little trickier. Let's take a page from the book of calculus. In calculus, the angle between intersecting curves is measured by zooming in to the infinitesimal level, at which point

33.2 Inversion

the angle between the curves becomes the angle between their tangent lines. A mapping that preserves angles between curves is called a *conformal map*. We will now show that inversion is a conformal map.

Theorem 33.6 *Let σ be the inversion across the circle $\mathcal{C}$ with center O and radius r. Let ℓ_1 and ℓ_2 be curves that intersect at some point P other than O. The curves may be both lines, both circles, or one of each. Let P be the intersection of ℓ_1 and ℓ_2. Then the angle between ℓ_1 and ℓ_2 at P is the same as the angle between $\sigma(\ell_1)$ and $\sigma(\ell_2)$ at $\sigma(P)$.*

Proof There are many cases here because one or both of the curves may pass through O. We will do the case where ℓ_1 and ℓ_2 are lines and leave the rest as exercises. ℓ_1 and ℓ_2 cannot both pass through O, for if they did, their intersection P would occur at O. It doesn't make sense to talk of the image of that point, which is why that case was prohibited in the statement of the theorem.

Suppose that ℓ_1 passes through O, but that ℓ_2 does not. Then σ will map ℓ_1 to itself and will map ℓ_2 to a circle that passes through O. In proving theorem 33.4, we found out that if F is the foot of the perpendicular to ℓ_2 from O, then $O\sigma(F)$ will be a diameter of $\sigma(\ell_2)$. If ℓ_1 and ℓ_2 intersect exactly at F, then ℓ_1 and ℓ_2 intersect at right angles, and in that case, the diameter of $\sigma(\ell_2)$ lies along the line ℓ_1. Thus the tangent line to the circle $\sigma(\ell_2)$ at $\sigma(F)$ again intersects $\sigma(\ell_1)$ at a right angle.

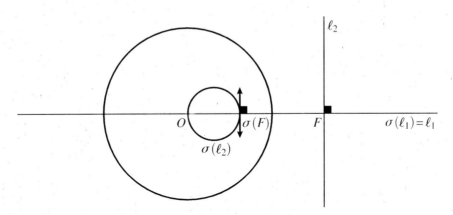

Figure 14. Conformality: a line through O and another that is perpendicular to it.

Generally, when ℓ_1 and ℓ_2 intersect at a point P other than F, then their angle of intersection is $\angle OPF$, and

$$\triangle OPF \sim \triangle O\sigma(F)\sigma(P),$$

so $\angle OPF \simeq \angle O\sigma(F)\sigma(P)$. Let Q be the center of the circle $\sigma(\ell_2)$. Both $Q\sigma(F)$ and $Q\sigma(P)$ are radii of the circle, so $\triangle Q\sigma(F)\sigma(P)$ is isosceles, and by the isosceles triangle theorem,

$$\angle Q\sigma(F)\sigma(P) \simeq \angle Q\sigma(P)\sigma(F).$$

Therefore $\angle Q\sigma(P)\sigma(F)$ and the angle between $\sigma(\ell_1)$ and $\sigma(\ell_2)$ are complementary to the same angle and so they are congruent.

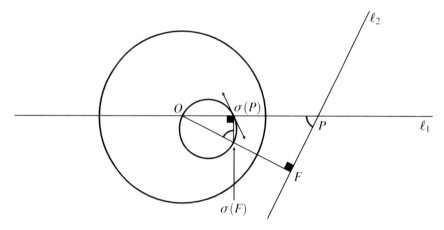

Figure 15. Conformality: a line through O and another intersecting line (not perpendicular to it).

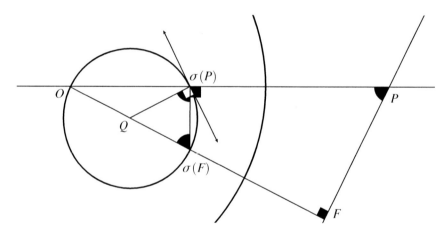

Figure 16. Close-up view near $\sigma(P)$: we see that $\angle Q\sigma(F)\sigma(P)$ is congruent to $\angle Q\sigma(P)\sigma(F)$ by the isosceles triangle theorem and that is congruent to the angle formed by $\sigma(\ell_1)$ and $\sigma(\ell_2)$ since they are complementary to the same angle.

Suppose that neither ℓ_1 nor ℓ_2 passes through O. In this case, $\overline{OP}$ splits the angle formed by ℓ_1 and ℓ_2 into two pieces. Let θ_1 be the angle between ℓ_1 and $\overline{OP}$ and let θ_2 be the angle between ℓ_2 and $\overline{OP}$. $\overline{OP}$ will also split the angle between $\sigma(\ell_1)$ and $\sigma(\ell_2)$. From our previous work, the angle between $\sigma(\ell_1)$ and $\overline{OP}$ is the same as the angle between ℓ_1 and $\overline{OP}$, and the angle between $\sigma(\ell_2)$ and $\overline{OP}$ is the same as the angle between ℓ_2 and $\overline{OP}$. Adding the pieces, the angle between $\sigma(\ell_1)$ and $\sigma(\ell_2)$ is the same as the angle between ℓ_1 and ℓ_2. □

That's some good news about angle measure. Unfortunately, we know that the news isn't so good when it comes to measuring distance. Does inversion have any kind of distance invariant?

33.2 Inversion

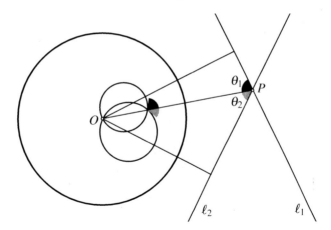

Figure 17. Conformality: the case of intersecting lines that do not pass through O. It can be transformed into two instances where one of the lines passes through O.

Yes, but to find it you have to use the similarity property of inversion. The invariant is called the *cross ratio*.

Definition 33.7 Let A, B, P, and Q be distinct points. The *cross ratio* of A, B, P, and Q, written $[AB, PQ]$, is the product of ratios

$$[AB, PQ] = \frac{|AP|}{|AQ|} \cdot \frac{|BQ|}{|BP|}.$$

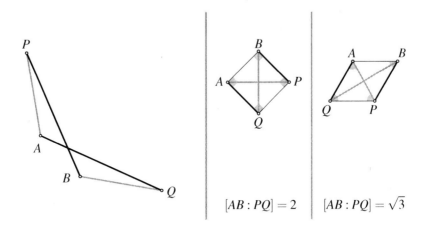

Figure 18. The cross ratio $[AB : PQ]$ is the product of $|AP|/|AQ|$ and $|BQ|/|BP|$.

Theorem 33.8 *The cross ratio is invariant under inversion. That is, for an inversion σ, and points A, B, P, and Q,*

$$[AB, PQ] = [\sigma(A)\sigma(B), \sigma(P)\sigma(Q)].$$

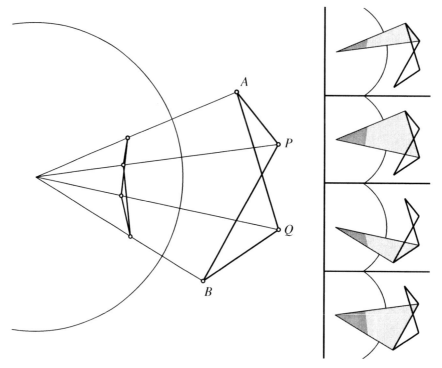

Figure 19. Ratios of sides in similar triangles to show that the cross ratio is an invariant of inversion.

Proof By the similarity property,

$$\frac{|\sigma(A)\sigma(P)|}{|AP|} = \frac{|O\sigma(P)|}{|OA|} \qquad \frac{|\sigma(B)\sigma(Q)|}{|BQ|} = \frac{|O\sigma(Q)|}{|OB|}$$

$$\frac{|\sigma(A)\sigma(Q)|}{|AQ|} = \frac{|O\sigma(Q)|}{|OA|} \qquad \frac{|\sigma(B)\sigma(P)|}{|BP|} = \frac{|O\sigma(P)|}{|OB|}.$$

Combining these in the right way allows for a lot of cancellation:

$$\frac{|\sigma(A)\sigma(P)|}{|AP|} \cdot \frac{|\sigma(B)\sigma(Q)|}{|BQ|} \cdot \frac{|AQ|}{|\sigma(A)\sigma(Q)|} \cdot \frac{|BP|}{|\sigma(B)\sigma(P)|}$$

$$= \frac{|O\sigma(P)|}{|OA|} \cdot \frac{|O\sigma(Q)|}{|OB|} \cdot \frac{|OA|}{|O\sigma(Q)|} \cdot \frac{|OB|}{|O\sigma(P)|} = 1.$$

Thus

$$\frac{|\sigma(A)\sigma(P)|}{|\sigma(A)\sigma(Q)|} \cdot \frac{|\sigma(B)\sigma(Q)|}{|\sigma(B)\sigma(P)|} = \frac{|AP|}{|AQ|} \cdot \frac{|BQ|}{|BP|},$$

and so

$$[\sigma(A)\sigma(B), \sigma(P)\sigma(Q)] = [AB, PQ]. \qquad \square$$

We will see the cross ratio again. It is an essential tool in non-Euclidean geometry.

33.3 Exercises

33.2. Let σ be the inversion across the circle $\mathcal{C}$ given by the equation $(x-1)^2 + (y+2)^2 = 4$, and let $P = (2, 2)$. What are the coordinates of $\sigma(P)$?

33.3. Let τ be a similarity mapping (so that a polygon and its image under τ are similar, as described in chapter 28). What is the relationship between the cross ratios $[AB, PQ]$ and $[\tau(A)\tau(B), \tau(P)\tau(Q)]$?

33.4. Let $A = (0, 0)$, $B = (2, 0)$, and $C = (1, 2)$, and let O be the circumcenter of $\triangle ABC$. Compute the cross ratio $[AB, CO]$.

33.5. Let $\square ABCD$ be a rectangle with side lengths $|AB| = x$, and $|BC| = y$. Compute the cross ratio $[AB, CD]$ in terms of x and y.

33.6. Consider the cyclic quadrilateral $\square APQR$ inscribed in the unit circle with $A = (\cos\theta, \sin\theta)$, $P = (1, 0)$, $Q = (0, -1)$, and $R = (-1, 0)$. Compute the cross ratio $[AP, QR]$ as a function of θ.

33.7. Verify that if two circles intersect at points P and Q, then their angle of intersection as measured at P is the same as the angle of intersection as measured at Q.

33.8. Let $\mathcal{C}_1$ and $\mathcal{C}_2$ be circles centered at O with radii r_1 and r_2 respectively. Let σ_1 be the inversion across $\mathcal{C}_1$ and σ_2 be the inversion across $\mathcal{C}_2$. If $|OP| = x$, what is $|O\sigma_1 \circ \sigma_2(P)|$ in terms of x, r_1, and r_2?

33.9. Let σ_1 be the inversion across the circle $x^2 + (y-1)^2 = 1$ and let σ_2 be the inversion across the circle $(x-1)^2 + y^2 = 1$. What is the domain of $\sigma_2 \circ \sigma_1$?

33.10. Let $\mathcal{C}_1$ and $\mathcal{C}_2$ be circles centered at O, with radii r_1 and r_2 respectively. Define

$$f(P) = \begin{cases} \sigma_1 \circ \sigma_2(P) & \text{if } P \neq O \\ O & \text{if } P = O. \end{cases}$$

Prove that f is a dilation (or the identity map).

33.11. There are $4! = 24$ permutations of the four letters in the cross ratio. Some give the same result; for instance

$$[PQ, AB] = \frac{|PA|}{|PB|} \cdot \frac{|QB|}{|QA|} = [AB, PQ]$$

but others do not. Determine which of the 24 permutations yield the same result.

33.12. Complete the proof that inversion is conformal. There are two cases: where both ℓ_1 and ℓ_2 are circles and where one is a circle and the other is a line.

33.13. Let r_1 and r_2 be rays with a common endpoint O. Let A, B, and C be points on r_1; let A', B', and C' be points on r_2; and let P be the intersection of AA' and CC'. Prove that B, B', and P are collinear if and only if

$$[AC, OB] = [A'C', OB'].$$

Compare to exercise 19.14.

34
Inversion II

Matrix and vector arithmetic is the natural language of isometries, but it does not do so well in describing inversion. For that, it is better to translate the problem into the language of complex arithmetic. We will start this chapter with a review of complex arithmetic. I assume that readers who have made it this far have some experience working with complex numbers—if not, then this cursory overview is probably not sufficient—our needs are minimal, but they are not non-existent.

34.1 Complex numbers, complex arithmetic

A complex number has the form $a + bi$ where a and b are real numbers and i is a solution to the equation $x^2 = -1$. The set of complex numbers $\mathbb{C}$ contains all the real numbers in the form $a + 0i$, but since the square of any real number is positive, i is not itself a real number. Thus $\mathbb{C}$ is an extension of the real numbers. There is a bijection between complex numbers and points (or vectors) in $\mathbb{R}^2$ via

$$a + bi \longleftrightarrow (a, b).$$

The correspondence is what allows us to translate problems in $\mathbb{R}^2$ into problems in $\mathbb{C}$. Why would we want to do that? The basic advantage of $\mathbb{C}$ over $\mathbb{R}^2$ is that $\mathbb{C}$ is a field: two numbers in it can be added, subtracted, multiplied, and (except in the case of 0) divided. In contrast, while the vectors of $\mathbb{R}^2$ are equipped with addition, subtraction, and scalar multiplication, there is no natural way to multiply or divide vectors. It is the multiplication and division operations that make it worth the effort.

Remark 34.1. Complex arithmetic. The sum, difference, product, and quotient of two complex numbers is another complex number. The calculations are:

$$(a + bi) + (c + di) = (a + c) + (b + d)i$$
$$(a + bi) - (c + di) = (a - c) + (b - d)i$$
$$(a + bi)(c + di) = ac + adi + bci + bdi^2 = (ac - bd) + (ad + bc)i$$
$$\frac{a + bi}{c + di} = \frac{a + bi}{c + di} \cdot \frac{c - di}{c - di} = \frac{ac + bd}{c^2 + d^2} + \frac{bc - ad}{c^2 + d^2}i.$$

The *complex conjugate* of $z = a + bi$ is $\bar{z} = a - bi$. The *norm* (or length or absolute value) of $z = a + bi$ is its distance from 0,

$$|z| = \sqrt{a^2 + b^2}.$$

The *argument* of z, arg(z), is the measure of the angle that it forms with the real axis (as measured in the counterclockwise direction), so

$$\tan(\arg(a + bi)) = b/a.$$

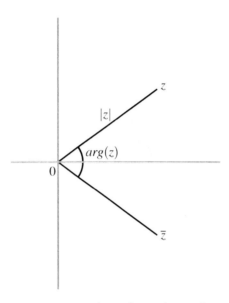

Figure 1. The argument, norm, and complex conjugate of a complex number.

While the standard presentation of a complex number is in the rectangular form $a + bi$, complex numbers can be expressed in a polar form: if $r = |z|$ and $\theta = \arg(z)$, then $a = r \cos \theta$ and $b = r \sin \theta$, so

$$a + bi = (r \cos \theta) + (r \sin \theta)i = r(\cos \theta + i \sin \theta).$$

For our purposes, the polar form is a stepping stone toward the goal of an exponential form. If you have only seen real-valued functions, then the trigonometric functions $\sin x$ and $\cos x$ probably seem vastly different from the exponential function e^x. For instance, $\sin x$ and $\cos x$ are bounded and periodic; the exponential function is neither bounded nor periodic. In the more expansive world of complex numbers, though, there are connections between the functions. The easiest way to see them is by looking at their Taylor series.

34.1.1 Taylor series: a quick review

Let $f(x)$ be a function whose derivatives at a point a are all defined. The nth Taylor polynomial of $f(x)$, expanded about the point a, is a specific degree n polynomial $p_n(x)$ that approximates $f(x)$ in a region around a. Its coefficients are calculated by matching the function value and the first n derivatives at a of $p_n(x)$ with those of $f(x)$. The derivatives at a give local information

34.1 Complex numbers, complex arithmetic

about the function around a (they tell us whether the function is increasing or decreasing, concave up or concave down, and so on). It makes sense that taking more derivatives would improve the approximation around a and perhaps extend the region for which the approximation is close. Taken to its natural extreme, then, if we want the best approximation, we let $n \to \infty$, and look at the Taylor series p_∞ that approximates $f(x)$. Matching derivatives gives the formula

$$p_\infty(x) = \sum_{n=0}^{\infty} \frac{p^{(n)}(a)}{n!}(x-a)^n.$$

Even with an infinite sum, there is no guarantee that $p_\infty(x)$ will be a good approximation of $f(x)$ as we move away from a (in fact, there is now the question of whether the series converges at all). Here's the good news: the Taylor series of e^x, $\sin x$, and $\cos x$ do converge to exactly the function value for all x, no matter what a is. The Taylor series expansions about $a = 0$ for them are

$$e^x = \sum_{n=0}^{\infty} \frac{1}{n!} x^n$$

$$\cos x = \sum_{n=0}^{\infty} \frac{(-1)^n}{(2n)!} x^{2n}$$

$$\sin x = \sum_{n=0}^{\infty} \frac{(-1)^n}{(2n+1)!} x^{2n+1}.$$

The formulas apply not just for real values of x, but for complex values as well.

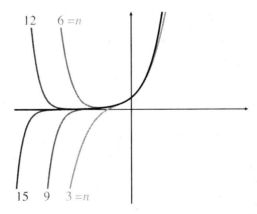

Figure 2. Taylor polynomial approximations to $f(x) = e^x$.

Exercise 34.1 Verify that the summations given for $\sin x$, $\cos x$, and e^x are the Taylor series (expanded about $a = 0$) for them (this is a standard problem in a calculus class).

Now let's see how that allows us to relate the sine and cosine functions to the exponential. Cosine is an even function and sine is an odd function, so if we take the series expansion of $e^{i\theta}$

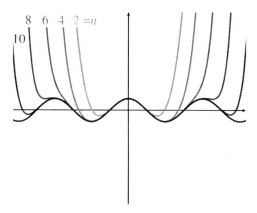

Figure 3. Taylor polynomial approximations to $f(x) = \cos x$.

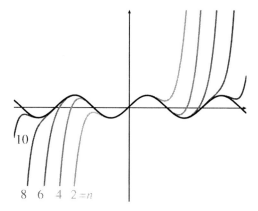

Figure 4. Taylor polynomial approximations to $f(x) = \sin x$.

and separate the even powers from the odd powers, we get

$$\begin{aligned}
e^{i\theta} &= \sum_{n=0}^{\infty} \frac{(i\theta)^n}{n!} \\
&= \sum_{n=0}^{\infty} \frac{(i\theta)^{2n}}{(2n)!} + \sum_{n=0}^{\infty} \frac{(i\theta)^{2n+1}}{(2n+1)!} \\
&= \sum_{n=0}^{\infty} \frac{i^{2n}\theta^{2n}}{(2n)!} + \sum_{n=0}^{\infty} \frac{i \cdot i^{2n}\theta^{2n+1}}{(2n+1)!} \\
&= \sum_{n=0}^{\infty} \frac{(-1)^n \theta^{2n}}{(2n)!} + i \sum_{n=0}^{\infty} \frac{(-1)^n \theta^{2n+1}}{(2n+1)!} \\
&= \cos\theta + i \sin\theta.
\end{aligned}$$

Therefore the polar form of a complex number z can be written in an exponential form

$$z = r(\cos\theta + i \sin\theta) = re^{i\theta}.$$

All the rules of exponents apply, so this is a powerful alternative to the rectangular form for a complex number.

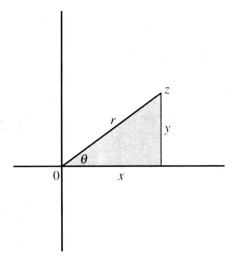

Figure 5. Three forms for a complex number. Rectangular: $z = x + iy$. Trigonometric: $z = r\cos\theta + ir\sin\theta$. Exponential: $z = re^{i\theta}$.

34.2 The geometry of complex arithmetic

Adding the complex number $z = a + bi$ to another complex number w has the effect of translating w by the vector $\langle a, b \rangle$. Subtracting z from w has a similar effect, but the translation is in the opposite direction. For multiplication and division it is best to look at the exponential form: write $z = re^{i\theta}$ and $w = se^{i\phi}$. Then

$$zw = re^{i\theta} \cdot se^{i\phi} = rse^{i(\theta+\phi)}.$$

The effect of multiplying by z, then, is to scale from the origin by r and to rotate counterclockwise around the origin by θ. Division works similarly:

$$w/z = se^{i\phi}/re^{i\theta} = (s/r)e^{i(\phi-\theta)},$$

but this time the scaling is by $1/r$ and the rotation by θ is in the clockwise direction. For this reason, some Euclidean isometries can be described naturally with complex arithmetic.

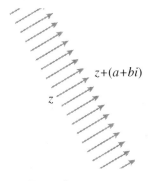

Figure 6. Translation.

The translation

$$t\begin{pmatrix}x\\y\end{pmatrix} = \begin{pmatrix}x+a\\y+b\end{pmatrix}$$

becomes

$$t(z) = z + (a+bi).$$

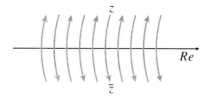

Figure 7. Reflection.

The reflection across the real axis (the x-axis)

$$s\begin{pmatrix}x\\y\end{pmatrix} = \begin{pmatrix}x\\-y\end{pmatrix}$$

becomes

$$s(z) = \bar{z}.$$

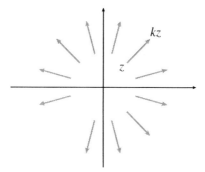

Figure 8. Dilation.

The dilation by k about the origin

$$d\begin{pmatrix}x\\y\end{pmatrix} = \begin{pmatrix}kx\\ky\end{pmatrix}$$

becomes

$$d(z) = kz.$$

34.2 The geometry of complex arithmetic

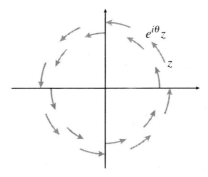

Figure 9. Rotation.

The rotation by θ about the origin

$$r\begin{pmatrix}x\\y\end{pmatrix} = \begin{pmatrix}\cos\theta & -\sin\theta \\ \sin\theta & \cos\theta\end{pmatrix}\begin{pmatrix}x\\y\end{pmatrix}$$

becomes

$$r(z) = e^{i\theta} \cdot z.$$

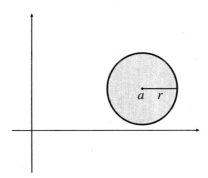

Figure 10. A circle.

For a complex number a and positive real number r, the equation $|z - a| = r$ describes a circle with center a and radius r.

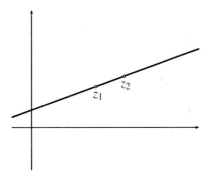

Figure 11. A line.

For two complex numbers z_1 and z_2, the function $r : \mathbb{R} \to \mathbb{C}$ defined by

$$r(t) = z_1 + t(z_2 - z_1)$$

describes the line through z_1 and z_2.

The point of this is to find a workable equation for inversion.

Theorem 34.2 *The inversion σ across $|z| = r$, the circle of radius r centered at the origin, is given by*

$$\sigma(z) = r^2/\bar{z}.$$

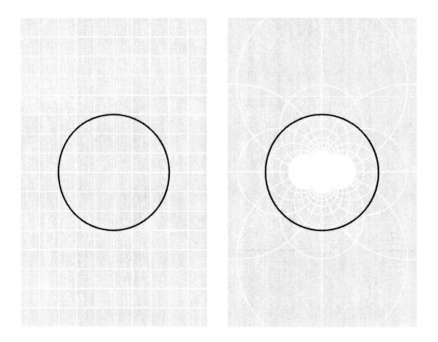

Figure 12. Inverting a rectangular grid.

Proof Write $z = Re^{i\theta}$. According to the definition of inversion, $\sigma(z)$ is on the ray from the origin passing through z and its distance from the origin is r^2/R. The points on this ray have an argument of θ. Therefore

$$\sigma(z) = \frac{r^2}{R}e^{i\theta} = \frac{r^2}{Re^{-i\theta}} = r^2/\bar{z}. \qquad \square$$

More generally, we can use a change of coordinates to find the equation of an inversion across a circle that is not centered at the origin.

34.3 Properties of the norm and conjugate

Corollary 34.3 *The inversion σ across $|z - a| = r$, the circle of radius r centered at a, is given by*

$$\sigma(z) = \frac{r^2}{\overline{z - a}} + a.$$

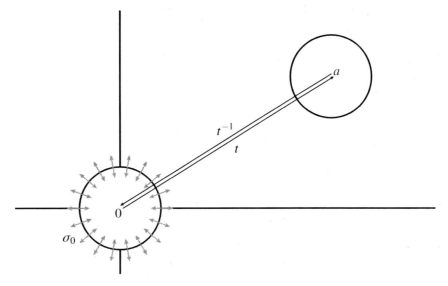

Figure 13. Use a change of coordinates to find the equation for inversion across an arbitrary circle.

Proof To use the previous formula, we need to change coordinates to reposition the origin at a. We can use the translation $t(z) = z + a$. If we label σ_0 as the inversion across the circle of radius r centered at the origin, then

$$\sigma(z) = t \circ \sigma_0 \circ t^{-1}(z) = t \circ \sigma_0(z - a) = t\left(\frac{r^2}{\overline{z - a}}\right) = \frac{r^2}{\overline{z - a}} + a. \qquad \square$$

34.3 Properties of the norm and conjugate

Some of our calculations in the coming chapters will rely on a few simple properties of the norm and the conjugate. They are listed in the following theorems. Their proofs are straightforward, and are left, for the most part, as exercises.

Theorem 34.4 *For complex numbers z, z_1, and z_2*

$$\overline{(\overline{z})} = z$$

$$\overline{z_1 + z_2} = \overline{z_1} + \overline{z_2}$$

$$\overline{z_1 - z_2} = \overline{z_1} - \overline{z_2}$$

$$z = re^{i\theta} \implies \overline{z} = re^{-i\theta}$$

$$\overline{z_1 \cdot z_2} = \overline{z_1} \cdot \overline{z_2}$$

$$\overline{z_1/z_2} = \overline{z_1}/\overline{z_2} \quad (if\ z_2 \neq 0).$$

Proof As an example, let's prove that $\overline{z_1 \cdot z_2} = \overline{z_1} \cdot \overline{z_2}$. Writing $z_1 = r_1 e^{i\theta_1}$ and $z_2 = r_2 e^{i\theta_2}$, then

$$\overline{z_1 \cdot z_2} = \overline{r_1 e^{i\theta_1} r_2 e^{i\theta_2}} = \overline{r_1 r_2 e^{i(\theta_1+\theta_2)}} = r_1 r_2 e^{-i(\theta_1+\theta_2)} = r_1 e^{-i\theta_1} r_2 e^{-i\theta_2} = \overline{z_1} \cdot \overline{z_2}. \qquad \square$$

Theorem 34.5 *For complex numbers z, z_1, and z_2*

$$z\bar{z} = |z|^2$$
$$|\bar{z}| = |z|$$
$$|z_1 \cdot z_2| = |z_1| \cdot |z_2|$$
$$|z_1/z_2| = |z_1|/|z_2| \quad (z_2 \neq 0)$$
$$|z_1 \pm z_2| \leq |z_1| + |z_2|.$$

Proof For example, consider the first. Write $z = re^{i\theta}$. Then

$$z\bar{z} = re^{i\theta} \cdot re^{-i\theta} = r^2 = |z|^2. \qquad \square$$

34.4 Exercises

34.2. Let r be the counterclockwise rotation by $\pi/2$ around the point $1 + 2i$. Write a complex equation describing r.

34.3. Let s be the reflection across the line $r(t) = t + (1 + t)i$. Write a complex equation describing s.

34.4. Let d be the dilation by a factor of 3 centered at the point $0 + i$. Write a complex equation describing d.

34.5. Let g be the glide reflection that reflects across the line $r(t) = t(1 + i)$ and maps the point 0 to $3 + 3i$. Write a complex equation describing g.

34.6. Write a complex equation for the circle that passes through the points 0, 1, and $2i$.

34.7. Find the equation for the inversion through the circle with radius 2 and center $1 + 5i$.

34.8. Let $\mathcal{C}_1$ and $\mathcal{C}_2$ be circles centered at the complex number a, with radii r_1 and r_2 respectively. Let σ_1 be the inversion across $\mathcal{C}_1$ and σ_2 be the inversion across $\mathcal{C}_2$. What is the complex equation of the transformation $\sigma_2 \circ \sigma_1$?

34.9. Let $\mathcal{C}_1$ be the circle with radius 1 centered at the complex number a_1 and let $\mathcal{C}_2$ be the circle with radius 1 centered at the complex number a_2. Let i_1 be the inversion across $\mathcal{C}_1$ and i_2 be the inversion across $\mathcal{C}_2$. Describe the fixed point(s) of the composition map $i_2 \circ i_1$ in terms of a_1 and a_2.

34.10. Let σ_1 be the inversion across the circle with radius 1 centered at the origin (the unit circle). Let σ_2 be the inversion across the circle with radius 2 centered at the origin. Describe the set of fixed points of $\sigma_1 \circ \sigma_2$.

34.4 Exercises

34.11. Verify the remaining properties of the conjugate (theorem 34.4).

34.12. Verify the remaining properties of the norm (theorem 34.5).
The remaining exercises deal with polar equations for conic sections.

34.13. Let a be a (real) constant. Show that the polar equation $r = a \cos \theta$ describes a circle. What is its radius and center?

34.14. Let d be a (real) constant. Show that the polar equation $r = d/(1 + \cos \theta)$ describes a parabola with directrix line $x = d$ and focus at the origin.

34.15. Let d and e be (real) constants, and suppose that $0 < e < 1$. Show that the polar equation $r = ed/(1 + e \cos \theta)$ describes an ellipse.

34.16. Let d and e be (real) constants, and suppose that $e > 1$. Show that the polar equation $r = ed/(1 + e \cos \theta)$ describes a hyperbola.

35

Applications of Inversion

In the last two chapters we came to some understanding of the workings of inversion. This will smooth the transition into non-Euclidean geometry. Before we make the transition, though, let's take a brief vacation, and look at two nice theorems that can be proved with the help of a well-placed inversion. Both involve a chain of mutually tangent circles. In the first, the chain of circles is inside a shape called an arbelos; in the second, it is wedged between two circles.

A typical inversion proof begins by applying an inversion to what appears to be a complicated picture. It transforms the complicated picture into a simpler one with more symmetry than the original. The symmetry makes the proof easy. The trick is to find the right inversion to start with. A good place to start looking is orthogonal circles.

35.1 Orthogonal circles

The angle between two intersecting circles is measured by the angle between their tangent lines.

Definition 35.1 Two intersecting circles $\mathcal{C}_1$ and $\mathcal{C}_2$ are *orthogonal* if the angle between them is a right angle.

Figure 1. A chain of pairwise orthogonal circles.

A reason for looking at orthogonal circles is

Theorem 35.2 *Suppose that $\mathcal{C}_1$ and $\mathcal{C}_2$ are orthogonal circles and that σ is the inversion across $\mathcal{C}_1$. Then $\sigma(\mathcal{C}_2) = \mathcal{C}_2$.*

Proof Orthogonal circles will intersect twice, and both points of intersection are fixed by σ (since they are on $\mathcal{C}_1$). So we know two points on $\sigma(\mathcal{C}_2)$ already, but what we want is its center. Label:

P_1, P_2 : the intersections of $\mathcal{C}_1$ and $\mathcal{C}_2$,
O_1, O_2 : the centers of $\mathcal{C}_1$ and $\mathcal{C}_2$, and
Q : the center of $\sigma(\mathcal{C}_2)$.

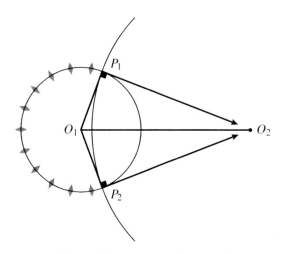

Figure 2. Inverting an orthogonal circle.

Because σ is conformal and the angle between $\mathcal{C}_1$ and $\mathcal{C}_2$ is a right angle, the angle between $\sigma(\mathcal{C}_2)$ and $\sigma(\mathcal{C}_1) = \mathcal{C}_1$ is too. Then Q is on the line through P_1 that is perpendicular to $O_1 P_1$ and on the line through P_2 that is perpendicular to $O_2 P_2$. Only one point meets those criteria, O_2. Therefore $\sigma(\mathcal{C}_2)$ is a circle centered at $Q = O_2$ and passing through P_1 and P_2, so $\sigma(\mathcal{C}_2)$ is $\mathcal{C}_2$. □

Though the theorem says that $\mathcal{C}_2$ is invariant, it does not say that the points of $\mathcal{C}_2$ are fixed. In fact, σ fixes only two points of $\mathcal{C}_2$, the points of intersection of $\mathcal{C}_1$ and $\mathcal{C}_2$. More subtly, while $\sigma(\mathcal{C}_2)$ has center O_2, $\sigma(O_2) \neq O_2$. If the orthogonal circles are going to play key roles in our proofs, we need to have some idea about how prevalent they are. The next two results address that issue.

Theorem 35.3 *Let $\mathcal{C}$ be a circle and P be a point outside $\mathcal{C}$. Then there is a unique circle centered at P that is orthogonal to $\mathcal{C}$.*

Proof It is easier to make this argument in reverse. Say that $\mathcal{C}$ has center O and radius r, and suppose that P is the center of a circle orthogonal to $\mathcal{C}$. What would its radius R be? Inside every pair of orthogonal circles is a right triangle. Two of its vertices are the centers of the circles; the

35.1 Orthogonal circles

third is one the points of intersection of the two circles. By the Pythagorean theorem,

$$r^2 + R^2 = |OP|^2 \implies R = [|OP|^2 - r^2]^{1/2}.$$

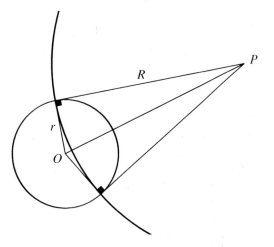

Figure 3. Finding the circle centered at P that is orthogonal to another circle. We can use the Pythagorean theorem to determine its radius.

Since P is outside $\mathcal{C}$, $|OP|$ is greater than r, so the equation has a solution. Because the Pythagorean theorem is bi-directional, the right triangle can be constructed with hypotenuse OP. In fact, it can be constructed on either side of OP. In both cases, the leg with length R is the radius of the one circle that is centered at P and orthogonal to $\mathcal{C}$. □

What about pairs of circles $\mathcal{C}_1$ and $\mathcal{C}_2$: are there circles that are orthogonal to both? Generally the answer is yes, but not always. The exception is that if $\mathcal{C}_1$ and $\mathcal{C}_2$ are concentric circles (that is, they have the same center), then there are no circles orthogonal to both $\mathcal{C}_1$ and $\mathcal{C}_2$. (This is actually a particularly good, rather than a particularly bad, case: if $\mathcal{C}_1$ and $\mathcal{C}_2$ are concentric, then it is the lines through their mutual center that will intersect both circles at right angles.) As long as $\mathcal{C}_1$ and $\mathcal{C}_2$ are not concentric, there are circles orthogonal to both. They are more scarce now, though, and the conditions for a point P to be the center of an orthogonal circle are more demanding. Label

- r_1 : the radius of $\mathcal{C}_1$,
- r_2 : the radius of $\mathcal{C}_2$,
- d_1 : the distance from the center of $\mathcal{C}_1$ to P,
- d_2 : the distance from the center of $\mathcal{C}_2$ to P.

Suppose $\mathcal{C}$ is a circle with radius r and center P that is orthogonal to $\mathcal{C}_1$ and $\mathcal{C}_2$. By the Pythagorean theorem,

$$r^2 + r_1^2 = d_1^2 \text{ and } r^2 + r_2^2 = d_2^2.$$

Solving for r^2 and setting them equal,

$$d_1^2 - r_1^2 = d_2^2 - r_2^2 \implies r_1^2 - r_2^2 = d_1^2 - d_2^2.$$

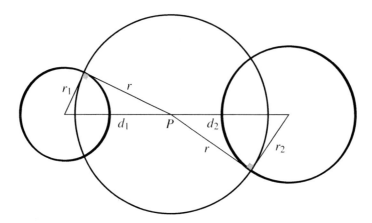

Figure 4. For P to be the center of a circle that is orthogonal to the other circles, the Pythagorean theorem requires that $r_1^2 - r_2^2 = d_1^2 - d_2^2$.

Since r_1 and r_2 are given by the circles $\mathcal{C}_1$ and $\mathcal{C}_2$, there are only certain combinations of d_1 and d_2 that satisfy this equation. The points that satisfy the condition form the *radical axis* of $\mathcal{C}_1$ and $\mathcal{C}_2$. What does the radical axis look like?

Theorem 35.4 *The radical axis of a pair of non-concentric circles is a line that is perpendicular to the line through the circles' centers.*

Proof Let O_1 and O_2 be the centers of the two circles. If the radical axis is a line perpendicular to $\overline{O_1 O_2}$ as claimed, then it intersects $\overline{O_1 O_2}$. Let's look for the intersection. To do so, set up a coordinate system measuring signed distance along $\overline{O_1 O_2}$. Center the coordinate system at O_1, so that O_1 is at coordinate 0, and label the corresponding coordinate for O_2 as α. For the point at coordinate x to be on the radical axis, it must satisfy

$$r_1^2 - r_2^2 = x^2 - (x - \alpha)^2 = x^2 - (x^2 - 2\alpha x + \alpha^2) = 2\alpha x - \alpha^2.$$

Solve for x to get $x = (r_1^2 - r_2^2 + \alpha^2)/(2\alpha)$. As long as the circles are not concentric, α will not be zero and the equation will have a (unique) solution. Therefore, exactly one point of $\overline{O_1 O_2}$ is on the radical axis, call it point Q. Let's label some distances too: $D_1 = |QO_1|$ and $D_2 = |QO_2|$. Label the line that passes through Q and is perpendicular to $O_1 O_2$ as ℓ. Since Q is on the radical axis,

$$D_1^2 - D_2^2 = r_1^2 - r_2^2.$$

We want to show that the other points that satisfy this condition are on ℓ (it is really an "if and only if" statement, which should be apparent in the proof). Take another point P and put $d_1 = |O_1 P|$, and $d_2 = |O_2 P|$.

In $\triangle O_1 QP$ and $\triangle O_2 QP$, the law of cosines gives

$$d_1^2 = D_1^2 + |PQ|^2 - 2D_1|PQ|\cos(\angle O_1 QP)$$
$$d_2^2 = D_2^2 + |PQ|^2 - 2D_2|PQ|\cos(\angle O_2 QP).$$

35.1 Orthogonal circles

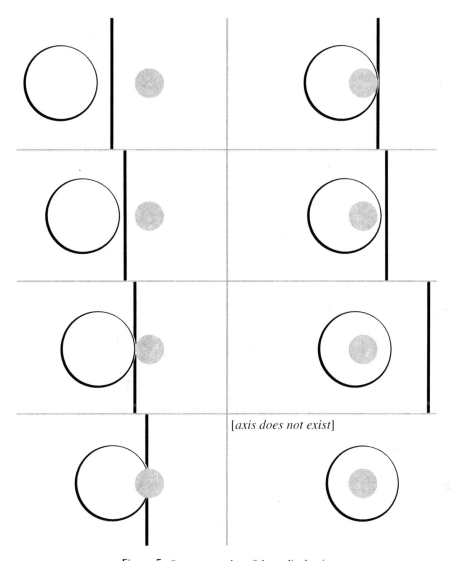

Figure 5. Some examples of the radical axis.

Therefore

$$d_1^2 - d_2^2 = \left[D_1^2 + |PQ|^2 - 2D_1|PQ|\cos(\angle O_1 QP) \right]$$
$$- \left[D_2^2 + |PQ|^2 - 2D_2|PQ|\cos(\angle O_2 QP) \right]$$
$$= (D_1^2 - D_2^2) - 2|PQ| \cdot \left[D_1 \cos(\angle O_1 QP) - D_2 \cos(\angle O_2 QP) \right].$$

For P to be on the radical axis, $d_1^2 - d_2^2$ needs to equal $r_1^2 - r_2^2$, but the term $(D_1^2 - D_2^2)$ already equals it. Therefore, P will be on the radical axis when

$$D_1 \cos(\angle O_1 QP) - D_2 \cos(\angle O_2 QP) = 0.$$

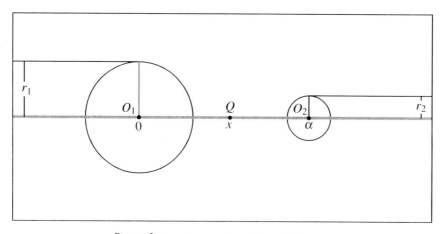

Figure 6. Locating a point on the radical axis.

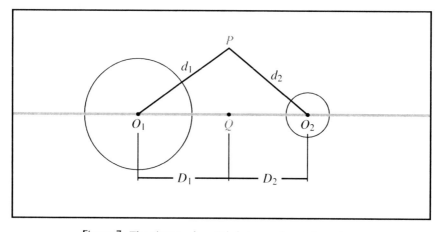

Figure 7. The picture when Q is between the circle centers.

There are two cases, depending on the position of Q relative to O_1 and O_2. If Q is between O_1 and O_2, then $\angle O_1 QP$ and $\angle O_2 QP$ are supplementary, so

$$\cos(\angle O_2 QP) = -\cos(\angle O_1 QP).$$

Then

$$D_1 \cos(\angle O_1 QP) - D_2 \cos(\angle O_2 QP)$$
$$= D_1 \cos(\angle O_1 QP) + D_2 \cos(\angle O_1 QP)$$
$$= (D_1 + D_2)\cos(\angle O_1 QP).$$

Since $D_1 + D_2 > 0$, the only way this can be zero is if $\cos(\angle O_1 QP) = 0$; that is, if $(\angle O_1 QP) = \pi/2$.

If Q is not between O_1 and O_2, then $\angle O_1 QP$ and $\angle O_2 QP$ are equal, so

$$D_1 \cos(\angle O_1 QP) - D_2 \cos(\angle O_2 QP)$$
$$= D_1 \cos(\angle O_1 QP) - D_2 \cos(\angle O_1 QP)$$
$$= (D_1 - D_2)\cos(\angle O_1 QP).$$

35.2 The arbelos

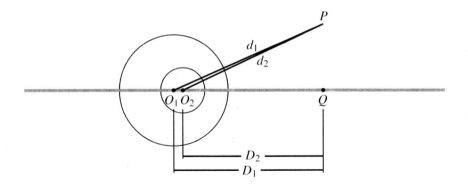

Figure 8. The picture when Q is not between the circle centers.

In this case, D_1 and D_2 cannot be equal because then the circles would be concentric. Therefore $D_1 - D_2 \neq 0$, so again $\cos(\angle O_1 QP) = 0$; that is, $(\angle O_1 QP) = \pi/2$. Either way, the angle at Q is a right angle. That places P on ℓ. □

Orthogonal circles will play a key role in our model for non-Euclidean geometry. We now turn to the first of the two main theorems of this chapter.

35.2 The arbelos

An *arbelos* is a shape built from three semicircles that is constructed as follows. Start with a semicircle with radius AC. Locate a point B between A and C. Now form two more semicircles on the same side of AC, with diameters AB and BC respectively. The resulting shape is an arbelos.

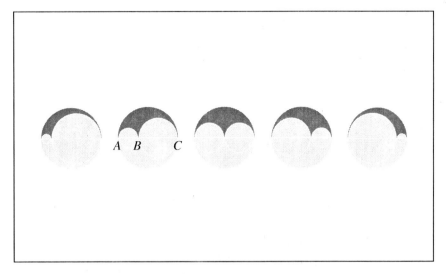

Figure 9. Five arbeloses (arbeli?).

The shape has a long history in classical geometry, going all the way back to the ancient Greeks. The name comes from them: apparently arbelos is a Greek word for a type of knife that was used by shoemakers. Its curved blade resembled the geometric shape that bears its name. Many interesting relationships have been found inside the arbelos, mainly in the form of hidden tangent circles. Our next theorem is in that vein. It presents a chain of mutually tangent circles. Its proof is easy once you have the right inversion.

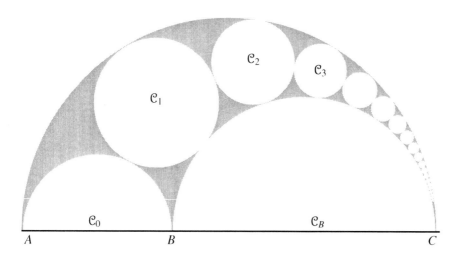

Figure 10. A chain of circles inside the arbelos.

Theorem 35.5 *Given an arbelos formed from three circles, with diameters AC, AB, and BC, where $A * B * C$, label*

$\mathcal{C}_A$: *the semicircle with radius AC*
$\mathcal{C}_B$: *the semicircle with radius BC*
$\mathcal{C}_0$: *the semicircle with radius AB.*

Then there is a circle $\mathcal{C}_1$ that is tangent to $\mathcal{C}_A$, $\mathcal{C}_B$, and $\mathcal{C}_0$; there is a circle $\mathcal{C}_2$ that is tangent to $\mathcal{C}_A$, $\mathcal{C}_B$, and $\mathcal{C}_1$; there is a circle $\mathcal{C}_3$ that is tangent to $\mathcal{C}_A$, $\mathcal{C}_B$, and $\mathcal{C}_2$; and in general, for any $n \geq 1$, there is a circle $\mathcal{C}_n$ that is tangent to $\mathcal{C}_A$, $\mathcal{C}_B$, and $\mathcal{C}_{n-1}$.

Proof From the discussion of orthogonal circles, since C is outside the circle $\mathcal{C}_0$, there is a circle centered at C that is orthogonal to $\mathcal{C}_0$. Let σ be the inversion across that circle. Then $\sigma(\mathcal{C}_0) = \mathcal{C}_0$ (but σ interchanges A and B). $\mathcal{C}_A$ and $\mathcal{C}_B$ pass through C, so σ inverts them to lines. Furthermore, since $\mathcal{C}_A$ and $\mathcal{C}_B$ are orthogonal to $\overline{AB}$, their images are perpendicular to $\sigma(\overline{AB}) = \overline{AB}$. That is, $\sigma(\mathcal{C}_A)$ is the line perpendicular to AB that passes through $\sigma(A) = B$, and $\sigma(\mathcal{C}_B)$ is the line perpendicular to AB that passes through $\sigma(B) = A$.

Now $\mathcal{C}_0 = \sigma(\mathcal{C}_0)$ is mutually tangent to two parallel lines, and in this configuration, it is easy to build a stack of circles on top of $\sigma(\mathcal{C}_0)$, each tangent to the lines $\sigma(\mathcal{C}_A)$ and $\sigma(\mathcal{C}_B)$ and the circle immediately below it. Apply σ again, sending the stack tumbling down: $\sigma(\mathcal{C}_A)$ back to $\mathcal{C}_A$, $\sigma(\mathcal{C}_B)$ back to $\mathcal{C}_B$, $\mathcal{C}_0$ staying put, but the circles stacked on top of it mapping to $\mathcal{C}_1$, $\mathcal{C}_2$, $\mathcal{C}_3$, ... □

35.3 Steiner's porism

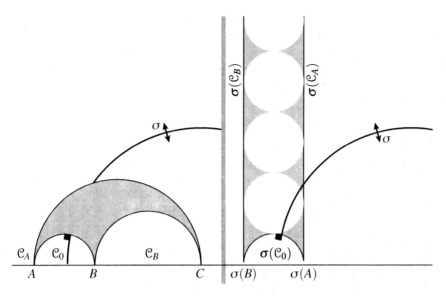

Figure 11. The vertical stack of circles inverts to a chain of mutually tangent circles in the arbelos.

35.3 Steiner's porism

Jakob Steiner was a nineteenth century geometer with an interest in inversion and a disdain for analytic geometry. This next chain of circles is named in his honor. Start with two circles $\mathcal{C}_A$ and $\mathcal{C}_B$, with $\mathcal{C}_B$ contained entirely in the interior of $\mathcal{C}_A$. We will build a chain of mutually tangent circles between $\mathcal{C}_A$ and $\mathcal{C}_B$. First choose a circle $\mathcal{C}_1$ that is tangent to both $\mathcal{C}_A$ and $\mathcal{C}_B$. Then

- There is a circle $\mathcal{C}_2$ that is tangent to $\mathcal{C}_A$, $\mathcal{C}_B$, and $\mathcal{C}_1$.
- There is a circle $\mathcal{C}_3$ that is tangent to $\mathcal{C}_A$, $\mathcal{C}_B$, and $\mathcal{C}_2$.
- In general, there is a circle that is tangent to $\mathcal{C}_A$, $\mathcal{C}_B$, and $\mathcal{C}_{n-1}$.

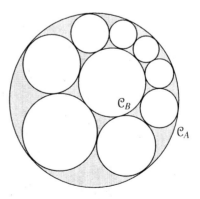

Figure 12. A Steiner chain between two circles.

This is similar to the chain of circles constructed in the arbelos. The difference is that eventually the circles inside this Steiner chain will loop around back to C_1. The natural question is: when the chain gets back around to C_1, will it join up perfectly? Will the last circle in the chain be tangent to C_1? That would be nice, but in general it does not happen. However, if conditions are so that the chain does close perfectly, then it will happen no matter what circle C_1 you use as the starting point. In other words, whether the chain closes perfectly depends only on C_A and C_B, not on C_1. I have made two claims that need proof.

Theorem 35.6 *(Steiner's porism) Suppose circle C_B is contained in the interior of C_A, and that a third circle C_1 in C_A is tangent to both C_A and C_B.*

(1) *There is a chain of circles $C_2, C_3, \ldots$, where each C_n is tangent to C_A, C_B, and C_{n-1}.*
(2) *If $\{C_1, C_2, \ldots, C_N\}$ is such a chain of circles and C_N is tangent to C_1, then for any such chain of circles $\{D_1, D_2, \ldots, D_N\}$, D_N will be tangent to D_1.*

Proof There is one case where these statements are obvious, when C_A and C_B are concentric. In that case, let O be the mutual centers of the circles, and let r_A and r_B be the respective radii. Then each circle in the chain C_i has its center on a circle halfway between C_A and C_B, on the circle with center O and radius $(r_A + r_B)/2$. All the circles in the chain are the same size with radii of $(r_A - r_B)/2$. Regarding the question of whether the chain will close up neatly, look at the angle θ at O that any one of the circles subtends. Slicing that angle in half creates a right triangle with an opposite side of $(r_A - r_B)/2$ and a hypotenuse of $(r_A + r_B)/2$. Therefore

$$\sin(\theta/2) = \frac{r_A - r_B}{r_A + r_B} \implies \theta = 2\sin^{-1}\left(\frac{r_A - r_B}{r_A + r_B}\right).$$

Then the question of whether the chain closes up is, is 2π divisible by θ? The answer depends only on r_A and r_B, not on where the starting circle C_1 is.

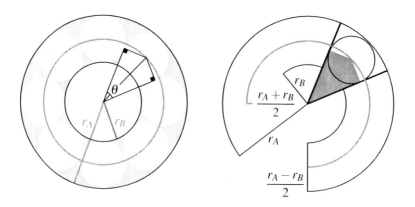

Figure 13. A Steiner chain between concentric circles.

If C_A and C_B are not concentric—that is, they have distinct centers O_A and O_B—then the circles in the chain will not all be the same size, as they are forced to adjust to fit in between

$\mathcal{C}_A$ and $\mathcal{C}_B$. That makes both claims more difficult to prove. Fortunately, there is an inversion to help, one that maps $\mathcal{C}_A$ and $\mathcal{C}_B$ to a concentric configuration.

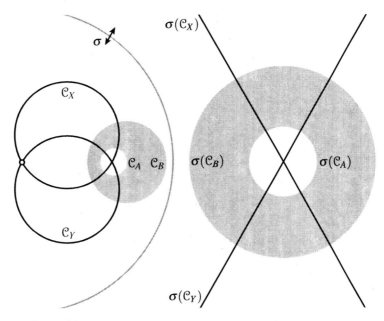

Figure 14. Inverting non-concentric circles to concentric ones.

We know that every point on the radical axis of $\mathcal{C}_A$ and $\mathcal{C}_B$ is the center of a circle that is orthogonal to $\mathcal{C}_A$ and $\mathcal{C}_B$. It is possible to choose two points on the radical axis so that the orthogonal circles $\mathcal{C}_X$ and $\mathcal{C}_Y$ intersect each other (it is an exercise to show this). Let σ be an inversion across a circle centered at one of the two intersections of $\mathcal{C}_X$ and $\mathcal{C}_Y$. Then $\sigma(\mathcal{C}_X)$ and $\sigma(\mathcal{C}_Y)$ are intersecting lines, and $\sigma(\mathcal{C}_A)$ and $\sigma(\mathcal{C}_B)$ are circles perpendicular to $\sigma(\mathcal{C}_X)$ and $\sigma(\mathcal{C}_Y)$. The only way that the lines $\sigma(\mathcal{C}_X)$ and $\sigma(\mathcal{C}_Y)$ can be perpendicular to the circles is if they both run along the diameters of the circles. Then their intersection must be at the centers of the circles. This applies to both circles $\sigma(\mathcal{C}_A)$ and $\sigma(\mathcal{C}_B)$. Therefore, $\sigma(\mathcal{C}_A)$ and $\sigma(\mathcal{C}_B)$ have the same center: they are concentric. We have just seen that we can construct a chain of circles between $\sigma(\mathcal{C}_A)$ and $\sigma(\mathcal{C}_B)$, and whether it closes depends only on $\sigma(\mathcal{C}_A)$ and $\sigma(\mathcal{C}_B)$. Apply σ again to such a chain of circles between $\sigma(\mathcal{C}_A)$ and $\sigma(\mathcal{C}_B)$. The result is a chain of circles between $\mathcal{C}_A$ and $\mathcal{C}_B$, and whether that chain closes up does not depend upon the location of the first circle in the chain. □

Exercise 35.1 This proof of Steiner's porism uses a pair of intersecting orthogonal circles. We had previously proved that the orthogonal circles exist, with their centers on the radical axis. Prove that it is possible to find a pair of orthogonal circles that intersect.

35.4 Exercises

35.2. Let $\mathcal{C}_1$ and $\mathcal{C}_2$ be two non-concentric circles. Prove that if the circles intersect, then their radical axis passes through the intersection point(s). Prove that if the circles don't intersect, then their radical axis lies outside both of them.

35.3. Given $A * B * C$, form an arbelos by removing half-circles with diameters AB and BC from the half-circle with diameter AC. Label the half-circles as $\mathcal{C}_0$, $\mathcal{C}_B$, and $\mathcal{C}_A$ (as in the proof of the arbelos chain). The line ℓ that passes through B and is perpendicular to AC intersects $\mathcal{C}_A$ at a point D. The circle with diameter BD intersects $\mathcal{C}_0$ and $\mathcal{C}_B$, label those points E and F. Prove that $\square BEDF$ is a rectangle.

35.4. Let σ be the inversion across circle $\mathcal{C}$ with center O, and let P be a point other than O. Find a compass and straightedge construction of $\sigma(P)$. Hint: the distance from O to $\sigma(P)$ is the key, and that is governed by the equation $|O\sigma(P)| = r^2/|OP|$. To get that ratio, consider a configuration of right triangles where $\triangle ABC$ has right angle at C and D is the foot of the altitude from C.

35.5. Given a circle $\mathcal{C}$ and two points P and Q in its interior that are not on the same diameter, give a compass and straightedge construction of the circle that passes through P and Q and is orthogonal to $\mathcal{C}$.

35.6. Let σ be an inversion across a circle $\mathcal{C}$ with center O. Let P be a point other than O and let $Q = \sigma(P)$. The line $\overline{PQ}$ intersects $\mathcal{C}$ twice. Let X be the intersection on the same side of O as P and Q. Let Y be the intersection on the other side. By definition,
$$\frac{|OQ|}{|OX|} = \frac{|OX|}{|OP|}.$$

Show that
$$\frac{|OQ| + |OX|}{|OQ| - |OX|} = \frac{|OX| + |OP|}{|OX| - |OP|}$$

and therefore that
$$\frac{|YQ|}{|XQ|} = \frac{|YP|}{|XP|}.$$

The remaining exercises prove Feuerbach's theorem, that for any triangle, the three excircles are tangent to the nine point circle. We will retain labels between problems and the results are cumulative, so they have to be worked (more or less) consecutively. This is essentially the proof in Pedoe's *A Course of Geometry for Colleges and Universities* [Ped70].

35.7. The first step is to eliminate an easy case where the general proof will not work. Suppose that $\triangle ABC$ is equilateral. Prove that the excircles are tangent to the nine-point circle at the midpoints of AB, AC, and BC.

35.8. Suppose that $\triangle ABC$ is not equilateral so that, for instance, $|AB| \neq |AC|$. Let $\mathcal{C}_B$ be the excircle that is in the interior of $\angle B$, and let $\mathcal{C}_C$ be the excircle that is in the interior of $\angle C$. Let O_B be the center of $\mathcal{C}_B$ and let O_C be the center of $\mathcal{C}_C$. Prove that A, O_B, and O_C are collinear.

35.9. Let F_B be the foot of the perpendicular to $\overline{BC}$ through O_B. Let F_C be the foot of the perpendicular to $\overline{BC}$ through O_C. Let s be the reflection across $\overline{O_B O_C}$. Let $B' = s(B)$,

35.4 Exercises

$C' = s(C)$, $F'_B = s(F_B)$, and $F'_C = s(F_C)$. Prove that:

- $\overline{B'C'}$ is tangent to $\mathcal{C}_B$ at F'_B and $\mathcal{C}_C$ at F'_C,
- the lines $\overline{BC}$, $\overline{O_1 O_2}$, $\overline{B'C'}$ are concurrent. Label their intersection as D,
- $B * A * C'$,
- $C * A * B'$.

35.10. Let G be the foot of the perpendicular from O_B to $\overline{AC}$, and label $G' = s(G)$. Show that

- $|BF_B| + |BG'| = |AB| + |BC| + |CA|$, and
- $|BF_B| = |BG'|$,

so that
$$|BF_B| = \frac{|AB| + |BC| + |AC|}{2}.$$
That is, $|BF_B|$ is the semiperimeter of $\triangle ABC$. So is $|CF_C|$. Therefore $|BF_B| = |CF_C|$. Hint: look for kites.

35.11. Let M be the midpoint of BC. Prove that M is also the midpoint of $F_B F_C$.

The key to this proof is an inversion σ across the circle that is centered at M and passes through F_B and F_C. $\mathcal{C}_B$ and $\mathcal{C}_C$ are orthogonal to the circle, so $\sigma(\mathcal{C}_B) = \mathcal{C}_B$ and $\sigma(\mathcal{C}_C) = \mathcal{C}_C$. The nine point circle, $\mathcal{C}_9$, passes through M, so $\sigma(\mathcal{C}_9)$ is a line. We will show that it is $\overline{B'C'}$. Since the line is tangent to $\mathcal{C}_B$ and $\mathcal{C}_C$ and inversion is conformal, $\mathcal{C}_9$ also is tangent to $\mathcal{C}_B$ and $\mathcal{C}_C$. The line $\sigma(\mathcal{C}_9)$ is determined by two points. We just need to find the right two points.

35.12. The first point. In $\triangle ABC$, let L be the foot of the altitude from A. L is one of the points on $\mathcal{C}_9$. Use the similar triangles $\triangle DAL$, $\triangle DO_C F_C$, and $\triangle DO_B F_B$ to show that
$$\frac{|LF_B|}{|LF_C|} = \frac{|AO_B|}{|AO_C|}.$$

35.13. Show that
$$\frac{|LF_B|}{|LF_C|} = \frac{|DF_B|}{|DF_C|}.$$
According to exercise 35.6, this implies that $\sigma(L) = D$, and, of course, D is on $\overline{B'C'}$.

35.14. The second point. Let M' be the midpoint of AB, a point on $\mathcal{C}_9$. Prove that the angle between MM' and the tangent line to $\mathcal{C}_9$ at M' is congruent to $\angle B$. Hint: let M'' be the midpoint of AC. Then $\angle MM''M'$ is inscribed in $\mathcal{C}_9$ and $\square BM'M''M$ is a parallelogram.

35.15. Since σ is conformal, $\sigma(\mathcal{C}_9)$ intersects $\sigma(\mathcal{C}_B) = \mathcal{C}_B$ at an angle congruent to $\angle B$. Using this, prove that $\sigma(M')$ is on $\overline{B'C'}$. With a similar argument $\sigma(M'')$ is on $B'C'$ as well.

Therefore $\sigma(\mathcal{C}_9) = \overline{B'C'}$ is tangent to $\sigma(\mathcal{C}_B) = \mathcal{C}_B$ and $\sigma(\mathcal{C}_C) = \mathcal{C}_C$. That shows that $\mathcal{C}_9$ is tangent to $\mathcal{C}_B$ and $\mathcal{C}_C$. We have made no mention of $\mathcal{C}_A$. But either $BC \neq AB$ or $BC \neq AC$, and we can then use the same argument with the pair of unequal sides.

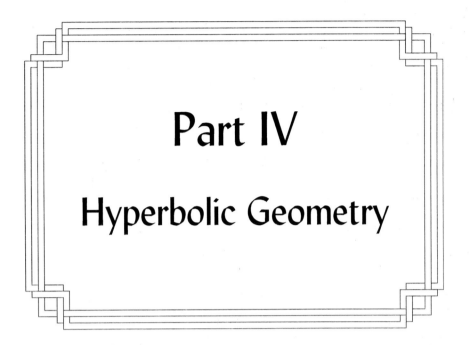

Part IV

Hyperbolic Geometry

The last part of this book is an introduction to hyperbolic geometry, the non-Euclidean neutral geometry. There are three major components.

- In chapters 36 and 37 we will try to get some sense of what to expect of hyperbolic geometry by looking at what would happen if Playfair's axiom were false.
- In chapters 38–41 we will build several models for hyperbolic geometry, starting with a partial model called the Beltrami pseudosphere, then moving to the upper half plane model, before ultimately settling on the Poincaré disk model. The calculations to get to the Beltrami pseudosphere model are difficult. They require multivariable calculus, linear algebra, and even some differential equations. You may want to skip straight ahead to the Poincaré disk model in chapter 41. In that chapter we verify that the Poincaré model satisfies all the axioms of neutral geometry, and so you can see how the Poincaré disk model works. I have included the chapters on the pseudosphere and the upper half plane to provide a better idea of why the Poincaré disk model works.
- The remaining chapters look at some elementary hyperbolic geometry in the Poincaré disk, deriving, for instance, hyperbolic analogs of the Pythagorean theorem and the circumference formula.

36

The Search for a Rectangle

The geometers who sought to prove Euclid's parallel axiom focused a lot of attention on rectangles. They are ubiquitous in Euclidean geometry, but without Euclid's parallel axiom, the situation is different. Let us see what happens when you let a rectangle into your world.

36.1 If there were a rectangle ...

By definition, a rectangle is a quadrilateral with four right angles. When we talk of a rectangle, we often talk of its *base* and *height*. There is ambiguity about which side to call the base and which to call the height, but the conversation would not make any sense if it weren't for this:

Lemma 36.1 *The opposite sides of a rectangle are congruent.*

Exercise 36.1 Prove the lemma (assuming that rectangles exist). Hint: the result is easy to prove if you use the converse of the alternate interior angle theorem, and the existence of a single rectangle implies the converse of the theorem. However, we have not yet established that. Consider the following approach: let $\square ABCD$ be a rectangle and suppose that $|BC| > |AD|$. Then there is a point E between B and C so that $AD \simeq BE$. Look for a contradiction involving $\triangle DCE$.

Now let's get to work.

Theorem 36.2 *Suppose that a rectangle R exists, with base b and height h. Then for an integer $n \geq 1$, there is a rectangle with base nb and height h.*

Proof By induction. The base case, when $n = 1$, is trivial because R meets the requirements. Now for the inductive step. Suppose that a rectangle with base nb and height h exists. Label it $\square ABCD$ where AB is the base. We need to build a rectangle with base $(n+1)b$ and height h. Two of its vertices are going to be A and D. The other two are E and F where

- $A * B * E$ and $|BE| = b$, and
- $D * C * F$ and $|CF| = b$.

By SASAS, □$BEFC$ is congruent to R, and therefore $\angle E$ and $\angle F$ are both right angles. That is, the interior angles of □$AEFD$ are right angles so □$AEFD$ is a rectangle. Its height is $|AD| = h$. Its base is

$$|AE| = |AB| + |BE| = nb + b = (n+1)b.$$

By induction, then, we can build nb-by-h rectangles.

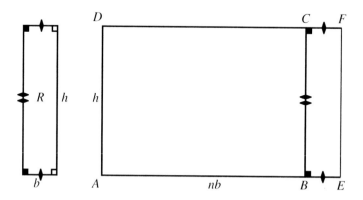

Figure 1. If there is a rectangle with base b and height h, then there is a rectangle with base nb and height h (a proof by induction). □

Theorem 36.3 *Suppose that a rectangle R exists, with base b and height h. Then for a positive real number x, there is a rectangle with base x and height h.*

Proof There is an integer n so that $nb > x$. By the previous result, there is a rectangle □$ABCD$ with base $|AB| = nb$ and height $|AD| = h$. Then there are points E and F so that

- $A * E * B$ and $|AE| = x$, and
- $D * F * C$ and $|DF| = x$.

The segment EF cuts □$ABCD$ into two pieces. We can show that both of those pieces are also rectangles. To see this, start by looking at their angle sums. As a corollary to the Saccheri-Legendre theorem, the angle sum of a quadrilateral is at most 2π. Therefore

$$(\angle A) + (\angle D) + (\angle AEF) + (\angle DFE) \leq 2\pi$$
$$(\angle B) + (\angle C) + (\angle BEF) + (\angle CFE) \leq 2\pi.$$

Since $(\angle A) = (\angle B) = (\angle C) = (\angle D) = \pi/2$, we get

$$(\angle AEF) + (\angle DFE) \leq \pi$$
$$(\angle BEF) + (\angle CFE) \leq \pi.$$

36.1 If there were a rectangle ...

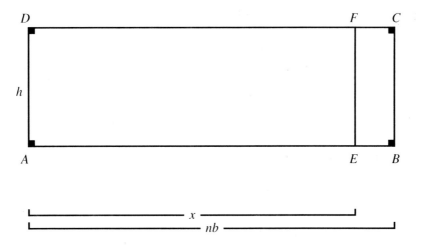

Figure 2. If there is a rectangle with base b and height h, then there is a rectangle with base x and height h. Extend the given rectangle to a new rectangle with base $nb > x$. Then divide that into two pieces, one with a base of x.

But

$$EA \simeq FD, \quad \angle A \simeq \angle D, \quad AD = DA, \quad \angle D \simeq \angle A, \text{ and } DF \simeq AE.$$

By SASAS, $\square AEFD$ is congruent to itself in a non-trivial way, so matching corresponding pieces, $\angle DFE \simeq \angle AEF$. Their supplements, $\angle CFE$ and $\angle BEF$, are also congruent.

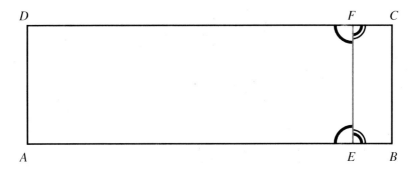

Figure 3. The segment EF cuts the rectangle $\square ABCD$ into two smaller pieces, both of which must also be rectangles.

Substitute this information into the pair of inequalities to get

$$2(\angle AEF) \leq \pi \implies (\angle AEF) \leq \pi/2$$
$$2(\angle BEF) \leq \pi \implies (\angle BEF) \leq \pi/2.$$

But $\angle AEF$ and $\angle BEF$ are supplementary, and so together sum to π. The only way this can happen is if $(\angle AEF) = (\angle BEF) = \pi/2$. That is, they are right angles, as are $\angle DFE$ and $\angle CFE$. Hence $\square AEFD$ and $\square BEFC$ are rectangles. In particular, $\square AEFD$ is a rectangle with base x and height h. □

Theorem 36.4 *Suppose that a rectangle R exists, with base b and height h. Then for positive real numbers x and y, there is a rectangle with base x and height y.*

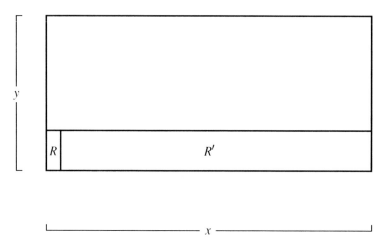

Figure 4. Use the previous result in both directions to build a rectangle with base x and height y.

Proof By the previous result, there is a rectangle R' with base x and height h. The roles of base and height are interchangeable, so that we can think of R' as a rectangle with base h and height x. Using the previous result, there is a rectangle R'' with base y and height x. Switch the roles of base and height in R'' back, so that R'' has a base x and a height y. □

The last three results tell us that if there is a rectangle in our geometry, then they are abundant. What are the implications of all of these rectangles?

Theorem 36.5 *If rectangles of all sizes exist, then the angle sum of a triangle is π.*

Proof There are two cases: first when the triangle is a right triangle, and second when it is not.

36.1 If there were a rectangle...

Case 1: Suppose that $\triangle ABC$ is a right triangle with right angle at C. There is a rectangle with base $|AC|$ and height $|BC|$. Label it $\square acbd$ so that ac is the base and cb is the height. The diagonal ab cuts the rectangle into two right triangles. By SAS, both are congruent to $\triangle ABC$. Therefore the angle sum of the two copies of $\triangle ABC$ equals the angle sum of a rectangle:

$$2((\angle A) + (\angle B) + (\angle C)) = 2\pi \implies (\angle A) + (\angle B) + (\angle C) = \pi.$$

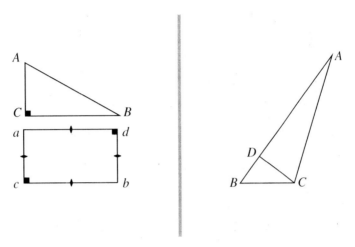

Figure 5. Given rectangles, the sum of the interior angles of the triangles must be π: (l) case 1, for right triangles. (r) case 2, for non-right triangles.

Case 2: Now suppose that $\triangle ABC$ is not a right triangle, and label it so that $\angle C$ is the largest angle. In that case, the altitude from C intersects AB between A and B. Label the intersection point as D. Then CD divides $\triangle ABC$ into two smaller right triangles: $\triangle ACD$ and $\triangle BCD$. From the previous part, both have an angle sum of π. They have right angles at D, so the other two angles sum to $\pi/2$. So

$$(\angle A) + (\angle ACD) = \pi/2 \quad \text{and} \quad (\angle BCD) + (\angle B) = \pi/2.$$

Add the equations to get

$$(\angle A) + (\angle ACD) + (\angle BCD) + (\angle B) = \pi/2 + \pi/2$$
$$(\angle A) + (\angle C) + (\angle B) = \pi. \qquad \square$$

Theorem 36.6 *If the angle sum of a triangle is π, then for a line ℓ and a point P not on ℓ, there is a unique line through P parallel to ℓ. That is, Playfair's axiom is true (and does not need to be taken as an axiom).*

Proof Begin by dropping the perpendicular from P to ℓ, and label its foot as Q_0. Let ℓ' be the line through P that is perpendicular to PQ_0. By the alternate interior angle theorem, it is parallel to ℓ. We need to show it is the only parallel. Suppose that ℓ'' is some other line through P. It forms an acute angle with PQ_0 on one of the sides of PQ_0 (and not on the other). On the acute side of PQ_0 we are going to build a sequence of isosceles triangles $\{\triangle PQ_{i-1}Q_i\}$.

(1) Locate Q_1 on ℓ so that $PQ_0 \simeq Q_0Q_1$. Then $\triangle PQ_0Q_1$ is isosceles. The angle at Q_0 is right, so if the angle sum of a triangle is π, then

$$\pi/2 + (\angle P) + (\angle Q_1) = \pi \quad \Longrightarrow \quad (\angle P) + (\angle Q_1) = \pi/2.$$

By the isosceles triangle theorem, $\angle P$ and $\angle Q_1$ are congruent, so they both measure $\pi/4$.

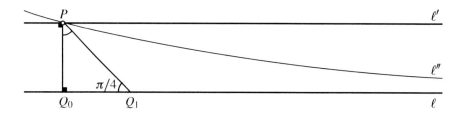

Figure 6. Constructing longer and longer isosceles triangles between ℓ and ℓ'. The first step.

(2) Now move farther out on ℓ, and locate Q_2 so that $Q_1Q_2 \simeq PQ_1$. The angle at Q_1 is $\pi - \pi/4 = 3\pi/4$, so if the angle sum of a triangle is π, then in $\triangle PQ_1Q_2$,

$$3\pi/4 + (\angle P) + (\angle Q_2) = \pi \quad \Longrightarrow \quad (\angle P) + (\angle Q_2) = \pi/4.$$

Because $\angle P$ and $\angle Q_2$ are congruent in this triangle, they each measure $\pi/8$. Therefore

$$(\angle Q_0PQ_2) = (\angle Q_0PQ_1) + (\angle Q_1PQ_2) = \pi/4 + \pi/8 = 3\pi/8.$$

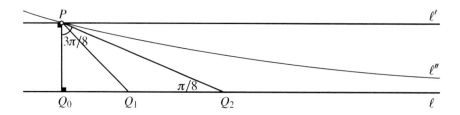

Figure 7. The second step.

36.2 The search for a rectangle

(3) We can continue in this way, building isosceles triangles $\triangle PQ_iQ_{i+1}$ with congruent sides PQ_i and Q_iQ_{i+1}. Measuring angles as we go:

i	$(\angle PQ_{i-1}Q_i)$	$(\angle Q_0PQ_i)$
1	$\pi/4$	$\pi/4$
2	$\pi/8$	$3\pi/8$
3	$\pi/16$	$7\pi/16$
4	$\pi/32$	$15\pi/32$
5	$\pi/64$	$31\pi/64$

In general,

$$(\angle Q_0PQ_i) = \frac{2^i - 1}{2^{i+1}}\pi.$$

As i approaches infinity, the angle approaches $\pi/2$. Now ℓ'' forms an acute angle with PQ_0. Therefore, for a large enough i, ℓ'' will lie between $\overrightarrow{PQ_0}$ and $\overrightarrow{PQ_i}$. By the crossbar theorem, ℓ'' will then intersect Q_0Q_i, and so ℓ'' will not be parallel to ℓ. That is, a line through P other than ℓ' will intersect ℓ. This means ℓ' is the unique parallel to ℓ through P. □

36.2 The search for a rectangle

Euclidean geometry is neutral geometry with one more axiom, Playfair's axiom, added. Before the discovery of non-Euclidean geometry, geometers were trying to prove that neutral geometry *is* Euclidean geometry by trying to prove Playfair's axiom from the axioms of neutral geometry. One way to do that would be to prove that there are rectangles in neutral geometry. Hence, much effort was spent in attempts to build a rectangle using only the axioms of neutral geometry. There isn't a construction that is going to just give you all four right angles right away, so the idea was to construct a quadrilateral that has some of the properties of a rectangle, and then to work from there. One such quasi-rectangle is called a Saccheri quadrilateral.

Definition 36.7 A convex quadrilateral $\square ABCD$ is called a *Saccheri quadrilateral* with *base* AB, *summit* CD, and *legs* BC and AD if:

- $\angle A$ and $\angle B$ are right angles, and
- segments BC and AD are congruent.

The angles $\angle C$ and $\angle D$ are called the *summit angles* of the Saccheri quadrilateral.

Saccheri quadrilaterals can be constructed in neutral geometry: start with a segment for the base, draw two rays from the endpoints of the base to form the two right angles, and then mark off congruent segments along them. The segments are the legs. Join their endpoints to form the summit and you have a Saccheri quadrilateral. A Saccheri quadrilateral seems to be about half the way to a rectangle because two of its angles are right, and one pair of opposite sides are congruent. The only Saccheri quadrilaterals in Euclidean geometry are rectangles because by the converse of the alternate interior angle theorem, the two summit angles must be right angles. With only the axioms of neutral geometry, we can't prove that the summit angles are right angles, but we can prove weaker statements. The first statement exploits the internal symmetry of a Saccheri quadrilateral (much like the proof of the isosceles triangle theorem).

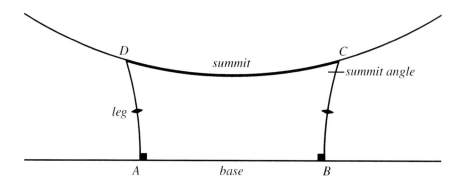

Figure 8. The parts of a Saccheri quadrilateral.

Theorem 36.8 *Let $\square ABCD$ be a Saccheri quadrilateral with base AB. Then the summit angles $\angle C$ and $\angle D$ are congruent.*

Proof Observe that

$$DA \simeq CB, \quad \angle A \simeq \angle B, \quad AB = BA, \quad \angle B \simeq \angle A, \quad \text{and} \quad BC \simeq AD.$$

By SASAS, $\square ABCD$ and $\square BADC$ are congruent. Matching the corresponding parts, $\angle C \simeq \angle D$.

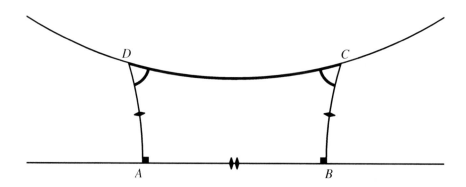

Figure 9. By SASAS, the summit angles of a Saccheri quadrilateral are congruent. □

36.2 The search for a rectangle

Theorem 36.9 *Let □ABCD be a Saccheri quadrilateral with base AB. Then the summit angles ∠C and ∠D cannot be obtuse.*

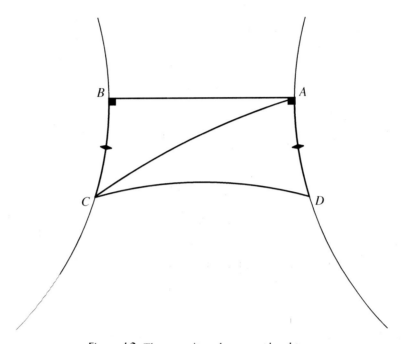

Figure 10. The summit angles cannot be obtuse.

Proof The diagonal AC divides the Saccheri quadrilateral into two triangles, and by the Saccheri-Legendre theorem the angle sum of each is at most π. Therefore the angle sum of □$ABCD$ is at most 2π. So

$$(\angle A) + (\angle B) + (\angle C) + (\angle D) \leq 2\pi$$
$$(\pi/2) + (\pi/2) + (\angle C) + (\angle D) \leq 2\pi$$
$$(\angle C) + (\angle D) \leq \pi.$$

We just proved that $\angle C$ and $\angle D$ are congruent, though, so they must measure at most $\pi/2$. □

If the summit angles are right, then the Saccheri quadrilateral is a rectangle. So in non-Euclidean geometry, the summit angles must be acute.

Definition 36.10 The *altitude* of a Saccheri quadrilateral is the line segment from the midpoint of its base to the midpoint of its summit.

Theorem 36.11 *Let □ABCD be a Saccheri quadrilateral with base AB and altitude EF (with E on the base and F on the summit). Then EF is perpendicular to the base AB and the summit CD.*

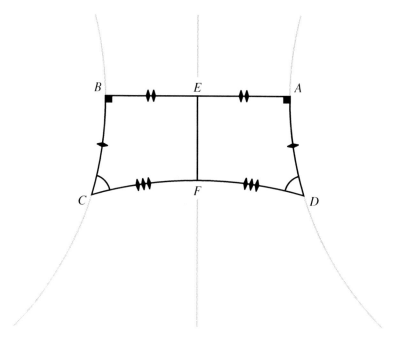

Figure 11. The segment EF is the altitude of $\square ABCD$. It is perpendicular to AB and CD.

Proof Compare the two quadrilaterals $\square AEFD$ and $\square BEFC$. Since $\square ABCD$ is Saccheri, $AD \simeq BC$ and $\angle A \simeq \angle B$, and since the summit angles of a Saccheri quadrilateral are congruent, $\angle D \simeq \angle C$. By the definition of an altitude, $AE \simeq BE$ and $DF \simeq CF$. By SASAS, then, $\square AEFD \simeq \square BEFC$. In particular, $\angle AEF$ is congruent to its own supplement, $\angle BEF$, and $\angle DFE$ is congruent to its own supplement, $\angle CFE$. They are both right angles, then, and so the altitude is perpendicular to both base and summit. □

Theorem 36.12 *Let $\square ABCD$ be a Saccheri quadrilateral with base AB and summit CD. Then $|CD| \geq |AB|$.*

Proof Assume instead that AB is greater than CD. Label E and F the midpoints of AB and CD, so that EF is the altitude. Then $|EB| > |FC|$ so we can locate a point G between E and B so that $EG \simeq FC$. Since $\square EFCG$ has right angles at E and F and a pair of congruent opposite sides, it is a Saccheri quadrilateral with base EF. As we saw above, the summit angle $\angle EGC$ cannot be obtuse, so its supplement cannot be acute. But $\triangle BGC$ has a right angle at B and an angle at G that isn't acute, so the angle sum of $\triangle BGC$ will exceed π, and this violates the Saccheri-Legendre theorem. □

Theorem 36.13 *Let $\square ABCD$ be a Saccheri quadrilateral with base AB. Let EF be the altitude, with E on AB and F on CD. Then the altitude EF cannot be longer than the legs AD and BC.*

36.2 The search for a rectangle

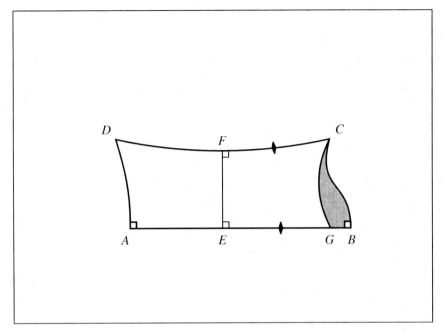

Figure 12. *The base cannot be longer than the summit.*

Proof Let's build a different Saccheri quadrilateral (see figure 13). Label:

F' : on $\overrightarrow{FE}$ so that $F * E * F'$ and $EF' \simeq EF$, and
C' : on $\overrightarrow{CB}$ so that $C * B * C'$ and $BC' \simeq BC$.

Now we have the congruences:

$$CB \simeq C'B, \quad BE = BE, \quad EF \simeq EF', \quad \angle CBE \simeq \angle C'BE, \text{ and } \angle BEF \simeq \angle BEF'.$$

By SASAS, $\square BEFC$ and $\square BEF'C'$ are congruent. Matching their corresponding pieces, $\angle F'$ is congruent to $\angle F$ and so is a right angle, and $F'C' \simeq FC$. Thus $\square CFF'C'$ is a Saccheri quadrilateral with base FF' and summit CC'. By the previous result, its base cannot be longer than the summit, and so

$$|EF| = \tfrac{1}{2}|FF'| \leq \tfrac{1}{2}|CC'| = |CB|.$$

We see, then, that EF, the altitude of $\square ABCD$ cannot be longer than the leg BC. □

Thinking more generally, what can we say about other perpendiculars that fall between the altitude and a leg?

Theorem 36.14 *Let $\square ABCD$ be a Saccheri quadrilateral with base AB and altitude EF. Let GH be another segment that connects the base to the summit, so that $E * G * B$, $F * H * C$, and GH is perpendicular to the base. Then*

$$|EF| \leq |GH| \leq |BC|.$$

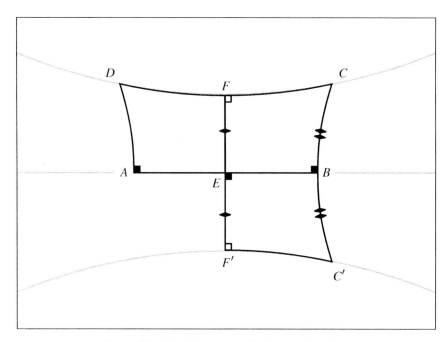

Figure 13. The altitude cannot be longer than a leg.

Proof We will prove the first inequality, and leave the second as an exercise. Suppose that $|EF| > |GH|$. Then there is a point K between E and F so that $EK \simeq GH$. Then $\square EGHK$ is a Saccheri quadrilateral with base EG. Now let's look at $\triangle KFH$. In it, $\angle F$ is right because the altitude is perpendicular to the summit; $\angle K$ is right or obtuse, since it is supplementary to a

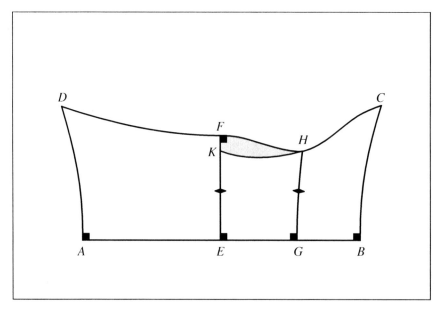

Figure 14. Given a segment GH that is between the leg BC and the altitude EF, $|GH| \geq |EF|$ (a proof by contradiction).

summit angle (remember that the summit angles cannot be obtuse). Then the sum of the angles in $\triangle KFH$ is greater than π, which contradicts the Saccheri-Legendre theorem. Therefore $|GH|$ cannot be less than $|EF|$. □

Exercise 36.2 Complete the proof of the theorem by showing that $|GH| \leq |BC|$.

If the first half of this chapter was a little jarring—it is a little crazy to think about a geometry without rectangles—I hope the second half gave some comfort. On the one hand, yes, things really are different in non-Euclidean geometry; but on the other, there is still order and structure. The next chapter is a more direct approach to understanding the order and structure: we will look at the behavior of parallel lines.

36.3 References

The general theme of this lesson, to start from a rectangle and to use it to establish Playfair's axiom, is taken from Wallace and West's *Roads to Geometry* [WW04].

36.4 Exercises

36.3. Prove that the diagonals of a Saccheri quadrilateral are congruent.

36.4. Prove that the diagonals of a Saccheri quadrilateral intersect one another at a point on its altitude.

36.5. Demonstrate that in neutral geometry it is possible to construct a convex quadrilateral whose four sides are all congruent.

36.6. Prove that if there is a triangle with an angle sum of π, then there must be a rectangle.

36.7. Let $\square ABCD$ be a convex quadrilateral so that $\angle A$ and $\angle B$ are right angles, and so that $\angle C \simeq \angle D$. Prove that $\square ABCD$ is a Saccheri quadrilateral.

36.8. Let $\square ABCD$ be a Saccheri quadrilateral with base AB and altitude EF (so that E is on the base and F is on the altitude). Prove that $|EC| \geq |FB|$.

36.9. Given $\triangle ABC$, let ℓ be the line that passes through the midpoints of AB and AC. Let F_B be the foot of the perpendicular to ℓ that passes through B, and let F_C be the foot of the perpendicular to ℓ that passes through C. Prove that $\square BCF_CF_B$ is a Saccheri quadrilateral with base F_BF_C.

36.10. Show that in non-Euclidean geometry, not every triangle has a circumcenter. The strategy is as follows. The circumcenter is the intersection of the perpendicular bisectors. Let P be a point not on the line ℓ. Let Q be the foot of the perpendicular to ℓ that passes through P. Let ℓ' be a line through Q that is parallel to ℓ but not perpendicular to PQ. Construct a triangle that has ℓ and ℓ' as perpendicular bisectors. Since the perpendicular bisectors don't even intersect, there cannot be a circumcenter.

36.11. Prove that in non-Euclidean geometry, a triangle has an incenter. (We shall see later, though, that some circles cannot be contained in any triangle. This is most easily seen in terms of area.)

37
Non-Euclidean Parallels

From here on out, let's assume that Playfair's axiom is false. That means that our geometry is not Euclidean. It means our triangle angle sums are strictly less than π. It means that Saccheri quadrilaterals are not rectangles, because rectangles do not exist, and so the summit angles are acute (not right). And other absurdities.

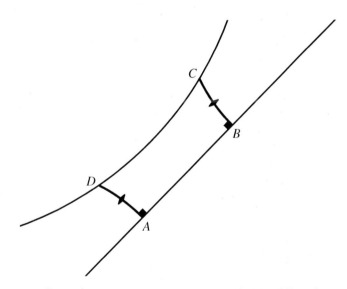

Figure 1. The base and summit of a Saccheri quadrilateral are on parallel lines.

It may have been implied, but in the last chapter we never proved that the base and summit of a Saccheri quadrilateral are parallel.

Exercise 37.1 Let $\square ABCD$ be a Saccheri quadrilateral with base AB and summit CD. Prove that $\overline{AB}$ and $\overline{CD}$ are parallel.

In this section, we will explore parallels more systematically. By the end of it, we should have a better sense of what to look for in a non-Euclidean geometry. To start, let's see that non-Euclidean behavior is pervasive, and not just an anomaly that occurs at a few isolated points.

Theorem 37.1 *Let ℓ be a line and let P be a point that is not on ℓ. Then there are infinitely many lines through P that are parallel to ℓ.*

Proof There are two pieces to this proof. First that there are always at least two parallels. Then that there are infinitely many.

Part 1, that there are (at least) two parallels. Drop the perpendicular from P to ℓ, and mark its intersection with ℓ as Q. Let ℓ' be the line through P that is perpendicular to PQ. According to the alternate interior angle theorem, ℓ' is parallel to ℓ. Now suppose that it is the only parallel. To get a contradiction, we are going to build a Saccheri quadrilateral and see that it would have to be a rectangle. Pick a point on ℓ other than Q and label it R. On the line through R that is perpendicular to ℓ, mark the point S so that $RS \simeq PQ$ (choose S on the same side of ℓ as P). Then $\square PQRS$ is a Saccheri quadrilateral with base QR and summit PS. According to exercise 37.1, $\overline{PS}$ and $\overline{QR} = \ell$ are parallel. But there is only one line through P that is parallel to ℓ, namely ℓ'. Therefore $\overline{PS} = \ell'$. In $\square PQRS$, the summit angle at P is a right angle; and the other summit angle must be congruent to it. Therefore all four angles of $\square PQRS$ are right angles and it is a rectangle. There are no rectangles in non-Euclidean geometry, so there must be more than one parallel to ℓ that passes through P.

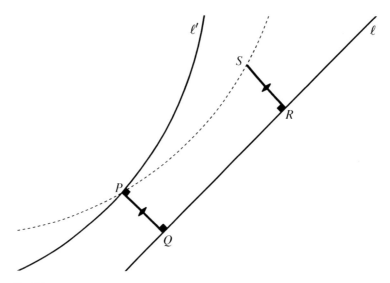

Figure 2. If there are no rectangles, then there must be at least two lines through P that are parallel to $\overline{QR}$.

Part 2, that there are infinitely many parallels. Let ℓ_1 and ℓ_2 be two lines through P that are parallel to ℓ. We can box-in P by surrounding it with more labeled points:

A, A' : on ℓ_1 so that $A * P * A'$,
B : on ℓ_2 and on the same side of PQ as A,
B' : on ℓ_2 and on the same side of PQ as A' (so $B * P * B'$).

Let C be a point on the segment AB. We will argue that $\overline{PC}$ is also parallel to ℓ. Suppose $\overline{PC}$ intersects ℓ at a point D. Then one of $\overrightarrow{PA}$, $\overrightarrow{PA'}$, $\overrightarrow{PB}$, or $\overrightarrow{PB'}$ would have to lie in the interior of

$\angle QPD$. According to the crossbar theorem, the ray would have to intersect QD. This cannot happen since ℓ_1 and ℓ_2 (which contain all four rays) are parallel to ℓ.

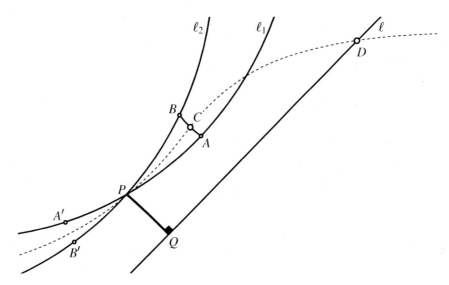

Figure 3. Given two parallels through P, we can find infinitely more between them. □

In Euclidean geometry, parallel lines are everywhere equidistant, so we can measure the distance between them with a single number. Non-Euclidean geometry is not so simple. Given parallel lines ℓ_1 and ℓ_2 in non-Euclidean geometry, the distance from a point P on ℓ_2 to the line ℓ_1 is measured as usual: drop a perpendicular from P to ℓ_1 and mark its foot (the intersection with ℓ_1) as F_P. Then the distance from P to ℓ_1 is $|PF_P|$.

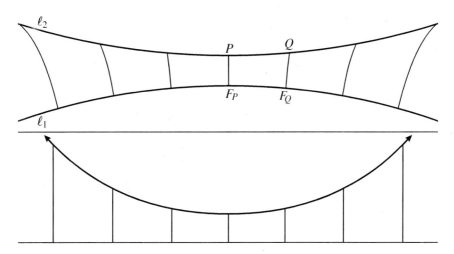

Figure 4. We can view the distance between parallel lines ℓ_1 and ℓ_2 as a function of the points on ℓ_2.

We can define a function $f : \ell_2 \to \mathbb{R}$ by setting $f(P)$ to be the distance from P to ℓ_1. The rest of this chapter investigates it. Before we go any further, we need the following lemma (note its similarity to the theorem in the last chapter that compared the length of the summit to the length of the base).

Lemma 37.2 *Let ℓ_1 and ℓ_2 be parallel lines, let P and Q be points on ℓ_2, and let F_P and F_Q be the respective feet of the perpendiculars on ℓ_1. Then $|F_P F_Q| < |PQ|$.*

Exercise 37.2 Prove the lemma.

Theorem 37.3 *The function f measuring the distance from points of ℓ_2 to the line ℓ_1 is continuous.*

Proof It is always possible to drop a perpendicular from a point P on ℓ_2 to the line ℓ_1, and it is always possible to measure the distance from P to its foot F_P, so the function f is always defined. For f to be continuous at the point P, we need a bit more: as a point Q on ℓ_2 approaches P, the value of $f(Q)$ must approach $f(P)$. To be more precise,

For every $\epsilon > 0$, there must be a $\delta > 0$ so that if $|PQ| < \delta$ then $|f(P) - f(Q)| < \epsilon$.

The verification of this hinges on a few simple geometric observations that follow from the triangle inequality.

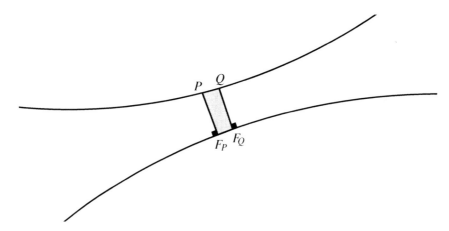

Figure 5. To show that f is continuous, consider two points P and Q that are close to one another.

- $|PF_P| \leq |PQ| + |QF_Q| + |F_P F_Q|$.
- $|QF_Q| \leq |PQ| + |PF_P| + |F_P F_Q| \implies |PF_P| \geq |QF_Q| - |PQ| - |F_P F_Q|$.
- Therefore

$$|QF_Q| - |PQ| - |F_P F_Q| \leq |PF_P| \leq |PQ| + |QF_Q| + |F_P F_Q|.$$

- Subtract $|QF_Q|$:

$$-|PQ| - |F_P F_Q| \leq |PF_P| - |QF_Q| \leq |PQ| + |F_P F_Q|.$$

- We don't know whether $|PF_P| > |QF_Q|$ or if it is the other way around. Either way, we can say that

$$\bigl||PF_P| - |QF_Q|\bigr| \leq |PQ| + |F_P F_Q|.$$

With that information, verifying continuity is straightforward. Fix $\epsilon > 0$ and put $\delta = \epsilon/2$. If $|PQ| < \delta = \epsilon/2$, then $|F_P F_Q| < \epsilon/2$ as well (according to lemma 37.2), and so

$$|f(P) - f(Q)| = \bigl||PF_P| - |QF_Q|\bigr| \leq |PQ| + |F_P F_Q| < \epsilon/2 + \epsilon/2 = \epsilon.$$

Therefore f is continuous at P. Since P was an arbitrary point on ℓ_2, f is continuous on all of ℓ_2. □

Theorem 37.4 *For a real number r, $|\{f^{-1}(r)\}| \leq 2$. That is, there are at most two points on ℓ_2 that are at a distance r from ℓ_1.*

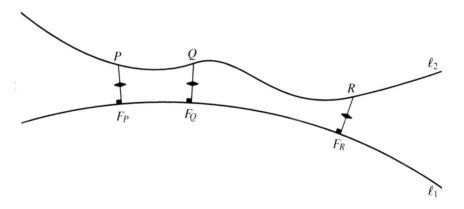

Figure 6. It is not possible for all of P, Q, and R to be the same distance from ℓ_1.

Proof Suppose that there is a real number r so that $|\{f^{-1}(r)\}| > 2$. Then there are (at least) three distinct points P, Q, and R on ℓ_2 with $P * Q * R$ so that $f(P) = f(Q) = f(R) = r$. If we label the feet of their respective perpendiculars as F_P, F_Q, and F_R, then

$$|PF_P| = |QF_Q| = |RF_R| = r.$$

Thus $\square F_P F_Q Q P$ and $\square F_Q F_R R Q$ are Saccheri quadrilaterals. They cannot be rectangles, though, so their summit angles $\angle PQF_Q$ and $\angle RQF_Q$ must be acute. But the angles are supplementary so they cannot both be acute. □

Theorem 37.5 *The function f has no local maxima and at most one local minimum.*

Proof Suppose that f has a local maximum at a point B. Then, since f is continuous, for a sufficiently small value positive value ϵ, there are points A and C on either side of B so that

$$f(A) = f(C) = f(B) - \epsilon.$$

Let F_A, F_B, and F_C be the feet of the perpendiculars to ℓ_1 from A, B, and C respectively. Then

$$|AF_A| = f(A) = f(C) = |CF_C|$$

so $\square F_A F_C CA$ is a Saccheri quadrilateral with base $F_A F_C$. The segment BF_B connects the summit to the base and is perpendicular to the base. We saw at the end of the last chapter that such segments cannot be longer than the legs. But the legs of this Saccheri quadrilateral are AF_A and CF_C, and they are shorter (by ϵ) than BF_B. This is a contradiction, and so f cannot have a local maximum.

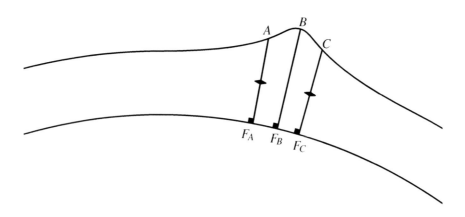

Figure 7. It is not possible for f to have a local maximum (as is suggested here at the point B).

The second part of the theorem follows immediately: if f had more than one minimum, then, because it is continuous, it would have a local maximum in between them. □

Theorem 37.6 *Let P and Q be points on ℓ_2 and suppose that $f(Q) > f(P)$. Then f approaches infinity along $\overrightarrow{PQ}$.*

It should be clear that the value of f continues to increase past Q. If at some point it began to decrease, then we would have found a local maximum, and we have just seen that f has none. But could it be that f continues to grow, but is bounded above? Is there an upper bound that f approaches in the limit but never surpasses? This is a more delicate question. The strategy of the proof is to build a wall of congruent Saccheri quadrilaterals that appears to contain $\overrightarrow{PQ}$, and then see why that cannot happen.

Proof Suppose that f does have an upper bound so that it approaches but never exceeds a number k. Starting from the point F_Q, the foot of the perpendicular from Q, mark off congruent segments along ℓ_1 on the same side of PF_P as $\overrightarrow{PQ}$. That is, label points F_i so that

$$F_Q = F_0 * F_1 * F_2 * F_3 * \cdots$$

and so that the segments $F_i F_{i+1}$ are congruent. Above each point F_i (on the same side of ℓ_1 as ℓ_2), erect a perpendicular segment $F_i P_i$ of length k. Then each $\Box F_i F_{i+1} P_{i+1} P_i$ is a Saccheri quadrilateral, and by SASAS they are all congruent.

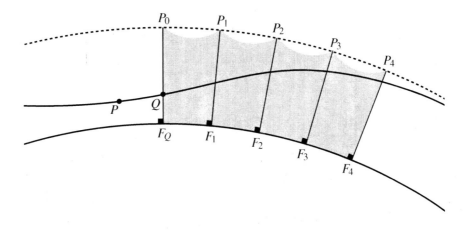

Figure 8. It is not possible for f to level off at some asymptotic value, as suggested here.

Their altitudes are shorter than their legs, by an amount that we can call ϵ (since the quadrilaterals are congruent, the discrepancy will be the same for all of them). Now let's get back to $\overrightarrow{PQ}$. For a while, it may run below the summits of the quadrilaterals we have assembled. But f has a limiting value of k along it, so eventually there is a point beyond which f is greater than $k - \epsilon$. Look beyond that point, at the interaction of $\overrightarrow{PQ}$ with one of the summits $P_n P_{n+1}$. Points P_n and P_{n+1} are on the opposite side of ℓ_2 from ℓ_1. But the altitude between them is on the same side of ℓ_2 as ℓ_1. This would imply that ℓ_2 intersects $P_n P_{n+1}$ twice. Two distinct lines can intersect only once, so this is a contradiction. $\square$

Looking in the other direction, if $f(Q) \geq f(P)$, then $f(P) < f(Q)$, and it is natural to wonder what happens as we progress along $\overrightarrow{QP}$. Does the distance to ℓ_2 continue to decrease, or does it at some point begin to increase again?

Definition 37.7 If f achieves a minimum, then ℓ_1 and ℓ_2 are called *divergent* parallels. If instead f approaches zero, then ℓ_1 and ℓ_2 are called *asymptotic* parallels.

If ℓ_1 and ℓ_2 are divergent parallels, then once $\overrightarrow{QP}$ has crossed over the minimum value, the value of f will begin to go up. Once that happens, by the previous result, f will go to infinity. Both types of parallels exist in non-Euclidean geometry. The base and summit of a Saccheri quadrilateral are divergent parallels. Asymptotic parallels are more elusive, but one of the exercises outlines a proof of their existence. Beyond that, there is one more obvious question: are there any parallels that are neither divergent nor asymptotic? Suppose ℓ_1 and ℓ_2 are not divergent. Then f does not achieve its minimum. It is easy then to pick points P and Q

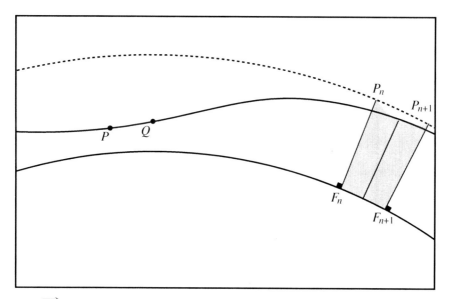

Figure 9. If $\overrightarrow{PQ}$ approached an asymptote, it would eventually intersect the summit of a constructed Saccheri quadrilateral more than once.

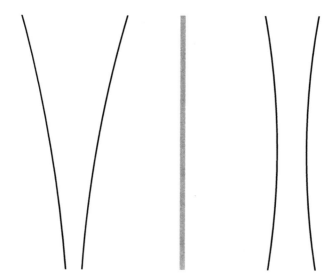

Figure 10. (l) asymptotic parallels; (r) divergent parallels.

so that $f(P) < f(Q)$, so f must continue to decrease past P as we move out along $\overrightarrow{QP}$. By definition, the lines are asymptotic if the value of f decreases to zero. Is it possible that the value of f is always decreasing, but never drops below some positive number k? No. The argument against a nonzero lower bound is essentially the same as the argument against an upper bound in the proof above.

Theorem 37.8 *If ℓ_1 and ℓ_2 are parallel, then they are either divergent or asymptotic.*

Exercise 37.3 Prove theorem 37.8.

The results of this chapter should give you some sense of the look of non-Euclidean geometry. But can such a geometry exist? If we want to answer YES, we need to find a model that meets all the requirements of neutral geometry, and has the non-Euclidean characteristics. In the next chapters, we will look at such models.

37.1 Exercises

37.4. Let r_1 and r_2 be two non-opposite rays with endpoint A. Label point B and C on r_1 and B' and C' on r_2 so that $A * B * C$ and $A * B' * C'$. Prove that the angle sum of $\triangle ACC'$ is less than the angle sum of $\triangle ABB'$.

37.5. Suppose that ℓ_1 and ℓ_2 are divergent parallels. Let P be the point on ℓ_2 that is the minimum distance from ℓ_1, and let Q be the foot of the perpendicular from P. Prove that QP intersects ℓ_2 at a right angle.

37.6. Let ℓ be a line and let P be a point that is not on ℓ. Prove that there is a line through P which is asymptotic to ℓ. Hint: let r_1 and r_2 be rays from P, one that intersects ℓ and one that is along a divergent parallel to ℓ. Let A be a point on r_1 and B be a point on r_2. A point Q on AB falls into one of two sets: either PQ is parallel to ℓ or it isn't. By Dedekind's axiom, there is a point Q_0 that separates those two sets. Show that PQ_0 is an asymptotic parallel.

37.7. Suppose that lines ℓ_1 and ℓ_2 intersect at a point O. Suppose that P and Q are on ℓ_2 so that $O * P * Q$, and that F_P and F_Q are the feet of the perpendiculars, from P and Q respectively, to ℓ_1. Prove that $|PF_P| < |QF_Q|$.

37.8. Let r_1 and r_2 be two rays with a common endpoint O. Prove that for a positive real number x, there is a point P on r_1 so that $d(P, r_2) > x$.

38

The Pseudosphere

In time, some geometers began to doubt that Playfair's axiom could be proved using only the axioms of neutral geometry. So they began to search for a model of a neutral geometry that would exhibit non-Euclidean behavior. In such a geometry parallel lines would have to bend away from one another. Viewed from a Euclidean perspective, the non-Euclidean plane would look warped. One idea was to use a curved surface as the model. In it, points would be modeled by the points on the surface, while lines will be modeled by geodesics on the surface. The distance between two points would be measured as the arc length of the geodesic connecting them, and the angle between two geodesics would be the angle between their tangent vectors at the point of intersection. The undefined terms *on*, *between*, and *congruent* would then be defined in the obvious way.

A geodesic is (roughly speaking) the shortest path between two points. For example, the geodesics on a plane are lines. Geodesics on a sphere are *great circles*, circles on the sphere whose centers coincide with the center of the sphere. Neither the plane nor the sphere works as a model for non-Euclidean geometry. The plane is flat and the sphere, while warped, seems to be warped in the wrong way because the great circles bend toward one another rather than away.

Figure 1. Some geodesics (great circles) on a sphere.

There are, however, some surfaces that exhibit the right kind of warping, those with constant negative Gaussian curvature. One is called the pseudosphere. The purpose of this chapter is to find equations for it. We will have to use differential geometry in order to do so, so this chapter

421

is advanced. My primary reference for this is Spivak's five volume Differential Geometry series, the calculations are in volume 3 [Spi99b].

We will get into the details as this chapter unfolds, but there are two parts to a surface of constant negative curvature: the constant part and the negative part. The curvature has to be constant so that the model will be "isotropic", that is it looks the same at every point. Without this requirement, the surface won't satisfy all of the neutral axioms. The curvature has to be negative so that parallels will bend away from one another as we would expect in a non-Euclidean geometry.

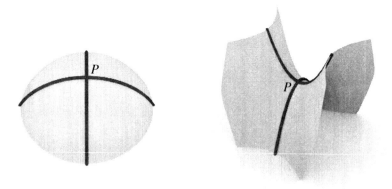

Figure 2. The curves are geodesics on the surfaces. On the left, P is a point of positive curvature. On the right, P is a point of negative curvature. For negatively curved surfaces, the geodesics bend away from each other.

38.1 Surfaces

There are several ways to describe a surface in $\mathbb{R}^3$. Perhaps the most elementary is when one variable can be written as a function of the other two such as $z = f(x, y)$. Because that form can be limiting, a more flexible approach is to parametrize the surface, writing coordinates as a function of two variables:

$$X(s, t) = (f(s, t), g(s, t), h(s, t)).$$

Given a surface S parametrized by $X(s, t)$, assuming that the surface is relatively smooth and that the parametrization function is relatively well-behaved, we can compute partial derivatives

$$X_s = \left.\frac{\partial X}{\partial s}\right|_{(s_0, t_0)} = \langle f_s(s_0, t_0), g_s(s_0, t_0), h_s(s_0, t_0) \rangle,$$

$$X_t = \left.\frac{\partial X}{\partial t}\right|_{(s_0, t_0)} = \langle f_t(s_0, t_0), g_t(s_0, t_0), h_t(s_0, t_0) \rangle$$

at a point $P = X(s_0, t_0)$ on S. The results are vectors that point in the direction of an infinitesimal change in s and an infinitesimal change in t. They form a basis for T_P, the *tangent plane* to S at P.

38.1 Surfaces

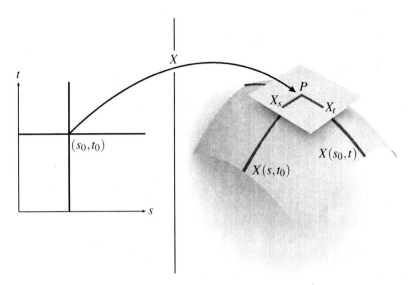

Figure 3. The parametrization X maps the st-domain space to the surface. The partial derivatives $X_s(s_0, t_0)$ and $X_t(s_0, t_0)$ form a basis for the tangent plane to the surface at P.

In the domain space, the line through (s_0, t_0) in the direction of the vector $\langle a, b \rangle$ can be parametrized as

$$\rho(w) = (s_0 + aw, t_0 + bw).$$

Then $X(\rho(w))$ is a curve on S,

$$X(\rho(w)) = \Big(f(s_0 + aw, t_0 + bw),\ g(s_0 + aw, t_0 + bw),\ h(s_0 + aw, t_0 + bw)\Big),$$

and the tangent vector at P can be computed by the chain rule:

$$\begin{aligned}
[X(\rho(w))]' &= \langle f_s \cdot a + f_t \cdot b,\ g_s \cdot a + g_t \cdot b,\ h_s \cdot a + h_t \cdot b \rangle \\
&= a \langle f_s, g_s, h_s \rangle + b \langle f_t, g_t, h_t \rangle \\
&= a X_s + b X_t.
\end{aligned}$$

Therefore, the vectors of the tangent plane T_P are all tangent vectors to parametrized curves on S that pass through P. So T_P is the best linear approximation to S in an infinitesimally small neighborhood around P.

Let's turn now to specifics. There are several surfaces that meet the constant negative curvature requirement. The one we are after, the pseudosphere, is relatively easy to work with because it is a surface of revolution. Start with a curve $\mathcal{C}$ in the yz-plane, parametrized as

$$r(t) = \big(0,\ y(t),\ z(t)\big).$$

It will simplify calculations if we place another restriction on the parametrization, and require that $\mathcal{C}$ be parametrized by arc length. That is, choose a parametrization that moves along $\mathcal{C}$ at a constant speed of 1, so that

$$\sqrt{[y'(t)]^2 + [z'(t)]^2} = 1.$$

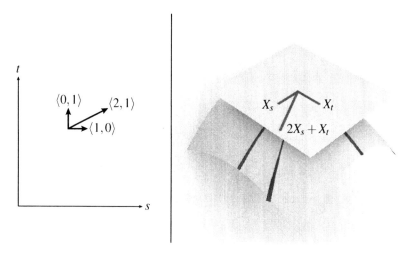

Figure 4. Any tangent vector in the tangent plane can be written as a linear combination of X_s and X_t.

Revolving $\mathcal{C}$ around the y-axis will generate a surface of revolution that can be parametrized as

$$X(s, t) = \langle z(t)\sin s, \; y(t), \; z(t)\cos s \rangle, \quad 0 \le s \le 2\pi.$$

At a point on this surface, the partial derivatives are

$$X_s = \langle z(t)\cos s, \; 0, \; -z(t)\sin s \rangle$$
$$X_t = \langle z'(t)\sin s, \; y'(t), \; z'(t)\cos s \rangle.$$

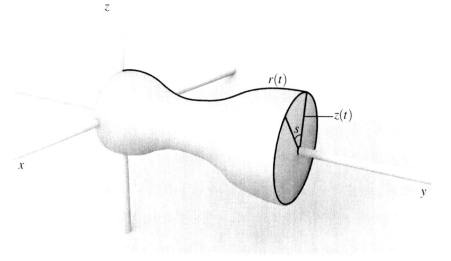

Figure 5. Form a surface by revolving the curve $r(t) = \langle 0, y(t), z(t) \rangle$ around the y-axis.

38.2 Maps between surfaces. The Gauss map

Because surfaces are now the objects of study, we consider maps between them. Let $f : S_1 \to S_2$ be a map from the points of a surface, S_1, to the points of another surface, S_2. Let P be a point on the first surface. The tangent plane T_P is the best linear approximation to S_1 around P. Likewise, on S_2, the tangent plane $T_{f(P)}$ is the best linear approximation to S_2 around $f(P)$. We want to define a linear map between the tangent planes $df : T_P \to T_{f(P)}$ that best approximates f near the points.

Let $X : D \to S_1$ be a parametrization of S_1 with $X(s_0, t_0) = P$. Every vector in T_P can be written as a linear combination $aX_s + bX_t$ and is therefore the tangent vector to the curve $X(\rho(w))$ where $\rho(w) = (s_0 + aw, t_0 + bw)$. The composition $f \circ X : D \to S_2$ is not likely to be a parametrization of S_2 since it will not typically be a bijection. Nevertheless, we can calculate the partial derivatives $(f \circ X)_s$ and $(f \circ X)_t$. A parametrized line $\rho(w)$ in D is mapped to a curve $f \circ X(\rho(w))$ on S_2, whose tangent vector at $f(P)$ is (by the chain rule)

$$[f \circ X(\rho(w))]' = (f \circ X)_s \cdot \frac{ds}{dw} + (f \circ X)_t \cdot \frac{dt}{dw}$$
$$= (f \circ X)_s \cdot a + (f \circ X)_t \cdot b.$$

We define the differential of f at P, $df : T_P \to T_{f(P)}$, by

$$df(aX_s + bX_t) = a(f \circ X)_s + b(f \circ X)_t.$$

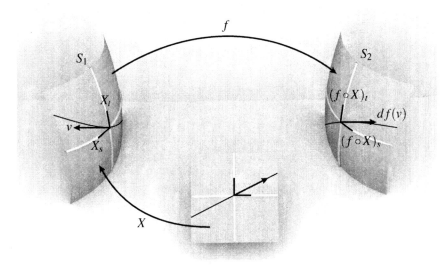

Figure 6. A map $f : S_1 \to S_2$ between two surfaces S_1 and S_2. The differential df is the map between tangent planes.

Gauss's idea was to use a particular map between surfaces, now called the Gauss map, to study curvature. The basic idea is simple and elegant. Imagine moving in a small neighborhood around a point P on a surface S, with an eye on the tangent plane as you do so. The tangent plane balances on the surface like a surfboard, and so as the surface bends, the tangent plane must turn. The more the tangent plane has to turn to keep up with the surface, the more highly curved the surface is. We can measure how fast the tangent plane is turning by looking at the rate of change of its unit normal vector (it is tempting to try to extend the surfboard analogy, and to call the unit normal vector the surfer on the surfboard, but that assumes the surfer is always perpendicular to the board).

Definition 38.1 Let $\mathbb{S}$ denote the sphere of radius one centered at the origin. For a point P on a smooth surface S, let $n(P)$ be the unit normal vector to S at P. If the tail of $n(P)$ is translated to the origin, then its head will be at a point $v(P)$ on $\mathbb{S}$. The *Gauss map* $v : S \to \mathbb{S}$ is defined by $P \mapsto v(P)$.

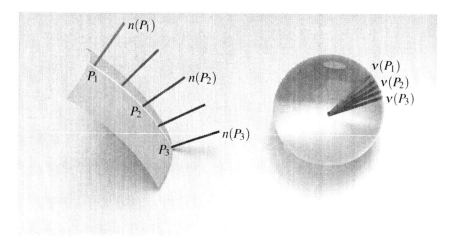

Figure 7. The Gauss map $v : S \to \mathbb{S}$.

The Gauss map may be complicated. However, the curvature at a point P only depends on what is happening near P, and there v can be replaced by its linear approximation dv. The differential of the Gauss map has the advantage over the differentials of other maps that the tangent planes T_P and $T_{v(P)}$ are parallel. Because they have the same vectors in them, $\{X_s, X_t\}$ is a basis of T_P and of $T_{v(P)}$. In it, dv can be written as a 2×2 matrix.

Let's now compute the Gauss map and its differential for our surface of revolution. To do that, we need to know the unit normal vector. The cross product of X_s and X_t is perpendicular to both of them, so it is a normal vector to the tangent plane:

$$N = X_s \times X_t$$

$$= \begin{vmatrix} i & j & k \\ z(t)\cos s & 0 & -z(t)\sin s \\ z'(t)\sin s & y'(t) & z'(t)\cos s \end{vmatrix}$$

$$= \left\langle y'(t)z(t)\sin s, \; -z(t)z'(t)\cos^2 s - z(t)z'(t)\sin^2 s, \; y'(t)z(t)\cos s \right\rangle$$

$$= \left\langle y'(t)z(t)\sin s, \; -z(t)z'(t), \; y'(t)z(t)\cos s \right\rangle.$$

38.2 Maps between surfaces. The Gauss map

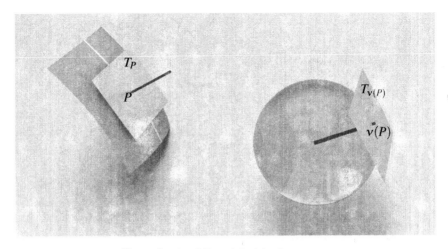

Figure 8. The differential of the Gauss map.

It may not be a unit vector, so we need to divide by its length

$$|N| = \sqrt{(y'(t)z(t)\sin s)^2 + (-z(t)z'(t))^2 + (y'(t)z(t)\cos s)^2}$$
$$= \sqrt{(y'(t)z(t))^2 + (-z(t)z'(t))^2}$$
$$= z(t)\sqrt{(y'(t))^2 + (z'(t))^2}$$
$$= z(t).$$

The last step takes advantage of the fact that $\mathcal{C}$ is parametrized by arc length. The unit normal is then

$$n = \frac{N}{|N|} = \langle y'(t)\sin s, \; -z'(t), \; y'(t)\cos s \rangle.$$

This calculation gives us the Gauss map as well. If $P = X(s, t)$, then

$$\nu(P) = \langle y'(t)\sin s, \; -z'(t), \; y'(t)\cos s \rangle.$$

Now let's find the $d\nu$ matrix. Compute the partial derivatives,

$$(\nu \circ X)_s = \langle y'(t)\cos s, \; 0, \; -y'(t)\sin s \rangle$$
$$(\nu \circ X)_t = \langle y''(t)\sin s, \; -z''(t), \; y''(t)\cos s \rangle.$$

We rewrite them as linear combinations of the basis elements X_s and X_t, where

$$X_s = \langle z(t)\cos s, \; 0, \; -z(t)\sin s \rangle$$
$$X_t = \langle z'(t)\sin s, \; y'(t), \; z'(t)\cos s \rangle.$$

The first is easy:

$$(v \circ X)_s = \langle y'(t)\cos s, \; 0, \; -y'(t)\sin s\rangle$$
$$= \frac{y'(t)}{z(t)} \cdot \langle z(t)\cos s, 0, -z(t)\sin s\rangle$$
$$= \frac{y'(t)}{z(t)} \cdot X_s.$$

The second requires a trick. Because $\mathcal{C}$ is parametrized by arc length,

$$[y'(t)]^2 + [z'(t)]^2 = 1.$$

If we differentiate with respect to t and solve for y'', we get

$$2y'(t)y''(t) + 2z'(t)z''(t) = 0 \implies y''(t) = -\frac{z'(t)z''(t)}{y'(t)}.$$

Substitute to get

$$(v \circ X)_t = \langle y''(t)\sin s, \; -z''(t), \; y''(t)\cos s\rangle$$
$$= \left\langle -\frac{z'(t)z''(t)}{y'(t)} \sin s, \; -z''(t), \; -\frac{z'(t)z''(t)}{y'(t)} \cos s \right\rangle$$
$$= -\frac{z''(t)}{y'(t)} \cdot \langle z'(t)\sin s, \; y'(t), \; z'(t)\cos s\rangle$$
$$= -\frac{z''(t)}{y'(t)} \cdot X_t.$$

Therefore

$$dv(aX_s + bX_t) = a(v \circ X)_s + b(v \circ X)_t$$
$$= a(y'/z)X_s + b(-z''/y')X_t.$$

In the $\{X_s, X_t\}$ basis, dv has the matrix form

$$dv\begin{pmatrix}a\\b\end{pmatrix} = \begin{pmatrix} y'/z & 0 \\ 0 & -z''/y' \end{pmatrix}\begin{pmatrix}a\\b\end{pmatrix}.$$

38.3 Gaussian curvature

The *Gaussian curvature* $K(P)$ of S at the point P is defined to be the determinant of the dv matrix at P. In our case

$$K(P) = -\frac{y'(t)}{z(t)} \cdot \frac{-z''(t)}{y'(t)} = -\frac{z''(t)}{z(t)}.$$

What is the geometric significance of that number?

Why we want K to be negative. The dv matrix is a symmetric matrix, which is obvious in our example, and true in general. Therefore it has real eigenvalues, and K, its determinant, is their product. If we apply the dv map to the circle of unit vectors $\langle \cos\theta, \sin\theta\rangle$, we get a set of vectors

38.3 Gaussian curvature

that trace out an ellipse. Because the dv matrix is symmetric, the eigenvectors are along the major and minor axes of the ellipse, and the corresponding eigenvalues are the signed distances from the origin to the ellipse along the axes. This is true in general, and easy to see in our example. The eigenvectors point in the directions of maximum and minimum curvature, then, and the corresponding eigenvalues are the curvatures in those directions (they are called the principal curvatures). If K is negative, then one of the eigenvalues is positive and the other is negative. So in the direction of one eigenvector, the surface will bend towards the normal vector $n(P)$; in the direction of the other, the surface will bend away from $n(P)$. Hence, negative K will make the surface bend so that lines on it diverge.

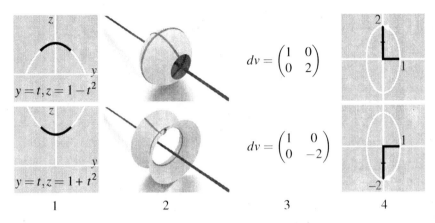

Figure 9. (1) A curve is (2) revolved around the y-axis. We calculate (3) the dv matrix, and can then (4) see its effects on the unit circle.

Why we want $|K|$ to be constant. The magnitude of $\det(dv)$ is the factor by which dv scales area. The dv map will transform a 1×1 square into a parallelogram with an area of $|\det(dv)|$, and since dv is linear, all other areas will follow suit. In our example, the parallelogram is a rectangle, so this is clear. In a small region (say of size ϵ) near P, the dv map is a good approximation of v, and so v also approximately scales area by $|\det(dv)|$. To get a sense of the scaling near P, consider the circle of points that are a distance of ϵ from P on the surface—a "circle" on the surface—and look at its circumference. In the plane that circumference is $2\pi\epsilon$, but on a curved surface it is not. The ratio depends on the curvature at P. Intuitively that is clear because if $|\det(dv)|$ is large, then v will wrap the circle farther around $\mathbb{S}$ than if it is small. For instance, consider the points P and Q shown in figure 10.

Both are points of positive curvature, but the curvature at P is larger than the curvature at Q. Let C_P and C_Q be circles of radius ϵ, centered at P and Q respectively. While both circles have the same radius, the circumference of C_P is smaller than the circumference of C_Q. As a model for neutral geometry, this would create the following problem. The circumference of a circle is the limit of perimeters of regular inscribed polygons. Since C_P and C_Q have the same radius, a regular n-gon inscribed in C_P is congruent to a regular n-gon inscribed in C_Q. Hence their perimeters would be the same, and so the circumferences of C_P and C_Q, as limits of them, should be the same (the Bertrand-Diquet-Puiseux theorem gives a precise relationship between the curvature and the circumference/radius ratio and that is a place to look if you want a more rigorous explanation [Spi99a]).

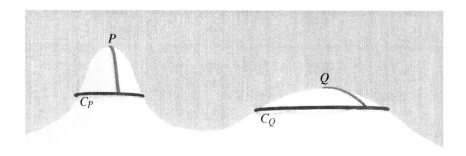

Figure 10. The curvature is greater at P than it is at Q. A circle centered at P of radius ϵ would have a smaller circumference than a circle centered at Q with the same radius.

38.4 The tractrix and pseudosphere

So it comes down to this: we want the curvature K to be a negative constant, say $K = -1$. That is,

$$-\frac{z''(t)}{z(t)} = -1 \implies z''(t) = z(t).$$

One solution to this differential equation (not the only one) is $z(t) = e^t$. What then, is $y(t)$? We get

$$(y'(t))^2 + (z'(t))^2 = 1$$
$$(y'(t))^2 + e^{2t} = 1$$
$$y'(t) = -\sqrt{1 - e^{2t}}$$
$$y(t) = \int -\sqrt{1 - e^{2t}}\, dt$$

(the solution curve comes in two pieces corresponding to the positive and negative roots and the negative root is the one that gives the standard piece). We leave evaluating the integral as an exercise.

Exercise 38.1 Show that

$$\int -\sqrt{1 - e^{2t}}\, dt = \ln\left(\frac{1 + \sqrt{1 - e^{2t}}}{e^t}\right) - \sqrt{1 - e^{2t}} + C.$$

Any value for C will work, so put $C = 0$. For the term inside the square root to be positive, t must be between $-\infty$ and zero. The result is a parametrization of $\mathcal{C}$. Let's go ahead and do one more step and write y directly in terms of z. It is easy since $z = e^t$:

$$y = \ln\left(\frac{1 + \sqrt{1 - z^2}}{z}\right) - \sqrt{1 - z^2}, \quad 0 < z \leq 1.$$

38.4 The tractrix and pseudosphere

The curve is called the *tractrix*. Its graph is shown in figure 11. Rotating the tractrix around the y-axis generates the pseudosphere, shown in figure 12.

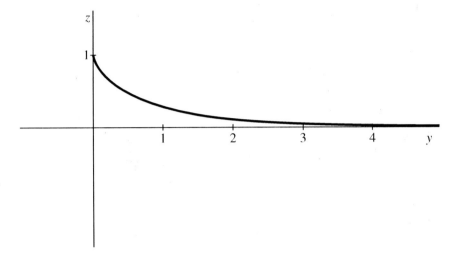

Figure 11. The graph of the tractrix (for positive y).

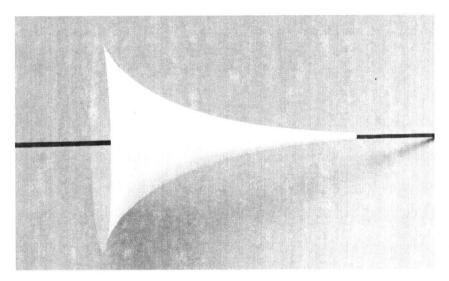

Figure 12. The pseudosphere.

The pseudosphere models a non-Euclidean neutral geometry locally, but globally there are some problems. One is apparent already: the pseudosphere ends at $y = 0$ where the tractrix has a vertical tangent. A neutral geometry plane cannot just stop like that. It is probably reasonable to hope that some other $K = -1$ surface (not a surface of revolution) would avoid this problem, but unfortunately this is not the case. Hilbert proved that no complete regular surface with constant negative curvature can be smoothly embedded in $\mathbb{R}^3$ [Spi99b]. In spite of its limitations, we will stick with the pseudosphere for a few more chapters. It will show the way to another model without the limitations.

38.5 Exercises

38.2. A torus can be realized as a surface of revolution formed by revolving a circle with radius a and center $(b, 0)$ around the y-axis (where $b > a$). Parametrize the surface.

38.3. Verify that the Gaussian curvature of the plane $z = 0$ is zero.

38.4. Show that the cylinder $x^2 + y^2 = 1$ has Gaussian curvature of zero.

38.5. Show that the Gaussian curvature of a sphere of radius r is $1/r^2$ (use spherical coordinates).

38.6. Compute the Gaussian curvature of the saddle surface $z = x^2 - y^2$ at the point $(0, 0, 0)$.

38.7. Let $P = (y_0, z_0)$ be a point on the tractrix. Let ℓ be the tangent line to the tractrix at P, and let $Q = (y_1, 0)$ be its intersection with the y-axis. Show that the distance from P to Q is 1 (so the distance from P to Q is independent of the coordinates of P).

39

Geodesics on the Pseudosphere

In the last chapter, we found a parametrization of the pseudosphere, a surface with constant negative curvature. The plan was to use that surface as a model for a non-Euclidean neutral geometry in which lines are represented by geodesics along the surface. The goal of this chapter is to calculate the geodesics.

39.1 Geodesics, the theory

Roughly, a geodesic between two points on a surface is the shortest path on the surface that connects them. To be more precise, we will talk about the calculus of variations. Let S be a surface parametrized by $X : D \to \mathbb{R}^3$ where

$$X(s, t) = (f(s, t), g(s, t), h(s, t))$$

and consider a parametrized curve $\rho(w)$ in the domain D (it can then be projected onto the surface S by X). We want to perturb $\rho(w)$ to see if the arc length of the curve on S increases or decreases. If the arc length always increases, no matter the variation, then $\rho(w)$ is a geodesic. What are the mechanics of perturbing?

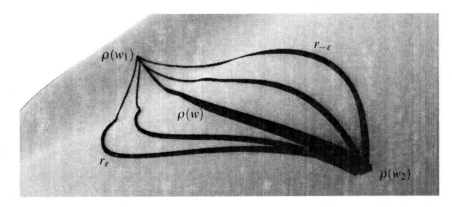

Figure 1. Some variations of the curve $\rho(w)$.

Definition 39.1 Let $\rho : [w_1, w_2] \to D$ be a parametrized curve in a domain D. A family of curves in D

$$\{r_v : [w_1, w_2] \longrightarrow D\}, \quad -\epsilon < v < \epsilon,$$

is a (smooth) *variation of ρ* if $R(w, v) = r_v(w)$ is a smooth map that satisfies:

- $R(w_1, v) = \rho(w_1)$ for all v; i.e., they all start at the same point,
- $R(w_2, v) = \rho(w_2)$ for all v; i.e., they all end at the same point,
- $R(w, 0) = \rho(w)$ (at $v = 0$, the original curve, unperturbed).

In general the length of a parametric curve $r(t)$ is calculated by integrating $|r'(t)|$. In our situation, the X map projects r_v of a variation onto the surface S, where we can calculate its arc length L as a function of v:

$$L(v) = \int_{w_1}^{w_2} |X(r_v(w))| \, dw = \int_{w_1}^{w_2} \sqrt{[X(r_v(w))]' \cdot [X(r_v(w))]'} \, dw.$$

For $\rho(w) = R(w, 0)$ to be a geodesic, $v = 0$ must be a local minimum of $L(v)$. This has to happen for every such variation.

Definition 39.2 Suppose that S is a surface parametrized by $X : D \longrightarrow \mathbb{R}^3$ and ρ is a parametrized curve in D. Then $X(\rho)$ is a *geodesic* of S if for every variation r_v of $\rho(w)$, $L(v)$, the length of $X(r_v)$, has a local minimum at $v = 0$.

Finding the local minimum across a perturbation requires locating the critical points, which requires computing the derivative with respect to v, of $L(v)$. We need to be careful with the variables: the integral is with respect to w, but the derivative is with respect to v. There is a trick that simplifies future calculations. The arc length integral involves a square root, which tends to make things difficult. There is, however, a related energy integral

$$\mathcal{E}(v) = \frac{1}{2} \int_{w_1}^{w_2} [X(r_v(w))]' \cdot [X(r_v(w))]' \, dw$$

that has the same critical points as L but not the square root (this is a commonly used trick in optimization problems; the calculations are not fundamentally different, but they are cleaned up). So to find the geodesics, we can solve the simple-looking equation

$$\left. \frac{d}{dv} \mathcal{E}(v) \right|_{v=0} = 0.$$

Alas, it is not as simple as it looks.

39.2 Geodesics, the calculations

In the last chapter, we parametrized the pseudosphere as a surface of revolution

$$X(s, t) = \big(z(t) \sin s, \ y(t), \ z(t) \cos s\big), \quad 0 \leq s \leq 2\pi,$$

39.2 Geodesics, the calculations

and found the equation for y in terms of z that would give the surface constant negative curvature. The result was the tractrix:

$$y = \ln\left(\frac{1+\sqrt{1-z^2}}{z}\right) - \sqrt{1-z^2}, \quad 0 < z \le 1.$$

The tractrix came to us in a parametrized form where $z = e^t$, for $-\infty < t \le 0$. We again need a parametrization of the tractrix, but this time a different one. Put $z(t) = 1/t$ with $1 \le t < \infty$. Then

$$\begin{aligned} y(t) &= \ln\left(\frac{1+\sqrt{1-(1/t)^2}}{1/t}\right) - \sqrt{1-(1/t)^2} \\ &= \ln\left(t + t\sqrt{1-(1/t)^2}\right) - \sqrt{1-(1/t)^2} \\ &= \ln\left(t + \sqrt{t^2-1}\right) - \sqrt{1-(1/t)^2}. \end{aligned}$$

In this parametrization, X is the map from $D = [0, 2\pi) \times [1, \infty)$ to $\mathbb{R}^3$ given by

$$X(s,t) = \left(\frac{1}{t}\sin s, \; \ln\left(t + \sqrt{t^2-1}\right) - \sqrt{1-(1/t)^2}, \; \frac{1}{t}\cos s\right).$$

Our goal is to find curves $\rho(w)$ in D so that $X(\rho(w))$ is a geodesic, i.e., curves that minimize energy. That calculation generates a pair of differential equations that $\rho(w)$ must satisfy. Solving the system will give the equations of the geodesics of the pseudosphere.

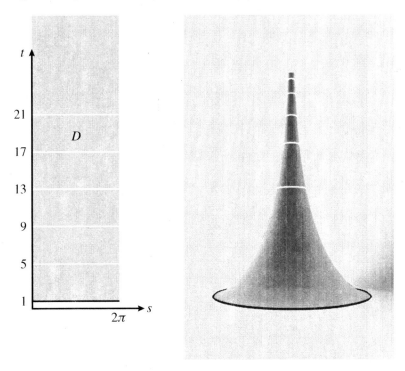

Figure 2. The domain D that is projected onto the pseudosphere by the parametrization X.

Given a variation $r_v(w)$, the energy integral is

$$\mathcal{E}(v) = \frac{1}{2}\int_{w_1}^{w_2} \frac{\partial X}{\partial w} \cdot \frac{\partial X}{\partial w}\, dw.$$

We need to find $\partial X/\partial w$. It may be helpful to review our variables. Because

$$X = \big(z(t)\sin s,\; y(t),\; z(t)\cos s\big),$$

X depends on s and t, which depend upon v and w (for the first calculation, we are setting up an integral to calculate the length of one curve in the variation, so the v variable is held constant). We have found y and z in terms of t, and it might be tempting to substitute now, but it is better to wait. We can use the chain rule to calculate

$$\frac{\partial X}{\partial w} = \Big\langle \frac{\partial}{\partial s}(z\sin s)\cdot\frac{\partial s}{\partial w} + \frac{\partial}{\partial t}(z\sin s)\cdot\frac{\partial t}{\partial w},\; \frac{dy}{dt}\frac{\partial t}{\partial w},$$

$$\frac{\partial}{\partial s}(z\cos s)\cdot\frac{\partial s}{\partial w} + \frac{\partial}{\partial t}(z\cos s)\cdot\frac{\partial t}{\partial w}\Big\rangle$$

$$= \Big\langle z\cos s\,\frac{\partial s}{\partial w} + \sin s\,\frac{dz}{dt}\frac{\partial t}{\partial w},\; \frac{dy}{dt}\frac{\partial t}{\partial w},\; -z\sin s\,\frac{\partial s}{\partial w} + \cos s\,\frac{dz}{dt}\frac{\partial t}{\partial w}\Big\rangle,$$

and then

$$\frac{\partial X}{\partial w}\cdot\frac{\partial X}{\partial w} = \Big[z\cos s\,\frac{\partial s}{\partial w} + \sin s\,\frac{dz}{dt}\frac{\partial t}{\partial w}\Big]^2 + \Big[\frac{dy}{dt}\frac{\partial t}{\partial w}\Big]^2 + \Big[-z\sin s\,\frac{\partial s}{\partial w} + \cos s\,\frac{dz}{dt}\frac{\partial t}{\partial w}\Big]^2.$$

$$= \Big[z^2\cos^2 s\Big(\frac{\partial s}{\partial w}\Big)^2 + 2z\sin s\cos s\,\frac{\partial s}{\partial w}\frac{dz}{dt}\frac{\partial t}{\partial w} + \sin^2 s\Big(\frac{dz}{dt}\frac{\partial t}{\partial w}\Big)^2\Big] + \Big[\frac{dy}{dt}\frac{\partial t}{\partial w}\Big]^2$$

$$+ \Big[z^2\sin^2 s\Big(\frac{\partial s}{\partial w}\Big)^2 - 2z\sin s\cos s\,\frac{\partial s}{\partial w}\frac{dz}{dt}\frac{\partial t}{\partial w} + \cos^2 s\Big(\frac{dz}{dt}\frac{\partial t}{\partial w}\Big)^2\Big].$$

Thanks to the Pythagorean identity $\sin^2 x + \cos^2 x = 1$, this simplifies to

$$\frac{\partial X}{\partial w}\cdot\frac{\partial X}{\partial w} = z^2\Big(\frac{\partial s}{\partial w}\Big)^2 + \Big(\frac{dz}{dt}\frac{\partial t}{\partial w}\Big)^2 + \Big(\frac{dy}{dt}\frac{\partial t}{\partial w}\Big)^2.$$

Now we can substitute y and z. We know $z = 1/t$, so $dz/dt = -1/t^2$. The dy/dt term is a little more difficult.

Lemma 39.3 *The derivative of*

$$y(t) = \ln(t + \sqrt{t^2 - 1}) - \sqrt{1 - (1/t)^2},\quad t \geq 1,$$

is $y'(t) = \sqrt{t^2 - 1}/t^2$, and thus $[y'(t)]^2 = t^{-2} - t^{-4}$.

Exercise 39.1 Verify lemma 39.3.

39.2 Geodesics, the calculations

Substituting,

$$\frac{\partial X}{\partial w} \cdot \frac{\partial X}{\partial w} = \frac{1}{t^2}\left(\frac{\partial s}{\partial w}\right)^2 + \frac{1}{t^4}\left(\frac{\partial t}{\partial w}\right)^2 + \left(\frac{1}{t^2} - \frac{1}{t^4}\right)\left(\frac{\partial t}{\partial w}\right)^2$$

$$= \frac{1}{t^2}\left[\left(\frac{\partial s}{\partial w}\right)^2 + \left(\frac{\partial t}{\partial w}\right)^2\right].$$

We have our energy function. For a variation $\{r_v(w)\}$ of ρ,

$$\mathcal{E}(v) = \frac{1}{2}\int_{w_1}^{w_2} \frac{1}{t^2}\left[\left(\frac{\partial s}{\partial w}\right)^2 + \left(\frac{\partial t}{\partial w}\right)^2\right] dw.$$

The geodesics will be critical points of the function. To find them, let's differentiate with respect to v (use the product rule):

$$\frac{d\mathcal{E}}{dv} = \frac{d}{dv}\left(\frac{1}{2}\int_{w_1}^{w_2} \frac{1}{t^2}\left[\left(\frac{\partial s}{\partial w}\right)^2 + \left(\frac{\partial t}{\partial w}\right)^2\right] dw\right)$$

$$= \frac{1}{2}\int_{w_1}^{w_2} \frac{\partial}{\partial v}\left(\frac{1}{t^2}\cdot\left[\left(\frac{\partial s}{\partial w}\right)^2 + \left(\frac{\partial t}{\partial w}\right)^2\right]\right) dw$$

$$= \frac{1}{2}\int_{w_1}^{w_2} \frac{1}{t^2}\left[2\frac{\partial s}{\partial w}\frac{\partial^2 s}{\partial w \partial v} + 2\frac{\partial t}{\partial w}\frac{\partial^2 t}{\partial w \partial v}\right] - \frac{2}{t^3}\frac{\partial t}{\partial v}\left[\left(\frac{\partial s}{\partial w}\right)^2 + \left(\frac{\partial t}{\partial w}\right)^2\right] dw$$

$$= \int_{w_1}^{w_2} \frac{1}{t^2}\frac{\partial s}{\partial w}\frac{\partial^2 s}{\partial w \partial v} + \frac{1}{t^2}\frac{\partial t}{\partial w}\frac{\partial^2 t}{\partial w \partial v} - \frac{1}{t^3}\left[\left(\frac{\partial s}{\partial w}\right)^2 + \left(\frac{\partial t}{\partial w}\right)^2\right]\frac{\partial t}{\partial v} dw.$$

The variation variable is v so we have very little control over it. Therefore, it makes sense to try to isolate the v terms as much as possible. The mixed partials are a challenge, but fortunately there is a trick.

Lemma 39.4

$$\int_{w_1}^{w_2} \frac{1}{t^2}\frac{\partial s}{\partial w}\frac{\partial^2 s}{\partial w \partial v} dw = \int_{w_1}^{w_2} \frac{1}{t^3}\left[2\frac{\partial s}{\partial w}\frac{\partial t}{\partial w} - t\frac{\partial^2 s}{\partial w^2}\right]\frac{\partial s}{\partial v} dw$$

$$\int_{w_1}^{w_2} \frac{1}{t^2}\frac{\partial t}{\partial w}\frac{\partial^2 t}{\partial w \partial v} dw = \int_{w_1}^{w_2} \frac{1}{t^3}\left[2\left(\frac{\partial t}{\partial w}\right)^2 - t\frac{\partial^2 t}{\partial w^2}\right]\frac{\partial t}{\partial v} dw.$$

Proof Both formulas are an application of integration by parts. We will do the first calculation, and leave the similar second calculation as an exercise. Put

$$U = \frac{1}{t^2}\cdot\frac{\partial s}{\partial w}, \qquad dU = \left(\frac{1}{t^2}\frac{\partial^2 s}{\partial w^2} - \frac{2}{t^3}\frac{\partial t}{\partial w}\frac{\partial s}{\partial w}\right) dw,$$

$$dV = \frac{\partial^2 s}{\partial w \partial v} dw, \qquad V = \frac{\partial s}{\partial v},$$

so that

$$\int_{w_1}^{w_2} \frac{1}{t^2} \frac{\partial s}{\partial w} \frac{\partial^2 s}{\partial w \partial v} \, dw = \left(\frac{1}{t^2} \frac{\partial s}{\partial w} \frac{\partial s}{\partial v} \right)\bigg|_{w_1}^{w_2} - \int_{w_1}^{w_2} \frac{\partial s}{\partial v} \left[\frac{1}{t^2} \frac{\partial^2 s}{\partial w^2} - \frac{2}{t^3} \frac{\partial t}{\partial w} \frac{\partial s}{\partial w} \right] dw.$$

The variation is fixed at both endpoints, so $\partial s/\partial v(w_1) = \partial s/\partial v(w_2) = 0$. The first term contributes nothing, so

$$\int_{w_1}^{w_2} \frac{1}{t^2} \frac{\partial s}{\partial w} \frac{\partial^2 s}{\partial w \partial v} \, dw = 0 - \int_{w_1}^{w_2} \frac{\partial s}{\partial v} \left[\frac{1}{t^2} \frac{\partial^2 s}{\partial w^2} - \frac{2}{t^3} \frac{\partial t}{\partial w} \frac{\partial s}{\partial w} \right] dw$$

$$= \int_{w_1}^{w_2} \frac{1}{t^3} \left[2 \frac{\partial t}{\partial w} \frac{\partial s}{\partial w} - t \frac{\partial^2 s}{\partial w^2} \right] \frac{\partial s}{\partial v} \, dw.$$

□

Exercise 39.2 Verify the second equation in the last lemma.

Substitute into $d\mathcal{E}/dv$ to get

$$\frac{d\mathcal{E}}{dv} = \int_{w_1}^{w_2} \frac{1}{t^3} \left[2\frac{\partial s}{\partial w}\frac{\partial t}{\partial w} - t\frac{\partial^2 s}{\partial w^2} \right] \frac{\partial s}{\partial v} + \frac{1}{t^3} \left[2 \left(\frac{\partial t}{\partial w} \right)^2 - t \frac{\partial^2 t}{\partial w^2} \right] \frac{\partial t}{\partial v}$$

$$- \frac{1}{t^3} \left[\left(\frac{\partial s}{\partial w} \right)^2 + \left(\frac{\partial t}{\partial w} \right)^2 \right] \frac{\partial t}{\partial v} \, dw$$

$$= \int_{w_1}^{w_2} \frac{1}{t^3} \left[2\frac{\partial s}{\partial w}\frac{\partial t}{\partial w} - t\frac{\partial^2 s}{\partial w^2} \right] \frac{\partial s}{\partial v} + \frac{1}{t^3} \left[\left(\frac{\partial t}{\partial w} \right)^2 - t \frac{\partial^2 t}{\partial w^2} - \left(\frac{\partial s}{\partial w} \right)^2 \right] \frac{\partial t}{\partial v} \, dw.$$

The critical points of $\mathcal{E}(v)$ occur when the integral is zero. Still, we know little about $\partial s/\partial v$ and $\partial t/\partial v$. There may be some variations where the $\partial s/\partial v$ and $\partial t/\partial v$ components cancel each other out. But geodesics occur only when energy is minimized for all variations. The only way to guarantee this regardless of variation is if when $v = 0$, at $\rho(v)$,

- $2\dfrac{ds}{dw}\dfrac{dt}{dw} - t\dfrac{d^2s}{dw^2} = 0$, and

- $\left(\dfrac{dt}{dw}\right)^2 - t\dfrac{d^2t}{dw^2} - \left(\dfrac{ds}{dw}\right)^2 = 0$,

the differential equations that describe a geodesic ρ. The next step is to put them into a more useful form (one that does not involve second derivatives) with the following lemmas.

Lemma 39.5 *If*

$$2\frac{ds}{dw}\frac{dt}{dw} - t\frac{d^2s}{dw^2} = 0,$$

then s'/t^2 is a constant.

39.2 Geodesics, the calculations

Proof Assuming as given that

$$-ts'' + 2t's' = 0,$$

use the quotient rule to compute the derivative of s'/t^2:

$$\left(\frac{s'}{t^2}\right)' = \frac{t^2 s'' - s' \cdot 2tt'}{(t^2)^2} = \frac{ts'' - 2s't'}{t^3} = 0.$$

Since the derivative of s'/t^2 is zero, s'/t^2 is a constant. □

Lemma 39.6 *If*

$$2\frac{ds}{dw}\frac{dt}{dw} - t\frac{d^2s}{dw^2} = 0 \quad \text{and} \quad \left(\frac{dt}{dw}\right)^2 - t\frac{d^2t}{dw^2} - \left(\frac{ds}{dw}\right)^2 = 0,$$

then $(s')^2 + (t')^2 = Ct^2$ for some constant C.

Proof Multiply both sides of the second equation by -1 to get $(s')^2 - (t')^2 + tt'' = 0$. Then add $(s')^2 + (t')^2$ to both sides and multiply by t':

$$(s')^2 - (t')^2 + tt'' = 0$$
$$2(s')^2 + tt'' = (s')^2 + (t')^2$$
$$2(s')^2 t' + tt't'' = [(s')^2 + (t')^2]t'.$$

According to the first equation, $2s't' = ts''$ so

$$s'(ts'') + tt't'' = [(s')^2 + (t')^2]t'$$
$$t[s's'' + t't''] = [(s')^2 + (t')^2]t'$$
$$[s's'' + t't'']/[(s')^2 + (t')^2] = t'/t.$$

Integrate with respect to w. On the right

$$\int \frac{t'}{t} dw = \ln(t) + C_1.$$

The left hand side is similar: put $u = (s')^2 + (t')^2$ so that $du = (2s's'' + 2t't'')dw$. Then

$$\int \frac{s's'' + t't''}{(s')^2 + (t')^2} dw = \int \frac{1/2}{u} du = \frac{1}{2}\ln|u| + C_2 = \frac{1}{2}\ln\left((s')^2 + (t')^2\right) + C_2.$$

Combining, we get

$$\frac{1}{2}\ln\left((s')^2 + (t')^2\right) + C_2 = \ln(t) + C_1$$
$$\ln\left((s')^2 + (t')^2\right) = 2\ln(t) + 2C_1 - 2C_2$$
$$(s')^2 + (t')^2 = e^{\ln(t^2) + 2C_1 - 2C_2}$$
$$(s')^2 + (t')^2 = Ct^2$$

where $C = e^{2C_1 - 2C_2}$. □

The constant C eventually gets wrapped into other constants, so there is no harm in assuming that $C = 1$ now. Doing that means

$$X(\rho)' \cdot X(\rho)' = \frac{1}{t^2}\left[(s')^2 + (t')^2\right] = \frac{1}{t^2} \cdot t^2 = 1.$$

The parametrized curve $X(\rho)$ on the pseudosphere moves at a constant speed of 1, so the curve is parametrized by arc length.

Now let's assume that a curve ρ satisfies the differential equations, and try to see what ρ will look like. Assuming $s'/t^2 = k$ for some constant k, write $s' = kt^2$. There is one case where this is easy: if $k = 0$, then $s' = 0$, so s is a constant, so ρ is a vertical line in D.

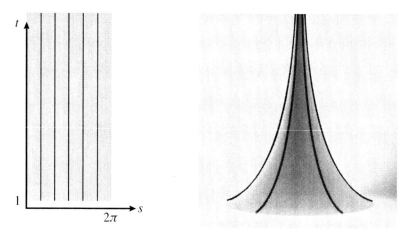

Figure 3. Geodesics on the pseudosphere that are projections of vertical lines.

If $k \neq 0$, the calculation is more difficult. Start by substituting $s' = kt^2$ into the second equation:

$$(s')^2 + (t')^2 = t^2 \implies (kt^2)^2 + (t')^2 = t^2$$
$$\implies (t')^2 = t^2 - k^2 t^4$$
$$\implies t' = \sqrt{t^2 - k^2 t^4} = t\sqrt{1 - k^2 t^2}.$$

Since s and t are functions of w, we can eliminate w to get a single differential equation relating s and t:

$$\frac{ds}{dt} = \frac{ds/dw}{dt/dw} = \frac{kt^2}{t\sqrt{1-k^2t^2}} = \frac{kt}{\sqrt{1-k^2t^2}}.$$

Integrating gives s in terms of t, the equations of the geodesics.

Lemma 39.7 *If $k \neq 0$, then*

$$\int \frac{kt}{\sqrt{1-k^2t^2}}\, dt = -\frac{1}{k}\sqrt{1-k^2t^2} + C.$$

39.2 Geodesics, the calculations

Proof Let $u = 1 - k^2t^2$ so $du = -2k^2t\,dt$ and

$$\int \frac{kt}{\sqrt{1-k^2t^2}}\,dt = \int \frac{-1/(2k)}{\sqrt{u}}\,du$$
$$= -\frac{1}{2k}\int u^{-1/2}\,du$$
$$= -\frac{1}{2k}2u^{1/2} + C$$
$$= -\frac{1}{k}\sqrt{1-k^2t^2} + C. \qquad \square$$

Therefore, if ρ is a geodesic, it satisfies

$$s = -\frac{1}{k}\sqrt{1-k^2t^2} + C \implies k(s-C) = -\sqrt{1-k^2t^2}.$$

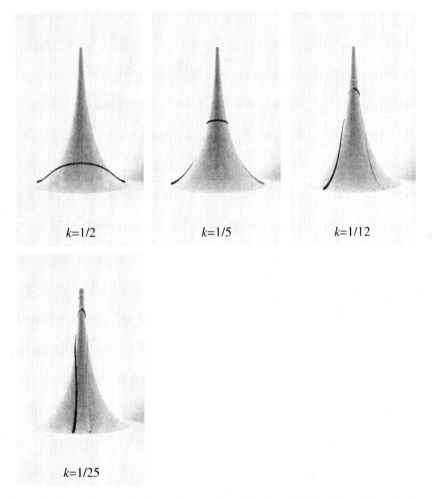

$k=1/2$ $k=1/5$ $k=1/12$

$k=1/25$

Figure 4. Geodesics on the pseudosphere that are the projections of half-circles. The projection of the entire half-circle is shown. The portion that lies in the domain D is shaded black. Extending the curves beyond D creates self-intersections that would be undesirable in a model for a non-Euclidean neutral geometry.

Square both sides:

$$k^2(s - C)^2 = 1 - k^2 t^2$$
$$k^2(s - C)^2 + k^2 t^2 = 1$$
$$(s - C)^2 + t^2 = (1/k)^2.$$

The curves in D that project to geodesics on the pseudosphere are circles that are centered on the s-axis (or at least the portions of them that lie in D).

The geodesics reveal another problem with the pseudosphere model. When the circles ρ reach the edge of D, the corresponding geodesics on the pseudosphere just stop. It is easy to get around that problem by extending D horizontally so that X wraps around the pseudosphere more than once. But then the geodesics cross over themselves, creating undesirable self-intersections.

39.3 References

Spivak discusses geodesics in the third volume of his *Comprehensive Introduction to Differential Geometry* [Spi99b], but my approach here more closely follows Thorpe's *Elementary Topics in Differential Geometry* [Tho79].

39.4 Exercises

39.3. The standard parameterization (in terms of sine and cosine) of the circle $(s - C)^2 + t^2 = (1/k)^2$ does not satisfy the differential equation for the arc length criterion or either of the differential equations for the geodesic criterion. The parametrization that works is

$$s(w) = \frac{1}{k} \tanh(w), \quad t(w) = \frac{1}{k} \operatorname{sech}(w).$$

Verify that the parametrization satisfies the differential equations (the sech(w) and tanh(w) functions are covered in chapter 44 if you are not familiar with them).

39.4. Show that lines are geodesics in the plane $z = 0$.

39.5. Show that the circle $r(t) = \langle \cos t, \sin t, k \rangle$ (k constant) is a geodesic on the cylinder $x^2 + y^2 = 1$. Show that the vertical line $r(t) = \langle 1, 0, t \rangle$ is also a geodesic on the cylinder.

39.6. Show that the equatorial circle $r(t) = \langle \cos t, \sin t, 0 \rangle$ is a geodesic on the sphere $x^2 + y^2 + z^2 = 1$.

40

The Upper Half Plane

In the last chapter, we parametrized the pseudosphere as

$$X(s,t) = \left(\frac{1}{t}\sin s,\ \ln\left(t + \sqrt{t^2 - 1}\right) - \sqrt{1 - \frac{1}{t^2}},\ \frac{1}{t}\cos s\right),$$

where $0 \leq s < 2\pi$ and $1 \leq t < \infty$, and found that its geodesics take the form $X(\rho)$ where ρ is one of two kinds of curves (restricted to the domain D): a vertical line, or a circle centered on the s-axis. In this chapter, we look at how we measure distance and angles on the pseudosphere. We will be able to relate them to measurements in the domain D of the parametrization. By moving away from the physical model of the pseudosphere, we will be able to abstract away many of the inherent problems of the pseudosphere model.

40.1 Distance

Let P and Q be two points in the domain D. The distance between their projections $X(P)$ and $X(Q)$ onto the pseudosphere is measured as the arc length of the geodesic segment $X(\rho(w))$ that connects them. Write $d_H(P, Q)$ for the distance. If $\rho(w_1) = P$ and $\rho(w_2) = Q$ (it is convenient to assume throughout that $w_1 < w_2$), then

$$d_H(P, Q) = \int_{w_1}^{w_2} \left|\frac{dX(\rho(w))}{dw}\right| dw = \int_{w_1}^{w_2} \sqrt{\frac{dX(\rho(w))}{dw} \cdot \frac{dX(\rho(w))}{dw}}\, dw.$$

In the last chapter, we calculated

$$\frac{dX}{dw} \cdot \frac{dX}{dw} = \frac{1}{t^2}\left[\left(\frac{ds}{dw}\right)^2 + \left(\frac{dt}{dw}\right)^2\right],$$

so

$$d_H(P, Q) = \int_{w_1}^{w_2} \frac{1}{t}\sqrt{\left(\frac{ds}{dw}\right)^2 + \left(\frac{dt}{dw}\right)^2}\, dw.$$

Let's do this calculation for both types of geodesic.

Calculation 1: Geodesics that come from vertical lines

Suppose that P and Q lie on the vertical line $s = C$. The line can be parametrized as

$$\rho(w) = \langle C, w \rangle \implies \rho'(w) = \langle 0, 1 \rangle,$$

and therefore

$$d_H(P, Q) = \int_{w_1}^{w_2} \frac{1}{w} \sqrt{(0)^2 + (1)^2} \, dw = \ln(w) \Big|_{w_1}^{w_2} = \ln(w_2) - \ln(w_1) = \ln(w_2/w_1).$$

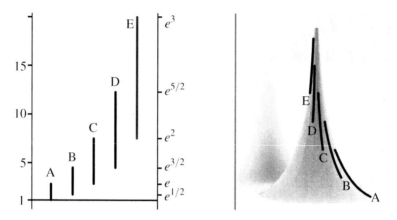

Figure 1. Segments of length 1 on the pseudosphere.

Calculation 2: Geodesics that come from circles   Suppose that P and Q lie on the circle $(s - h)^2 + t^2 = r^2$. The circle can be parametrized as

$$\rho(w) = \langle r \cos w + h, r \sin w \rangle \implies \rho'(w) = \langle -r \sin w, r \cos w \rangle,$$

and therefore

$$\begin{aligned}
d_H(P, Q) &= \int_{w_1}^{w_2} \frac{1}{r \sin w} \sqrt{(-r \sin w)^2 + (r \cos w)^2} \, dw \\
&= \int_{w_1}^{w_2} \frac{1}{r \sin w} \cdot r \, dw \\
&= \int_{w_1}^{w_2} \csc w \, dw \\
&= -\ln |\csc w + \cot w| \Big|_{w_1}^{w_2} \\
&= -\ln |\csc w_2 + \cot w_2| + \ln |\csc w_1 + \cot w_1| \\
&= \ln \left| \frac{\csc w_1 + \cot w_1}{\csc w_2 + \cot w_2} \right| \\
&= \ln \left| \frac{(1/\sin w_1) + (\cos w_1 / \sin w_1)}{(1/\sin w_2) + (\cos w_2 / \sin w_2)} \right|.
\end{aligned}$$

40.1 Distance

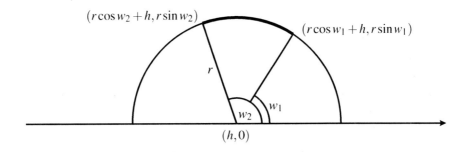

Figure 2. Calculating distance along geodesics that are the images of circles.

To simplify further, multiply the numerator and denominator of the fraction by $\sin w_1 \sin w_2$ to get

$$d_H(P, Q) = \ln \left[\frac{\sin w_2 + \sin w_2 \cos w_1}{\sin w_1 + \sin w_1 \cos w_2} \right] = \ln \left[\frac{\sin w_2 (1 + \cos w_1)}{\sin w_1 (1 + \cos w_2)} \right]$$

(all the terms in the logarithm are positive since the points are on the top half of the circle). That is an acceptable answer, but there is another way to present it that has important geometric implications. In addition to $P = (r \cos w_1 + h, r \sin w_1)$ and $Q = (r \cos w_2 + h, r \sin w_2)$, label two more points on the circle ρ: $A = (r + h, 0)$ and $B = (-r + h, 0)$, and consider the cross ratio of the four:

$$[PQ : AB] = \frac{|PA|}{|QA|} \cdot \frac{|QB|}{|PB|}.$$

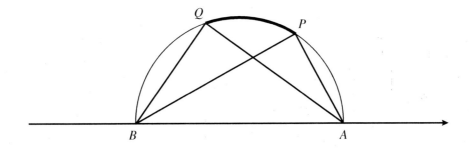

Figure 3. Distance can also be calculated using the cross ratio.

The four segment lengths are easy to calculate. For instance,

$$|PA| = \sqrt{[(r\cos w_1 + h) - (r+h)]^2 + [r\sin w_1 - 0]^2}$$
$$= \sqrt{(r\cos w_1 - r)^2 + (r\sin w_1)^2}$$
$$= \sqrt{r^2\cos^2 w_1 - 2r^2\cos w_1 + r^2 + r^2\sin^2 w_1}$$
$$= \sqrt{2r^2 - 2r^2\cos w_1}$$
$$= \sqrt{2}r\sqrt{1-\cos w_1},$$

and similarly,

$$|QA| = \sqrt{2}r\sqrt{1-\cos w_2}, \quad |PB| = \sqrt{2}r\sqrt{1+\cos w_1}, \quad \text{and} \quad |QB| = \sqrt{2}r\sqrt{1+\cos w_2}.$$

The cross ratio is then

$$[PQ:AB] = \frac{\sqrt{2}r\sqrt{1-\cos w_1}}{\sqrt{2}r\sqrt{1-\cos w_2}} \cdot \frac{\sqrt{2}r\sqrt{1+\cos w_2}}{\sqrt{2}r\sqrt{1+\cos w_1}} = \sqrt{\frac{(1-\cos w_1)(1+\cos w_2)}{(1-\cos w_2)(1+\cos w_1)}}.$$

Multiply the numerator and denominator inside the square root by $(1+\cos w_1)(1+\cos w_2)$ to get

$$[PQ:AB] = \sqrt{\frac{(1-\cos^2 w_1)(1+\cos w_2)^2}{(1-\cos^2 w_2)(1+\cos w_1)^2}}$$
$$= \sqrt{\frac{\sin^2 w_1(1+\cos w_2)^2}{\sin^2 w_2(1+\cos w_1)^2}}$$
$$= \frac{\sin w_1(1+\cos w_2)}{\sin w_2(1+\cos w_1)}.$$

We just saw that distance on the pseudosphere is the natural logarithm of the reciprocal of this. Since $\ln(1/x) = -\ln(x)$, $|\ln(1/x)| = |\ln(x)|$. We want distance to be positive, so we can write

$$d_H(P, Q) = |\ln([PQ:AB])|,$$

and our excursion into differential geometry is starting to come back to classical geometry.

40.2 Angle measure

There are two cases to consider depending upon whether the intersecting geodesics are the projections of two circles, or of one circle and one vertical line. Throughout this section, we will look at the first case, and leave the other as an exercise. Parametrize two intersecting circles, one with radius r_1 and center $(h_1, 0)$ and the other with radius r_2 and center $(h_2, 0)$, as

$$\rho_1(w) = (s_1(w), t_1(w)) = (r_1\cos w + h_1, r_1\sin w)$$
$$\rho_2(w) = (s_2(w), t_2(w)) = (r_2\cos w + h_2, r_2\sin w)$$

and suppose that their intersection is at $\rho_1(w_1) = \rho_2(w_2)$. The angle between the curves $X(\rho_1)$ and $X(\rho_2)$ is measured as the angle θ between their tangent vectors $[X(\rho_1)]'$ and $[X(\rho_2)]'$ at the

40.2 Angle measure

point of intersection. The dot product gives a way to calculate the angle:

$$\cos\theta = \frac{X(\rho_1)' \cdot X(\rho_2)'}{|X(\rho_1)'||X(\rho_1)'|}.$$

The pieces of the dot product are spread across the next few lemmas. There is significant overlap between these calculations and those of the last chapter, but I choose to be methodical rather than efficient.

We parametrized the pseudosphere as $X = (z \sin s, y, z \cos s)$ where

- $z = 1/t$ so $z' = -1/t^2$, and
- $y = \ln(t + \sqrt{t^2 - 1}) - \sqrt{1 - (1/t^2)}$, so $y' = \sqrt{t^2 - 1}/t^2$.

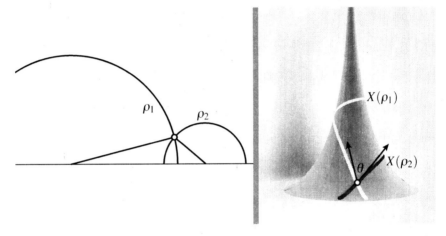

Figure 4. The angle θ between the projected curves $X(\rho_1)$ and $X(\rho_2)$ is the angle between their tangent vectors.

We will need the calculations

$$\frac{\partial X}{\partial s} = \langle z \cos s, 0, -z \sin s \rangle$$

$$\frac{\partial X}{\partial t} = \langle z' \sin s, y', z' \cos s \rangle$$

$$\frac{\partial X}{\partial s} \cdot \frac{\partial X}{\partial s} = z^2 \cos^2 s + 0 + z^2 \sin^2 s = z^2 = 1/t^2$$

$$\frac{\partial X}{\partial s} \cdot \frac{\partial X}{\partial t} = zz' \sin s \cos s + 0 - zz' \sin s \cos s = 0$$

$$\frac{\partial X}{\partial t} \cdot \frac{\partial X}{\partial t} = (z')^2 \sin^2 s + (y')^2 + (z')^2 \cos^2 s = (z')^2 + (y')^2$$

$$= (1/t^4) + (1/t^2 - 1/t^4) = 1/t^2$$

(the dot products of the partials are the coefficients of the "first fundamental form" of X).

Lemma 40.1

$$|X(\rho_1)'| = r_1/t_1 \quad \text{and} \quad |X(\rho_2)'| = r_2/t_2.$$

Proof Let's look at the first of these (the second is no different). By the chain rule,

$$[X(\rho_1)]' = \frac{\partial X}{\partial s} \cdot \frac{ds_1}{dw} + \frac{\partial X}{\partial t} \cdot \frac{dt_1}{dw}.$$

Then

$$\begin{aligned}
|X(\rho_1)'|^2 &= X(\rho_1)' \cdot X(\rho_1)' \\
&= \left(\frac{\partial X}{\partial s} \cdot \frac{ds_1}{dw} + \frac{\partial X}{\partial t} \cdot \frac{dt_1}{dw}\right) \cdot \left(\frac{\partial X}{\partial s} \cdot \frac{ds_1}{dw} + \frac{\partial X}{\partial t} \cdot \frac{dt_1}{dw}\right) \\
&= \left(\frac{ds_1}{dw}\right)^2 \frac{\partial X}{\partial s} \cdot \frac{\partial X}{\partial s} + 2\left(\frac{ds_1}{dw}\frac{dt_1}{dw}\right)\frac{\partial X}{\partial s} \cdot \frac{\partial X}{\partial t} + \left(\frac{dt_1}{dw}\right)^2 \frac{\partial X}{\partial t} \cdot \frac{\partial X}{\partial t} \\
&= (-r_1 \sin w)^2 \cdot \frac{1}{t_1^2} + 0 + (r_1 \cos w)^2 \cdot \frac{1}{t_1^2} \\
&= r_1^2/t_1^2.
\end{aligned}$$

Therefore $|X(\rho_1)'| = r_1/t_1$. □

Lemma 40.2

$$X(\rho_1)' \cdot X(\rho_2)' = \frac{r_1 r_2}{t_1 t_2}(\sin w_1 \sin w_2 + \cos w_1 \cos w_2).$$

Proof At the point of intersection, $t_1 = t_2$. Write t for the shared value. Now we calculate

$$\begin{aligned}
X(\rho_1)' \cdot X(\rho_2)' &= \left(\frac{\partial X}{\partial s}\frac{ds_1}{dw} + \frac{\partial X}{\partial t}\frac{dt_1}{dw}\right) \cdot \left(\frac{\partial X}{\partial s}\frac{ds_2}{dw} + \frac{\partial X}{\partial t}\frac{dt_2}{dw}\right) \\
&= \left(\frac{ds_1}{dw}\frac{ds_2}{dw}\right)\frac{\partial X}{\partial s} \cdot \frac{\partial X}{\partial s} + \left(\frac{ds_1}{dw}\frac{dt_2}{dw} + \frac{ds_2}{dw}\frac{dt_1}{dw}\right)\frac{\partial X}{\partial s} \cdot \frac{\partial X}{\partial t} \\
&\quad + \left(\frac{dt_1}{dw}\frac{dt_2}{dw}\right)\frac{\partial X}{\partial t} \cdot \frac{\partial X}{\partial t} \\
&= (-r_1 \sin w_1)(-r_2 \sin w_2)\frac{1}{t^2} + 0 + (r_1 \cos w_1)(r_2 \cos w_2)\frac{1}{t^2} \\
&= \frac{r_1 r_2}{t^2}\left[\sin w_1 \sin w_2 + \cos w_1 \cos w_2\right] \\
&= \frac{r_1 r_2}{t_1 t_2}\left[\sin w_1 \sin w_2 + \cos w_1 \cos w_2\right].
\end{aligned}$$
□

Lemma 40.3 *Let θ denote the angle between $X(\rho_1)$ and $X(\rho_2)$. Then*

$$\cos\theta = \sin w_1 \sin w_2 + \cos w_1 \cos w_2.$$

40.2 Angle measure

Proof Substitute, using the previous two lemmas. At the intersection of ρ_1 and ρ_2,

$$\cos\theta = \frac{X(\rho_1)' \cdot X(\rho_2)'}{|X(\rho_1)'||X(\rho_2)'|}$$

$$= \frac{(r_1 r_2 / t_1 t_2)(\sin w_1 \sin w_2 + \cos w_1 \cos w_2)}{(r_1/t_1)(r_2/t_2)}$$

$$= \sin w_1 \sin w_2 + \cos w_1 \cos w_2. \qquad \square$$

We're not quite done. Let's now compare θ to the angle between ρ_1 and ρ_2 before projecting onto the pseudosphere.

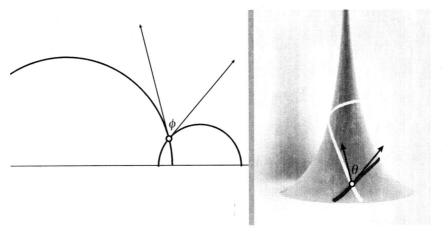

Figure 5. We need to compare the angle ϕ between curves in D and the angle θ between their projections on the pseudosphere.

Theorem 40.4 *Let ϕ be the angle between ρ_1 and ρ_2, and let θ be the angle between $X(\rho_1)$ and $X(\rho_2)$. Then $\phi = \theta$.*

Proof We can calculate $\cos\phi$ with a dot product:

$$\cos\phi = \frac{\rho_1'(w_1) \cdot \rho_2'(w_2)}{|\rho_1'(w_1)||\rho_2'(w_2)|}.$$

The calculations are easier this time:

$$\rho_1'(w_1) = \langle -r_1 \sin w_1, r_1 \cos w_1 \rangle$$
$$\rho_2'(w_2) = \langle -r_2 \sin w_2, r_2 \cos w_2 \rangle,$$

so

$$\rho_1'(w_1) \cdot \rho_2'(w_2) = r_1 r_2 \sin w_1 \sin w_2 + r_1 r_2 \cos w_1 \cos w_2$$
$$|\rho_1'(w_1)| = \sqrt{(-r_1 \sin w_1)^2 + (r_1 \cos w_1)^2} = r_1$$
$$|\rho_2'(w_2)| = \sqrt{(-r_2 \sin w_2)^2 + (r_2 \cos w_2)^2} = r_2.$$

Therefore
$$\cos\phi = \frac{r_1 r_2 \sin w_1 \sin w_2 + r_1 r_2 \cos w_1 \cos w_2}{r_1 r_2}$$
$$= \sin w_1 \sin w_2 + \cos w_1 \cos w_2,$$

the same value as $\cos\theta$. Because θ and ϕ measure between 0 and π and have the same cosine value they must be the same. □

So while the X projection distorts distance, it does not alter the angles between curves.

40.3 Extending the domain

We have seen that there are some problems with the pseudosphere model. But locally the pseudosphere model exhibits the right behavior. The problems are at the boundary: in the s direction, the domain wraps all the way around every 2π; in the t direction, there is a cutoff at $t = 1$. What if we took the good local behavior derived from the pseudosphere, pulled it back to the domain, and then extended the domain without regard for the constraints placed on it by the pseudosphere? The result is called the upper half plane (UHP) model for hyperbolic geometry. In the UHP, the undefined terms of neutral geometry are interpreted as follows.

Point: The points are represented by points in the set $\{(s, t) \in \mathbb{R}^2 | t > 0\}$.

Line: Lines are represented by either vertical lines, $s = C$, or circles centered on the horizontal axis, $(s - C)^2 + t^2 = r^2$. In each case, t is restricted to be greater than zero.

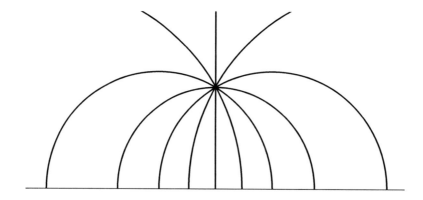

Figure 6. Some lines through a point in the upper half plane.

On: A point is on a line if its coordinates satisfy the equation that defines the line.

Between: Suppose that P, Q, and R are points on a line represented by $s = C$, so that $P = (p, C)$, $Q = (q, C)$, and $R = (r, C)$ for positive real numbers p, q, and r. The point Q is between P and R if the number q is between p and r (either $p < q < r$ or $p > q > r$). Suppose that P, Q, and R are points on a line represented by $(s - C)^2 + t^2 = r^2$, so

40.3 Extending the domain

that they can be written in the form

$$P = (r\cos\theta_P + C, r\sin\theta_P)$$
$$Q = (r\cos\theta_Q + C, r\sin\theta_Q)$$
$$R = (r\cos\theta_R + C, r\sin\theta_R),$$

where θ_P, θ_Q, and θ_R are between 0 and π. Point Q is between P and R if the number θ_Q is between θ_P and θ_R (either $\theta_P < \theta_Q < \theta_R$ or $\theta_P > \theta_Q > \theta_R$).

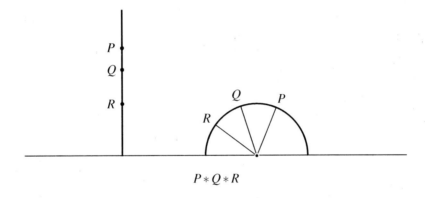

$P * Q * R$

Figure 7. A triple of points along lines represented by (l) a vertical line, and (r) an orthogonal circle. In both cases, $P * Q * R$.

Congruent (segments): If P and Q are points on a line represented by $s = C$, then the distance from $P = (p, C)$ to $Q = (q, C)$ is given by

$$d_H(P, Q) = |\ln(q/p)|.$$

As one of the points, say P, approaches the s-axis along the $s = C$ line, the distance $d_H(P, Q)$ approaches infinity, which is why t values must be greater than zero. If P and Q are points on a line represented by $(s - C)^2 + t^2 = r^2$, then the distance from P to Q is given by

$$d_H(P, Q) = |\ln([PQ, AB])|.$$

where $A = (C + r, 0)$, $B = (C - r, 0)$, and $[PQ, AB]$ denotes the cross ratio. Two segments PQ and RS are congruent if $d_H(P, Q) = d_H(R, S)$.

Congruent (angles): The angle measure of $\angle ABC$ is the same as the measure of the angle between the tangent lines to the curves that represent $\vec{BA}$ and $\vec{BC}$ at the intersection point B. Two angles are congruent if they have the same angle measure.

The UHP model is a valid model for a non-Euclidean neutral geometry! It satisfies all the axioms of neutral geometry, but not Playfair's axiom. In the next chapter, we will look at a related model called the Poincaré disk model, and verify that all the axioms are met.

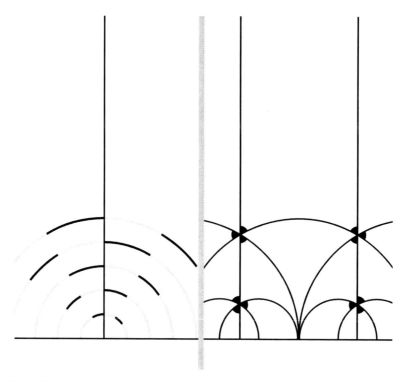

Figure 8. (l) Some congruent segments. (r) Some congruent angles, measuring $\pi/3$.

40.4 Exercises

40.1. Let ρ_1 be a circle centered on the s-axis, let ρ_2 be a vertical line which intersects ρ_1, and let ϕ be the angle of intersection of the two curves. Show that $X(\rho_1)$ and $X(\rho_2)$ intersect at the same angle (this is the other half of the argument that dealt with the case of two intersecting circles).

40.2. Let P and Q be points in the UHP at coordinates $P = (1, 3)$, $Q = (1, 5)$. What is $d_H(P, Q)$?

40.3. Let P and Q be points in the UHP at coordinates $P = (1, 2)$, $Q = (3, 4)$. What is the equation of the circle that is centered on the x-axis that passes through P and Q?

40.4. Let P and Q be points in the UHP, at coordinates $P = (1, 1)$ and $Q = (-1, 1)$. What are the two locations (i.e., coordinates) for the point R on the y-axis so that the altitude from R of $\triangle PQR$ is congruent to the side PQ?

40.5. Let P be a point in the UHP, and let Q be a point on the vertical line through P. Prove that $d_H(P, Q)$ approaches infinity as Q approaches the x-axis.

40.6. Let $\mathcal{C}$ be a circle centered on the x-axis and let P be a fixed point on $\mathcal{C}$. If Q is a point on $\mathcal{C}$, show that as Q approaches the x-axis (along $\mathcal{C}$), $d_H(P, Q)$ approaches infinity.

40.7. Let $P = (0, 1)$ and $Q = (0, 2)$. Find the coordinates of R so that $\angle PQR$, as measured in the UHP model, is a right angle.

40.4 Exercises

40.8. Let $P = (0, 1)$ and $Q = (1, 2)$. Find the coordinates of R so that $\angle PQR$, as measured in the UHP model, is a right angle.

40.9. Let P and Q be points in the UHP at coordinates $P = (1, 2)$, $Q = (3, 4)$. What is $d_H(P, Q)$?

40.10. Suppose that $P = (x_1, y)$ and $Q = (x_2, y)$ are two points in the UHP. What are the coordinates of the midpoint of PQ (using the d_H distance formula)?

40.11. Let r be a positive real number. What is the equation of the set of points that are a distance r from $(0, 1)$ (using the d_H distance formula)?

40.12. Associate the points of the UHP with the complex numbers z such that $Im(z) > 0$. Show that if $a, b, c,$ and d are real numbers such that $ad - bc > 0$, then the function

$$f(z) = \frac{az + b}{cz + d}$$

maps a point of the UHP to another point of the UHP.

41
The Poincaré disk

At the end of the last chapter, we found the upper half plane, and I claimed that it is a valid model for a non-Euclidean neutral geometry. We did not verify the claim, nor will we. Rather, we will move on to another model, the Poincaré disk model.

41.1 To the Poincaré disk model

The points of the Poincaré disk model are the points inside the unit circle. For future calculations, it is often easier to think in terms of the complex plane:

$$D = \{(x, y) \in \mathbb{R}^2 \,|\, x^2 + y^2 < 1\} = \{z \in \mathbb{C} \,|\, |z| < 1\}.$$

Fortunately, the geometry of the upper half plane transfers to the Poincaré disk via a fairly straightforward mapping.

Theorem 41.1 *Let $\rho(z)$ be the reflection of the complex plane across the real axis. Let $\sigma(z)$ be the inversion across the circle with radius $\sqrt{2}$ and center at the complex coordinate i. Then $\sigma \circ \rho$ maps the UHP to D.*

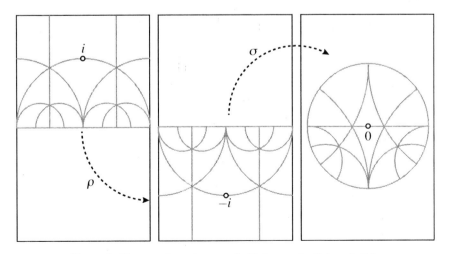

Figure 1. The map from the upper half plane to the Poincaré disk.

455

Proof To start, we find the equations of ρ and σ. The first is easy: $\rho(z) = \bar{z}$. The second is not too much more complicated. The general form for an inversion across a circle with radius r and center a is

$$\sigma(z) = \frac{r^2}{\overline{z-a}} + a.$$

With $r = \sqrt{2}$ and $a = i$, this becomes

$$\sigma(z) = \frac{2}{\overline{z-i}} + i = \frac{2}{\bar{z}+i} + i = \frac{2 + i(\bar{z}+i)}{\bar{z}+i} = \frac{i\bar{z}+1}{\bar{z}+i}.$$

The composition $\sigma \circ \rho$ is then

$$\sigma \circ \rho(z) = \sigma(\bar{z}) = \frac{iz+1}{z+i}.$$

Now look at the points along the boundary of the UHP, those of the form $z = x + 0i$. They are mapped to points whose norm is

$$|\sigma \circ \rho(x)| = \left[\sigma \circ \rho(x) \cdot \overline{\sigma \circ \rho(x)}\right]^{1/2}$$
$$= \left[\frac{ix+1}{x+i} \cdot \overline{\left(\frac{ix+1}{x+i}\right)}\right]^{1/2}$$
$$= \left[\frac{ix+1}{x+i} \cdot \frac{-ix+1}{x-i}\right]^{1/2}$$
$$= \left[\frac{x^2+1}{x^2+1}\right]^{1/2}$$
$$= 1.$$

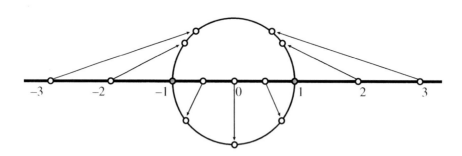

Figure 2. The image of the real axis in the map from the UHP to D.

So the image of a point on the boundary of the UHP is a point on ∂D, the boundary of D. What about points in the interior of the UHP? The reflection flips them to the lower half plane, fixing the real axis. The inversion maps the two sides of the real axis to the two "sides" of ∂D, the inside and the outside. Together, $\sigma \circ \rho(UHP)$ is either the set of points inside ∂D or the set of points outside of it. The easiest way to figure out which is to test a point.

For instance,
$$\sigma \circ \rho(i) = \frac{i \cdot i + 1}{i + i} = 0.$$
So $\sigma \circ \rho$ maps the points of the UHP to the points inside ∂D. □

While $\sigma \circ \rho$ is one of the standard maps from the UHP to the Poincaré disk, it is not the only one. You could do it with a single inversion. An advantage of $\sigma \circ \rho$ over a single inversion is that it preserves, rather than reverses, orientation.

41.2 Interpreting "undefineds" in the Poincaré disk

As a composition of a reflection and an inversion, $\sigma \circ \rho$ is particularly well-behaved. Both σ and ρ are conformal, and both preserve the cross ratio. The foundations of measurement, inherited from the pseudosphere and then extended to the UHP model, will transfer quite naturally to the Poincaré disk model.

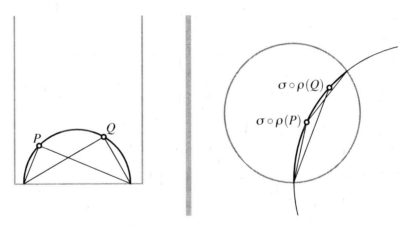

Figure 3. (l) The cross ratio in the UHP. (r) The cross ratio in D.

Points. The points are represented by the complex numbers in the set
$$D = \left\{ z \in \mathbb{C} \big| |z| < 1 \right\}.$$

Lines. Lines in the UHP are represented by portions of vertical lines and circles that are perpendicular to the real axis. Consider the image under $\sigma \circ \rho$ of one such vertical line or circle, X. First it is reflected, and $\rho(X) = X$ (although the individual points of X are not fixed). Then it is inverted: if X passes through i, the center of the inverting circle, then $\rho(X)$ will also pass through i, and so be inverted to a line. The line will pass through the origin since $\sigma \circ \rho(i) = 0$. If X does not pass through i, then neither will $\rho(X)$, and so it will be inverted to a circle. X is perpendicular to the boundary of the UHP, the real axis. Since $\sigma \circ \rho$ is conformal, $\sigma \circ \rho(X)$ is perpendicular to the boundary of the Poincaré disk, ∂D. Therefore, lines in the Poincaré disk model are represented as the restriction to D of one of two Euclidean shapes: either a line through the origin, or a circle orthogonal to ∂D.

On. A point is on a line ℓ if it is on (in the Euclidean sense) the orthogonal circle or line through the origin that represents ℓ.

Between. Suppose that P, Q, and R are points on a line represented by a line through the origin. Then Q is between P and R if it is between them in the Euclidean sense. If instead P, Q, and R are points on a line represented by an orthogonal circle with center O, then Q is between P and R if it is in the interior of $\angle POR$.

Congruence (of segments). In the UHP, distance is the logarithm of a cross ratio. But the cross ratio (and hence its logarithm) is an invariant of the reflection map ρ and the inversion σ. Therefore, in the Poincaré disk, the distance between points A and B is

$$d_H(A, B) = |\ln[AB, PQ]|,$$

where P and Q are the intersections of ∂D with the line or circle that represents $\overline{AB}$. The length of a segment AB is the distance between its endpoints, and two segments are congruent if they have the same length.

Figure 4. Some lines in the Poincaré disk model.

Congruence (of angles). In the UHP, angle measure is the Euclidean angle measure between tangent lines. Since the reflection ρ and the inversion σ are conformal, this is how angles are measured in D as well. The measure of $\angle BAC$ is defined as follows. Let r_B be the Euclidean ray along the tangent line to $\overline{AB}$ (at A) that points in the same direction as $\overrightarrow{AB}$. Let r_C be the Euclidean ray along the tangent line to $\overline{AC}$ (at A) which points in the same direction as $\overrightarrow{AC}$. Then $(\angle BAC)$ is the Euclidean measure of the angle between r_B and r_C. Two angles are congruent if they have the same angle measure.

41.3 Verifying the axioms

Let's verify that the model satisfies the axioms of a neutral geometry. Many of them are obvious, so we can focus our attention on the ones that aren't. We have been using $\mathbb{R}^2$ (and to a lesser

41.3 Verifying the axioms

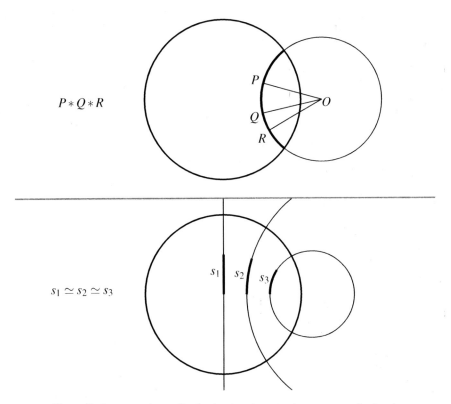

Figure 5. Interpretations of order (top) and segment congruence (bottom).

extent $\mathbb{C}$) as a model for Euclidean geometry. Now we have a subset of $\mathbb{R}^2$ (or $\mathbb{C}$) as a model for a non-Euclidean geometry. An object such as a circle that is orthogonal to the unit circle has one interpretation in the Euclidean model—it's a circle—and another in the non-Euclidean model—it's a line. This can lead to confusion, so you do have to be careful.

The axioms of incidence

In 1: There is a unique line on any two distinct points.
In 2: There are at least two points on any line.
In 3: There exist at least three points that do not all lie on the same line.

The second and third axioms of incidence are clearly true in the Poincaré disk model. Verifying the first requires more effort. To begin, let's find an equation that describes the circles that are orthogonal to the unit circle.

Lemma 41.2 *The equation of the circle with center at coordinates (h, k) that is orthogonal to ∂D is*

$$x^2 - 2xh + y^2 - 2yk = -1.$$

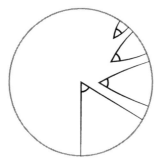

Figure 6. Four congruent angles in the Poincaré disk model.

Proof In chapter 35, we saw that as long as (h, k) is outside the unit circle, there will be one circle centered at (h, k) that is orthogonal to ∂D. We are given its center. To find its radius r, we need some labels:

O : the origin,
Q : the center of the orthogonal circle, at coordinates (h, k),
P : one of the points of intersection of the two circles.

Then $\triangle OPQ$ is a right triangle with legs of length $|OP| = 1$ and $|PQ| = r$, and hypotenuse of length $|OQ| = \sqrt{h^2 + k^2}$. By the Pythagorean theorem,

$$1 + r^2 = h^2 + k^2 \implies r^2 = h^2 + k^2 - 1.$$

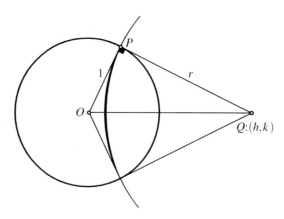

Figure 7. Finding the radius of the orthogonal circle through P.

Now substitute into the general form for a circle and simplify:

$$(x - h)^2 + (y - k)^2 = r^2$$
$$x^2 - 2xh + h^2 + y^2 - 2yk + k^2 = h^2 + k^2 - 1$$
$$x^2 - 2xh + y^2 - 2yk = -1.$$
□

41.3 Verifying the axioms

The lemma provides the analytic framework needed to verify that the Poincaré disk satisfies the first axiom of incidence.

Theorem 41.3 *In the Poincaré disk model, there is a unique line through two distinct points.*

Proof Let $P = (a, b)$ and $Q = (c, d)$ be distinct points in the Poincaré disk D. There are two possibilities, since there are two ways to represent lines, either as orthogonal circles or as lines through the origin. Suppose that P and Q are on the same line through the origin. Then the ratios b/a and d/c are the same. More properly (so we do not divide by zero), if $ad = bc$, then P and Q are on the same line through the origin, and the line is unique.

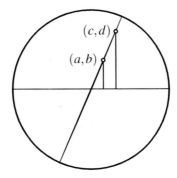

Figure 8. A line through two points, the "line through the origin" case.

Suppose that $ad \neq bc$. In that case P and Q are not on the same line through the origin, so we need to show that they are on an orthogonal circle. The general form for such a circle is

$$x^2 - 2xh + y^2 - 2yk = -1$$

where (h, k) is its center. For P and Q to be on that circle, h and k have to satisfy a system of equations:

$$a^2 - 2ah + b^2 - 2bk = -1 \quad \Longrightarrow \quad ah + bk = \tfrac{1}{2}(1 + a^2 + b^2)$$
$$c^2 - 2ch + d^2 - 2dk = -1 \qquad\qquad ch + dk = \tfrac{1}{2}(1 + c^2 + d^2).$$

In matrix form,

$$\begin{pmatrix} a & b \\ c & d \end{pmatrix} \begin{pmatrix} h \\ k \end{pmatrix} = \frac{1}{2} \begin{pmatrix} 1 + a^2 + b^2 \\ 1 + c^2 + d^2 \end{pmatrix}.$$

This matrix equation (and hence the system of equations) has a unique solution since the determinant $ad - bc$ is nonzero. So, if $ad \neq bc$, then there is exactly one orthogonal circle through P and Q. The equation indicates that if P and Q are on a line through the origin, then it is not possible for them also to be on an orthogonal circle. For if they are on a line through the origin, then $ad - bc = 0$, and the resulting system of equations to find the center of the orthogonal circle is inconsistent. Thus, two distinct points in D are either on exactly one line

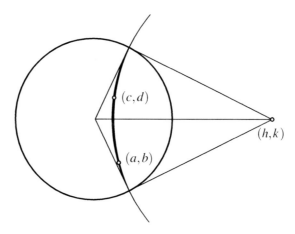

Figure 9. A line through two points, the "orthogonal circle" case.

through the origin, or on exactly one orthogonal circle (but never both), so there is a unique line through any two distinct points. □

The axioms of order

Or 1: If $A * B * C$, then the points A, B, and C are distinct collinear points, and $C * B * A$.

Or 2: For two distinct points B and D, there are points A, C, and E, such that $A * B * D$, $B * C * D$, and $B * D * E$.

Or 3: Of three distinct points on a line, exactly one lies between the other two.

Or 4: *The plane separation axiom.* Given a line ℓ and points A, B, and C that are not on ℓ. If A and B are on the same side of ℓ and A and C are on the same side of ℓ, then B and C are on the same side of ℓ. If A and B are not on the same side of ℓ and A and C are not on the same side of ℓ, then B and C are on the same side of ℓ.

None of the axioms of order are particularly troubling. The axioms of congruence and continuity are a different matter.

The axioms of congruence

Cg1: *The segment construction axiom.* If A and B are distinct points and if A' is any point, then for each ray r with endpoint A', there is a unique point B' on r such that $AB \simeq A'B'$.

Cg2: Segment congruence is reflexive (every segment is congruent to itself), symmetric (if $AA' \simeq BB'$ then $BB' \simeq AA'$), and transitive (if $AA' \simeq BB'$ and $BB' \simeq CC'$, then $AA' \simeq CC'$).

Cg3: *The segment addition axiom.* If $A * B * C$ and $A' * B' * C'$, and if $AB \simeq A'B'$ and $BC \simeq B'C'$, then $AC \simeq A'C'$.

Cg4: *The angle construction axiom.* Given $\angle BAC$ and $\overrightarrow{A'B'}$, there is a unique ray $\overrightarrow{A'C'}$ on a given side of the line $\overline{A'B'}$ such that $\angle BAC \simeq \angle B'A'C'$.

Cg5: Angle congruence is reflexive (every angle is congruent to itself), symmetric (if $\angle A \simeq \angle B$, then $\angle B \simeq \angle A$), and transitive (if $\angle A \simeq \angle B$ and $\angle B \simeq \angle C$, then $\angle A \simeq \angle C$).

Cg6: *The side angle side (SAS) axiom.* In $\triangle ABC$ and $\triangle A'B'C'$ if $AB \simeq A'B'$, $\angle B \simeq \angle B'$, and $BC \simeq B'C'$, then $\angle A \simeq \angle A'$.

41.3 Verifying the axioms

Of these axioms, I think that only *Cg2* and *Cg5* are easy to verify. The rest deserve careful consideration. We will verify *Cg1* using a distance function along a line, and in the process see the correspondence between a line and $\mathbb{R}$ that verifies the two axioms of continuity (listed below) as well.

The axioms of continuity

Ct1: *Archimedes' axiom.* If AB and CD are two segments, there is a positive integer n such that n congruent copies of CD constructed end-to-end from A along $\overrightarrow{AB}$ will pass beyond B.

Ct2: *Dedekind's axiom.* Let $\mathbb{S}_<$ and $\mathbb{S}_\geq$ be two nonempty subsets of a line ℓ satisfying: $\mathbb{S}_< \cup \mathbb{S}_\geq = \ell$; no point of $\mathbb{S}_<$ is between two points of $\mathbb{S}_\geq$; and no point of $\mathbb{S}_\geq$ is between two points of $\mathbb{S}_<$. Then there is a unique point O on ℓ such that for two other points P_1 and P_2 with $P_1 \in \mathbb{S}_<$ and $P_2 \in \mathbb{S}_\geq$ then $P_1 * O * P_2$

Let's verify the segment construction axiom. Since we have defined congruence in terms of distance, this basically is a question of measuring distances from a point A along $\overrightarrow{AB}$ using the cross ratio formula. The measurement is a straightforward calculation involving a little algebra and trigonometry.

Lemma 41.4 *Let A and B be points on a line that is represented by an orthogonal circle $\mathcal{C}$ that intersects ∂D at points P and Q. Suppose that $\mathcal{C}$ is parametrized as*

$$R(t) = (h + r\cos t, k + r\sin t).$$

If t_A, t_B, t_P, and t_Q are real numbers so that $R(t_A) = A$, $R(t_B) = B$, $R(t_P) = P$, and $R(t_Q) = Q$, then

$$d_H(A, B) = \frac{1}{2} \left| \ln \frac{(1 - \cos(t_A - t_P))(1 - \cos(t_B - t_Q))}{(1 - \cos(t_B - t_P))(1 - \cos(t_A - t_Q))} \right|.$$

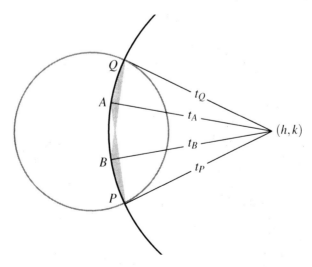

Figure 10. Distance between points on a line represented by an orthogonal circle.

Proof Point A has coordinates $(h + r\cos t_A, k + r\sin t_A)$ and point P has coordinates $(h + r\cos t_P, k + r\sin t_P)$. The Euclidean distance from A to P is

$$|AP| = \sqrt{[h + r\cos t_A - h - r\cos t_P]^2 + [k + r\sin t_A - k - r\sin t_P]^2}$$
$$= \sqrt{r^2(\cos t_A - \cos t_P)^2 + r^2(\sin t_A - \sin t_P)^2}$$
$$= r\sqrt{\cos^2 t_A - 2\cos t_A \cos t_P + \cos^2 t_P + \sin^2 t_A - 2\sin t_A \sin t_P + \sin^2 t_P}$$
$$= r\sqrt{2 - 2\cos t_A \cos t_P - 2\sin t_A \sin t_P}$$
$$= \sqrt{2}r\sqrt{1 - (\cos t_A \cos t_P + \sin t_A \sin t_P)}$$
$$= \sqrt{2}r\sqrt{1 - \cos(t_A - t_P)}.$$

(The last step uses the subtraction formula for cosine.) Similarly,

$$|BP| = \sqrt{2}r\sqrt{1 - \cos(t_B - t_P)}$$
$$|AQ| = \sqrt{2}r\sqrt{1 - \cos(t_A - t_Q)}$$
$$|BQ| = \sqrt{2}r\sqrt{1 - \cos(t_B - t_Q)}.$$

Putting the pieces together,

$$d_H(A, B) = \left|\ln\left(\frac{|AP|}{|AQ|} \cdot \frac{|BQ|}{|BP|}\right)\right|$$
$$= \left|\ln\left(\frac{\sqrt{2}r\sqrt{1 - \cos(t_A - t_P)}}{\sqrt{2}r\sqrt{1 - \cos(t_A - t_Q)}} \cdot \frac{\sqrt{2}r\sqrt{1 - \cos(t_B - t_Q)}}{\sqrt{2}r\sqrt{1 - \cos(t_B - t_P)}}\right)\right|$$
$$= \left|\ln\sqrt{\frac{(1 - \cos(t_A - t_P))(1 - \cos(t_B - t_Q))}{(1 - \cos(t_A - t_Q))(1 - \cos(t_B - t_P))}}\right|$$
$$= \frac{1}{2}\left|\ln\frac{(1 - \cos(t_A - t_P))(1 - \cos(t_B - t_Q))}{(1 - \cos(t_A - t_Q))(1 - \cos(t_B - t_P))}\right|. \quad \square$$

There is a similar formula if $\overline{AB}$ is represented by a (Euclidean) line through the origin.

Lemma 41.5 *Let A and B be points on a line that is represented by a (Euclidean) line ℓ through the origin that intersects ∂D at points $P = (c, d)$ and $Q = (-c, -d)$. Suppose that ℓ is parametrized as $R(t) = (ct, dt)$, so that $R(1) = P$ and $R(-1) = Q$. If t_A and t_B are real numbers so that $R(t_A) = A$ and $R(t_B) = B$, then*

$$d_H(A, B) = \left|\ln\frac{(1 - t_A)(1 + t_B)}{(1 + t_A)(1 - t_B)}\right|.$$

41.3 Verifying the axioms

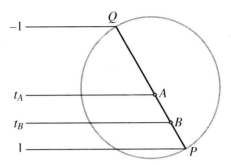

Figure 11. Distance between points on a line represented by a line through the origin.

Exercise 41.1 Prove lemma 41.5.

Exercise 41.2 Let A be a point that is a Euclidean distance of r from the origin O. Prove that

$$d_H(O, A) = \ln\left(\frac{1+r}{1-r}\right).$$

Those are the computational components that we need for the next result.

Theorem 41.6 *For a ray $\overrightarrow{AB}$ and a positive real number x, there is a unique point X on $\overrightarrow{AB}$ so that $d_H(A, X) = x$.*

For the existence part of the proof, we will show that distance as measured from A is a continuous function. We will show that it is zero at A, approaches infinity as we move away from A, and hence has every value in between along the way. For the uniqueness part, we will show that the distance is a one-to-one function along the ray, increasing as we move away from A. Therefore, there is only one point on the ray at a given distance from A. In both parts, we will consider only the case where $\overline{AB}$ is represented by an orthogonal circle. The case of the line through the origin is similar, and is left as an exercise.

Proof Label P and Q, the intersections of $\overline{AB}$ with ∂D, so that $\overrightarrow{AB}$ points toward P and away from Q. Let X be a point on $\overrightarrow{AB}$. Because A, X, P, and Q are all on the same half of the orthogonal circle that represents $\overrightarrow{AB}$, it is possible to give a parametrization $r(t)$ of the circle so that

$$r(t_A) = A, \quad r(t_X) = X, \quad r(t_P) = P, \text{ and } r(t_Q) = Q,$$

where $0 < t_Q < t_A < t_X < t_P < \pi$. Then we can use the distance formula from the earlier lemma without the absolute value signs:

$$d_H(A, X) = \frac{1}{2}\ln\frac{(1-\cos(t_A - t_P))(1-\cos(t_X - t_Q))}{(1-\cos(t_X - t_P))(1-\cos(t_A - t_Q))}.$$

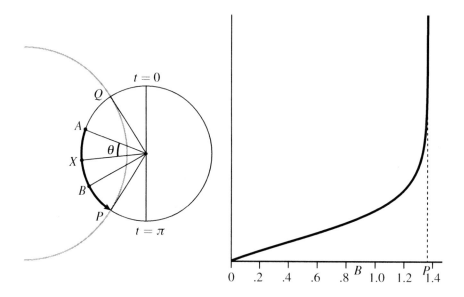

Figure 12. The distance d_H from A, as a function of θ (measured in radians).

Existence. $d_H(A, X)$ is a continuous function of t_X, since

- the cosine terms are continuous for all real numbers
- as long as X is in D, $t_X \neq t_P$, so $\cos(t_X - t_P)$ is strictly less than 1; likewise for the term $\cos(t_A - t_Q)$, so the denominators in the fractions are not zero
- the factors inside the logarithm are positive, and the logarithm is a continuous function for all positive reals.

As should be expected, when $X = A$, $t_A = t_X$ and so

$$d_H(A, A) = \frac{1}{2} \ln \frac{(1 - \cos(t_A - t_P))(1 - \cos(t_A - t_Q))}{(1 - \cos(t_A - t_P))(1 - \cos(t_A - t_Q))} = \frac{1}{2} \ln(1) = 0.$$

At the other end of the spectrum, as X approaches P, t_X approaches t_P, so the term $1 - \cos(t_X - t_P)$ in the denominator approaches zero, and thus $d_H(A, X)$ approaches infinity. Since d_H is continuous, it must take on every number between zero and infinity. That is, there is a point on $\overrightarrow{AB}$ that is a distance x from A for every positive real number x.

Uniqueness. We can determine whether a function is increasing by looking at its derivative. Before differentiating, we use the laws of logarithms to simplify:

$$d_H(A, X) = \frac{1}{2}\big[\ln(1 - \cos(t_A - t_P)) + \ln(1 - \cos(t_X - t_Q)) \\ - \ln(1 - \cos(t_X - t_P)) - \ln(1 - \cos(t_A - t_Q))\big].$$

We want to see what happens as X moves, so t_X is the variable and the other three angles, t_A, t_P, and t_Q, do not change. The derivative with respect to t_X is

$$(d_H)' = \frac{1}{2}\left[\frac{\sin(t_X - t_Q)}{1 - \cos(t_X - t_Q)} - \frac{\sin(t_X - t_P)}{1 - \cos(t_X - t_P)}\right].$$

41.3 Verifying the axioms

This can be simplified with the trigonometric identity $\sin\theta/(1 - \cos\theta) = \cot(\theta/2)$:

$$(d_H)' = \frac{1}{2}\left[\cot\left(\frac{t_X - t_Q}{2}\right) - \cot\left(\frac{t_X - t_P}{2}\right)\right]$$

$$= \frac{1}{2}\left[\cot\left(\frac{t_X - t_Q}{2}\right) + \cot\left(\frac{t_P - t_X}{2}\right)\right].$$

Since $0 < t_Q < t_X < t_P < \pi$, the terms $(t_X - t_Q)/2$ and $(t_P - t_X)/2$ are between zero and $\pi/2$, and in this interval the cotangent function is positive. Thus, $(d_H)'$ is positive, so $d_H(A, X)$ is increasing on $\overrightarrow{AB}$. It is therefore a one-to-one function on $\overrightarrow{AB}$, so no two distinct points on $\overrightarrow{AB}$ can be the same distance from A. □

Exercise 41.3 Complete the proof of the previous theorem by considering the other case where the line containing $\overrightarrow{AB}$ is modeled by a line through the origin.

Corollary 41.7 *For a segment AB and a ray $\overrightarrow{CD}$, there is a unique point E on $\overrightarrow{CD}$ so that $AB \simeq CE$.*

Proof Let AB be a segment and $\overrightarrow{CD}$ be a ray in the Poincaré disk. According to the previous result, there is a unique point E on $\overrightarrow{CD}$ so that $|CE| = |AB|$ and then $CE \simeq AB$. □

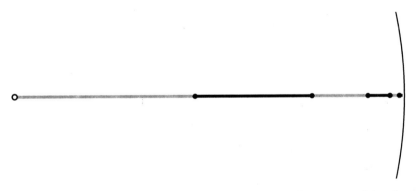

Figure 13. End-to-end segments of length 1 from the center of D out toward the edge.

We can use the distance function to establish a bijective correspondence between the points of a geometric line and the points of the real number line (taking positive distances on one ray from A, and the negatives of the distance from A on the opposite ray). This means that D will satisfy the two axioms of continuity, Dedekind's and Archimedes', as well. Next, the segment addition axiom. The key here is that distance along a line is additive.

Theorem 41.8 *(Segment addition, measured version) If $A * B * C$, then $|AC| = |AB| + |BC|$.*

Proof Let P and Q be the intersections of the orthogonal circle or line representing $\overline{AC}$ with ∂D. Choose the labels so that $\overrightarrow{AC}$ points in the direction of P not Q. Then (here the subscript

H is used to distinguish distance in the Poincaré model from Euclidean distance),

$$|AB|_H + |BC|_H = \ln \frac{|AP||BQ|}{|BP||AQ|} + \ln \frac{|BP||CQ|}{|CP||BQ|}$$

$$= \ln \frac{|AP||BQ||BP||CQ|}{|BP||AQ||CP||BQ|}$$

$$= \ln \frac{|AP||CQ|}{|CP||AQ|}$$

$$= |AC|_H. \qquad \square$$

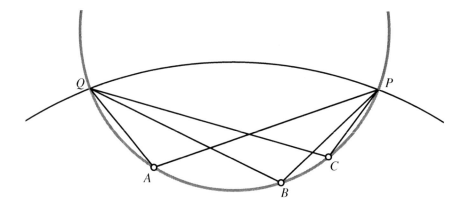

Figure 14. We can use the cross ratio to verify the measured version of the segment addition axiom.

Corollary 41.9 *The segment addition axiom holds in the Poincaré disk model.*

Proof Suppose that $A * B * C$ and $A' * B' * C'$ and that $AB \simeq A'B'$ and $BC \simeq B'C'$. Then

$$|A'C'| = |A'B'| + |B'C'| = |AB| + |BC| = |AC|,$$

and so $AC \simeq A'C'$. $\qquad \square$

Next, the angle construction axiom. Angle congruence is defined in terms of angle measure, which is measured by the Euclidean angle between tangent vectors. The next lemma helps with that measurement.

Lemma 41.10 *Let X be a point in the Poincaré disk, and let ℓ be a Euclidean line through X. Either there is a unique orthogonal circle that passes through X and has tangent line ℓ, or ℓ passes through the origin.*

Proof Label the coordinates of X as (x_0, y_0) and parametrize the line ℓ as

$$R(t) = (x_0 + at, y_0 + bt)$$

41.3 Verifying the axioms

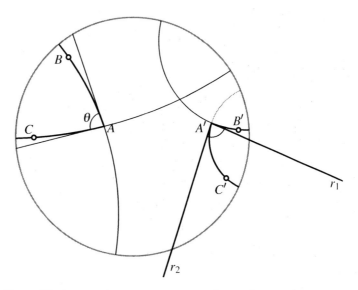

Figure 15. Constructing a congruent copy of an angle.

for some real numbers a and b (that are not both zero). Suppose that ℓ does not pass through the origin. We will show that ℓ is the tangent line at X to some orthogonal circle. To know that the circle exists, we need to find its center. The first observation is that the radial line from the circle's center out to X would have to be perpendicular to its tangent line ℓ. We can parametrize the perpendicular radial line as

$$R_\perp(t) = (x_0 + bt, y_0 - at)$$

(the lines are perpendicular because the dot product of their direction vectors is $\langle a, b \rangle \cdot \langle b, -a \rangle = ab - ba = 0$). For a point $Q = R_\perp(\tau)$ on $\ell_\perp$ other than X there is a unique circle that is centered at Q, passes through X, and has tangent line ℓ. Its radius is

$$|QX| = \sqrt{(x_0 + b\tau - x_0)^2 + (y_0 - a\tau - y_0)^2} = |\tau|\sqrt{a^2 + b^2}.$$

In general, the circle will not be an orthogonal circle. For that to happen, the circle's center and radius must satisfy the relation $r^2 = h^2 + k^2 - 1$. That is,

$$\tau^2(a^2 + b^2) = (x_0 + b\tau)^2 + (y_0 - a\tau)^2 - 1$$
$$a^2\tau^2 + b^2\tau^2 = x_0^2 + 2bx_0\tau + b^2\tau^2 + y_0^2 - 2ay_0\tau + a^2\tau^2 - 1$$
$$0 = x_0^2 + 2bx_0\tau + y_0^2 - 2ay_0\tau - 1$$
$$2ay_0\tau - 2bx_0\tau = x_0^2 + y_0^2 - 1$$
$$\tau(2ay_0 - 2bx_0) = x_0^2 + y_0^2 - 1$$
$$\tau = (x_0^2 + y_0^2 - 1)/(2(ay_0 - bx_0)).$$

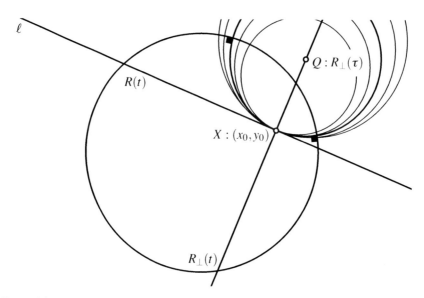

Figure 16. Many circles are tangent to $R(t)$ at X, but only one is an orthogonal circle.

The denominator is not zero: since $R(t)$ does not pass through the origin, $ay_0 \neq bx_0$ (if ay_0 and bx_0 were equal, then $R(-y_0/b)$ or $R(-x_0/a)$ would pass through the origin). Therefore the equation always has a unique solution and so there is an orthogonal circle with ℓ as its tangent line. □

Corollary 41.11 *Given $\angle BAC$ and $\overrightarrow{A'B'}$, there is a unique ray $\overrightarrow{A'C'}$ on a given side of the line $\overline{A'B'}$ such that $\angle BAC \simeq \angle B'A'C'$.*

Proof Given $\angle BAC$ with $(\angle BAC) = \theta$ and $\overrightarrow{A'B'}$ in the Poincaré disk, choose a side of $\overline{A'B'}$. Let r_1 be the Euclidean tangent ray to the orthogonal circle or line through the origin that represents $\overrightarrow{A'B'}$. On the chosen side of r_1, there is a unique ray r_2 so that the angle between r_1 and r_2 has a Euclidean measure of θ. According to the last theorem, there is either an orthogonal circle through A' with it as its tangent or the line containing r_2 passes through the origin. Either way, r_2 is the tangent vector to something that represents a line in the Poincaré disk model. Now return to the non-Euclidean side of the story. Let C' be a point on the (non-Euclidean) line so that $\overrightarrow{A'C'}$ points in the same direction as the tangent vector r_2. Then $\overrightarrow{A'C'}$ is the unique ray on the chosen side of $\overline{A'B'}$ so that $\angle B'A'C'$ is congruent to $\angle BAC$. □

The last tricky axiom on the list is the SAS axiom. To verify it, we will use a common technique of the Poincaré disk model and shift calculations to the origin where they are generally easier. In order to freely move objects around we have to use isometries. Soon we will have a whole arsenal of isometries, but for now, we will make do with what we have.

41.3 Verifying the axioms

Lemma 41.12 *Let $B = \rho e^{i\theta}$ be a point in the Poincaré disk other than the origin. Let σ be the inversion across the circle $\mathcal{C}$ with radius $r = \sqrt{1 - \rho^2}/\rho$ and center $a = (1/\rho)e^{i\theta}$. Then σ fixes the boundary of the Poincaré disk and maps B to the origin.*

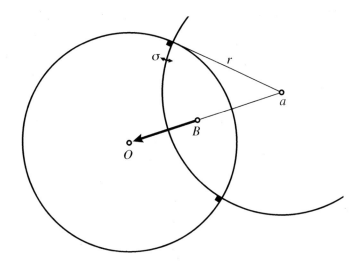

Figure 17. A hyperbolic isometry that maps the point B to the origin O.

Proof The first claim is that $\sigma(\partial D) = \partial D$. For the circle $\mathcal{C}$,

$$r^2 = \frac{1 - \rho^2}{\rho^2} = \frac{1}{\rho^2} - 1 \quad \text{and} \quad h^2 + k^2 = \frac{1}{\rho^2}.$$

Therefore $r^2 = h^2 + k^2 - 1$. That is the condition for $\mathcal{C}$ to be orthogonal to ∂D. So $\mathcal{C}$ and ∂D are orthogonal circles, and we have seen in chapter 35 that orthogonal circles are invariant under inversion. Therefore $\sigma(\partial D) = \partial D$.

The second claim is that $\sigma(B) = 0$. The equation for σ is

$$\sigma(z) = \frac{r^2}{\overline{z} - \overline{a}} + a = \frac{\frac{1}{\rho^2} - 1}{\overline{z} - \frac{1}{\rho}e^{-i\theta}} + \frac{1}{\rho}e^{i\theta}.$$

Put $z = \rho e^{i\theta}$:

$$\sigma(\rho e^{i\theta}) = \frac{\frac{1}{\rho^2} - 1}{\rho e^{-i\theta} - \frac{1}{\rho}e^{-i\theta}} + \frac{1}{\rho}e^{i\theta} = \frac{\frac{1}{\rho^2} - 1}{-\rho e^{-i\theta}(-1 + \frac{1}{\rho^2})} + \frac{1}{\rho}e^{i\theta} = -\frac{1}{\rho}e^{i\theta} + \frac{1}{\rho}e^{i\theta} = 0.$$

□

What's the advantage of the map σ? Suppose ℓ is a line that passes through B. Then ℓ may be represented by an orthogonal circle or by a line through the origin, but $\sigma(\ell)$ will always be represented by a line through the origin. It is generally easier to work with lines through the origin than orthogonal circles.

Exercise 41.4 Let σ be defined as above and let ℓ be a line through B. Prove that $\sigma(\ell)$ is represented by a line that passes through the origin.

We can now verify the SAS axiom.

Theorem 41.13 *If $\triangle ABC$ and $\triangle A'B'C'$ are two triangles in the Poincaré disk with $AB \simeq A'B'$, $\angle B \simeq \angle B'$, and $BC \simeq B'C'$, then $\angle A \simeq \angle A'$.*

Proof Given $\triangle ABC$ and $\triangle A'B'C'$ with $AB \simeq A'B'$, $\angle B \simeq \angle B'$, and $BC \simeq B'C'$, we want to show that, as the SAS axiom claims, $\angle A \simeq \angle A'$. We have just seen that there is an inversion σ that maps B to the origin. It has the effect of straightening the lines BA and BC so that they are represented by lines through the origin. We aren't done repositioning $\triangle ABC$ yet. Let r be the rotation that lines up $\overrightarrow{AB}$ along the positive real axis (or the identity if $\overrightarrow{AB}$ is along the positive real axis). Let s be the reflection across the real axis if C is below the real axis and the identity if C is already above the real axis. Let $f = s \circ r \circ \sigma$. f does not alter distances (as measured by the cross ratio) or angle measures (as measured by their tangent lines). There is an equivalent map f' for A', B', and C'.

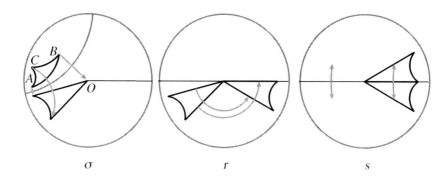

Figure 18. Verifying the SAS axiom with a sequence of isometries.

Since $AB \simeq A'B'$, $f(A)$ and $f'(A')$ are the same distance from the origin along the real axis, so $f(A) = f'(A')$. Furthermore, $\angle f(B) \simeq \angle f'(B')$, and $f(C)$ and $f'(C')$ are on the same side of the real axis. Thus $\overrightarrow{Of(C)}$ and $\overrightarrow{Of'(C')}$ are the same. On it, $f(C)$ and $f'(C')$ are the same distance from O. Therefore $f(C) = f'(C')$. There is only one orthogonal circle that passes through the points $f(A) = f'(A')$ and $f(C) = f'(C')$. We can then conclude

$$\angle A \simeq \angle f(A) = \angle f'(A') \simeq \angle A'. \qquad \square$$

Exercise 41.5 Demonstrate that the Poincaré disk model does not satisfy Playfair's axiom, and hence is not a model for Euclidean geometry.

The Poincaré disk is a valid model for a neutral non-Euclidean geometry called hyperbolic geometry. It turns out that there are only two neutral geometries, Euclidean geometry and hyperbolic geometry, though we will not prove it.

41.4 Exercises

41.6. Let $P = (0, 1/3)$ and $Q = (1/3, 0)$. What is the equation of the orthogonal circle that models the line $\overline{PQ}$?

41.7. Let $P = (1/2, 0)$ and $Q = (0, 1/2)$. What are the coordinates of the points where the orthogonal circle that models $\overline{PQ}$ intersects ∂D? The points on ∂D are not points in the geometry, but rather like points at infinity.

41.8. Give an isometry that maps $Q = (a, 0)$ to the origin.

41.9. Give an isometry that maps $P = (1/2, 1/2)$ to the origin.

41.10. Calculate the hyperbolic distance from $P = (-1/4, 0)$ to $Q = (1/2, 0)$.

41.11. Calculate the hyperbolic distance from $P = (1/2, 0)$ to $Q = (0, 1/2)$.

41.12. What is the measure of $\angle PQR$ where $P = (0, 0)$, $Q = (a, 0)$, and $R = (0, a)$ in terms of a? What happens to its measure as a approaches 1 (so Q and R approach ∂D)?

41.13. What is the angle sum of $\triangle ABC$ where $A = (0, 0)$, $B = (0, 0.5)$, and $C = (0.5, 0)$?

41.14. Verify that the Poincaré disk model satisfies the plane separation axiom.

41.15. Find a single inversion σ that maps the points of the UHP to the points of the Poincaré disk.

42

Hyperbolic Reflections

In chapters 22–27 we developed a theory of Euclidean isometries, using reflections as building blocks. It is possible to mimic that development in the hyperbolic world, and we will do that to an extent. This time we will cut it short: we will look at reflections and compositions of two reflections, but not at compositions of three reflections.

Definition 42.1 A bijection f of the Poincare disk D is a *hyperbolic isometry* if, for points A and B,

$$|f(AB)|_H = |AB|_H.$$

Large chunks of our study of Euclidean isometries really relied on neutral, not Euclidean, geometry. We only proved those results in the Euclidean context, but the proofs themselves carry over fine to hyperbolic geometry. So, we will not re-prove them here.

Theorem 42.2 *(Properties of hyperbolic isometries)*

- *The composition of two hyperbolic isometries is a hyperbolic isometry. The identity map is a hyperbolic isometry. The inverse of a hyperbolic isometry is a hyperbolic isometry.*
- *A hyperbolic isometry preserves segment and angle congruence.*
- *A hyperbolic isometry preserves the relations of incidence and order.*
- *If an isometry f fixes distinct points A and B, then it fixes all the points of $\overline{AB}$.*
- *If an isometry f fixes non-collinear points A, B, and C, then it fixes all points (it is the identity isometry).*

Following the pattern established in the study of Euclidean isometries, let's start with reflections.

Definition 42.3 Define the *hyperbolic reflection* s across a line ℓ as follows. For a point P on l, set $s(P) = P$. For a point P that is not on ℓ, there is a unique line passing through P that is perpendicular to ℓ. On this line, there is one other point that is the same (hyperbolic) distance from ℓ as P on the opposite side of ℓ from P. Set $s(P)$ to be this point.

Hyperbolic reflections retain a lot of the structure of Euclidean reflections, as collected in the next theorem. The Euclidean proofs carry over with minimal modification, so we will not take the time to re-prove them.

Theorem 42.4 *(Properties of hyperbolic reflections)*

- *A hyperbolic reflection is an isometry.*
- *If an isometry fixes all the points of a line but is not the identity, then it must be a reflection.*
- *Every hyperbolic isometry can be written as either a reflection, a composition of two reflections, or a composition of three reflections.*

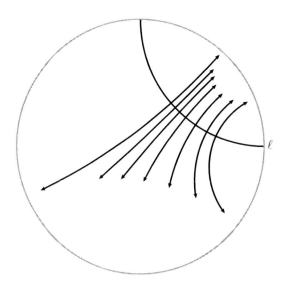

Figure 1. Reflections across a line ℓ.

Let's try to get a better sense of how hyperbolic reflections are manifested in the Poincaré disk model.

Theorem 42.5 *Let s be a hyperbolic reflection across a line ℓ. If ℓ is represented by a line through the origin, then s is the Euclidean reflection across the line. If ℓ is represented by an orthogonal circle, then s is the Euclidean inversion across the circle.*

Proof In either case, it is clear that the points on ℓ are fixed, as required. Let A be a point that is not on ℓ. There are two things to check.

That the line through $s(A)$ and A is perpendicular to ℓ. The line through A and $s(A)$ intersects ℓ at a point B. Let C be another point on ℓ. Then s fixes B and C. It may be represented as a reflection or as an inversion, but either way it is conformal so

$$\angle ABC \simeq s(\angle ABC) \simeq \angle s(A)BC.$$

The angles are congruent and supplementary so they are right angles, and the line through $s(A)$ and A is perpendicular to ℓ.

That the points $s(A)$ and A are equidistant from ℓ. Let B be the intersection of ℓ with $\overline{As(A)}$. Let P and Q be the intersections of $\overline{As(A)}$ with ∂D. Because ∂D is orthogonal to ℓ, whether s is a reflection or an inversion, s maps ∂D to itself. Likewise, $\overline{As(A)}$ is orthogonal to

Hyperbolic Reflections

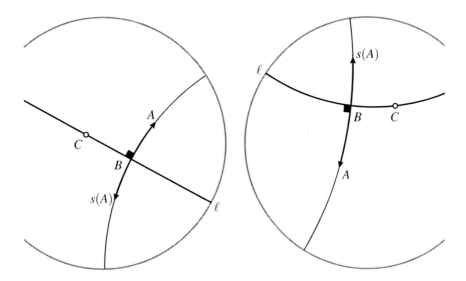

Figure 2. The two types of hyperbolic reflections: (l) reflection across a line through the origin, (r) inversion across an orthogonal circle.

ℓ so s maps $\overline{As(A)}$ to itself. Points P and Q are the only points on both ∂D and $\overline{As(A)}$; since s does not fix them, it must interchange them: $s(P) = Q$ and $s(Q) = P$. Furthermore, since B is on ℓ, it is fixed, so $s(B) = B$. Therefore

$$|s(A)B|_H = \left| \ln \left(\frac{|s(A)P|}{|BP|} \cdot \frac{|BQ|}{|s(A)Q|} \right) \right|$$
$$= \left| \ln \left(\frac{|s(A)s(Q)|}{|s(B)s(Q)|} \cdot \frac{|s(B)s(P)|}{|s(A)s(P)|} \right) \right|$$
$$= \left| \ln \left(\frac{|AQ|}{|BQ|} \cdot \frac{|BP|}{|AP|} \right) \right|$$

(for a reflection, the four segment lengths are the same; for an inversion, the segment lengths are not the same, but the cross ratio is). Continuing,

$$|s(A)B|_H = \left| \ln \left(\frac{|BQ|}{|AQ|} \cdot \frac{|AP|}{|BP|} \right)^{-1} \right| = \left| -\ln \left(\frac{|AP|}{|BP|} \cdot \frac{|BQ|}{|AQ|} \right) \right| = |AB|_H.$$

So A and $s(A)$ are the same distance from ℓ. □

In the next chapters, isometries play a fundamental role in calculations. Because of that, we will need more than a synthetic description of them. We will need to know the equations that describe them.

Theorem 42.6 *Let s be a reflection across a line ℓ. (i) If ℓ is represented by a line through the origin and ℓ forms an angle of θ with the real axis, then $s(z) = e^{2i\theta}\bar{z}$. (ii) If ℓ is represented by an orthogonal circle with center a, then $s(z) = (a\bar{z} - 1)/(\bar{z} - \bar{a})$.*

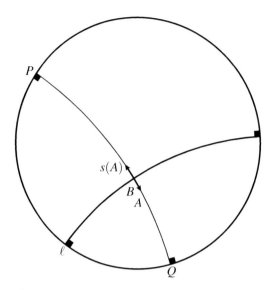

Figure 3. Point B is the midpoint of the segment from A to $s(A)$.

Proof (i) We can calculate this equation via a change of coordinates: (1) rotate by $-\theta$, (2) reflect across the real axis, (3) rotate back by θ. Here it is all together:

$$z \stackrel{(1)}{\mapsto} e^{-i\theta}z \stackrel{(2)}{\mapsto} \overline{e^{-i\theta}z} = e^{i\theta}\bar{z} \stackrel{(3)}{\mapsto} e^{i\theta}e^{i\theta}\bar{z} = e^{2i\theta}\bar{z}.$$

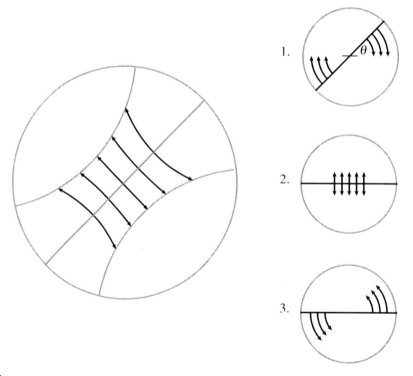

Figure 4. Finding the equation of a reflection across a line through the origin: (1) rotate, (2) reflect, (3) rotate back.

Hyperbolic Reflections

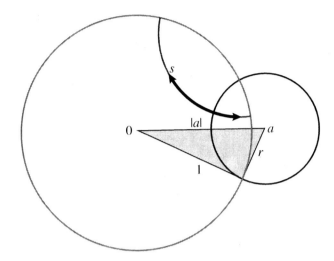

Figure 5. Finding the formula for an inversion. The orthogonal circle that is centered at the point $a = h + ik$ has radius $h^2 + k^2 - 1 = a\bar{a} - 1$.

(ii) In chapter 34 we derived a general formula for the inversion across a circle with radius r and center a:

$$s(z) = \frac{r^2}{\overline{z - a}} + a.$$

The circle is now orthogonal to the unit circle, so its radius r and center $a = h + ik$ are related by the equation

$$r^2 = h^2 + k^2 - 1 \implies r^2 = a\bar{a} - 1.$$

Therefore

$$s(z) = \frac{a\bar{a} - 1}{\overline{z - a}} + a = \frac{a\bar{a} - 1}{\bar{z} - \bar{a}} + \frac{a\bar{z} - a\bar{a}}{\bar{z} - \bar{a}} = \frac{a\bar{z} - 1}{\bar{z} - \bar{a}}. \qquad \square$$

We have grappled with two representations of lines. There is a unifying form that describes reflections in either case.

Theorem 42.7 *Let $s(z)$ be a hyperbolic reflection. Then $s(z)$ can be written in the form*

$$s(z) = \frac{A\bar{z} - \overline{B}}{B\bar{z} - \overline{A}}$$

where $A\overline{A} - B\overline{B} = 1$.

Proof If $s(z)$ is a reflection across a line represented by a line through O, then it has the form $s(z) = e^{2i\theta}\bar{z}$. Put $A = e^{i(\theta+\pi/2)}$ and $B = 0$. Then, because $e^{i\pi} = -1$,

$$\frac{A\bar{z} - \overline{B}}{B\bar{z} - \overline{A}} = \frac{e^{i(\theta+\pi/2)}\bar{z} - 0}{0\bar{z} - e^{-i(\theta+\pi/2)}}$$

$$= \frac{e^{i\theta}e^{i\pi/2}\bar{z}}{-e^{-i\theta}e^{-i\pi/2}}$$

$$= -e^{i\pi} \cdot e^{2i\theta}\bar{z}$$

$$= e^{2i\theta}\bar{z}$$

$$= s(z),$$

and

$$A\overline{A} - B\overline{B} = e^{i(\theta+\pi/2)} \cdot e^{-i(\theta+\pi/2)} - 0 = 1.$$

If, on the other hand, $s(z)$ is a reflection across an orthogonal circle, then it has the form $s(z) = (a\bar{z} - 1)/(\bar{z} - \bar{a})$. Put $A = a/r$ and $B = 1/r$, where r is the radius of the orthogonal circle. Then r is a real number, equal to its complex conjugate, so

$$\frac{A\bar{z} - \overline{B}}{B\bar{z} - \overline{A}} = \frac{(a/r)\bar{z} - (1/r)}{(1/r)\bar{z} - \overline{(a/r)}} = \frac{a\bar{z} - 1}{\bar{z} - \bar{a}} = s(z),$$

and since $r^2 = a\bar{a} - 1$,

$$A\overline{A} - B\overline{B} = \frac{a}{r}\frac{\bar{a}}{r} - \frac{1}{r}\frac{1}{r} = \frac{1}{r^2}(a\bar{a} - 1) = 1.$$

$\square$

While all hyperbolic reflections take this form, not all maps of this form are hyperbolic reflections. There is an easy check.

Theorem 42.8 *A map*

$$s(z) = \frac{A\bar{z} - \overline{B}}{B\bar{z} - \overline{A}}$$

with $A\overline{A} - B\overline{B} = 1$ *is a hyperbolic reflection if and only if B is a real number.*

Proof $\Longleftarrow$ First suppose that B is a real number. If $B = 0$, write A in its exponential form $A = re^{i\theta}$. Then

$$s(z) = \frac{A\bar{z}}{-\overline{A}} = \frac{re^{i\theta}\bar{z}}{-re^{-i\theta}} = -e^{2i\theta}\bar{z} = e^{i\pi}e^{2i\theta}\bar{z} = e^{2i(\theta+\pi/2)}\bar{z},$$

so $s(z)$ takes the form of a reflection across a line through the origin. If $B \neq 0$ (and remember, since it is real $B = \overline{B}$), then

$$s(z) = \frac{A\bar{z} - \overline{B}}{B\bar{z} - \overline{A}} = \frac{(A/B)\bar{z} - 1}{\bar{z} - \overline{(A/B)}}.$$

Hyperbolic Reflections

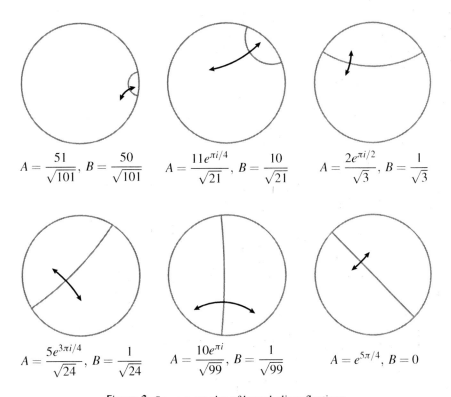

Figure 6. Some examples of hyperbolic reflections.

Because
$$(A/B) \cdot \overline{(A/B)} = \frac{A \cdot \overline{A}}{B \cdot \overline{B}} = (1 + B\overline{B})/B\overline{B} > 1,$$

A/B is a point outside the unit circle. Therefore $s(z)$ takes the form of an inversion across an orthogonal circle whose center is A/B.

$\implies$ Suppose that $s(z)$ is a hyperbolic reflection. It has many fixed points inside the Poincaré disk, so there are solutions in D to the equation

$$z = \frac{A\overline{z} - \overline{B}}{B\overline{z} - \overline{A}} \implies z(B\overline{z} - \overline{A}) = A\overline{z} - \overline{B} \implies Bz\overline{z} + \overline{B} = A\overline{z} + \overline{A}z.$$

The right hand side of the last equation is the sum of a complex number and its conjugate so it is a real number. What about the left hand side? Write $B = b_1 + ib_2$. Then

$$\begin{aligned} Bz\overline{z} + \overline{B} &= (b_1 + ib_2)z\overline{z} + (b_1 - ib_2) \\ &= b_1 z\overline{z} + ib_2 z\overline{z} + b_1 - ib_2 \\ &= b_1(z\overline{z} + 1) + b_2(z\overline{z} - 1)i. \end{aligned}$$

Inside the Poincaré disk, $z\overline{z}$ is strictly less than one, so the only way that this can be a real number is if b_2 is zero, so B is a real number. $\square$

In the next chapter, we will use the formula to investigate the composition of two hyperbolic reflections. This will provide us with a classification of orientation-preserving hyperbolic isometries.

42.1 Exercises

42.1. Show that if a hyperbolic isometry f fixes two distinct points A and B, then it fixes all the points of $\overline{AB}$. Show that if f fixes three non-collinear points, then it is the identity. (Verify that the proofs we gave earlier still hold in hyperbolic geometry.)

42.2. Retrace the steps in the proof of the Euclidean three reflections theorem (theorem 23.4) to verify that it is valid in hyperbolic geometry as well.

42.3. Let $f(z) = (A\bar{z} - b)/(b\bar{z} - \overline{A})$ where b is a real number and $A\overline{A} - b^2 = 1$. Show by direct calculation that $f \circ f$ is the identity map.

42.4. Let $f(z) = (A\bar{z} - \overline{B})/(B\bar{z} - \overline{A})$ where $A\overline{A} - B\overline{B} = 1$. Show by direct calculation that if B is not a real number, then $f \circ f$ is not the identity map. Hint: investigate $f \circ f(0)$.

42.5. Let $f(z) = (A\bar{z} - b)/(b\bar{z} - \overline{A})$ where b is a nonzero real number and $A\overline{A} - b^2 = 1$. Then f is a reflection across a hyperbolic line represented by an orthogonal circle. Find the coordinates of the center of the orthogonal circle in terms of A and b.

42.6. Find the equation for the reflection across the line that passes through the points at coordinates 0 and $0.25 + .5i$.

42.7. Find the equation for the reflection across the line that passes through the points at coordinates 0.5 and $0.5i$.

42.8. Let $s(z) = (A\bar{z} - \overline{B})/(B\bar{z} - \overline{A})$ where $A = \sqrt{2}$ and $B = 1$. What is the equation of the line that is fixed by s?

42.9. Let $s(z) = (A\bar{z} - \overline{B})/(B\bar{z} - \overline{A})$ where $A = 2i/\sqrt{3}$ and $B = 1/\sqrt{3}$. Let $O = 0 + 0i$. What is the midpoint of the segment $Os(O)$?

42.10. Find a reflection that maps the point $0.5 + 0i$ to $0 + 0i$.

42.11. Find the equation of the reflection that swaps the points $x + 0i$ and $0 + 0i$.

42.12. Find a reflection that maps the point $2/3 + 0i$ to $1/3 + 0i$.

43

Orientation-Preserving Hyperbolic Isometries

In the last chapter, we found the general form for a hyperbolic reflection in the Poincaré disk. As a consequence of the three reflections theorem, every orientation-preserving hyperbolic isometry can be written as a composition of two of them. By composing two reflections in general form, we can find the general form for an orientation-preserving isometry.

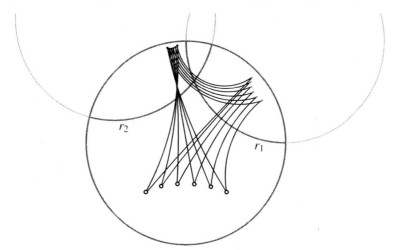

Figure 1. A composition of two reflections.

Theorem 43.1 *An orientation-preserving isometry of the Poincaré disk has the form*

$$f(z) = \frac{Ez + F}{\overline{F}z + \overline{E}}$$

where $E\overline{E} - F\overline{F} = 1$.

Proof Start with two reflections s_1 and s_2, given by

$$s_1(z) = \frac{A\overline{z} - b}{b\overline{z} - \overline{A}} \quad \text{and} \quad s_2(z) = \frac{C\overline{z} - d}{d\overline{z} - \overline{C}},$$

483

where $A\overline{A} - b^2 = 1$ and $C\overline{C} - d^2 = 1$, and b and d are real numbers. Their composition is

$$s_1 \circ s_2(z) = \frac{A\overline{\left(\dfrac{C\overline{z} - d}{d\overline{z} - \overline{C}}\right)} - b}{b\overline{\left(\dfrac{C\overline{z} - d}{d\overline{z} - \overline{C}}\right)} - \overline{A}}$$

$$= \frac{A\left(\dfrac{\overline{C}z - d}{dz - C}\right) - b}{b\left(\dfrac{\overline{C}z - d}{dz - C}\right) - \overline{A}}$$

$$= \frac{A(\overline{C}z - d) - b(dz - C)}{b(\overline{C}z - d) - \overline{A}(dz - C)}$$

$$= \frac{A\overline{C}z - Ad - bdz + bC}{b\overline{C}z - bd - \overline{A}dz + \overline{A}C}$$

$$= \frac{(A\overline{C} - bd)z + (bC - Ad)}{(b\overline{C} - \overline{A}d)z + (\overline{A}C - bd)}.$$

With the substitutions $E = A\overline{C} - bd$ and $F = bC - Ad$, this takes the desired form

$$s_1 \circ s_2(z) = \frac{Ez + F}{\overline{F}z + \overline{E}}.$$

The second part of the theorem, that $E\overline{E} - F\overline{F} = 1$, is a straightforward calculation:

$$\begin{aligned}
E\overline{E} - F\overline{F} &= (A\overline{C} - bd)\overline{(A\overline{C} - bd)} - (bC - Ad)\overline{(bC - Ad)} \\
&= (A\overline{C} - bd)(\overline{A}C - bd) - (bC - Ad)(b\overline{C} - \overline{A}d) \\
&= A\overline{A}C\overline{C} - Ab\overline{C}d - \overline{A}bCd + b^2d^2 - b^2C\overline{C} + \overline{A}bCd + Ab\overline{C}d - A\overline{A}d^2 \\
&= A\overline{A}C\overline{C} + b^2d^2 - b^2C\overline{C} - A\overline{A}d^2 \\
&= A\overline{A}(C\overline{C} - d^2) - b^2(C\overline{C} - d^2) \\
&= A\overline{A} - b^2 \\
&= 1.
\end{aligned}$$

$\square$

There are two points worth mentioning. First, while it is easy enough to calculate $E\overline{E} - F\overline{F}$ and to see that it is equal to one, it may not be clear why you would think to calculate that expression in the first place. The fact that reflections have a similar requirement might be a clue. Beyond that, there is a nice correspondence between orientation-preserving isometries and 2×2 matrices that is explored in the exercises. In the correspondence, $E\overline{E} - F\overline{F}$ is the determinant.

Second, although the theorem shows that every orientation-preserving hyperbolic isometry can be written as a function of this form, it does not answer the question of whether every function of this form is a hyperbolic isometry. In fact, it is true that every function of this form is a hyperbolic isometry.

43.1 An important example

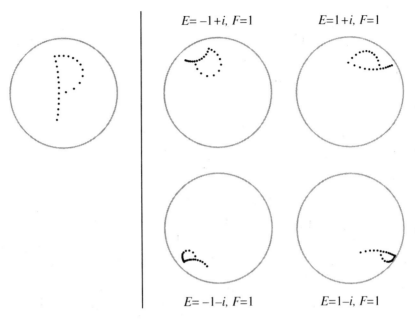

Figure 2. Four orientation-preserving isometries of the letter P (for Poincaré).

Exercise 43.1 Show that if $|z| < 1$ and $f(z) = (Ez + F)/(\overline{E}z + \overline{F})$ with $E\overline{E} - F\overline{F} = 1$, then $|f(z)| < 1$ (that is $f(z) \cdot \overline{f(z)} < 1$). Therefore f is a function from the Poincaré disk to itself.

Exercise 43.2 Show that if $f(z) = (Ez + F)/(\overline{E}z + \overline{F})$ with $E\overline{E} - F\overline{F} = 1$, then f is a bijection of the Poincaré disk. (This can be done by showing that f has an inverse that is defined on D.)

Exercise 43.3 Show that for a function of the form $f(z) = (az + b)/(cz + d)$, with $ad - bc \neq 0$,

$$f(z) - f(w) = \frac{ad - bc}{(cz + b)(cw + d)} \cdot (z - w).$$

Exercise 43.4 Show that a function of the form $f(z) = (az + b)/(cz + d)$ with $ad - bc \neq 0$, preserves the cross ratio. That is, for four points z, w, s, and t in D,

$$[f(z)f(w) : f(s)f(t)] = [zw : st].$$

43.1 An important example

It is generally easier to do calculations in the Poincaré disk if they are "at the origin" where lines are represented by Euclidean lines rather than orthogonal circles. For instance, $|AB|_H$ is easier to calculate if $\overline{AB}$ passes through the origin, and $(\angle ABC)$ is easier to calculate if B is located at the origin. To take advantage of this, we need an isometry that will map an arbitrary point to the origin.

Theorem 43.2 *Let α be a point in D. The map*

$$f(z) = \frac{z - \alpha}{1 - \overline{\alpha}z}$$

is a hyperbolic isometry that maps α to zero.

Proof Of course

$$f(\alpha) = \frac{\alpha - \alpha}{1 - \overline{\alpha}\alpha} = 0.$$

To see that f is an isometry, let's see how it can be written in the form $(Ez + F)/(\overline{F}z + \overline{E})$ where $E\overline{E} - F\overline{F} = 1$. To find the right E and F put $\rho = 1/\sqrt{1 - \alpha\overline{\alpha}}$, and let $E = \rho$ and $F = -\rho\alpha$. Because ρ is a real number, $\rho = \overline{\rho}$. Substituting and simplifying,

$$\frac{Ez + F}{\overline{F}z + \overline{E}} = \frac{\rho z - \rho\alpha}{-\overline{\rho\alpha}z + \rho} = \frac{\rho(z - \alpha)}{\rho(-\overline{\alpha}z + 1)} = \frac{z - \alpha}{1 - \overline{\alpha}z},$$

and

$$\begin{aligned}
E\overline{E} - F\overline{F} &= \rho \cdot \rho - (-\rho\alpha)\overline{(-\rho\alpha)} \\
&= \rho^2 - \rho^2 \alpha\overline{\alpha} \\
&= \rho^2(1 - \alpha\overline{\alpha}) \\
&= [1/(1 - \alpha\overline{\alpha})] \cdot (1 - \alpha\overline{\alpha}) \\
&= 1.
\end{aligned}$$

□

43.2 Classification by fixed points

Orientation-preserving Euclidean isometries are classified by their fixed points:

- no fixed points: translation,
- one fixed point: rotation,
- all points fixed: identity.

Let us undertake a similar analysis of the fixed points of orientation-preserving hyperbolic isometries.

Theorem 43.3 *Other than the identity, a map $f(z) = (Ez + F)/(\overline{F}z + \overline{E})$ with $E\overline{E} - F\overline{F} = 1$ has either one or two fixed points in $\mathbb{C}$. If $f(z)$ has one fixed point, then it is either at the origin or on the unit circle ∂D. If $f(z)$ has two fixed points, then either both are on ∂D, or one is inside D and one is outside of it.*

Proof A fixed point, z, of f will satisfy

$$\frac{Ez + F}{\overline{F}z + \overline{E}} = z.$$

43.2 Classification by fixed points

Solving for z,

$$Ez + F = z(\overline{F}z + \overline{E})$$
$$Ez + F = \overline{F}z^2 + \overline{E}z$$
$$0 = \overline{F}z^2 + (\overline{E} - E)z - F. \quad (1)$$

As long as $\overline{F} \neq 0$ (this case is dealt with later), the quadratic formula gives the solutions:

$$z = \frac{-(\overline{E} - E) \pm \sqrt{(\overline{E} - E)^2 + 4F\overline{F}}}{2\overline{F}}$$

$$= \frac{(E - \overline{E}) \pm \sqrt{\overline{E}^2 - 2E\overline{E} + E^2 + 4F\overline{F}}}{2\overline{F}}$$

$$= \frac{(E - \overline{E}) \pm \sqrt{\overline{E}^2 + 2E\overline{E} + E^2 - 4E\overline{E} + 4F\overline{F}}}{2\overline{F}}$$

$$= \frac{(E - \overline{E}) \pm \sqrt{(E + \overline{E})^2 - 4}}{2\overline{F}}.$$

The discriminant reveals how many solutions this equation has. There is just one solution (a double root) if it is zero. This occurs when

$$(E + \overline{E})^2 - 4 = 0$$
$$(E + \overline{E})^2 = 4$$
$$E + \overline{E} = \pm 2.$$

Otherwise, there will be two solutions.

Where are the fixed points located? Let a and b denote the two fixed points (whether or not they are distinct). Comparing the factored and non-factored forms of the quadratic equation (1),

$$\overline{F}z^2 + (\overline{E} - E)z - F = \overline{F}(z - a)(z - b).$$

The constant terms match up as $-F = \overline{F}ab$ and so the product of the roots is $ab = -F/\overline{F}$. Therefore

$$|a||b| = |ab| = |-F/\overline{F}| = \sqrt{\left(-\frac{F}{\overline{F}}\right)\overline{\left(-\frac{F}{\overline{F}}\right)}} = \sqrt{\frac{F}{\overline{F}} \cdot \frac{\overline{F}}{F}} = 1.$$

Either both $|a|$ and $|b|$ are 1, or one is less than 1 and the other greater. That is, either both a and b are on ∂D, or one is inside D and the other is outside it.

What happens when $\overline{F}$ is zero? In that case $F = 0$, and equation (1) becomes

$$(\overline{E} - E)z = 0.$$

The value E cannot be a real number, for if it were, f would be the identity map. Therefore $\overline{E} - E \neq 0$, and so there is a single fixed point at $z = 0$. □

Now that we have some understanding of the arrangements of fixed points, we can classify the orientation-preserving hyperbolic isometries.

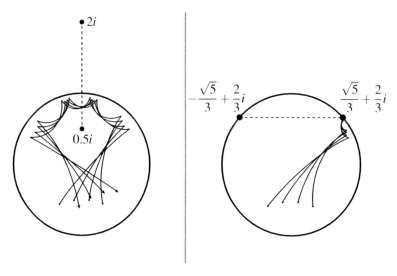

Figure 3. The paths of nine points under repeated iteration of hyperbolic isometries. (l) $E = 0.8 + i$, $F = 0.8$, with a single fixed point inside D; (r) $E = 1.5 + i$, $F = 1.5$, with two fixed points on ∂D.

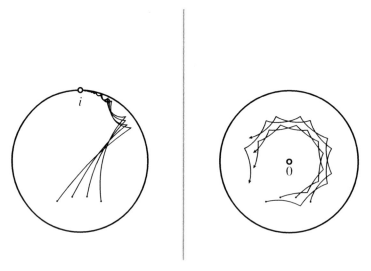

Figure 4. (l) $E = 1 + i$, $F = 1$, with a single fixed point on ∂D ($E + \overline{E} = 2$); (r) $E = e^{i\pi/8}$, $F = 0$, with a single fixed point at the origin.

Definition 43.4 Let $f(z)$ be a hyperbolic isometry. Then $f(z)$ is called:

- *elliptic* if it has a fixed point in D,
- *parabolic* if it has a single fixed point on ∂D,
- *loxodromic* if it has two fixed points on ∂D.

Each is a composition of two reflections. We conclude with a few comments about how fixed point behavior relates to the lines of reflection. Elliptic isometries are the result of a

43.2 Classification by fixed points

composition of reflections across intersecting lines. The fixed point is their point of intersection. Elliptic isometries are the hyperbolic equivalent of rotations. Parabolic isometries are the result of a composition of reflections across asymptotically parallel lines. The fixed point on ∂D is where the two lines "meet at infinity". Loxodromic isometries are the result of a composition of two reflections across divergent parallels. Divergent parallels share a unique mutual perpendicular and the two fixed points are where it intersects ∂D.

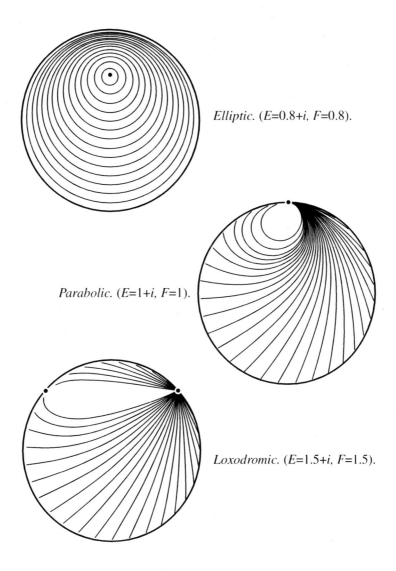

Elliptic. ($E=0.8+i$, $F=0.8$).

Parabolic. ($E=1+i$, $F=1$).

Loxodromic. ($E=1.5+i$, $F=1.5$).

Figure 5. Flow lines for the three types of orientation-preserving isometry indicating the motions of points in relation to the fixed point(s).

43.3 References

I highly recommend *Indra's Pearls* [MSW02] for a fascinating study of fractional linear transformations.

43.4 Exercises

43.5. A hyperbolic isometry is a special case of a more general form of mapping known as a fractional linear transformation,

$$f(z) = \frac{az+b}{cz+d}, \quad ad - bc \neq 0.$$

They are closely connected to the set of 2×2 invertible matrices

$$\begin{pmatrix} a & b \\ c & d \end{pmatrix}, \quad ad - bc \neq 0$$

via the correspondence

$$\Phi\left(\frac{az+b}{cz+d}\right) = \begin{pmatrix} a & b \\ c & d \end{pmatrix}.$$

Verify that $\Phi(f \circ g) = \Phi(f) \cdot \Phi(g)$ and that $\Phi(f^{-1}) = \Phi(f)^{-1}$.

43.6. Classify the isometry

$$f(z) = \frac{Ez + F}{\overline{F}z + \overline{E}}$$

where $E = 0.5 + i$ and $F = 0.5$. Find the fixed point(s) of f.

43.7. Classify the isometry

$$f(z) = \frac{Ez + F}{\overline{F}z + \overline{E}}$$

where $E = 2 + i$ and $F = \sqrt{2} + i\sqrt{2}$. Find the fixed point(s) of f.

43.8. Let s_1 and s_2 be reflections across divergent parallels, and let ℓ be the mutual perpendicular to their lines of reflection. Prove that the fixed points of $s_1 \circ s_2$ are at the intersections of ℓ with ∂D.

43.9. Find the equation of a parabolic isometry whose fixed point is at 1.

43.10. Find the equation of an elliptic isometry whose fixed point is at $0.5 + 0.5i$. Hint: consider a change of coordinates that will move $0.5 + 0.5i$ to the origin.

43.11. Find the equation of the reflection across the line that passes through $P : 0.2$ and $Q : 0.6i$.

43.12. Let $P = (1/3) + (1/3)i$ and $Q = 1/3 + 0i$. What is the hyperbolic distance from to P to Q?

43.13. Consider $\square ABCD$ where $A = a + 0i$, $B = 0 + ai$, $C = -a + 0i$, and $D = 0 - ai$. What is the measure of an interior angle of the quadrilateral (in terms of a)?

43.4 Exercises

43.14. Show that every map of the form $f(z) = (Ez + F)/(\overline{E}z + \overline{F})$ with $E\overline{E} - F\overline{F} = 1$ is a hyperbolic isometry.

43.15. Prove that in the Poincaré disk, circles centered at the point $O = 0 + 0i$ are modeled by Euclidean circles $r(\theta) = c \cos \theta + i \cdot c \sin \theta$ where c is a real number between 0 and 1. What is the relationship between the radius of the hyperbolic circle and c?

43.16. Let τ be an orientation-preserving hyperbolic isometry and let $\mathcal{C}$ be a (hyperbolic) circle centered at the origin $O = 0 + 0i$. According to the previous problem, $\mathcal{C}$ is modeled by a Euclidean circle. Since isometries preserve distances, we know that $\tau(\mathcal{C})$ is also a (hyperbolic) circle. Give an argument for why $\tau(\mathcal{C})$ is also modeled by a Euclidean circle. Because of this, we say that "circles look like circles" in the Poincaré disk model, but be warned: the center of $\tau(\mathcal{C})$ is not $\tau(O)$.

44

The Six Hyperbolic Trigonometric Functions

The hyperbolic trigonometric functions are often first introduced in a calculus class, apparently to provide some more examples for differentiation. But they play a key role in the trigonometry of the hyperbolic plane (though it is rarely as simple as, for instance, replacing sines with hyperbolic sines). The purpose of this chapter is to review some of their basic properties, in preparation for the hyperbolic trigonometry of the next chapter.

Definition 44.1 The hyperbolic sine of x, written $\sinh x$, and the hyperbolic cosine of x, written $\cosh x$, are defined in terms of the exponential function as

$$\sinh x = \frac{e^x - e^{-x}}{2} \quad \text{and} \quad \cosh x = \frac{e^x + e^{-x}}{2}.$$

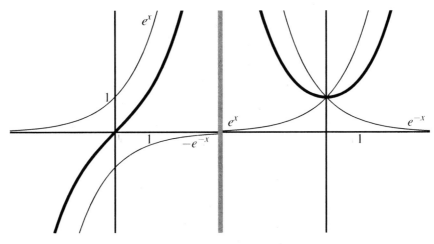

Figure 1. (l) $y = \sinh(x)$ and (r) $y = \cosh(x)$.

From the definitions, it is easy to calculate their Taylor series, which are similar to the Taylor series of the regular sine and cosine.

Theorem 44.2 *The Taylor series for the hyperbolic sine and hyperbolic cosine, expanded about the point 0, are*

$$\sinh x = \sum_{n=0}^{\infty} \frac{x^{2n+1}}{(2n+1)!}, \quad \text{and} \quad \cosh x = \sum_{n=0}^{\infty} \frac{x^{2n}}{(2n)!}.$$

Proof This is a straightforward calculation using the series expansion of the exponential function.

$$\sinh x = \frac{1}{2}(e^x - e^{-x})$$
$$= \frac{1}{2}\left[\sum_{n=0}^{\infty} \frac{x^n}{n!} - \sum_{n=0}^{\infty} \frac{(-x)^n}{n!}\right]$$
$$= \frac{1}{2}\left[\left(1 + x + \frac{x^2}{2!} + \frac{x^3}{3!} + \cdots\right) - \left(1 - x + \frac{x^2}{2!} - \frac{x^3}{3!} + \cdots\right)\right]$$
$$= \frac{1}{2}\left[2x + 2\frac{x^3}{3!} + 2\frac{x^5}{5!} + \cdots\right]$$
$$= x + \frac{x^3}{3!} + \frac{x^5}{5!} + \cdots$$
$$= \sum_{n=0}^{\infty} \frac{x^{2n+1}}{(2n+1)!}.$$

The calculation for $\cosh x$ is similar. □

The remaining four hyperbolic trigonometric functions are defined in terms of the hyperbolic sine and hyperbolic cosine, by analogy with the regular trigonometric functions.

Definition 44.3

$$\tanh x = \frac{\sinh x}{\cosh x} = \frac{e^x - e^{-x}}{e^x + e^{-x}}, \quad \coth x = \frac{\cosh x}{\sinh x} = \frac{e^x + e^{-x}}{e^x - e^{-x}},$$

$$\operatorname{sech} x = \frac{1}{\cosh x} = \frac{2}{e^x + e^{-x}}, \quad \operatorname{csch} x = \frac{1}{\sinh x} = \frac{2}{e^x - e^{-x}}.$$

As with the regular trigonometric functions, $\cosh x$ and $\operatorname{sech} x$ are even functions, while the other four are odd. As with regular trigonometric functions, there are many identities involving them. They parallel their non-hyperbolic counterparts enough so that the regular trigonometric identities can motivate the search for hyperbolic identities.

Theorem 44.4 *(Pythagorean identities)*

$$\cosh^2 x - \sinh^2 x = 1$$
$$\operatorname{sech}^2 x = 1 - \tanh^2 x$$
$$\operatorname{csch}^2 x = \coth^2 x - 1.$$

The Six Hyperbolic Trigonometric Functions

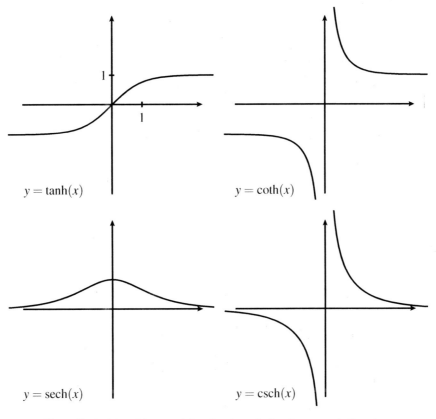

Figure 2. Graphs of the remaining four hyperbolic trigonometric functions.

Proof We will prove the first.

$$\cosh^2 x - \sinh^2 x = \left(\frac{e^x + e^{-x}}{2}\right)^2 - \left(\frac{e^x - e^{-x}}{2}\right)^2$$

$$= \frac{e^{2x} + 2 + e^{-2x}}{4} - \frac{e^{2x} - 2 + e^{-2x}}{4}$$

$$= \frac{4}{4} = 1.$$

The second can be derived by dividing by $\cosh^2 x$ and the third by dividing by $\sinh^2 x$. □

Theorem 44.5 *(Addition formulas)*

$$\sinh(x + y) = \sinh x \cosh y + \cosh x \sinh y$$
$$\cosh(x + y) = \cosh x \cosh y + \sinh x \sinh y.$$

Proof For the first of these (the second is similar),

$$\begin{aligned}
\sinh x \cosh y &+ \cosh x \sinh y \\
&= \frac{e^x - e^{-x}}{2} \cdot \frac{e^y + e^{-y}}{2} + \frac{e^x + e^{-x}}{2} \cdot \frac{e^y - e^{-y}}{2} \\
&= \frac{e^x e^y + e^x e^{-y} - e^{-x} e^y - e^{-x} e^{-y}}{4} + \frac{e^x e^y - e^x e^{-y} + e^{-x} e^y - e^{-x} e^{-y}}{4} \\
&= \frac{2 e^x e^y - 2 e^{-x} e^{-y}}{4} \\
&= \frac{e^{x+y} - e^{-x-y}}{2} \\
&= \sinh(x + y).
\end{aligned}$$
□

The addition formulas unlock a lot of other identities, listed below. Each is a simple calculation, so the proofs are just outlined.

Corollary 44.6 *(Double angle formulas)*

$$\begin{aligned}
\sinh(2x) &= 2 \sinh x \cosh x, \\
\cosh(2x) &= \cosh^2 x + \sinh^2 x \\
&= 1 + 2 \sinh^2 x \\
&= 2 \cosh^2 x - 1.
\end{aligned}$$

Proof Write $2x$ as $x + x$ and use the addition formulas. □

Corollary 44.7 *(Power reduction formulas)*

$$\begin{aligned}
\sinh^2 x &= \frac{\cosh(2x) - 1}{2} \\
\cosh^2 x &= \frac{\cosh(2x) + 1}{2}.
\end{aligned}$$

Proof Solve for $\sinh^2 x$ and $\cosh^2 x$ in the second and third variations of the double angle formula for $\cosh x$. □

Corollary 44.8 *(Half angle formulas)*

$$\begin{aligned}
\sinh^2(x/2) &= \frac{\cosh x - 1}{2} \\
\cosh^2(x/2) &= \frac{\cosh x + 1}{2}.
\end{aligned}$$

Proof Substitute $x/2$ for x in the power reduction formulas. □

The inverses of the hyperbolic trigonometric functions will appear in upcoming calculations. Since the hyperbolic trigonometric functions are defined in terms of exponentials, we should expect that their inverses would be defined in terms of logarithms, and that is the case. Four of the hyperbolic trigonometric functions are one-to-one, and so have proper inverses, but both $\cosh x$ and $\text{sech}\, x$ are not one-to-one, so to invert them, we have to restrict them to a suitable domain where they are, usually $[0, \infty)$.

Theorem 44.9 *(Inverse hyperbolic trigonometric functions)*

$$\sinh^{-1} x = \ln\left[x + \sqrt{x^2 + 1}\right] \qquad \cosh^{-1} x = \ln\left[x + \sqrt{x^2 - 1}\right]$$

$$\tanh^{-1} x = \frac{1}{2} \ln\left[\frac{1+x}{1-x}\right] \qquad \coth^{-1} x = \frac{1}{2} \ln\left[\frac{x+1}{x-1}\right]$$

$$\text{sech}^{-1} x = \ln\left[\frac{1 + \sqrt{1 - x^2}}{x}\right] \qquad \text{csch}^{-1} x = \ln\left[\frac{1 + \sqrt{1 + x^2}}{x}\right].$$

Proof To find the inverses, switch the roles of x and y and solve for y. For instance, the calculation of the inverse hyperbolic sine is as follows. Put

$$x = \sinh y = \frac{e^y - e^{-y}}{2}.$$

Then

$$e^y - e^{-y} = 2x$$
$$e^y - 2x - e^{-y} = 0$$
$$e^{2y} - 2xe^y - 1 = 0,$$

and from there the quadratic formula gives

$$e^y = \frac{2x \pm \sqrt{4x^2 + 4}}{2} = x \pm \sqrt{x^2 + 1}.$$

There are seemingly two roots, but e^y has to be positive, so

$$e^y = x + \sqrt{x^2 + 1} \implies y = \ln(x + \sqrt{x^2 + 1}).$$

The remaining inverses can be found in the same manner. □

The next chapter is full of heavy calculations involving these hyperbolic trigonometric functions. The properties and identities will help us to battle through them.

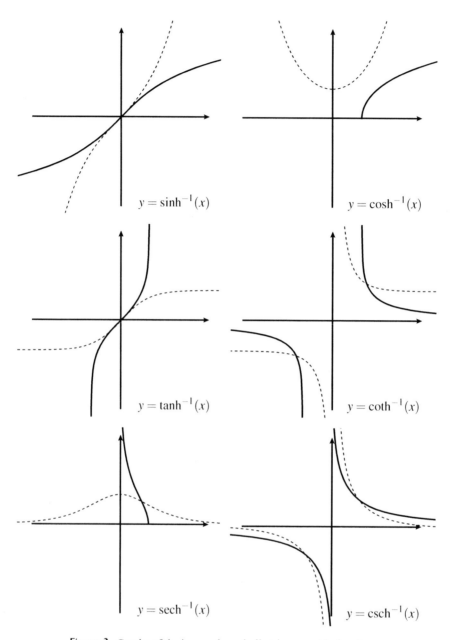

Figure 3. Graphs of the inverse hyperbolic trigonometric functions.

44.1 Exercises

44.1. Verify that the Taylor series for the hyperbolic cosine (expanded about zero) is

$$\cosh x = \sum_{n=0}^{\infty} \frac{x^{2n}}{(2n)!}.$$

44.1 Exercises

44.2. Verify the addition formula for hyperbolic cosine.

44.3. Verify the formula for the inverse of the hyperbolic tangent function.

44.4. Compute the derivatives of the hyperbolic trigonometric functions.

44.5. Compute the derivatives of the inverse hyperbolic trigonometric functions.

44.6. Show that the hyperbolic distance from the origin to a point P at complex coordinate z is
$$|OP|_H = 2\tanh^{-1}(|z|).$$

45

Hyperbolic Trigonometry

Suppose $\triangle ABC$ and $\triangle A'B'C'$ have congruent corresponding angles:

$$\angle A \simeq \angle A', \qquad \angle B \simeq \angle B', \qquad \angle C \simeq \angle C'.$$

In Euclidean geometry, this would mean that the triangles are similar. It would mean that, while the triangles might not be congruent, one would be just a scaled version of the other. In hyperbolic geometry, the situation is different. With the right isometries, we can translate A' to A, rotate around A to put B' on $\overrightarrow{AB}$, and then (if necessary) reflect across $\overline{AB}$ to put C' on $\overrightarrow{AC}$. The resulting image of $\triangle A'B'C'$ (call it $\triangle AB''C''$) is congruent to $\triangle A'B'C'$, but overlays $\triangle ABC$. Unless $B'' = B$ and $C'' = C$, either $A * B * B''$ and $A * C * C''$ or $A * B'' * B$ and $A * C'' * C$.

Exercise 45.1 Explain why these are the only two possibilities. Why, for instance, is it not possible to have $A * B * B''$ and $A * C'' * C$?

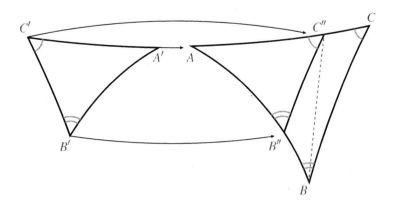

Figure 1. Attempting to create similar but non-congruent triangles.

Either way, look at $\square BB''C''C$. Since $\angle B \simeq \angle B''$ and $\angle C \simeq \angle C''$, its interior angles sum to

$$(\angle B) + (\pi - (\angle B)) + (\angle C) + (\pi - (\angle C)) = 2\pi.$$

501

But the diagonal BC'' divides the quadrilateral into two triangles, and in the hyperbolic world, their angle sums would each be less than π and so the angle sum of $\square BB''C''C$ would have to be less than 2π, contradicting the previous calculation. The only possible conclusion is that in hyperbolic geometry, there are no similar triangles other than congruent ones. In hyperbolic geometry, AAA is a congruence theorem. That has implications for trigonometry. In Euclidean geometry, the trigonometric functions are defined as ratios of sides but it is because of similarity that those ratios are constant. Otherwise, the trigonometric functions would not be well-defined. Things are going to be different in hyperbolic geometry. This chapter gives a glimpse of hyperbolic trigonometry.

45.1 Pythagorean theorem

The first piece of the puzzle is a hyperbolic analog of the Pythagorean theorem, an equation that relates the lengths of the sides of a right triangle. The proof of the theorem relies on a half-angle formula for the hyperbolic tangent.

Lemma 45.1

$$\tanh^2(x/2) = \frac{\cosh x - 1}{\cosh x + 1}.$$

Exercise 45.2 Prove the lemma.

Now on to the Pythagorean theorem.

Theorem 45.2 *(The hyperbolic Pythagorean theorem) Let $\triangle ABC$ be a right triangle with right angle at C. Label the lengths of the sides of this triangle as $a = |BC|$, $b = |AC|$, and $c = |AB|$. Then*

$$\cosh c = \cosh a \cdot \cosh b.$$

Proof The triangle is congruent to another triangle whose legs are along the positive real and imaginary axes. We can get from the first to the second by

- translating C to the origin,
- rotating about the origin to put B on the positive real axis, and
- flipping across the real axis (if necessary) to put A on the positive imaginary axis.

Any relationship between the sides of this more conveniently aligned triangle will also hold for the general triangle, so there is no loss of generality in assuming that our triangle is so aligned in the first place. Suppose that point A is located at $0 + iy$ and that point B is located at $x + 0i$. Then the lengths a and b can be calculated with the hyperbolic distance formula, easily since the lines pass through the origin (see exercise 41.2):

$$a = \ln\left(\frac{1+x}{1-x}\right) \quad \text{and} \quad b = \ln\left(\frac{1+y}{1-y}\right).$$

45.1 Pythagorean theorem

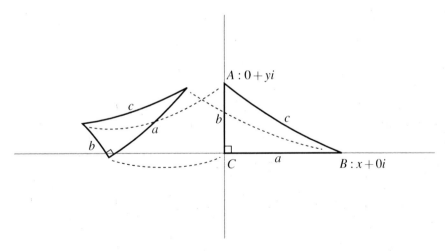

Figure 2. Proof of the hyperbolic Pythagorean theorem: positioning the triangle to measure the legs.

The easiest way to measure c is to translate B to the origin. In chapter 43 we found an isometry that would move any point to the origin. Here it takes the form

$$f(z) = \frac{z - (x + 0i)}{1 - \overline{(x + 0i)}z} = \frac{z - x}{1 - xz}.$$

Then

$$f(iy) = \frac{iy - x}{1 - xyi} \implies |f(iy)| = \frac{\sqrt{x^2 + y^2}}{\sqrt{1 + x^2 y^2}}.$$

Therefore

$$c = |Of(A)| = \ln\left(\left(1 + \frac{\sqrt{x^2 + y^2}}{\sqrt{1 + x^2 y^2}}\right) \Big/ \left(1 - \frac{\sqrt{x^2 + y^2}}{\sqrt{1 + x^2 y^2}}\right)\right).$$

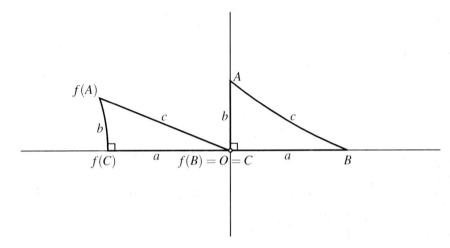

Figure 3. Proof of the hyperbolic Pythagorean theorem: positioning the triangle to measure the hypotenuse.

At this point, we could substitute a, b, and c into the equation given in the statement of the theorem. After simplification, the two sides would be the same. But, while that verifies the equation, it doesn't motivate the equation.

The inverse hyperbolic tangent is

$$\tanh^{-1}(x) = \frac{1}{2}\ln\left(\frac{1+x}{1-x}\right).$$

Therefore

$$a = 2\tanh^{-1}(x) \implies x = \tanh(a/2)$$
$$b = 2\tanh^{-1}(y) \implies y = \tanh(b/2)$$
$$c = 2\tanh^{-1}\left(\frac{\sqrt{x^2+y^2}}{\sqrt{1+x^2y^2}}\right) \implies \frac{x^2+y^2}{1+x^2y^2} = \tanh^2(c/2).$$

Let's work on the last equation. On the right side, using lemma 45.1,

$$\tanh^2(c/2) = \frac{\cosh c - 1}{\cosh c + 1}.$$

On the left side (again using the lemma),

$$\frac{x^2+y^2}{1+x^2y^2} = \frac{\tanh^2(a/2) + \tanh^2(b/2)}{1 + \tanh^2(a/2)\tanh^2(b/2)} = \frac{\dfrac{\cosh a - 1}{\cosh a + 1} + \dfrac{\cosh b - 1}{\cosh b + 1}}{1 + \dfrac{\cosh a - 1}{\cosh a + 1} \cdot \dfrac{\cosh b - 1}{\cosh b + 1}}.$$

Multiply the numerator and denominator by $(\cosh a + 1)(\cosh b + 1)$, expand and simplify:

$$\frac{x^2+y^2}{1+x^2y^2} = \frac{(\cosh a - 1)(\cosh b + 1) + (\cosh a + 1)(\cosh b - 1)}{(\cosh a + 1)(\cosh b + 1) + (\cosh a - 1)(\cosh b - 1)}$$
$$= \frac{\cosh a \cosh b + \cosh a - \cosh b - 1 + \cosh a \cosh b - \cosh a + \cosh b - 1}{\cosh a \cosh b + \cosh a + \cosh b + 1 + \cosh a \cosh b - \cosh a - \cosh b + 1}$$
$$= \frac{2\cosh a \cosh b - 2}{2\cosh a \cosh b + 2}$$
$$= \frac{\cosh a \cosh b - 1}{\cosh a \cosh b + 1}.$$

So

$$\frac{\cosh c - 1}{\cosh c + 1} = \frac{\cosh a \cosh b - 1}{\cosh a \cosh b + 1},$$

and

$$(\cosh c - 1)(\cosh a \cosh b + 1) = (\cosh c + 1)(\cosh a \cosh b - 1)$$
$$\cosh a \cosh b \cosh c - \cosh a \cosh b + \cosh c - 1$$
$$= \cosh a \cosh b \cosh c + \cosh a \cosh b - \cosh c - 1$$
$$2\cosh c = 2\cosh a \cosh b$$
$$\cosh c = \cosh a \cosh b. \qquad \square$$

45.2 Sine and cosine in a hyperbolic triangle

In the trigonometry of Euclidean geometry, the trigonometric functions describe relationships between the angles and sides of a right triangle. It would be nice to see if there are analogous relationships in hyperbolic geometry. Is there, for instance, a way to relate the cosine of an angle to the lengths of the adjacent side and the hypotenuse? Yes, there is. We begin with a little calculation.

Lemma 45.3
$$\frac{\tanh(x/2)}{\tanh^2(x/2) + 1} = \frac{\tanh x}{2}.$$

Exercise 45.3 Use lemma 45.1 to verify this lemma.

Theorem 45.4 *(Cosine in a hyperbolic triangle) Let $\triangle ABC$ be a right triangle with right angle at C. Label the lengths of the sides of this triangle as $a = |BC|$, $b = |AC|$, and $c = |AB|$. Then*
$$\cos A = \frac{\tanh b}{\tanh c}.$$

Proof As in the proof of the hyperbolic Pythagorean theorem, we will focus on triangles that are conveniently positioned, with A at the origin and C along the positive real axis. Any triangle can be put into that position with a composition of isometries that first translate A to the origin, and then rotate to move C to the positive real axis. Without loss of generality then, we may focus on triangles that are so positioned. Let z be the complex number that represents the point B.

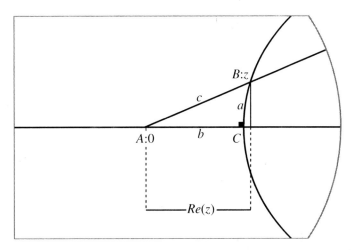

Figure 4. The cosine relationship in hyperbolic trigonometry: position the triangle so one vertex is at the origin.

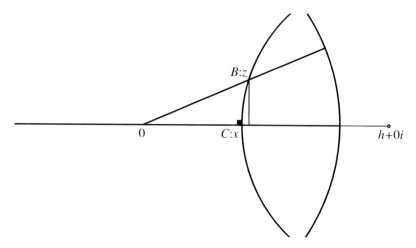

Figure 5. The cosine relationship in hyperbolic trigonometry: we need to find the orthogonal circle that passes through B and C.

Then it is easy to read off, from the Euclidean triangle, that

$$\cos A = \frac{\operatorname{Re}(z)}{|z|}.$$

The goal is to write both $\operatorname{Re}(z)$ and $|z|$ in terms of the hyperbolic lengths b and c. For the second

$$c = 2\tanh^{-1}(|z|) \implies |z| = \tanh(c/2).$$

For the first, let x denote the complex number that represents the point C (of course, it is on the real axis). Then

$$b = 2\tanh^{-1}(x) \implies x = \tanh(b/2).$$

The hyperbolic line through B and C is represented by the orthogonal circle through x and z. Its equation provides the connection between x and z that will allow us to find $\operatorname{Re}(z)$ in terms of b and c.

The center of the orthogonal circle is on the real axis at a point $h + 0i$. Its radius is $\sqrt{h^2 - 1}$, and so its equation (we use the complex form for the equation of a circle described in chapter 34) is

$$|w - h| = \sqrt{h^2 - 1}$$

that can be simplified to

$$(w - h)\overline{(w - h)} = h^2 - 1$$
$$(w - h)(\overline{w} - h) = h^2 - 1$$
$$w\overline{w} - \overline{w}h - wh + h^2 = h^2 - 1$$
$$w\overline{w} - (\overline{w} + w)h = -1.$$

45.2 Sine and cosine in a hyperbolic triangle

Since B and C are on the circle, x and z are solutions. Substitute x and solve for h (remember that x is real):

$$x^2 - 2xh = -1 \implies h = \frac{x^2 + 1}{2x}.$$

Substitute z and solve for its real part:

$$z\bar{z} - (\bar{z} + z)h = -1$$
$$|z|^2 - 2\text{Re}(z)h = -1$$
$$\text{Re}(z) = \frac{|z|^2 + 1}{2h}.$$

Therefore

$$\text{Re}(z) = \frac{|z|^2 + 1}{2\left(\frac{x^2+1}{2x}\right)} = (|z|^2 + 1) \cdot \frac{x}{x^2+1}.$$

Using lemma 45.3,

$$\cos(A) = \frac{\text{Re}(z)}{|z|}$$
$$= \frac{|z|^2 + 1}{|z|} \cdot \frac{x}{x^2+1}$$
$$= \frac{\tanh^2(c/2) + 1}{\tanh(c/2)} \cdot \frac{\tanh(b/2)}{\tanh^2(b/2) + 1}$$
$$= \frac{2}{\tanh c} \cdot \frac{\tanh b}{2}$$
$$= \frac{\tanh b}{\tanh c}. \qquad \square$$

With the cosine done, the sine is easier. The goal is the same, to try to write the sine of an angle in a right triangle as a function of the lengths of the opposite side and the hypotenuse.

Theorem 45.5 *(Sine in a hyperbolic triangle)* Let $\triangle ABC$ be a right triangle with right angle at C. Label the lengths of the sides of this triangle as $a = |BC|$, $b = |AC|$, and $c = |AB|$. Then

$$\sin A = \frac{\sinh a}{\sinh c}.$$

Proof Unlike the proofs of the last two theorems, which involved geometry, this one is pure algebra, using the identity $\cosh^2 x - \sinh^2 x = 1$ repeatedly, the hyperbolic Pythagorean

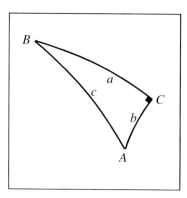

Figure 6. The sine relationship in hyperbolic trigonometry: labels.

theorem for the steps marked (∗) and the cosine relationship for the step marked (∗∗):

$$\begin{aligned}
\sin^2 A &= 1 - \cos^2 A \\
&= 1 - \frac{\tanh^2 b}{\tanh^2 c} \quad (**) \\
&= 1 - \frac{\sinh^2 b / \cosh^2 b}{\sinh^2 c / \cosh^2 c} \\
&= 1 - \frac{\sinh^2 b \cosh^2 c}{\sinh^2 c \cosh^2 b} \\
&= 1 - \frac{\sinh^2 b (\cosh^2 a \cosh^2 b)}{\sinh^2 c \cosh^2 b} \quad (*) \\
&= 1 - \frac{\sinh^2 b \cosh^2 a}{\sinh^2 c} \\
&= \frac{\sinh^2 c}{\sinh^2 c} - \frac{\sinh^2 b \cosh^2 a}{\sinh^2 c} \\
&= \frac{\cosh^2 c - 1}{\sinh^2 c} - \frac{\sinh^2 b \cosh^2 a}{\sinh^2 c} \\
&= \frac{(\cosh^2 a \cosh^2 b) - 1}{\sinh^2 c} - \frac{\sinh^2 b \cosh^2 a}{\sinh^2 c} \quad (*) \\
&= \frac{\cosh^2 a \cosh^2 b - 1 - \sinh^2 b \cosh^2 a}{\sinh^2 c} \\
&= \frac{\cosh^2 a (\cosh^2 b - \sinh^2 b) - 1}{\sinh^2 c} \\
&= \frac{\cosh^2 a - 1}{\sinh^2 c} \\
&= \frac{\sinh^2 a}{\sinh^2 c}.
\end{aligned}$$

Take a square root of both sides ($0 \le A \le \pi/2$, $a > 0$, and $c > 0$, so $\sin A$, $\sinh a$, and $\sinh c$ are positive) to get

$$\sin A = \frac{\sinh a}{\sinh c}.$$

□

45.3 Circumference of a hyperbolic circle

For our final result, let's find the relationship between the radius and the circumference of a circle. As in the Euclidean case, the circumference is a limit of perimeters of inscribed regular polygons. We find two limits before proving the theorem.

Lemma 45.6

$$\lim_{n \to \infty} n \sin \frac{x}{n} = x \quad \text{and} \quad \lim_{n \to \infty} n \sinh \frac{x}{n} = x.$$

Proof We will verify the first. The second is left as an exercise. In chapter 15, the radian system for angle measurement was set up so that

$$\pi = \lim_{n \to \infty} n \sin\left(\frac{\pi}{n}\right).$$

The formula we want is close to this. It is essentially just a change of variables: put $m = \pi n/x$ (so $n = mx/\pi$). As n approaches infinity, then so does m, and so

$$\lim_{n \to \infty} n \sin\left(\frac{x}{n}\right) = \lim_{m \to \infty} \frac{mx}{\pi} \sin\left(\frac{x}{mx/\pi}\right)$$
$$= \frac{x}{\pi} \lim_{m \to \infty} m \sin\left(\frac{\pi}{m}\right)$$
$$= \frac{x}{\pi} \cdot \pi$$
$$= x.$$

□

Exercise 45.4 Verify the second identity (L'Hôpital's rule works well).

Theorem 45.7 *(Hyperbolic circumference) A circle with radius r has circumference $C = 2\pi \sinh r$.*

Proof Let $A_1 A_2 \cdots A_n$ be a regular n-gon inscribed in a circle with radius r. Let O be its center and let B be the midpoint of $A_1 A_2$. Then $\triangle O A_1 A_2$ is isosceles; segment OB cuts it in half, into two congruent (by SSS) triangles. Thus $\angle OBA_1$ and $\angle OBA_2$ are congruent supplements, so $\triangle OA_1 B$ is a right triangle, and we use trigonometry. The polygon $A_1 A_2 \cdots A_n$ is built from $2n$ congruent copies of the triangle. Its perimeter, P_n, is $2n \cdot |A_1 B|$, and

$$(\angle A_1 O B) = 2\pi/(2n) = \pi/n.$$

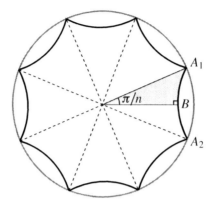

Figure 7. The circumference is the limit of perimeters of inscribed regular polygons.

The sine relationship in the right triangle tells us that

$$\sin(\pi/n) = \frac{\sinh|A_1B|}{\sinh r}$$
$$\sinh(|A_1B|) = \sin(\pi/n)\sinh r$$
$$\sinh(P_n/2n) = \sin(\pi/n)\sinh r.$$

Figure 8. The approximating polygon for $n = 4$, 8, and 12.

What happens as n goes to infinity? On the left, zero; on the right, zero; so $0 = 0$, which is comforting, but not useful. However, suppose we multiply both sides by n. Then

$$n\sinh((P_n/2)/n) = n\sin(\pi/n)\cdot\sinh r.$$

Using the previous lemma on limits, as n approaches infinity and P_n approaches the circumference C,

$$C/2 = \pi\sinh r \implies C = 2\pi\sinh r. \qquad \square$$

45.4 On a small scale

The formulas of this chapter confirm that Euclidean geometry and hyperbolic geometry are very different. But, when viewed on a small enough scale, they are not that different. Take, for instance, the hyperbolic Pythagorean theorem,

$$\cosh c = \cosh a \cosh b.$$

In that form, it doesn't look like its Euclidean counterpart. But let's rewrite it using the Taylor series for the hyperbolic cosine:

$$1 + \frac{c^2}{2!} + \frac{c^4}{4!} + \cdots = \left(1 + \frac{a^2}{2!} + \frac{a^4}{4!} + \cdots\right)\left(1 + \frac{b^2}{2!} + \frac{b^4}{4!} + \cdots\right)$$

$$= 1 + \frac{a^2}{2!} + \frac{b^2}{2!} + \frac{a^4}{4!} + \frac{a^2 b^2}{2! 2!} + \frac{b^4}{4!} + \cdots$$

If the values of a, b, and c are small enough, the fourth powers and higher are very small, and make little contribution to the overall expression. Ignore them and we end up with (approximately) the Euclidean Pythagorean theorem:

$$1 + \frac{c^2}{2!} \approx 1 + \frac{a^2}{2!} + \frac{b^2}{2!}$$

or

$$c^2 \approx a^2 + b^2.$$

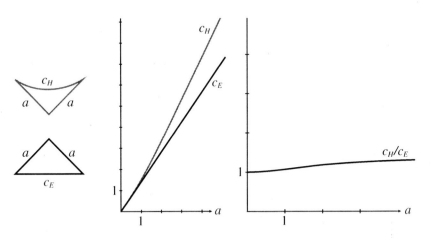

Figure 9. The hypotenuses c_E (Euclidean) and c_H (hyperbolic) of an isosceles right triangle as a function of the length of its legs.

Likewise, with the circumference formula

$$C = 2\pi \sinh r = 2\pi \left(r + \frac{r^3}{3!} + \frac{r^5}{5!} + \cdots\right).$$

For small values of r, the r^3 and higher power terms make only a small contribution to the overall circumference, and so

$$C \approx 2\pi r.$$

This is true in general. Euclidean and hyperbolic geometry are different globally but at a small scale they are hard to tell apart.

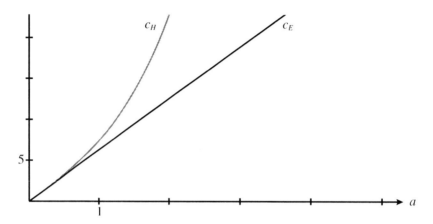

Figure 10. The circumferences c_E (Euclidean) and c_H (hyperbolic) of a circle as a function of its radius a.

45.5 Exercises

45.5. Compute directly the lengths of the sides of the right triangle with vertices at coordinates 0, 0.5, and $0.5i$. Show that they satisfy the hyperbolic Pythagorean theorem.

45.6. A convex quadrilateral $\square ABCD$ has $A = 0.25 + 0i$, $B = 0 + 0i$, $C = 0 + 0.25i$ and right angles at A, B, and C. What are the coordinates of D?

45.7. Let $\square ABCD$ be as in the last exercise. Compute the lengths of the sides $|AB|$ (which is the same as $|BC|$) and $|AD|$ (which is the same as $|CD|$).

45.8. Let $\square ABCD$ be as in the last exercise. Compute the lengths of the diagonals $|AC|$ and $|BD|$.

45.9. Let $\square ABCD$ be as in the last exercise. Compute $(\angle D)$.

45.10. Let $\square EFGH$ be the Saccheri quadrilateral whose summit vertices are $E = -(1/3) + (1/3)i$ and $F = (1/3) + (1/3)i$ and whose base is on the real axis. What are the coordinates of G and H?

45.11. Let $\square EFGH$ be as in the last exercise. What are $|EF|$ and $|GH|$?

45.12. Let $\square EFGH$ be as in the last exercise. What is the measure of the summit angle $\angle E$ (which is the same as the measure of $\angle F$)?

45.5 Exercises

45.13. Given an isosceles right triangle with legs of length x, what are the measures of its non-right angles as a function of x?

45.14. Let $\triangle ABC$ be the equilateral triangle whose center is at the origin $0 + 0i$ and has one vertex at $a + 0i$ (where $0 < a < 1$). Compute $(\angle A)$ as a function of a.

45.15. Let $\triangle ABC$ be the equilateral triangle described in the last exercise. Compute the length of one of its sides, say $|AB|$, as a function of a.

45.16. Find an equation that relates the angle sum of an equilateral triangle to the perimeter of its circumcircle.

46
Hyperbolic Area

When we developed Euclidean area in chapter 30, we described it as a positive real number associated with each polygon so that

- congruent polygons would have the same area,
- if a polygon could be broken into smaller pieces, the area of that polygon would be the sum of the areas of the pieces, and
- the area of a rectangle would be given by the formula $A = bh$.

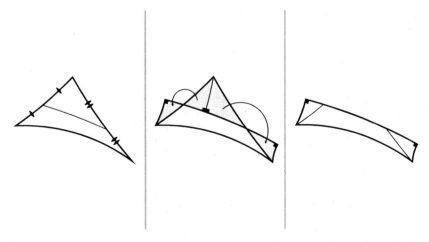

Figure 1. A triangle can be cut and the pieces rearranged to form a Saccheri quadrilateral. The triangle and the Saccheri quadrilateral should have the same area.

Hyperbolic area should retain the first two properties, but the last one isn't going to work because there are no rectangles. Hyperbolic area will have to be built from some other shape. The obvious choice is a triangle (and you could argue that it would have been better to build up Euclidean areas from triangles as well). The area of a Euclidean triangle is $A = bh/2$ but the formula doesn't work in the hyperbolic realm (a specific example is investigated in the exercises). So where can we look to find the formula for the area of a hyperbolic triangle? We will go back to the pseudosphere.

46.1 Area on the pseudosphere

In multivariable calculus the area $\mathscr{A}$ of a parametrized surface $\Omega = X(D)$ can be calculated as the double integral

$$\mathscr{A} = \iint_\Omega 1\, d\sigma = \iint_D |N|\, dt\, ds$$

where $|N|$ is the length of the normal vector. We have parametrized the pseudosphere by

$$X(s,t) = \left(\frac{\sin s}{t},\ \ln(t + \sqrt{t^2 - 1}) - \sqrt{1 - \frac{1}{t^2}},\ \frac{\cos s}{t} \right),$$

where $0 \leq s \leq 2\pi$ and $1 \leq t < \infty$. The restriction on t is essential, but the $[0, 2\pi]$ restriction on s is there just to ensure that the domain wraps around the pseudosphere only once. That horizontal constraint is fairly restricting but we can ignore it, so long as we understand that this will cause X to wrap around the pseudosphere repeatedly.

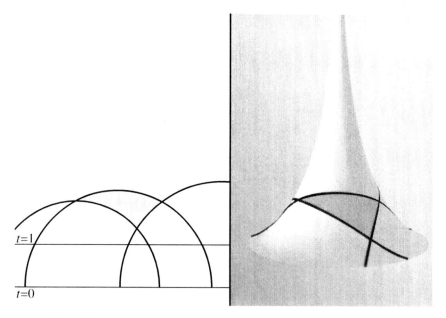

Figure 2. The projection of a triangle from the UHP to the pseudosphere.

The N vector is the cross product of the partial derivatives,

$$X_s = \left\langle \frac{\cos s}{t},\ 0,\ -\frac{\sin s}{t} \right\rangle$$

$$X_t = \left\langle -\frac{\sin s}{t^2},\ \frac{\sqrt{t^2 - 1}}{t^2},\ -\frac{\cos s}{t^2} \right\rangle$$

46.1 Area on the pseudosphere

(the calculation of $\partial y/\partial t$ in X_t was an exercise in chapter 39). Then

$$N = \begin{vmatrix} i & j & k \\ \dfrac{\cos s}{t} & 0 & -\dfrac{\sin s}{t} \\ -\dfrac{\sin s}{t^2} & \dfrac{\sqrt{t^2-1}}{t^2} & -\dfrac{\cos t}{t^2} \end{vmatrix}$$

$$= \left\langle \dfrac{\sin s}{t} \cdot \dfrac{\sqrt{t^2-1}}{t^2},\ \dfrac{\sin s}{t} \cdot \dfrac{\sin s}{t^2} + \dfrac{\cos s}{t}\dfrac{\cos s}{t^2},\ \dfrac{\cos s}{t} \cdot \dfrac{\sqrt{t^2-1}}{t^2} \right\rangle$$

$$= \left\langle \dfrac{\sin s \sqrt{t^2-1}}{t^3},\ \dfrac{1}{t^3},\ \dfrac{\cos s \sqrt{t^2-1}}{t^3} \right\rangle,$$

and so

$$|N| = (N \cdot N)^{1/2}$$
$$= \left[\dfrac{\sin^2 s(t^2-1) + 1 + \cos^2 s(t^2-1)}{t^6} \right]^{1/2}$$
$$= \dfrac{[(t^2-1)+1]^{1/2}}{t^3}$$
$$= 1/t^2.$$

Substitution yields a pleasant integral:

$$\mathscr{A} = \iint_D \dfrac{1}{t^2}\,ds\,dt.$$

We are not interested in the surface area of the pseudosphere as a whole. We want to know the areas of hyperbolic triangles on it. For that, the tricky part is to set up the proper domain D to cover them. We proceed in several steps.

Lemma 46.1 *Let*

$$D = \{(s,t) \mid (s^2+t^2) < r^2,\ 0 \le a \le s \le b,\ t \ge 1\},$$

an approximately quadrilateral region in the UHP. Then the area of $X(D)$ is

$$\mathscr{A} = (b-a) + (\sin^{-1}(a/r) - \sin^{-1}(b/r)).$$

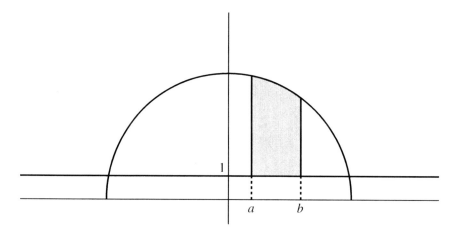

Figure 3. The first domain, with $0 \leq a \leq b$.

Proof This is a straightforward calculation.

$$\mathcal{A} = \iint_D \frac{1}{t^2} \, ds\, dt$$

$$= \int_a^b \int_1^{\sqrt{r^2-s^2}} \frac{1}{t^2} \, dt\, ds$$

$$= \int_a^b -\frac{1}{t} \Big|_1^{\sqrt{r^2-s^2}} \, ds$$

$$= \int_a^b 1 - \frac{1}{\sqrt{r^2-s^2}} \, ds$$

$$= \int_a^b 1 \, ds - \int_a^b \frac{1}{\sqrt{r^2-s^2}} \, ds$$

$$= (b-a) - \int_a^b \frac{1}{\sqrt{r^2-s^2}} \, ds.$$

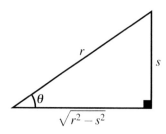

Figure 4. A trigonometric substitution.

46.1 Area on the pseudosphere

Use trigonometric substitution: put $s = r\sin\theta$, so that $\sqrt{r^2 - s^2} = r\cos\theta$ and $ds = r\cos\theta\, d\theta$. Then

$$\mathscr{A} = (b-a) - \int_{s=a}^{s=b} \frac{r\cos\theta}{r\cos\theta} d\theta$$

$$= (b-a) - \left(\theta\Big|_{s=a}^{s=b}\right)$$

$$= (b-a) - \left(\sin^{-1}(s/r)\Big|_a^b\right)$$

$$= (b-a) - \left[\sin^{-1}(b/r) - \sin^{-1}(a/r)\right]$$

$$= (b-a) + \sin^{-1}(a/r) - \sin^{-1}(b/r). \qquad \square$$

The previous lemma ended with two inverse sine functions. They can be rewritten in a more useful form.

Lemma 46.2 *Let D be the same region as in the previous lemma, but now label its top-most vertices*

$$A = (a, \sqrt{r^2 - a^2}) \text{ and } B = (b, \sqrt{r^2 - b^2}).$$

Then

$$\mathscr{A} = (b-a) + (\pi - (\angle A) - (\angle B)),$$

where $(\angle A)$ and $(\angle B)$ are computed by measuring the angles between the vertical lines and the tangent lines to D at A and B.

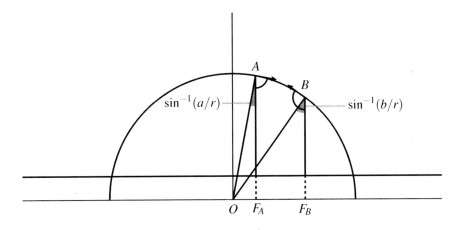

Figure 5. Relating the area to the measures of $\angle A$ and $\angle B$.

Proof We start with labels: let O be the origin, $F_A = (a, 0)$, and $F_B = (b, 0)$. The right triangle $\triangle OF_A A$ has hypotenuse r and a side of length a, so

$$\sin(\angle OAF_A) = \frac{a}{r} \implies (\angle OAF_A) = \sin^{-1}(a/r).$$

This angle is complementary to $\angle A$, so $\sin^{-1}(a/r) = \pi/2 - (\angle A)$. The behavior at B is different. Again

$$\sin(\angle OBF_B) = \frac{b}{r} \implies (\angle OBF_B) = \sin^{-1}(b/r),$$

but this time $\sin^{-1}(b/r) = (\angle B) - \pi/2$. Substituting into the formula for $\mathscr{A}$,

$$\mathscr{A} = (b-a) + (\pi/2 - (\angle A)) - ((\angle B) - \pi/2)$$
$$= (b-a) + (\pi - (\angle A) - (\angle B)). \qquad \square$$

In the first two lemmas, we put restrictions on D that are not really necessary: that a and b have to be positive, and that the circle must be centered at the origin. Now let's remove them in the next two lemmas.

Lemma 46.3 *If*

$$D = \{(s,t) \mid (s^2 + t^2) < r^2,\ a \leq s \leq b,\ t \geq 1\},$$

then $\mathscr{A} = (b-a) + (\pi - (\angle A) - (\angle B))$.

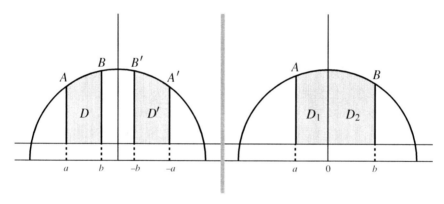

Figure 6. Extending to other types of domains: (l) $a \leq b \leq 0$ and (r) $a \leq 0 \leq b$.

Proof Step 1. Suppose that $a < b \leq 0$. The reflection across the vertical axis will transform D into a congruent shape D', mapping A to A' and B to B'. The area of $X(D')$ is the same as the area of $X(D)$, though, and $\angle A \simeq \angle A'$ and $\angle B \simeq \angle B'$, so

$$\mathscr{A}(D) = \mathscr{A}(D')$$
$$= ((-a) - (-b)) + (\pi - (\angle B') - (\angle A'))$$
$$= (b-a) + (\pi - (\angle A) - (\angle B)).$$

Step 2. Suppose that $a < 0 < b$. Then D can be broken up into two smaller regions, D_1 from $x = a$ to $x = 0$, and D_2 from $x = 0$ to $x = b$. We have

$$\mathscr{A}(D_1) = (0 - a) + (\pi - (\angle A) - \pi/2) = -a + \pi/2 - (\angle A)$$
$$\mathscr{A}(D_2) = (b - 0) + (\pi - \pi/2 - (\angle B)) = b + \pi/2 - (\angle B),$$

46.1 Area on the pseudosphere

and so

$$\mathcal{A}(D) = \mathcal{A}(D_1) + \mathcal{A}(D_2)$$
$$= (-a + \pi/2 - (\angle A)) + (b + \pi/2 - (\angle B))$$
$$= (b - a) + (\pi - (\angle A) - (\angle B)). \qquad \square$$

Lemma 46.4 *If*

$$D = \{(s,t) \mid (s-h)^2 + t^2 < r^2, \ a \leq s \leq b, \ t \geq 1\},$$

then $\mathcal{A} = (b - a) + (\pi - (\angle A) - (\angle B))$.

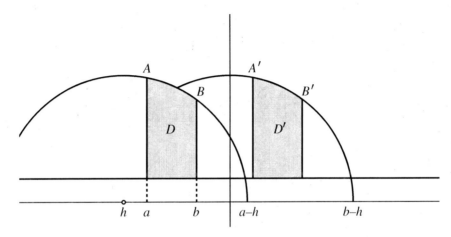

Figure 7. Extending to other types of domains: circles that are not centered at the origin.

Proof The horizontal translation by $-h$ shifts D to a congruent domain D' whose bounding circle is centered at the origin, mapping A to A' and B to B'. It does not alter area: the area of $X(D)$ is the same as the area of $X(D')$. Nor does it alter angle measures, so $(\angle A) = (\angle A')$ and $(\angle B) = (\angle B')$. Therefore

$$\mathcal{A}(D) = \mathcal{A}(D')$$
$$= ((b-h) - (a-h)) + (\pi - (\angle A') - (\angle B'))$$
$$= (b-a) + (\pi - (\angle A) - (\angle B)). \qquad \square$$

Now we can cut apart and piece together domains to find the area of a triangle.

Theorem 46.5 *For* $\triangle ABC$ *in the UHP, the area of* $X(\triangle ABC)$ *is*

$$\mathcal{A} = \pi - (\angle A) - (\angle B) - (\angle C).$$

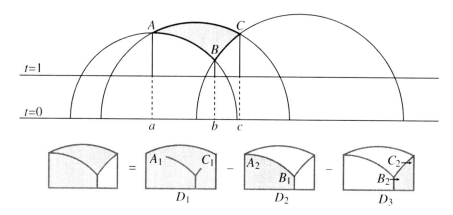

Figure 8. Calculating the area of a hyperbolic triangle.

Proof There are a few configurations. We will look at just one, as illustrated and labeled in figure 8. In this case, the area of $X(\triangle ABC)$ is

$$\mathscr{A}(\triangle ABC) = \mathscr{A}(D_1) - \mathscr{A}(D_2) - \mathscr{A}(D_3)$$

where

$$\mathscr{A}(D_1) = (c - a) + [\pi - (\angle A_1) - (\angle C_1)]$$
$$\mathscr{A}(D_2) = (b - a) + [\pi - (\angle A_2) - (\angle B_1)]$$
$$\mathscr{A}(D_3) = (c - b) + [\pi - (\angle B_2) - (\angle C_2)],$$

and so

$$\begin{aligned}\mathscr{A}(\triangle ABC) &= (c - a) + [\pi - (\angle A_1) - (\angle C_1)] \\ &\quad - (b - a) - [\pi - (\angle A_2) - (\angle B_1)] \\ &\quad - (c - b) - [\pi - (\angle B_2) - (\angle C_2)] \\ &= [(\angle A_2) - (\angle A_1)] + [(\angle B_2) + (\angle B_1)] + [(\angle C_2) - (\angle C_1)] - \pi \\ &= -(\angle A) + [2\pi - (\angle B)] - (\angle C) - \pi \\ &= \pi - (\angle A) - (\angle B) - (\angle C).\end{aligned}$$

□

46.2 Areas of polygons in the Poincaré disk

The UHP model inherits its metrical properties from the pseudosphere, so $\triangle ABC$ in the UHP should also have area $\pi - (\angle A) - (\angle B) - (\angle C)$. The map $\sigma \circ \rho$ from the UHP to D is conformal, so the image of $\triangle ABC$ in D will have the same angles. Therefore, the pseudosphere area calculation carries over directly to suggest an area formula in the Poincaré disk.

Definition 46.6 The area of $\triangle ABC$ is $\pi - (\angle A) - (\angle B) - (\angle C)$.

In spite of all the calculations that have led to this point, this is still (I think) a little surprising. We are saying that triangle areas are calculated entirely in terms of angle sums! The

46.2 Areas of polygons in the Poincaré disk

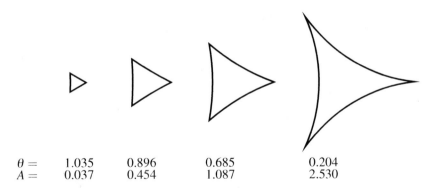

$\theta =$	1.035	0.896	0.685	0.204
$A =$	0.037	0.454	1.087	2.530

Figure 9. Hyperbolic areas of equilateral triangles, given by the formula $\mathcal{A} = \pi - 3\theta$.

term $\pi - (\angle A) - (\angle B) - (\angle C)$ is called the *defect* of the triangle. In Euclidean geometry, it would always be zero, so the defect measures how far the triangle deviates from a Euclidean triangle. This formula is saying that small triangles (in the sense of area) are the ones with small defects. That jibes with the general principal of the last lesson, that Euclidean and hyperbolic geometry look approximately the same for small objects, but diverge from one another for bigger objects. It also means that no hyperbolic triangle can have an area greater than π! Surprising or not, it turns out that the formula meets the requirements of an area function.

In chapter 31, we found a formula for the area of a general simple polygon, using an inductive argument. It was based on the idea that a simple polygon can be split into smaller pieces, and then the area of the whole is the sum of the areas of the pieces (every simple polygon can be triangulated and its area is the sum of the areas of the triangles). The key to the argument was that every simple polygon with more than three sides has a diagonal (theorem 31.1) that breaks the polygon into two smaller pieces. The argument for the existence of a diagonal works just fine in hyperbolic geometry.

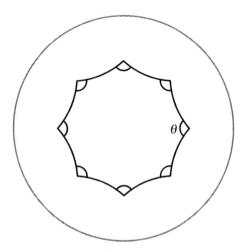

Figure 10. The area of a regular octagon with interior angle θ is $\mathcal{A} = 6\pi - 8\theta$.

Exercise 46.1 Check the proof of theorem 31.1 to make sure that it is still valid in hyperbolic geometry.

Theorem 46.7 *The area of a simple polygon $\mathcal{P} = P_1 P_2 \cdots P_n$ is*

$$(n-2)\pi - \sum_{i=1}^{n} (\angle P_i).$$

Proof As in the Euclidean case, we proceed by induction on the number of sides of $\mathcal{P}$. The base case consists of polygons of three sides: triangles. For them, the formula matches the defect formula. Now assume that polygons with n or fewer sides have areas that are given by the formula, and consider an $(n+1)$-sided polygon $\mathcal{P} = P_1 P_2 \cdots P_{n+1}$. It has a diagonal cutting $\mathcal{P}$ into two smaller pieces $\mathcal{P}_A$ and $\mathcal{P}_B$. To simplify the notation, let's relabel the vertices of $\mathcal{P}$ so that

$$\mathcal{P}_A = P_1 P_2 \cdots P_j$$
$$\mathcal{P}_B = P_1 P_j P_{j+1} \cdots P_{n+1}.$$

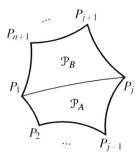

Figure 11. To prove the area formula, we use an inductive proof, cutting the polygon into two smaller pieces with a diagonal.

Between $\mathcal{P}_A$ and $\mathcal{P}_B$, there are $n+3$ sides (the $n+1$ original sides and the diagonal that is counted twice because it occurs in both pieces). The first piece, $\mathcal{P}_A$, has j of the sides. The other piece, $\mathcal{P}_B$, must have the remaining $n+3-j$ sides. According to the inductive hypothesis,

$$\mathcal{A}(\mathcal{P}_A) = (j-2)\pi - \left[(\angle P_j P_1 P_2) + \sum_{i=2}^{j-1} (\angle P_i) + (\angle P_{j-1} P_j P_1) \right]$$

$$\mathcal{A}(\mathcal{P}_B) = (n+1-j)\pi - \left[(\angle P_1 P_j P_{j+1}) + \sum_{i=j+1}^{n+1} (\angle P_i) + (\angle P_{n+1} P_1 P_j) \right].$$

Then

$$\mathcal{A}(\mathcal{P}) = \mathcal{A}(\mathcal{P}_A) + \mathcal{A}(\mathcal{P}_B)$$

$$= (n-1)\pi - \left[(\angle P_1) + \sum_{i=2}^{j-1}(\angle P_i) + (\angle P_j) + \sum_{i=j+1}^{n+1}(\angle P_i) \right]$$

$$= ((n+1) - 2)\pi - \sum_{i=1}^{n+1}(\angle P_i),$$

as needed. By induction, this is a valid formula for any simple polygon. □

Now that we have this formula for the area of a general polygon, we can check that it satisfies the requirements of an area function.

Exercise 46.2 Verify that the formula assigns to every simple polygon a positive real number. Verify that the formula assigns the same number to two congruent polygons.

The last requirement, that the area of the whole is the sum of the areas of the pieces, is more complicated to verify. The next exercise confirms it in a few specific cases.

Exercise 46.3 Let D be a point in the interior of $\triangle ABC$. Prove that

$$\mathcal{A}(\triangle ABC) = \mathcal{A}(\triangle ABD) + \mathcal{A}(\triangle BCD) + \mathcal{A}(\triangle ACD).$$

Given a convex quadrilateral $\square ABCD$, let $A * E * B$ and $C * F * D$. Prove that

$$\mathcal{A}(\square ABCD) = \mathcal{A}(\square AEFD) + \mathcal{A}(\square BEFC).$$

46.3 Area of a circle

To end the chapter, let's use the formula for the area of a polygon to find the area of a circle. The formula depends upon two things. One is lemma 45.6, that as n approaches infinity, $n \sin(x/n)$ approaches x. The other is a kind of hyperbolic cofactor identity:

Lemma 46.8 *Let α and β be the two acute angles in a right triangle and let a be the length of the side opposite α. Then*

$$\cos \alpha = \sin \beta \cdot \cosh a.$$

Exercise 46.4 Prove the lemma. Hint: to get started, rewrite the cosine term in terms of hyperbolic tangents.

The derivation of the area formula for circles is like the derivation of the circumference formula: we approximate the circle with regular inscribed polygons and then take a limit as the number of sides approaches infinity. The area formula would seem to make this easy, since it only depends on the interior angles and all the interior angles of a regular polygon are congruent. It turns out that there are some computational complications with this approach, and so we will instead do something a little less direct.

Theorem 46.9 *The area of a circle with radius r is*

$$\mathscr{A} = 2\pi(\cosh r - 1).$$

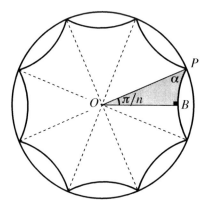

Figure 12. Use inscribed regular polygons to approximate the area inside a circle.

Proof Write $\mathscr{A}$ for the area of a circle with radius r. Let $\mathscr{A}_n$ denote the area of a regular n-gon $\mathcal{P}_n$ inscribed in the circle. It can be broken up into $2n$ right triangles each of which is congruent to $\triangle OPB$, where O is the center of the circle, P is one of the vertices of the n-gon, and B is the right angle. The angle at P is the key. Let's say it has a measure of α. Then each interior angle of $\mathcal{P}_n$ has a measure of 2α and therefore

$$\mathscr{A}_n = (n-2)\pi - \sum_{i=1}^{n} 2\alpha = (n-2)\pi - 2n\alpha.$$

Rather than trying to find the value of α, we solve the equation for α:

$$2n\alpha = (n-2)\pi - \mathscr{A}_n \implies \alpha = \pi/2 - \pi/n - \mathscr{A}_n/2n.$$

The trick is to look not at α itself, but instead at $\cos \alpha$. There are two ways we can work with that expression. One way is to use the cofactor identity in lemma 46.8. In $\triangle OPB$, the angle at O measures π/n. Therefore

$$\cos \alpha = \sin(\pi/n) \cosh(|OB|).$$

The other way is via the traditional cofactor identity:

$$\cos \alpha = \cos\left(\frac{\pi}{2} - \frac{\pi}{n} - \frac{\mathscr{A}_n}{2n}\right)$$
$$= \cos\left(\frac{\pi}{2} - \left(\frac{\pi}{n} + \frac{\mathscr{A}_n}{2n}\right)\right)$$
$$= \sin\left(\frac{\pi}{n} + \frac{\mathscr{A}_n}{2n}\right).$$

46.4 Exercises

Setting them equal we get

$$\sin(\pi/n)\cosh(|OB|) = \sin\left(\frac{\pi}{n} + \frac{\mathscr{A}_n}{2n}\right).$$

As n approaches infinity, both sides of the equation approach zero, but if we multiply by n, then

$$n\sin(\pi/n)\cosh(|OB|) = n\sin\left(\frac{\pi + \mathscr{A}_n/2}{n}\right).$$

As n approaches infinity,

- $|OB|$ approaches r,
- $\mathscr{A}_n$ approaches $\mathscr{A}$,
- the small angle formula describes what happens to the sine terms.

The result is

$$\pi \cdot \cosh(r) = \pi + \mathscr{A}/2,$$

and we can solve for $\mathscr{A}$:

$$\mathscr{A}/2 = \pi \cosh r - \pi \implies \mathscr{A} = 2\pi(\cosh r - 1). \qquad \square$$

The formula looks different from the Euclidean one. Look, though, at what happens when it is rewritten using the Taylor series for hyperbolic cosine:

$$\mathscr{A} = 2\pi\left(\left(\sum_{n=0}^{\infty} \frac{r^{2n}}{(2n)!}\right) - 1\right)$$

$$= 2\pi\left(\left(1 + \frac{r^2}{2!} + \frac{r^4}{4!} + \cdots\right) - 1\right)$$

$$= 2\pi\left(\frac{r^2}{2} + \frac{r^4}{24} + \cdots\right)$$

$$= \pi r^2 + \frac{\pi r^4}{12} + \cdots.$$

For small values of r, the later terms in the series contribute very little to the value of $\mathscr{A}$. That is, when $r \approx 0$, $\mathscr{A} \approx \pi r^2$. Once again, the hyperbolic world appears to be similar to the Euclidean world when it is viewed at a small enough scale.

46.4 Exercises

46.5. A torus is formed by revolving the circle

$$(x - b)^2 + y^2 = a^2, \quad 0 < a < b,$$

around the y-axis (see exercise 38.2 in lesson 38). Compute the surface area of the torus.

46.6. Let $\triangle ABC$ be a right triangle with vertices $A = 0$, $B = 0.5$, and $C = 0.5i$. First consider AB to be its base, and calculate the lengths of its base and height. Then consider BC as the base, and calculate the lengths of the base and height. Observe that the values of $bh/2$ are not the same.

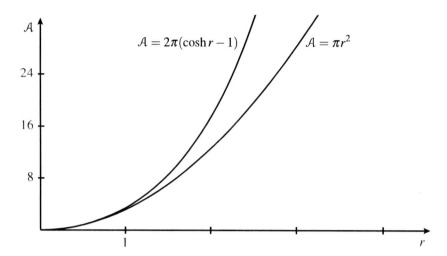

Figure 13. Comparisons of the areas of hyperbolic and Euclidean circles as a function of their radii.

46.7. Compute the hyperbolic area of the triangle with vertices $A = 0$, $B = 0.5$, and $C = 0.5i$.

46.8. In the proof of the formula for the area of a triangle, we considered the case where the vertex B is below AC. Show that the formula is valid if B is above AC.

46.9. Let R be a regular quadrilateral inscribed in a circle of radius r. Find a formula for the area of R as a function of r.

46.10. Prove that there are circles that cannot be contained in the interior of any triangle.

46.11. What is the area of the convex quadrilateral $\square ABCD$ with $A = 0.25 + 0i$, $B = 0 + 0i$, $C = 0 + 0.25i$ and right angles at A, B, and C (see exercise 45.6)?

46.12. What is the area of the Saccheri quadrilateral $\square EFGH$ that has summit vertices $E = (-1/3) + (1/3)i$ and $F = (1/3) + (1/3)i$ and its base along the real axis (see exercise 45.10)?

46.13. Suppose that $\triangle ABC$ is decomposed into a collection of smaller triangles $\{T_i\}$. Explain why the area of $\triangle ABC$ is the sum of their areas. You will need to use Euler's formula for planar polygons, that $v - e + f = 1$, where v is the total number of vertices, e is the total number of edges, and f is the total number of faces in the decomposition.

47
Tiling

In this final chapter, we will draw one more contrast between the geometries of the Euclidean and hyperbolic planes. We have approached hyperbolic geometry along a traditional path, but I hope that this chapter can lead you into a more modern study of the subject. We will look at the topic of tiling by regular polygons.

Definition 47.1 A *tiling* of the plane (either Euclidean or hyperbolic) is a collection of simple polygons $\{P_i\}$ that

- Cover the entire plane (that is, for any point Q, there is a polygon P_i so that Q is either on P_i or in its interior), and
- Overlap only on the edges (for two distinct polygons P_i and P_j, $\text{int}(P_i) \cap \text{int}(P_j) = \emptyset$).

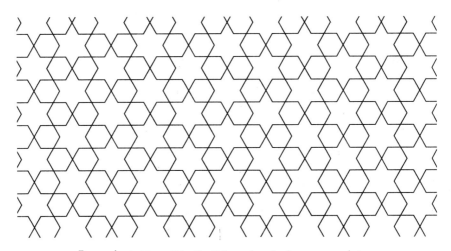

Figure 1. A tiling of the Euclidean plane by hexagons and stars.

The study of planar tilings is both fascinating and elusive. The definitive text is *Tilings and Patterns* by Grunbaum and Shephard [GS87]. However, we have only modest goals here: our focus is on planar tilings $\{\mathcal{P}_i\}$ where all the $\mathcal{P}_i$ are congruent regular polygons that are arranged so that when two polygons are adjacent they share either an entire side or only a vertex. Let's start with Euclidean tilings. We are all familiar with tilings by equilateral triangles, squares, and regular hexagons. But when you try to tile by regular pentagons, it doesn't work. Around each

529

vertex, three pentagons is not enough, but four is too many. The problem is the interior angles: for a tiling they need to evenly divide 2π, and that doesn't happen for the pentagon. This is the restriction in general.

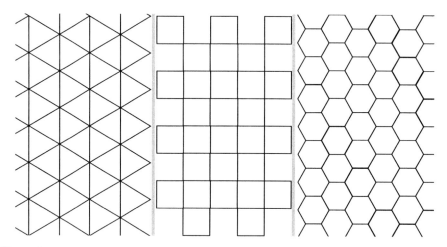

Figure 2. Tilings of the Euclidean plane by equilateral triangles, squares, and regular hexagons.

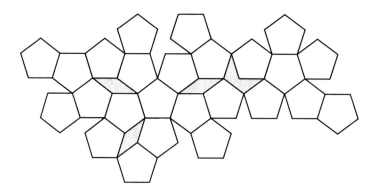

Figure 3. A (failed) attempt to tile the Euclidean plane with regular pentagons.

Theorem 47.2 *The only regular polygons that tile the Euclidean plane are equilateral triangles, squares, and regular hexagons.*

Proof This proof is easier if you measure angles using degrees rather than radians, so that's what we will do here. We can calculate the interior angles of a regular n-gon $\mathcal{P}_n$ by dividing it into isosceles triangles (that all share a vertex at the center O of $\mathcal{P}_n$). In any of the triangles, the angle at O measures $360°/n$. Then the other two angles in those triangles measure

$$\frac{180° - 360°/n}{2}.$$

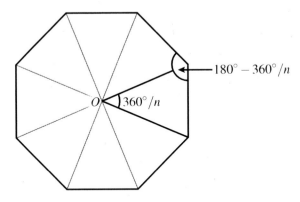

Figure 4. Measuring the interior angles of a Euclidean regular polygon.

An interior angle of $\mathcal{P}_n$ is built from two of them, so the measure of an interior angle of a regular n-gon is $\theta_n = 180° - 360°/n$. Let's look at some values for small n.

n	θ_n
3	$60°$
4	$90°$
5	$108°$
6	$120°$
7	$128.57°$ (approx.)
8	$135°$
9	$140°$
10	$144°$

For $\mathcal{P}_n$ to tile the plane, θ_n has to divide $360°$ evenly. Of the first ten, only $n = 3, 4$, and 6 work. For $n > 10$,

$$120° < \theta_n < 180°.$$

There are no divisors of 360 between 120 and 180, so no polygon with a larger number of sides will tile. Therefore tilings only occur when $n = 3, 4$, or 6. □

All this depends on the relationship between the measures of angles in an isosceles triangle, a relationship provided by the fact that in Euclidean geometry, the angle sum of a triangle is $180°$. In hyperbolic geometry, the situation is different. The angle measures of small regular polygons are not that different from their Euclidean counterparts, but for larger polygons, the deviation becomes large.

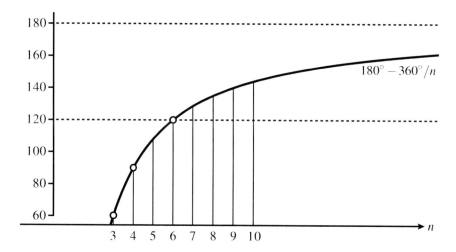

Figure 5. The measure of the interior angle of a Euclidean regular n-gon as a function of n.

Lemma 47.3 *Let θ_n be the measure of an interior angle of a regular n-gon $\mathcal{P}_n$ with sides of length a. Then*

$$\sin(\theta_n/2) = \frac{\cos(\pi/n)}{\cosh(a/2)}.$$

This is an immediate consequence of lemma 46.8 from chapter 46, a lemma on cofactors that states that the acute angles α and β of a right triangle are related by $\sin(\beta) = \cos(\alpha)/\cosh(A)$, where A is the length of the side opposite angle α.

Exercise 47.1 Prove the lemma.

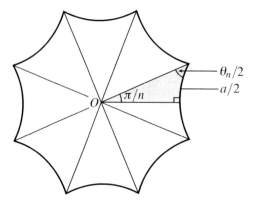

Figure 6. Measuring the interior angles of a hyperbolic regular polygon.

Tiling

Theorem 47.4 *There is a tiling of the hyperbolic plane by regular n-gons for every $n \geq 3$.*

The measures of the interior angles must evenly divide 2π. In the hyperbolic case we can control the interior angle measures by adjusting the size of the polygon. The formula in the preceding lemma directly relates polygon size to angle measure.

Proof Let θ_n denote the angle measure of an interior angle of a regular n-gon. It is a function of the radius a of the polygon. As a approaches zero and as a approaches infinity,

$$\lim_{a \to 0} \sin(\theta_n/2) = \lim_{a \to 0} \cos(\pi/n)/\cosh(a/2)$$
$$= \cos(\pi/n)$$
$$= \sin(\pi/2 - \pi/n)$$
$$= \sin((\pi - 2\pi/n)/2),$$
$$\lim_{a \to \infty} \sin(\theta_n/2) = \lim_{a \to \infty} \cos(\pi/n)/\cosh(a/2)$$
$$= 0$$
$$= \sin(0/2).$$

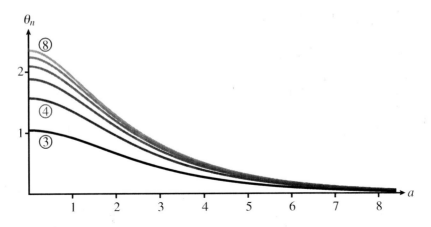

Figure 7. The measure of the interior angle of a hyperbolic regular n-gon as a function of a for $n = 3, 4, 5, 6, 7,$ and 8.

Therefore, as a approaches zero, θ_n approaches $\pi - 2\pi/n$, and as a approaches infinity, θ_n approaches zero. This happens continuously, so as a grows, θ_n takes on every value between $\pi - 2\pi/n$ and zero. No matter what n is, there are some values between $\pi - 2\pi/n$ and zero that divide 2π evenly. Therefore, for every $n \geq 3$, there are some n-gons that are just the right size to tile the hyperbolic plane. □

So Euclidean and hyperbolic tilings offer different types of flexibility. In Euclidean geometry, tiling is only possible for certain n (3, 4, and 6), but the size of the pieces of the tiling are not important. In hyperbolic geometry, tiling is possible for all n, but only at certain sizes (some examples are shown in figure 8).

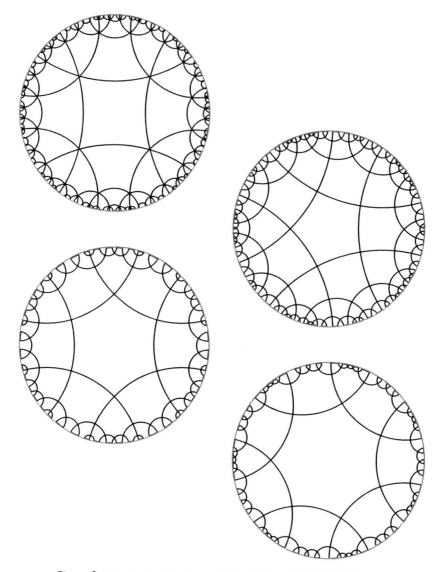

Figure 8. Hyperbolic tiling by regular 4-, 5-, 6-, and 7-sided polygons.

That brings us to the end of this book. Where do you go from here? Modern geometry takes planar geometries and applies them to more general contexts, to manifolds and beyond. For instance, two dimensional compact, smooth, orientable manifolds can all be realized as quotients of the plane by $4n$-gon tilings. When $n = 1$, the quotient is a torus, and since the Euclidean plane can be tiled by 4-gons (squares) the torus can be given a Euclidean structure. When $n > 1$, the quotients are higher genus surfaces, n-holed surfaces. These surfaces do not inherit a Euclidean structure, because $4n$-gons do not tile the Euclidean plane when $n > 1$. But $4n$-gons tile the hyperbolic plane, and so the surfaces inherit a hyperbolic structure. William Thurston and others have extended these ideas to the geometry of three dimensional manifolds. Three dimensions are vastly more difficult, and so the resulting study is much deeper and more sophisticated. For further reading in hyperbolic geometry, I recommend Thurston's *Three-Dimensional Geometry and Topology* [Thu97], Week's *The Shape of Space* [Wee02], Ratcliffe's *Foundations of Hyperbolic Manifolds* [Rat94], and Conway, Burgiel, and Goodman-Strauss's *The Symmetries of Things* [CBGS08].

47.1 Exercises

47.2. There is a tiling of the hyperbolic plane by regular quadrilaterals where six quadrilaterals come together at each vertex. Give a numerical approximation of the length of the side of the quadrilateral in the tiling.

47.3. Suppose that $\mathcal{P}$ is a regular n-gon that tiles the hyperbolic plane so that v polygons come together at each vertex. What is the area of $\mathcal{P}$ (in terms of n and v)?

Bibliography

[Bie04] Jürgen Bierbrauer, *Finite geometries: MA5980*, Lecture notes distributed on World Wide Web, 2004.

[Bog] Alexander Bogomolny, *Napoleon's theorem by transformation*, distributed on World Wide Web, www.cut-the-knot.org/Curriculum/Geometry/NapoleonByTransformation.shtml.

[CBGS08] John H. Conway, Heidi Burgiel, and Chaim Goodman-Strauss, *The symmetries of things*, 1st ed., AK Peters Ltd., Wellesley, Mass., 2008.

[CG67] H.S.M. Coxeter and Samuel L. Greitzer, *Geometry revisited*, 1st ed., Random House, New York, 1967.

[Con] John Conway, *Trilinear vs barycentric coordinates*, Correspondence, distributed on World Wide Web, Currently available at mathforum.org/kb/message.jspa?messageID=1091956.

[Cox64] H.S.M. Coxeter, *Projective geometry*, 1st ed., Blaisdell Publishing Co., New York, 1964.

[Cox69] _____, *Introduction to geometry*, 2nd ed., John Wiley and Sons, Inc., New York, 1969.

[CR41] Richard Courant and Herbert Robbins, *What is mathematics? : an elementary approach to ideas and methods*, 1st ed., Oxford University Press, London, 1941.

[Dur92] John R. Durbin, *Modern algebra: An introduction*, 3rd ed., John Wiley and Sons, Inc., New York, 1992.

[Edg90] Gerald A. Edgar, *Measure, topology, and fractal geometry*, 1st ed., Springer-Verlag, New York, 1990.

[Euc56] Euclid, *The thirteen books of euclid's elements*, 2nd ed., Dover Publications, New York, 1956, Translated from the text of Heiberg, with introduction and commentary by Sir Thomas L. Heath.

[Euc02] _____, 1st ed., Green Lion Press, Santa Fe, New Mexico, 2002.

[Gre08] Marvin J. Greenberg, *Euclidean and non-euclidean geometries: Development and history*, 4th ed., W.H. Freeman and Company, New York, 2008.

[Grü03] Branko Grünbaum, *Are your polyhedra the same as my polyhedra?*, Discrete and Computational Geometry: The Goodman-Pollack Festschrift (2003).

[GS87] Branko Grünbaum and G. C. Shephard, *Tilings and patterns*, 1st ed., W.H. Freeman and Company, New York, 1987.

[Hil50]   David Hilbert, *The foundations of geometry*, reprint edition ed., The Open Court Publishing Company, La Salle, Illinois, 1950, Translated by E.J. Townsend.

[Kim]   Clark Kimberling, *Encyclopedia of triangle centers - etc*, distributed on World Wide Web, faculty.evansville.edu/ck6/encyclopedia /ETC.html.

[Mas91]   William S. Massey, *A basic course in algebraic topology*, Springer-Verlag, New York, 1991.

[Moi74]   Edwin E. Moise, *Elementary geometry from an advanced standpoint*, 2nd ed., Addison Wesley Publishing Company, Reading, Massachusetts, 1974.

[MSW02]   David Mumford, Caroline Series, and David Wright, *Indra's pearls: The vision of Felix Klein*, 1st ed., Cambridge University Press, Cambridge, 2002.

[Ped70]   Daniel Pedoe, *A course of geometry for colleges and universities*, 1st ed., Cambridge University Press, London, 1970.

[Rat94]   John G. Ratcliffe, *Foundations of hyperbolic manifolds*, 1st ed., Springer-Verlag, New York, 1994.

[Spi99a]   Michael Spivak, *A comprehensive introduction to differential geometry, vol. 2*, 3nd ed., Publish or Perish, Inc., Houston, Texas, 1999.

[Spi99b]   _____, *A comprehensive introduction to differential geometry, vol. 3*, 3rd ed., Publish or Perish, Inc., Houston, Texas, 1999.

[Tho79]   John A. Thorpe, *Elementary topics in differential geometry*, 1st ed., Springer-Verlag, New York, 1979.

[Thu97]   William P. Thurston, *Three-dimensional geometry and topology*, 1st ed., Princeton University Press, Princeton, New Jersey, 1997.

[Wee02]   Jeffrey R. Weeks, *The shape of space*, 2nd ed., Marcel Dekker, New York, 2002.

[WW04]   Edward C. Wallace and Stephen F. West, *Roads to geometry*, 3rd ed., Pearson Education, Inc., Upper Saddle River, New Jersey, 2004.

Index

AA similarity, 129
AAA, 49
AAASS, 101–103
AAS triangle congruence, 43
AASAS, 97–100
Absolute geometry, 7
Acute angle, 69
Acute triangle, 74
Addition rule
 for cosine, 233–234
 for hyperbolic cosine, 495
 for hyperbolic sine, 495
 for sine, 234–235
Adjacent interior angles, 40
Alternate interior angle theorem, 41, 44, 109
Alternate interior angles, 40
Altitude, 181, 193, 209
Angle, 19
 bisector, 44, 71, 162, 186
 interior, 19
Angle addition theorem, 46, 71
Angle bisector, 44, 71, 162, 186
Angle construction axiom, 28, 470
Angle measure, 70–71
 in the Poincaré disk, 468–470
 in the UHP, 446–450
Angle subtraction theorem, 45
Angle sum of a triangle, 73, 110, 400–401
Arbelos, 387–388
Arc, 136
Arc length, 151, 156–157
Archimedes' axiom, 57
Area, 313–319, 346
 of a polygon (determinant formula), 334–336
 and barycentric coordinates, 348, 349
 hyperbolic, 515–527
 of a circle, 338
 of a hyperbolic circle, 527
 of a hyperbolic polygon, 524–525
 of a hyperbolic triangle, 522
 of a parallelogram, 315–317
 of a polygon, 329–337
 of a regular polygon, 336–337
 of a trapezoid, 319
 of a triangle, 317–319
 of a triangle (determinant formula), 331–334
 of hyperbolic circle, 525
 on the pseudosphere, 516–522
Argument (of a complex number), 370
ASA, 31
ASASA, 97–100
Asymptotic parallel, 417
Automorphism, 240
Axiom
 of congruence, 28
 of continuity, 57
 of incidence, 11
 of order, 12
 Playfair's, 107

Barycentric coordinates, 341–352
 and area, 348
 and area, 349
 of circumcenter, 350
 of excenters, 352
 of incenter, 351
 of orthocenter, 352
Between, 12
Bijection, 239
Bisector (of an angle), 44, 71, 162, 186
Brocard point, 198

Cartesian plane, 225–226
Central angle, 135
Centroid, 184–186, 191
 trilinear coordinates, 220

539

Ceva's theorem, 200–205, 328, 344
Change of coordinates, 285–289
Chord, 135
Chord-chord theorem, 144
Circle, 135–146, 228
    arc, 136
    area, 338
    center, 135
    central angle, 135
    chord, 135
    complex equation, 375
    diameter, 135
    intersections, 137–141
    orthogonal, 381–387
    parametrization, 236
    radius, 135
    semicircle, 136
Circumcenter, 180–181, 191, 197
    barycentric coordinates, 350
    trilinear coordinates, 215
Circumcircle, 181
    and the law of sines, 327
Circumference, 149–158
    in hyperbolic geometry, 509–510
Cofunction identities, 236
Collapsing compass, 165
Collinear, 11
Compass and straightedge, 160–161
Complementary angles, 72
Complex arithmetic, 369–370
    circle, 375
    dilation, 374
    inversion, 376–377
    line, 376
    reflection, 374
    rotation, 375
    translation, 374
Complex conjugate, 370, 377
Complex number, 369–370
    argument, 370
    conjugate, 377
    exponential form, 371
    norm, 370, 378
    trigonometric form, 371
Concurrence, 179
Conformal, 363–364
Congruence
    in the Poincaré disk, 458
    of polygons, 89
Conic section
    ellipse, 237, 265, 379
    hyperbola, 238, 379
    parabola, 237, 265, 379

Constructions, 159–177
Convex polygon, 89–92
Coordinate, 225–226
Cosh, 493–497
Cross ratio, 365, 444–446
    and inversion, 365
Crossbar theorem, 21–22
Curvature, 421, 428–429
Cyclic polygon, 92

Dedekind's axiom, 57
Diagonal of a polygon, 93
Diameter, 135
Dilation, 293–300
    analytic equations, 296
    complex equation, 374
Distance, 51
    Euclidean, 226–228
    in neutral geometry, 53–62
    in the Poincaré disk, 463–467
    in the UHP, 443–446
    signed, 204
Divergent parallel, 417
Double angle formulas, 236
    in hyperbolic trigonometry, 496
Dyadic number, 57, 58–61

Ellipse, 237, 265, 379
Elliptic isometry, 487
Equilateral, 34
Equilateral polygon, 92
Equilateral triangle, 72, 171
Euclid, 9
Euclid's fifth postulate, 9, 107–108
Euclid's postulates, 9
Euclidean tiling, 530–531
Euler line, 191–193, 197
Excenter, 200
    barycentric coordinates, 352
    trilinear coordinates, 222
Excircle, 200
Explementary angles, 88
Exterior angle, 42
Exterior angle theorem, 42–43, 44, 78

Fano's geometry, 2–4
Feuerbach's theorem, 209, 392–393
Fifth postulate, 9, 107
Fixed point, 243–245
    of hyperbolic isometry, 486–487
Foot of a perpendicular, 101, 112
Fractional linear transformation, 490

Index

Gauss map, 425–428
Gaussian curvature, 428–429
Geodesic, 421, 433–442
Gergonne point, 210
Glide reflection, 273–280
Green's theorem, 336, 339

Half angle formulas, 237
   in hyperbolic trigonometry, 496
Half-turn, 265, 304, 311
Heron's formula, 324–326
Hilbert, 10
HL right triangle congruence, 78
Hyperbola, 238, 379
Hyperbolic area, 515–527
   of a circle, 525–527
   of a polygon, 524–525
   of a triangle, 522
Hyperbolic geometry
   circumference, 509–510
   cosine, 505–507
   sine, 507–509
Hyperbolic isometry, 475, 483–489
   elliptic, 487
   loxodromic, 487
   parabolic, 487
Hyperbolic Pythagorean theorem, 502–504
Hyperbolic reflections, 475–482
   equations of, 477–481
Hyperbolic tiling, 531–533
Hyperbolic trigonometric functions, 493–497
Hyperbolic trigonometric identities, 494–497
Hyperbolic trigonometry
   addition formulas, 495
   double angle formulas, 496
   half-angle formulas, 496
   inverse functions, 497
   power reduction formulas, 496

Identities
   hyperbolic trigonometric, 494–497
Incenter, 186–188
   barycentric coordinates, 351
Incidence, 11
   in the Poincaré disk, 457
Inscribed angle, 141
Inscribed angle theorem, 141–144
Intersecting (lines), 12
Inverse hyperbolic trigonometric functions, 497
Inversion, 353–366
   and similarity, 359
   is conformal, 363–364
   and cross ratio, 365

   and orthogonal circles, 381
   complex equation, 376–377
   image of a circle, 361–362
   image of a line, 360
Isometry, 239–247
   analytic equations, 245, 289
   and congruence, 241–242
   and incidence and order, 242–243
   complete classification, 280
   composition of, 239–240
   glide reflection, 273–280
   hyperbolic, 475, 483–489
   orientation, 267–271
   orientation-preserving/reversing, 270
   reflection, 249–255
   rotation, 257, 261–263
   translation, 257–261
Isosceles, 34
Isosceles triangle theorem, 34

Jordan curve theorem, 85

Kite, 103
Klein bottle, 272
Koch snowflake, 152

Law of cosines, 321–324
Law of sines, 319–320, 328
Law of tangents, 328
Lemoine point, 328
Lever, 341
Line
   complex equation, 376
   Euclidean equation of, 228–230
Line segment, 13
Lines
   in the Poincaré disk, 457
Loxodromic isometry, 487

Möbius strip, 268
Major arc, 136
Mass, 342–345
Median, 184, 209
Menelaus's theorem, 205–207
Midpoint, 54, 55, 193
   formula, 235
Miguel point, 198
Minor arc, 136
Mobile, 341

Nagel point, 207–209
Napoleon's theorem, 305–307
Neutral geometry, 7

Nine point circle, 193–197, 308–310
Norm (of a complex number), 370, 378

Obtuse angle, 69
Obtuse triangle, 74
Opposite ray, 14
Order, 12–16
   in the Poincaré disk, 457
Orientation, 267–271
Orientation-preserving/reversing, 270
Orthic triangle, 189, 199
Orthocenter, 181–183, 191, 193, 197
   barycentric coordinates, 352
   trilinear coordinates, 217
Orthogonal circle, 381–387
   and inversion, 381
   equation, 459
Orthonormal frame, 267

Pappus's theorem, 210
Parabola, 237, 265, 379
Parabolic isometry, 487
Parallel, 12, 401–403
   asyptotic, 417
   divergent, 417
   Euclidean, 112
   non-Euclidean, 411–419
Parallel projection, 116–124
Parallelogram, 95, 115–116
   area, 315–317
Parametrized surface, 422
Pasch's lemma, 20–21
Penrose tiles, 327
Perimeter, 83, 149–150
Perpedicular line, 69
Perpendicular, 162–164
Perpendicular bisector, 70, 161, 179–180
Perpendicular line, 101
Pi, 150, 153–156
Plane separation axiom, 12
Playfair's axiom, 10, 107
Poincaré disk, 455–473
Polygon, 81–93
   area (determinant formula), 334–336
   diagonal, 93
   diagonal (existence), 329–331
   interior, 88–92
   simple, 82
Polygons
   similar, 125
Power reduction formulas, 236
   in hyperbolic trigonometry, 496
Product-to-sum formulas, 237

Pseudosphere, 421–431, 433–442
Pythagorean theorem, 130–131, 320–321
   hyperbolic, 502–504

Quadrilateral congruence
   AAASS, 101–103
   AASAS, 97–100
   ASASA, 97–100
   SASAS, 97–100
   SSSSA, 100–101
Quadrilateral congruence theorems, 97

Radians, 157
Radical axis, 384–387
Radius, 135
Ray, 13
Rectangle, 95, 110–112, 397–400
Reflection, 249–255
   analytic equations, 254–255, 263
   complex equation, 374
   hyperbolic, 475–482
Reflex angle, 88
Regular pentagon, 172–177
Regular polygon, 92
   construction, 170
Rhombus, 95
Right angle, 67–69
Right triangle, 74
Rotation, 257, 261–263
   analytic equations, 263–264
   complex equation, 375
   half-turn, 265

Saccheri quadrilateral, 403–409
   altitude, 405–406
   base, 403, 406, 407
   leg, 403, 406
   summit, 403, 406, 407
   summit angle, 403, 404
Saccheri-Legendre theorem, 73, 74–76
SAS, 30
SAS axiom, 28
SAS similarity, 127
SASAS, 97–100
Scalene, 34
Scalene triangle theorem, 76, 78
Secant-secant formula, 146
Segment addition axiom, 28, 467–468
Segment addition theorem, 61
Segment construction axiom, 28
Segment subtraction theorem, 45
Semicircle, 136
Semiperimeter, 324

Index

Side angle side axiom, 28
Signed distance, 204, 211
Similarity, 125–130, 291
   and inversion, 359
Similarity mapping, 291–293
   and congruence, 293
   and incidence and order, 292–293
Simple polygon, 82
Simpson line, 198
Sinh, 493–497
Square, 95, 172
SSA, 49
SSS, 47
SSS similarity, 130
   SSSSA, 100, 101
Star polygon, 93
Steiner's porism, 389–391
Stereographic projection, 353–358
Sum-to-product formulas, 237
Supplementary angles, 38–39
Surface of revolution, 423
Symmedians, 328

Tangent plane, 422
Taylor series, 370–373
   for hyperbolic trigonometric functions, 493
Thales' theorem, 144
Three reflections theorem, 252–254
Tiling, 529–533
Torus, 272, 432
Tractrix, 431
Translation, 257–261
   complex equation, 374
Transversals, 40
Trapezoid, 95
   area, 319
Triangle, 20
   AA similarity, 129
   angle sum, 400–401
   area, 317–319

   area (determinant formula), 331–334
   SAS similarity, 127
   similarity, 127–130
   SSS similarity, 130
Triangle congruence
   AAS, 43
   ASA, 31
   HL, 78
   SAS, 30
   SSS, 47
Triangle inequality, 77–78
Trigonometry, 230–238
   addition rules, 233–235
   cofunction identities, 236
   double angle formulas, 236
   half angle formulas, 237
   identities, 236–237
   power reduction formulas, 236
   product-to-sum formulas, 237
   sum-to-product formulas, 237
Trilinear coordinates, 211–221, 346

UHP, 443–451
   angle measure, 446–450
   distance, 443–446
   model, 450–451
Unit circle, 230–231
Upper half plane, 443–451
   angle measure, 446–450
   distance, 443–446
   model, 450–451

Variation, of a curve, 433
Varignon's theorem, 303–305
Vector, 260, 281–284, 342
   arithmetic properties, 284
   norm, 284
Vertical angles, 39–40

Young's geometry, 5